FLOW ANALYSIS WITH SPECTROPHOTOMETRIC AND LUMINOMETRIC DETECTION

This book is part of the collection: Handbooks in Analytical Science

FLOW ANALYSIS WITH SPECTROPHOTOMETRIC AND LUMINOMETRIC DETECTION

ELIAS A. G. ZAGATTO
University of São Paulo, Brazil

CLÁUDIO C. OLIVEIRA
University of Maringá, Brazil

ALAN TOWNSHEND
University of Hull, U.K.

PAUL J. WORSFOLD
University of Plymouth, U.K.

ELSEVIER

AMSTERDAM • BOSTON • HEIDELBERG • LONDON • NEW YORK • OXFORD
PARIS • SAN DIEGO • SAN FRANCISCO • SINGAPORE • SYDNEY • TOKYO

Elsevier
225 Wyman Street, Waltham, Massachusetts 02451, USA
The Boulevard, Langford Lane, Kidlington, Oxford OX5 1GB, UK
Radarweg 29, PO Box 211, 1000 AE Amsterdam, The Netherlands

First edition 2012

Library of Congress Cataloging-in-Publication Data
Worsfold, Paul.
Flow analysis with spectrophotometric and luminometric detection / Paul Worsfold. – 1st ed.
p. cm.
ISBN 978-0-12-385924-2 (hardback)
1. Instrumental analysis. 2. Flow injection analysis. 3. Spectrophotometry.
4. Luminescence spectroscopy. I. Title.
QD79.I5W67 2012
543'.55–dc23
2011044013

British Library Cataloguing in Publication Data
A catalogue record for this book is available from the British Library

Printed and bound in US

12 13 14 10 9 8 7 6 5 4 3 2 1

ISBN: 978-0-12-385924-2

Contents

Preface

Flow analysis is now firmly established as an important branch of analytical chemistry. It is an excellent tool for solution management, providing reproducible conditions for physical and chemical sample treatment within a controlled and reproducible environment. Most detection techniques can be coupled with flow analysis but spectrophotometry and luminescence are the most commonly used ones and hence are the focus for this monograph.

The aim is to provide a comprehensive flow analysis handbook that encompasses theoretical and practical aspects, discusses all modes of flow analysis and considers in detail the special strategies available for flow manipulation. In this regard, the concept of multi-commutation receives particular emphasis. Flow analysis in its various forms is now widely used but its potential has not yet been fully exploited. One aspiration for this monograph is that it will provide users with the enthusiasm and the tools to unlock that potential. The flow system (instrument) offers a number of strategies relying on different flow manipulations but most have not been exploited in commercially available flow analysers. Another goal of the monograph is therefore to highlight these strategies (opportunities) in order to stimulate the manufacture of appropriate hardware. It is expected to be a useful resource for users, students and anyone who needs to adapt a given traditional analytical procedure to a flow-based method.

The monograph has nine chapters. The first two provide an introduction and a historical perspective. This is followed by fundamental aspects and interaction of radiation of the flowing sample. Chapter five discusses the various types of flow analysers and chapter six considers the specific components that make up a flow analyser. This is followed by chapters on flow manipulation and strategies for sample handling. It concludes with a short chapter on future trends.

The production of this monograph has been a longstanding ambition for Elias Zagatto, and the other authors have encouraged and assisted him along the way. All of the authors would like to acknowledge assistance from professional colleagues and institutions and thank family members for their support throughout the period from its original conception to its final delivery.

Elias A.G. Zagatto
University of São Paulo, Brazil

Cláudio C. Oliveira
University of Maringá, Brazil

Alan Townshend
University of Hull, U.K.

Paul J. Worsfold
University of Plymouth, U.K.

Foreword

Flow analysis techniques date to over eighty years ago, but modern analytical flow techniques began in the 1950s with the introduction of segmented flow analysis, followed about two decades later by flow injection analysis. Numerous books have been written over the years on flow analysis in general and flow injection analysis in particular. The most widely used detection systems employ flow cells utilising attenuation or radiation of light. This is the first book to focus on these important detection systems and methods, i.e., spectrophotometry, turbidimetry and nephelometry, and techniques based on fluorescence, chemiluminescence and bioluminescence. It is intended to be complementary to existing monographs.

Written by pioneers and leading practitioners in the field, it serves as both a reference book and a textbook on flow analysis principles and the fundamentals of the optical measurement techniques. It begins with a comprehensive historical record, in Chapter 2, of the techniques, including the pioneering work on flow injection analysis performed in the laboratory of Professor Zagatto and his colleagues at the Centre for Nuclear Energy in Agriculture, University of São Paulo, in collaboration with Jaromir Ruzicka during his extended visit there. The authors then describe and discuss commonalities of flow systems, giving in Chapter 3 a detailed tutorial on types of flow, sample dispersion, system configurations and types of detector responses. The principles of the optical measurement techniques in flowing systems are detailed in Chapter 4, which includes an excellent overview of the Schlieren effect and how to minimise it. The different types of flow analysers are comprehensively discussed in Chapter 5, followed by the instrumentation for implementing them in Chapter 6. Chapter 7 focusses on strategies for unsegmented flow analysers, such as stream splitting, merging zones and stopped flow.

The important use of flow systems for sample handling is described in Chapter 8, including in-line sampling, dilution, preconcentration, calibration, gas diffusion and dialysis and solvent extraction, among others. This chapter also describes special strategies we may use for kinetic, catalytic and enzymatic analysis. The authors end with a projection, in Chapter 9, of future trends in flow analysis. The many references are complete, with titles of all references given, which will be of great value to readers wishing to follow up on specific articles. This well-organised, comprehensive and detailed monograph will be of value for the student, the researcher and the practitioner of flow analysis methods, particularly utilising optical detection.

Gary D. Christian
University of Washington, U.S.A.

CHAPTER

1

Introduction

The pathway of true wisdom can be identified by three things:
There must be love,
it must be practical,
it can be followed by anyone. **Paulo Coelho**

OUTLINE

Flow Analysis – the art of performing analytical chemistry in flowing streams – is a branch of analytical chemistry that has experienced significant growth in recent decades. The concepts, methods and applications, as well as the overall acceptance of flow-based analytical procedures, have advanced rapidly because the flow analyser is an excellent tool for solution management [1] and provides reproducible conditions for sample handling. Flow analysers have unique characteristics such as reproducible timing, low consumption of samples and reagents, ease of operation, no need for glassware such as beakers, volumetric flasks and pipettes, excellent analytical figures of merit and reduced human intervention in the measurement process.

1.1. GENERAL

In flow analysis, a discrete aliquot of an aqueous sample is precisely selected and introduced into the flow manifold (Fig. 1.1). A reproducible sample zone is then established and pushed towards the detector by the carrier/wash stream. While being transported through the analytical path (usually narrow-bore tubing), the sample zone undergoes in-line physical and chemical treatment involving, *e.g.*, dilution, reagent addition, and analyte separation/concentration. The steps related to the specific analytical application are therefore efficiently accomplished under nonequilibrium yet reproducible conditions. During passage of the sample zone through the detector, a transient signal is generated and recorded as a peak that reflects the analyte concentration in the sample.

Almost all detection techniques can be coupled with flow analysis; even the analytical balance [2] and the microscope [3] have been used as detectors. It is however a fact that *spectrophotometry*[1] is the most common detection techniques used in flow analysis, as it usually requires extensive manual solution handling. Moreover, the flow-through cuvette is permanently located in the detector unit, thus maintaining the detection geometry. This leads to an improvement in the measurement repeatability compared with batch-wise analytical procedures, in which the cuvette is removed from the instrument between measurements.

Several modes of flow analysis have been developed during the last half century [4], the most important being segmented flow analysis, flow injection analysis and sequential injection analysis. In the early 1980s, it was stated that "*the majority of chemical assays are still performed manually, with methods and tools (pipettes, beakers, volumetric flasks, etc) designed in the nineteenth century, and this time- and labour-consuming approach forms a contrast to other technical areas which have undergone amazing development*

[1]In this book, the italicised word *spectrophotometry* (and derived expressions) is used in a broader context, encompassing UV–visible spectrophotometry, fluorimetry, turbidimetry, chemi- and bio-luminometry and related techniques.

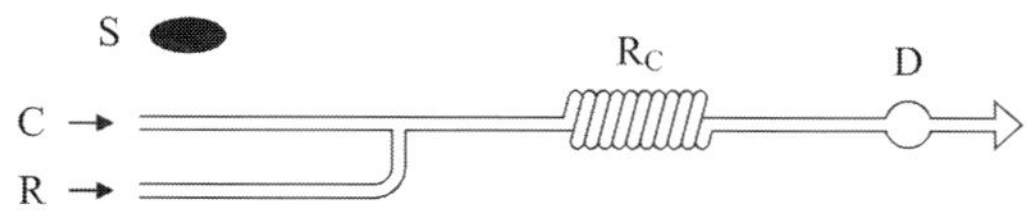

FIGURE 1.1 **The flow analyser.** The figure represents a flow system involving only three streams: sample (S), carrier/wash (C) and reagent (R). R_C and D specify the coiled reactor and the flow-through detector.

within the last two decades" [5]. This quotation referred to chemical assays not taking advantage of flow analysis and is still valid, especially in the context of some batch-wise procedures involving classical techniques. Consequently, there is a general view that *spectrophotometric* analytical procedures are old-fashioned. This is not the case however and one of the goals of this monograph is to clarify this misconception.

This monograph covers the different modes of flow analysis, as well as more generic aspects, including applications relying on *spectrophotometry.* Emphasis is given to flow-through detectors based on attenuation of radiation (spectrophotometry and turbidimetry) or radiation emission (fluorimetry, chemiluminescence and bioluminescence) by a flowing sample.

Although monitoring of a flowing sample is inherent to techniques such as mass spectrometry, flame atomic absorption spectrometry and chromatography, even when the instrument is operated manually these techniques are not considered to belong to the field of flow analysis by the analytical community. Therefore, a deeper discussion of these techniques, as well as capillary electrophoresis and micro-fluidic systems, is not undertaken.

This monograph is complementary to existing publications devoted to general and specific aspects of flow analysis (Appendix 1.1). In relation to flow systems with spectrophotometric detection, the quotation at the beginning of the chapter could be modified to:

> The ideal flow analyser can be identified by three things:
> There must be careful design,
> it must be practical,
> it can be assembled by anyone.

1.2. THE ADVENT OF FLOW ANALYSIS

The laboratory scene in Fig. 1.2 (upper) caricatures a situation of extreme overload in a traditional laboratory devoted to high-throughput analysis. An increase in the demand for analyses often leads to the development of novel concepts, instruments, methods and procedures, increasing the capacity of the laboratory; this in turn leads to a further increase in the demand for chemical assays, and so on. The synergistic chemical analysis/analytical chemistry dichotomy is often the driver for

FIGURE 1.2 Overload in a traditional routine analysis laboratory (upper). Conceptual diagram of a modern "laboratory" for chemical analyses (lower).

innovation [6]. In this context, flow analysis emerged in the nineteen-fifties to address the need for high sample throughput in clinical laboratories.

1.3. THE DEVELOPMENT OF FLOW ANALYSIS

With the continuous evolution of flow analysis, the capacity of analytical laboratories expanded greatly, advanced flow analysers became commercially available, and *spectrophotometry* experienced a renaissance [7]. Significant improvements in *spectrophotometry* have been observed in recent years [8,9], through developments in novel light sources, sensors (including bio-sensors), detectors, light transmission (*e.g.*, fibre optics technology), in-line detection approaches (*e.g.*, solid-phase spectrophotometry) and strategies for data treatment.

In view of the wide acceptance of flow analysis [10] and the increasing availability of flow-based procedures for real sample analysis [11], the situation hinted at in Fig. 1.2 (upper) has undergone a significant improvement, approaching that shown in Fig. 1.2 (lower), which depicts a "laboratory" exploiting flow analysis for *in vivo* assays at the beginning of the third millennium. Nowadays, modern *spectrophotometric* flow-based techniques are a competitive alternative to other modern analytical techniques and are also amenable to miniaturisation [12].

1.4. MAIN FEATURES OF FLOW ANALYSIS

Flow analysis has often been referred to as an analytical technique, but this is not strictly true, as it is an advanced procedure for carrying out automated chemical assays. The cornerstone features inherent to flow injection analysis, namely sample insertion, controlled dispersion and reproducible timing [5], are considered here in a broader context, in order to encompass the different modes of flow analysers.

1.4.1. Sample Insertion

A 5–500 μl aliquot of the aqueous sample is precisely selected (Fig. 1.3a) and introduced into the flow manifold. This is especially relevant for the analysis of biological fluids, cell tissues, blood sera, dew and other volume-limited samples, as well as for *in vivo* assays. Moreover, sampling strategies relying on mini-probes become more practical. In spite of the low sample volume, reliable results are obtained even for very low analyte concentrations.

The *characteristic mass* (analyte mass yielding 0.01 absorbance), often considered in atomic absorption spectrometry and related techniques as an indicator of sensitivity, is also relevant in the context of flow-based analytical procedures involving UV–visible spectrophotometry [7].

As a small sample aliquot is used, the reagent consumption tends to be low, thus minimising the need for special laboratory facilities and waste generation. If really necessary, waste treatment can be carried out in-line, because flow analysis can provide a suitable strategy for handling highly concentrated and toxic reagents. This feature supports the desirable trend toward "environmentally friendly" or "green" analytical methods [13].

The selected sample aliquot is inserted into a flow manifold (Fig. 1.3b) and pushed forwards by the carrier/wash stream flowing through narrow-bore (typically 0.3–1.0 mm i.d.) tubing. During sample handling inside the analytical path, there is no physical contact between the sample and the external environment (and *vice versa*). Analyte losses leading to biased results and/or to indoor environmental pollution are therefore

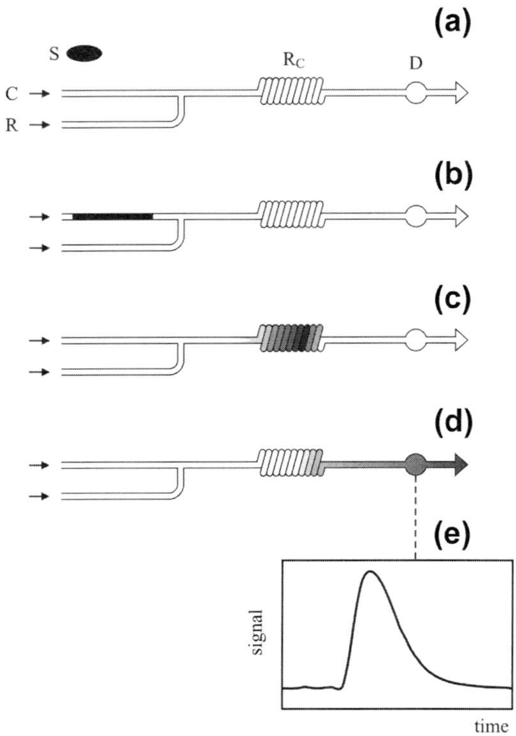

FIGURE 1.3 Didactic representation of sample processing inside a flow analyser. It refers to the simplest analyser designed in the confluence configuration. *Adapted from P.J. Worsfold, A. Townshend, C.F. Poole (Eds), Encyclopedia of Analytical Science, 2nd Edn, Oxford, 2005, v.3, p.24, E.A.G. Zagatto, P.J. Worsfold, Flow Analysis: Overview, with permission from Elsevier (Ref. [4]).*

avoided, and sample contamination, which might degrade the detection limit, is minimised. The analytical path therefore can be regarded as a "clean room" inside which the sample is manipulated. This closed status also minimises drawbacks associated with the use of hazardous and/or volatile reagents.

1.4.2. Sample Dispersion

While being transported through the analytical path, the sample zone undergoes continuous dispersion as well as sudden variations in concentration at the points where confluent streams are added (Fig. 1.3c), and control of these processes is often the key to system design. It is possible to manipulate dispersion, keeping it low when sample integrity needs to be preserved, or otherwise adjusting it in order to, *e.g.*, achieve an analytical signal within the dynamic range of the detector.

As a consequence of dispersion, interaction of the sample with the reagents is improved, allowing precise analytical results to be obtained. Moreover, concentration gradients are established along the flowing sample, and their exploitation expands the application range of analytical methods [14]. The deleterious influence of some undesirable concentration gradients on the analytical signal, especially those related to Schlieren noise (changes in refractive index) [15], can be efficiently circumvented.

Passage of the sample zone through the analytical path is fast, the mean sample residence time typically being 10–180 s. During this time, it participates in the steps inherent to the specific analytical application. Conditions for sample handling are strictly maintained from one sample to the next, and this feature is particularly noteworthy in applications involving slurry processing, in-line sample digestion, micellar media, monitoring of suspensions and exploitation of renewable sensors [16]. Reproducibility of sample handling may not be achieved in some circumstances, *e.g.*, in the analysis of sample batches with high matrix variability. However, different strategies for circumventing matrix effects, *e.g.*, standard addition [17] or matrix matching [18], can usually be efficiently accomplished in the flow analyser.

Another important feature arising from the short mean sample residence time is the rapid acquisition of the analytical signal, which is particularly attractive for process control, screening procedures (selection of high priority samples), one-off samples and studies involving concentration-orientated feedback mechanisms. On the other hand, if a relatively long time interval is needed for sample handling, specific strategies for increasing the mean residence time, *e.g.*, stream segmentation or sample stopping, can be exploited. These strategies have often been reported in relation to, *e.g.*, high-sensitivity analytical procedures involving relatively slow chemical reactions (see 7.5) and phosphorimetric detection (see 4.1.4).

A corollary of the short sample residence time in the analytical path is the short wash time. Intermingling of successive samples, leading to pronounced carryover effects, is only a minor issue, and the next sample can be introduced without a long delay. This permits high sampling rates, typically 30–300 h^{-1}, which can be exploited for, *e.g.*, repetitive measurements, simultaneous determinations, titrations, standard additions and industrial quality control.

1.4.3. Reproducible Timing

Rigid timing control is essential, as most of the steps related to sample handling are time dependent. Precise timing is inherent to flow analysis, allowing flow-based analytical procedures to be carried out under transient conditions, often without attaining chemical equilibrium. Operations such as dialysis, gas diffusion and ion exchange are easily and efficiently accomplished in-line, and the tedious wait to achieve equilibrium is not necessary. External timing control through the operation of

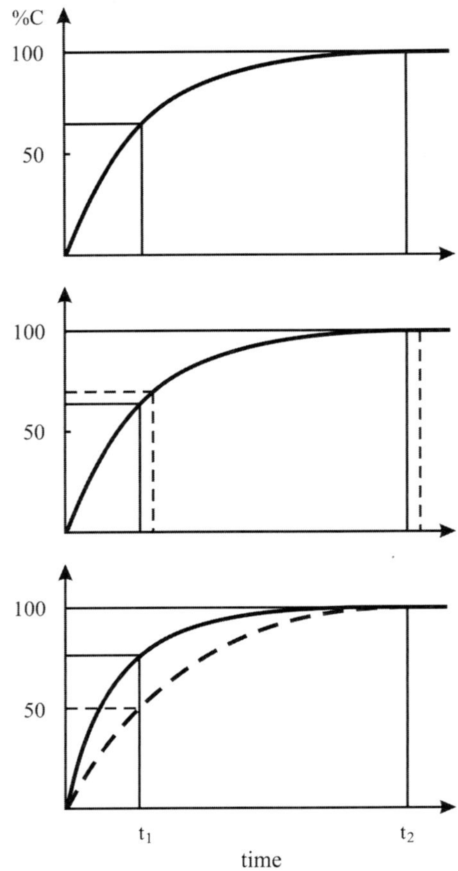

FIGURE 1.4 **Didactic representation of temporal variation in the concentration of a monitored chemical species.** Top, centre and bottom = no modification (ideal situation), modification in timing, and modification in reaction kinetics, respectively; %C = relative concentration of the monitored species; t_1 and t_2 = instants corresponding to situations without and with attainment of equilibrium.

manifold valves and/or pumps is also valuable, especially in relation to procedures exploiting time-based injections, re-sampling and sample stopping.

During the fast passage of the sample through the analytical path, some processes inherent to the specific analytical method may not reach completion, and this feature makes it easier to exploit partial yet reproducible development of chemical reactions, monitor unstable chemical species, and implement catalytic methods of analysis. The favourable characteristics of flow-based methods that do not require chemical equilibrium to be reached have often been emphasised, and can be exploited to expand the potential and application range of analytical procedures.

However, this feature should be considered with care, as there are some processes that should always be quantitative. In fact, chemical equilibrium should be attained for in-line masking or separation processes. Moreover, analytical procedures relying on incomplete development of certain processes tend to be less rugged, and thus more susceptible to poor accuracy, as illustrated in Fig. 1.4, which shows the increase in concentration of a chemical species monitored over time. At instant t_1, formation of this species is not complete, and this situation is often associated with flow analysis; instant t_2 refers to complete reaction, inherent to batch analysis. Examination of Fig. 1.4 reveals that variations in timing (centre) or reaction kinetics (bottom) may result in poor accuracy. The quality of the pumping device and the need for sample/reagent matrix matching may then be limiting factors in the reliability of results.

1.4.4. Other Attractive Features

1.4.4.1. Transient Analytical Signal

Passage of the processed sample through the flow cell results in a transient recorded peak, features of which such as height, area, or width are ideally proportional to analyte concentration (Fig.1.3d). Monitoring the detector output in the presence (analytical signal) and absence (baseline) of the sample is an important diagnostic parameter for assessing the correct operation of the system.

It is interesting to report that different fluid elements of a flowing sample are associated with specific residence times and hence different sample handling conditions. Monitoring the entire flowing sample zone then provides a number of successive measurements reflecting these different conditions. After proper selection, these successive measurements can be taken into account for multi-parametric determinations [19].

A unique characteristic of flow analysis is the low susceptibility to instrumental drift, and this is a consequence of considering the transient signals as the basis for measurement [20]. This is evident in Fig. 1.5, where recorded peak heights (and areas) are maintained regardless of the pronounced baseline drift. This feature is especially relevant in process analysis where continuous monitoring of the same environment is usually required.

1.4.4.2. Low Susceptibility to Biased Results

In flow analysis, random errors due to operator intervention are significantly reduced, and traditional glassware is less intensively used. Attention should however be paid to the possibility of systematic errors, which generally increase when a batch-wise analytical method is carried

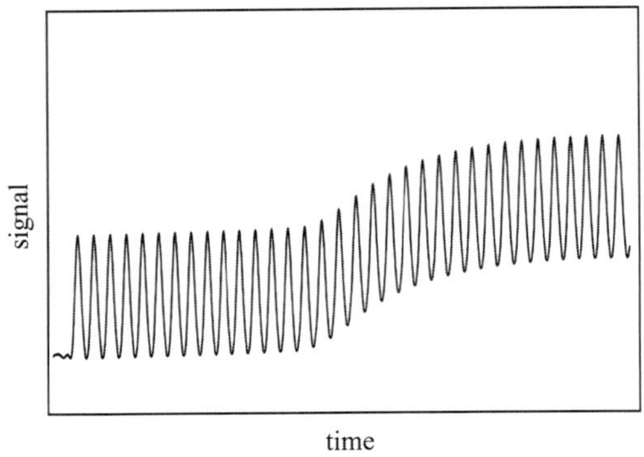

FIGURE 1.5 Didactic representation of a recorder output related to process monitoring in a situation of pronounced baseline drift.

out in a mechanised manner. Regarding this aspect, detecting and circumventing potential sources of inaccuracy can be efficiently accomplished in-line [21].

Reliable results are obtained because the sample is manipulated inside a closed environment which minimises analyte losses and/or sample contamination. Better precision is achieved with flow methods as compared with batch methods due to the reproducible timing and ease of automation.

1.4.4.3. *Improved System Design*

Other attractive features relate to the possibility of designing modular flow systems, and this aspect is a consequence of the high versatility associated with flow analysis. In view of the flexible design of the flow manifold and the high versatility of the flow system, several approaches for sample handling have been proposed, as discussed in Chapters 7 and 8.

1.5. AIMS AND SCOPE OF THE MONOGRAPH

Due to the intrinsic characteristics of flow analysis mentioned above, more and more attention is being given to this family of techniques. This is reflected in the numerous articles, books, reviews, and conference proceedings that have been published on flow analysis (see *e.g.*, Appendices 1.1 and 1.2). Internet databases, web pages of prominent researchers, tutorials, as well as standard, recommended, and/or official methods have recently been highlighted [22]. The ultimate objectives of this book are therefore to provide a sound scientific foundation for those interested in flow-based techniques and to familiarise a wide community of potential end-users, *e.g.*, researchers, students and technical staff, with the practical applications of *spectrophotometric* flow analysis, with an emphasis on environmental, biomedical, agronomical, food, industrial and pharmaceutical applications, as well as to encourage the analytical community, especially instrument manufacturers, developers and practitioners, to better exploit this important branch of analytical chemistry.

References

[1] F.J. Krug, H. Bergamin-Filho, E.A.G. Zagatto, Commutation in flow injection analysis, Anal. Chim. Acta 179 (1986) 103.
[2] A.O. Jacintho, M.A.Z. Arruda, E.A.G. Zagatto, B.F. Reis, Analytical balance as a detector in flow analysis, Anal. Chim. Acta 258 (1992) 129.
[3] J. Ruzicka, C.H. Pollema, K.M. Scudder, Jet ring cell. A tool for flow-injection spectroscopy and microscopy on a renewable solid support, Anal. Chem. 65 (1993) 3566.
[4] E.A.G. Zagatto, P.J. Worsfold, Flow analysis: overview, in: P.J. Worsfold, A. Townshend, C.F. Poole (Eds.), second ed., Encyclopedia of Analytical Science, vol. 3 Elsevier, Oxford, 2005, p. 24.
[5] J. Ruzicka, E.H. Hansen, Flow Injection Analysis, J. Wiley & Sons, New York, 1981.
[6] J.F. Tyson, Analysis: What Analytical Chemists Do, Royal Society of Chemistry, London, 1988.
[7] E.A.G. Zagatto, The renaissance of spectrophotometry, J. Flow Injection Anal. 19 (2002) 1.
[8] L. Sommer, Analytical Absorption Spectrophotometry in the Visible and Ultraviolet: The Principles, Elsevier, Amsterdam, 1989.
[9] C.L. Bashford, D.A. Harris (Eds.), Spectrophotometry and Spectrofluorimetry: A Practical Approach, IRL Press, Oxford, Washington DC, 1987.
[10] E.A.G. Zagatto, B.F. Reis, C.C. Oliveira, R.P. Sartini, M.A.Z. Arruda, Evolution of the commutation concept associated with the development of flow analysis, Anal. Chim. Acta 400 (1999) 249.
[11] T. Takayanagi, FIA bibliography (51), J. Flow Injection Anal. 26 (2009) 64.
[12] J. Ruzicka, From beaker chemistry to programmable microfluidics, Collect. Czech. Chem. Commun. 70 (2005) 1737.
[13] M. de la Guardia, J. Ruzicka, Towards environmentally conscientious analytical-chemistry through miniaturization, containment and reagent replacement, Analyst 120 (1995) 17N.
[14] D. Betteridge, B. Fields, Construction of pH gradients in flow injection analysis and their potential use for multielement analysis in a single sample bolus, Anal. Chem. 50 (1978) 654.
[15] A.C.B. Dias, E.P. Borges, E.A.G. Zagatto, P.J. Worsfold, A critical examination of the components of the Schlieren effect in flow analysis, Talanta 68 (2006) 1076.
[16] J. Ruzicka, L. Scampavia, From flow injection to bead injection, Anal. Chem. 71 (1999) 257A.
[17] J.J. Harrow, J. Janata, Heterogeneous samples in flow-injection systems: part. 2. standard addition, Anal. Chim. Acta 174 (1985) 123.
[18] M.F. Gine, H. Bergamin-Filho, B.F. Reis, R.L. Tuon, Intelligent flow-injection inductively-coupled plasma system for matrix matching, Anal. Chim. Acta 234 (1990) 207.
[19] P.R. Fortes, S.R.P. Meneses, E.A.G. Zagatto, A novel flow-based strategy for implementing differential kinetic analysis, Anal. Chim. Acta 572 (2006) 316.
[20] W.E. van der Linden, Flow injection analysis in on-line process control, Anal. Chim. Acta 179 (1986) 91.
[21] E.A.G. Zagatto, F.R.P. Rocha, P.B. Martelli, B.F. Reis, Detecting and circumventing sources of inaccuracy in flow analysis, Pure Appl. Chem. 73 (2001) 45.

[22] S.J. Chalk, Flow analysis and the internet – databases, instrumentation and resources, in: M. Trojanowicz (Ed.), Advances in Flow Analysis, Wiley-VCH, Weinheim, 2008, p. 321 (Chapter 12).

Appendix 1.1. Important Monographs Related to Flow Analysis

J.K. Foreman, P.B. Stockwell, Automatic Chemical Analysis, Ellis Horwood, Chichester, 1975.

W.B. Furman, Continuous-Flow Analysis. Theory and Practice, Marcel Dekker, New York, 1976.

J. Ruzicka, E.H. Hansen, Flow Injection Analysis, J. Wiley & Sons, New York, 1981.

W. Coakley, Handbook of Automated Analysis. Continuous Flow Techniques, Marcel Dekker, New York, 1981.

M. Valcarcel, M.D. Luque de Castro, Flow Injection Analysis. Principles and Applications, Ellis Horwood, Chichester, 1987.

J. Ruzicka, E.H. Hansen, Flow Injection Analysis, second ed. Wiley-Interscience, New York, 1988.

B. Karlberg, G.E. Pacey, Flow Injection Analysis. A Practical Guide, Elsevier, Amsterdam, 1989.

M. Valcarcel, M.D. Luque de Castro, Automatic Methods of Analysis, Elsevier, Amsterdam, 1989.

J.L. Burguera (Ed.), Flow Injection Atomic Spectroscopy, Marcel Dekker, New York, 1989.

R.D. Schmid (Ed.), Flow Injection Analysis (FIA) Based on Enzymes or Antibodies, VCH Pub., Weinheim, 1991.

M. Valcarcel, M.D. Luque de Castro, Non-chromatographic Continuous Separation Techniques, Royal Society of Chemistry, Cambridge, 1991.

Z. Fang, Flow Injection Separation and Preconcentration, VCH Pub., Weinheim, 1993.

Z. Fang, Flow Injection Atomic Absorption Spectrometry, Wiley, Somerset NJ, USA, 1995.

J.M. Calatayud, Flow Injection Analysis of Pharmaceuticals: Automation in the Laboratory, Taylor & Francis, London, 1996.

A. Sanz-Medel, Flow Analysis with Atomic Spectrometric Detectors, Elsevier, Amsterdam, 1999.

M. Trojanowicz, Flow Injection Analysis: Instrumentation and Applications, World Scientific, London, 2000.

M. Trojanowicz (Ed.), Advances in Flow Analysis, Wiley-VCH, Weinheim, 2008.

S.D. Kolev, I.D. McKelvie (Eds.), Advances in Flow Injection Analysis and Related Techniques, Wilson and Wilson's Comprehensive Analytical Chemistry, vol. 54, Elsevier, Amsterdam, 2008.

A. Cerda, V. Cerda, An Introduction to Flow Analysis, Universitat de les Illes Balears, Mallorca, Spain, 2009.

Appendix 1.2. Journal Special Issues Dedicated to Conferences on Flow Analysis

International Conferences on Flow Analysis, Anal. Chim. Acta 114 (1980), 145(1983), 179 (1986), 214 (1988), 261 (1992), 308 (1995), 366 (1998), 438 (2001), 499 (2003), 600 (2007). 668 (2010).

International Conference on Flow Analysis (and related techniques), Talanta 52 (2000), 58(2002), 66 (2004), 68 (2005), 77 (2008), 79 (2009).

CHAPTER

2

Historical View

The whole of science is nothing more than refinement of everyday thinking. **A. Einstein**

OUTLINE

It is somewhat surprising that the monitoring of flowing solutions, now routine in Analytical Chemistry, started at the beginning of the twentieth century in the context of Physical Chemistry. At that time more information on novel chemical reactions was required, especially for the evaluation of reaction rate constants. To this end, visual measurements using static solutions were initially used, but they did not provide complete information on chemical kinetics, especially when the reaction times involved were too short. The need for rapid solution mixing and in-line detection quickly became evident, and flow-based strategies were seen as the best way to achieve these objectives. The landmark article by Hartridge & Roughton [1] highlighted the state-of- the-art of flow analysis as early as 1923, emphasising its potential for monitoring purposes. These flow systems relied on pressurised air as the fluid propeller and on detection with the naked-eye but, even so, the influence of the flow pattern on mixing conditions, as well as the vortex movement of fluids, were thoroughly investigated. Better results were obtained when solutions at high flow rates were accommodated inside narrow bore tubing (Fig. 2.1). These features were important in the context of the development of the *stopped-flow analyser.*

In 1940, a stopped-flow apparatus with spectrophotometric detection was designed for the determination of fast reaction rate constants [2]. The sample and the reagent streams were merged at the inlet of a mixing chamber positioned inside the spectrophotometer. A motor driven syringe produced a sudden and high flow rate that improved homogenisation of the reaction medium. The stream was then stopped and the formation of the reaction products monitored. During solution insertion (GO period), turbulence was approached, thus improving solution mixing. Successive measurements were performed during the STOP period which was dependent on the reaction kinetics and the required analytical sensitivity. The development of the stopped-flow analyser for chemical analysis and/or reaction rate evaluation has been reviewed by Chance [3].

In the context of developments in spectrophotometry, the commercial availability of the Beckman DU photoelectric colorimeter in the 1940's should be highlighted. Its original design dates from 1941, and it was produced until 1976 [4]. Although the instrument did not provide facilities for automated scanning and recording, it became "the universal spectrophotometer" during the science boom after World War II.

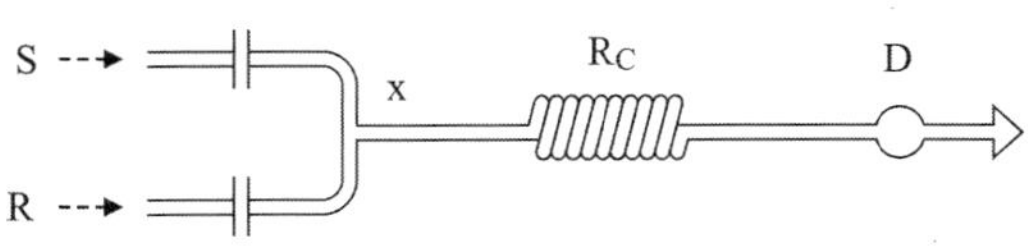

FIGURE 2.1 Didactic representation of a stopped-flow analyser. S = sample; R = reagent; R_C = reactor (often a transmission line); D = flow-through detector; x = special confluent joint for attaining turbulent mixing and vortex formation; dashed arrows, || = devices for simultaneous and sudden introduction of sample and reagent solutions (GO period), such as hammers, pistons and valves. For details, see text.

Its general acceptance was the driving force for the inception of different strategies for flow-based analytical procedures [5]. These ranged from simple devices for feeding/washing the flow cell with samples to more sophisticated systems where the sample was pumped through narrow-bore tubing and, after proper treatment, reached the spectrophotometric flow-through detector where a steady signal was quantified.

2.1. EARLY DEVELOPMENTS

In order to perform repetitive assays, simple systems with mechanical feeding/washing of the flow cell were conceived in the early nineteen fifties, and a wide variety of flow spectrophotometers, turbidimeters and nephelometers were already commercially available in 1955 [6]. This approach is still exploited, mainly in connection with process analyses.

Another relevant feature common in the 1950's was the adaptation of the spectrophotometer in order to permit measurements of temperature-dependent absorbance [7]. The sample was initially immersed in a temperature-controlled water-bath for thermal pre-equilibration. Thereafter, a sample aliquot was selected and placed inside a specially designed temperature-controlled cell where absorbance was measured. In the more sophisticated versions, the instrument was furnished with a peristaltic pump for propelling the sample solution towards the cell and then to waste.

There were situations where the analyte needed to be converted into a more detectable species, and thus reagent addition was often required. These steps were efficiently accomplished in a flow system, as demonstrated in the determination of chlorine in waters [8]. A potassium iodide solution merged with the sample flowing stream, and the mixture passed through a coil, so that enough time for the stoichiometric release of iodine was achieved, and the liberated species was electrochemically determined.

In the above-mentioned applications, a steady situation was approached during passage of the sample through the detector, resembling what is now called the *"infinite volume situation"* [9]. As successive samples were analysed, the introduction of a sample was used to expel the previous one from the system. However, quantitative filling of the flow cell with the sample to be monitored was not always attained, especially for smaller sample volumes. Consequently, portions of the previous sample were a potential source of cross contamination. In order to circumvent these drawbacks, a large sample volume, and hence a high sampling time, was required. As this parameter was also dependent on the geometry and volume of the analytical path (including the flow-through detector), it became evident in the early 1950's that system dimensions needed to be reduced.

Limitations associated with carryover and large sample volumes became less severe when a washing solution was aspirated between successive samples. This led to the concept of the *sample carrier stream*. This stream (also called *background solution* [10]) was introduced after the sample aliquot and pushed it towards the detector, also acting as a wash solution. The required sample volume was thereby drastically reduced. This manifold architecture was a driving force for the inception of segmented flow analysis.

2.2. SEGMENTED FLOW ANALYSIS

Segmented flow analysis was conceived in the early 1950's [11] by Skeggs[1] (Fig. 2.2) as a logical consequence of the early developments in flow analysis. Automation at the Veterans Administration Hospital in Cleveland (Ohio, USA) was needed, and Skeggs considered robotic systems at that time too cumbersome to be practical. Therefore, he began to perfect the art of doing chemistry in flowing streams. To this end, his expertise in haemodialysis was also valuable [12].

Segmentation of the main flowing streams by adding air bubbles was the main innovation. It is interesting to note that Skeggs initially tried to develop a flow system without addition of air bubbles, but it was difficult to avoid the inlet of air during replacement of sample and carrier solutions. Skeggs recognised the benefits of adding air bubbles and decided to exploit segmentation. It is however somewhat surprising that air-segmentation was not emphasised in his landmark article *"An automatic method for colorimetric analysis"* [13] which focussed on urea determination in whole blood or blood serum using in-line dialysis. No comments on the addition and removal of air bubbles appeared in the summary or introductory section of this article, nor in its brief in Chemical Abstracts [14].

The development of air-segmented flow analysis was restricted to a single company (Technicon Corporation Inc.), owner of the main patents until the mid 1970's. The first air-segmented system marked with the *AutoAnalyzer®* trade name is shown in Fig. 2.3.

The expression *continuous flow analysis* and the corresponding acronym CFA have often been used to specify this mode of flow analysis, in accordance with official recommendations [15]. A historical survey

[1]Skeggs and Skeggs Jr are written interchangeably in the scientific literature. In this monograph, Skeggs is used.

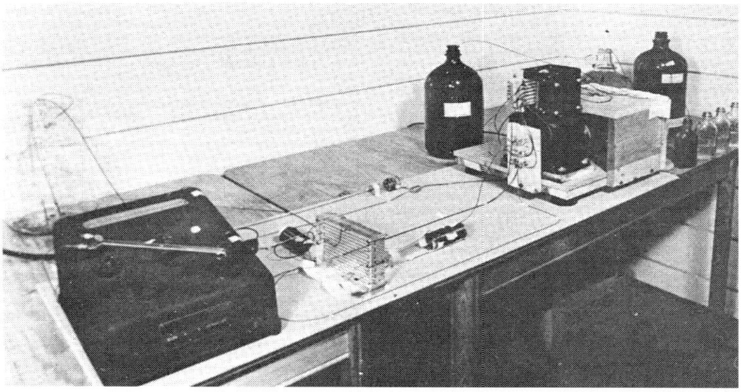

FIGURE 2.2 **The first prototype of a single-channel segmented flow analyser from May 1951.** *Reproduced with permission of Hindawi Publishing Corporation from "R. Stanley, Journal of Automatic Chemistry, 6 (1984) 175".*

FIGURE 2.3 **A first-generation single-channel AutoAnalyzer® (Technicon Instruments, *circa* 1957).** *Reproduced with permission of Hindawi Publishing Corporation from "R. Stanley, Journal of Automatic Chemistry, 6 (1984) 175".*

reveals however that the expression has also been associated with unsegmented flow analysis [16–19]. In the present monograph therefore the expression *continuous flow analysis* - or CFA - is not used to indicate segmented flow analysis.

Expansion of the original AutoAnalyzer concept resulted in the multichannel AutoAnalyzer dedicated to simultaneous determinations. For clinical analyses, the multiple, simultaneously recorded peaks can be combined to build up a *multiple analysis chart* [20], which has been recognised as a very important tool for clinical diagnosis.

The AutoAnalyzer underwent fast development and played a dominant role in the field of routine chemical assays for several decades, especially in relation to clinical chemistry. Nowadays, segmented flow analysis is very important in the context of large-scale analysis and applied worldwide in different fields [21].

In the segmented flow analyser (Fig. 2.4), a sampling arm successively selects the sample or the carrier/wash solution to be aspirated towards the detector, thus establishing the main aqueous stream. A convergent stream of air is thereafter added to promote segmentation.

Stream sectioning into a number of small segments reduces broadening of the flowing sample, and plays a beneficial role in the mixing process, reducing intermixing of the sample with the carrier stream and between successive samples.

During sample transport towards the detector, the analytical steps required by the specific application are performed in-line under reproducible conditions. Precise timing is involved; therefore chemical equilibrium is not always reached. Immediately before detection, the flow stream passes through a de-bubbler to remove the air bubbles. Sample passage through the detector results in a change in the monitored signal, which is recorded. The difference between the baseline and the maximum signal is related to the analyte concentration in the sample.

Segmented flow analysis relies on three cornerstone features: sample aspiration, stream segmentation and reproducible conditions due to precise timing.

The maximum signal is associated with the less dilute portion of the flowing sample and is often referred to as the plateau region. The recorded peak shape shows a tendency towards this steady state situation, as well as a slight axial dispersion of the sample zone. As segmentation is involved, a small sample aliquot is enough for the sample to reach the detector with its central portion almost undispersed. All of the above-mentioned features have a positive influence on sample throughput.

The segmented flow analyser is very robust but lacks versatility because its only commuting element is the sampling arm. Moreover, the

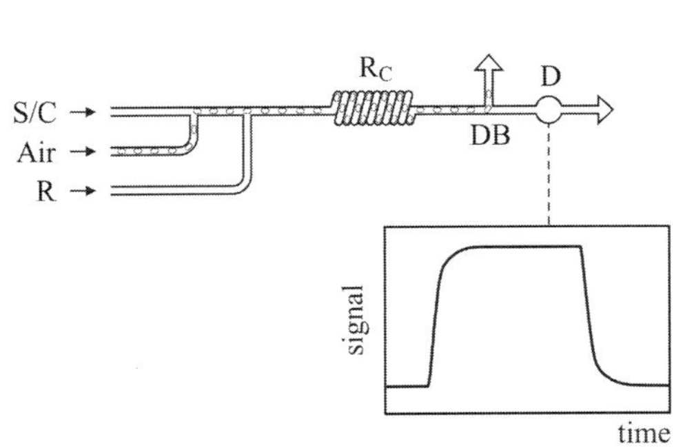

FIGURE 2.4 **Flow diagram of a single-channel segmented flow analyser and the associated recorder tracing.** S/C = sample/carrier wash stream; Air = air; R = reagent; R_C = coiled reactor; DB = de-bubbler; D = detector; arrows = sites where pumping is applied.

possibilities for system miniaturisation are limited, because the integrity of the air bubbles cannot be maintained inside very narrow bore tubing. A critical evaluation of segmented-flow analysers is given elsewhere [22].

During the 1960's and 1970's, the design of air-segmented flow systems continued to evolve and different kinds of pumps, tubing, flow-through detectors, and devices for specific in-line operations such as filtration, heating, dialysis, liquid-liquid extraction, ion-exchange and evaporation became commercially available [23]. In this period, more than one hundred million samples were assayed in clinical laboratories using air-segmented flow systems [24] and about 5000 papers were published [25], mostly dealing with methodological developments. However, few conceptual advances were made during this period, and the most significant achievements were: electronic rather than mechanical timing of the sampler, a rapidly moving sampling arm [26], a bubble-gating flow cell through which the air bubbles were allowed to pass [27], and a computer monitored instrument for analytical curve regeneration [28].

After the intensive development of air-segmented flow analysis, some successful analytical procedures without stream segmentation were proposed, mainly in connection with enthalpimetric [29,30], chemiluminometric [31] or spectrophotometric [32,33] detection. In these systems the time domain involved was generally short, thus air addition and removal was not performed. Either the sample or the wash solution was continuously aspirated towards the main channel and measurements were made under an almost steady state situation. Without air-segmentation, however, sample throughput was impaired and sample changing was cumbersome.

It did not take long for several researchers [16,34–40] to realise that insertion of a sample plug into a continuously moving unsegmented carrier stream was beneficial for overcoming some of the problems associated with flow segmentation, the dimensions of the analytical path, sample/reagent consumption and sample throughput. This strategy is considered the essence of flow injection analysis.

The development of chromatography also played an important role in the initial development of flow injection analysis, as pumps, tubing, flow-through detectors and accessories typical of chromatography systems were commercially available at the beginning of the 1970's.

Here, a parallel between the inception of segmented flow and flow injection analysis (section 2.3) can be drawn. Skeggs did not want air bubbles and incidentally realised the improved performance of systems with segmentation; on the other hand, Ruzicka & Hansen did want air bubbles and incidentally realised the improved performance of systems without segmentation [41]. Both were correct.

2.3. FLOW INJECTION ANALYSIS

The initial development of flow injection analysis was strongly influenced by projects at the Centre for Nuclear Energy in Agriculture, University of São Paulo (Brazil) sponsored by the International Atomic Energy Agency (IAEA), which required a high number of chemical analyses of natural waters and plant materials. In 1974/5, J. Ruzicka spent a year in Brazil as an IAEA expert, and the seminal article *"Flow Injection Analyses. Part I. A new concept of fast continuous flow analysis"* [16] was prepared there. This article relied on previous experiments carried out at the Technical University of Denmark in collaboration with E.H. Hansen [41]. Flow injection analysis was first applied to large-scale analysis at the analytical laboratories headed by H. Bergamin F° in Brazil, and the presence of J.W.B. Stewart (also an IAEA expert), S.S. Jørgensen and the staff of the host laboratory played an important role in this development. Requirements inherent to large-scale, routine analysis for solving practical problems, mainly in relation to agriculture, culminated in several novel approaches to

FIGURE 2.5 A flow injection system for the potentiometric determination of ammonium in Kjeldahl plant digests used in Brazil in 1976. The flow setup is supported by LEGO® blocks and includes a needleless syringes for manual sample injection, an air-gap ammonium electrode (in white), a model 8511 Polymetron peristaltic pump and a model 64 Radiometer pH meter (recorder not shown). For experimental details, see Ref. [43].

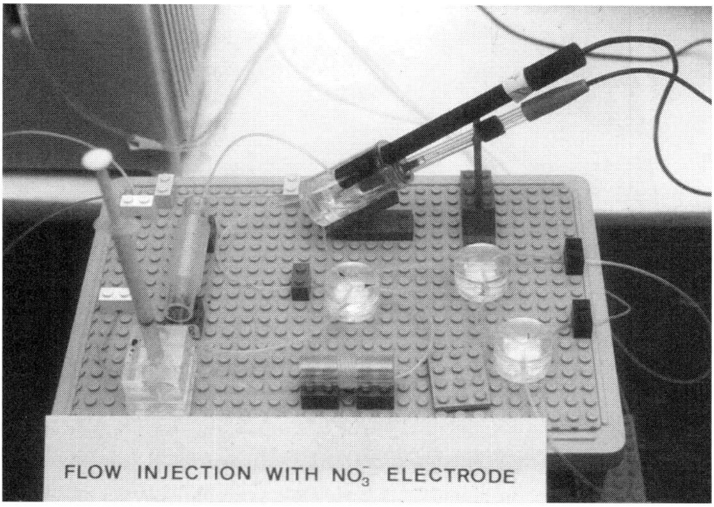

FIGURE 2.6 **A flow injection system for the potentiometric determination of nitrate in natural waters.** For experimental details see Ref. [52].

flow injection analysis [42]. Figs 2.5 and 2.6 show some of the first home-made flow injection systems developed in Brazil for routine assays.

The classical 10-article series [16–18,43–49] discussing in detail theoretical, methodological and practical aspects of flow injection analysis, as well as emphasising its potential, limitations, applications and trends, was published during the 1970's. Concomitantly, other early contributions from different countries appeared in the literature [50], and those prepared by the research team headed by M. Valcarcel in Spain [51] should be highlighted.

The flow injection analyser is operated as follows. A sample volume is precisely selected and introduced as a plug into an unsegmented flowing carrier stream, producing a fuzzy but very reproducible sample zone that undergoes continuous dispersion while being transported towards the detector. During sample transport through the analytical path, the steps inherent in the specific analytical application such as sample dilution, reagent addition and analyte concentration/separation, take place in a reproducible manner as a consequence of the fixed flow geometry and reproducible timing. When the sample passes through the detector, a transient signal is generated and recorded as a peak which is proportional to the analyte concentration in the sample.

One can then infer that flow injection analysis relies on three cornerstone features: sample injection, controlled dispersion and reproducible timing [50]. A typical flow injection analyser and the related recorder output are shown in Fig. 2.7.

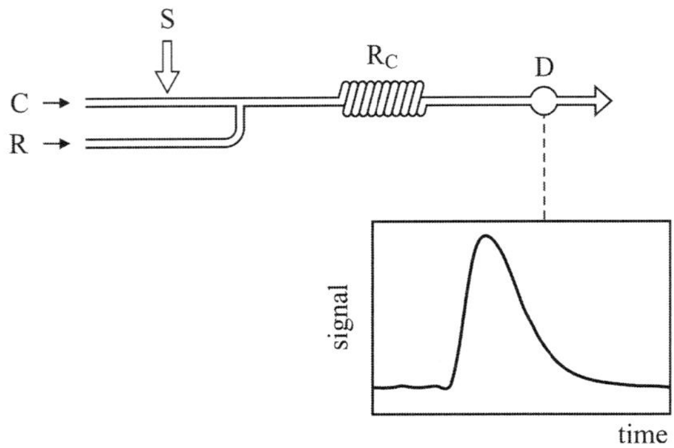

FIGURE 2.7 Flow diagram of a typical flow injection analyser and the associated recorder tracing. S = sample; C = sample/wash carrier stream; R = reagent; R_C = coiled reactor; D = detector; arrows = sites where pumping is applied.

As the sample plug is intercalated into the carrier/wash stream, the sample zone does not pass through the pumping device, so the time interval between sample introduction and detection is reduced. The steady situation corresponding to the infinite sample volume [9] is usually not required, thus only a small volume of sample is used. In view of the precise timing and the low mean sample residence time in the system, chemical equilibrium may not be attained; this feature is more relevant than in segmented flow analysis. As stream segmentation is not exploited, limitations related to increasing total flow rate inside the analytical path and to thermal expansion/contraction of the air-bubbles are less pronounced, and system miniaturisation can be more efficiently achieved. These aspects are beneficial for improving sample throughput and reducing reagent consumption. Sample broadening tends to be more pronounced, therefore exploitation of concentration gradients becomes easier [53,54].

As the unsegmented streams are almost incompressible, they are efficiently manipulated, and this aspect led to enhanced system versatility. In addition, commutation became more efficient, allowing different processes such as stream re-directing, flow reversal, addition/removal of manifold components, re-sampling and stream stopping to be efficiently accomplished. Consequently, several innovative approaches such as merging zones [55,56], zone trapping [57], zone slicing [58], zone sampling [59], stream splitting [44] and sample stopping [60,61], were proposed. Exploitation of these approaches in combination with chemometrics allowed reliable results to be obtained in a more efficient way [62].The features mentioned above led to an expansion of the application range of flow injection analysis.

As a consequence, a tendency to disregard segmented flow analysis became evident in the early 1980's. The dominance of unsegmented relative to segmented flow analysis became so pronounced that the phrase *Flow injection analysis - the end of the beginning? Segmented-flow analysis - the beginning of the end?* appeared in the scientific literature [63]. The number of researchers in the field of flow injection analysis experienced a marked growth, the related instrumentation became more accessible, and many novel strategies were proposed. Exploitation of segmented streams however did not cease, and relevant contributions also appeared in the literature [23].

In 1985, mono-segmented flow analysis was proposed [64] as a means of achieving extended sample incubation times without excessive sample dispersion. The sample was inserted between two air bubbles into an unsegmented carrier stream; therefore the innovation combined the favourable characteristics of both segmented and unsegmented flow systems. Further development revealed other potential applications, especially with regard to relatively slow chemical reactions, flow titrations, sample introduction to atomic absorption spectrometers, liquid-liquid extraction and multi-site detection (Chapters 7 and 8). This innovation was also referred to as *segmental flow injection analysis* [65].

Computer facilities started to be exploited in flow analysis at the end of the 1960's, mainly in relation to clinical chemistry [66]. Computerised segmented flow analysers resulted in enhanced solution management and system operation, as well as medical diagnosis [12,67]. Regarding flow injection analysis, a fully computer-controlled system was designed in 1980 [68], and programs for data treatment and instrument control were published [50,61]. The improved performance of unsegmented flow systems exploiting computer-assisted commutation has been critically discussed [69–71].

By the end of the 1980's, the presence of microcomputers in the analytical laboratories was widespread. In addition, there was a drive to develop more versatile systems able to perform different kinds of analysis without requiring frequent re-configuration of the manifold architecture. Consequently, there was a tendency to have several active, discrete computer-controlled devices in the manifold of flow analysers. Using these devices increased system flexibility, mainly the ability to handle fluidic zones, generate flow gradients and provide commuting facilities.

At that time, it was recognised that flow injection analysis, although a powerful tool for processing, manipulation and automation of chemical assays in the laboratory, was sometimes less suitable for use in the industrial environment due to the need for complex flow manifolds [72]. In addition, the diversity of industrial and *in situ* assays led to intensive studies on the potential of variable flow rates, especially sinusoidal flows [73,74].

These aspects, as well as the need for simpler systems, can be considered the driving forces for the proposal and development of sequential injection analysis.

2.4. SEQUENTIAL INJECTION ANALYSIS

Sequential injection analysis was conceived in 1990 by Ruzicka & Marshall [75] as a novel approach to process analysis and laboratory assays, as well as to the development of sensors. A flow diagram of the simplest analyser is shown in Fig. 2.8.

The heart of the system is a computer-controlled multi-way valve that reproducibly selects aliquots of sample and reagent solutions to be sequentially aspirated towards a holding coil, thus establishing a stack of zones. Thereafter, the flow is reversed and the sample/reagents mixture is directed towards the detector. Mixing between the zones is improved because, on transporting the zone stack, the zones penetrate each other. Analogous to segmented flow and flow injection systems, the steps inherent in a specific application take place under reproducible conditions inside the analytical path, which now includes the holding coil. When the sample passes through the detector, the resulting transient signal is recorded as a peak, the height of which is proportional to the analyte concentration in the sample. The recorded peak shape also reflects the sample/reagent interactions that are inherent to the specific application involved. As a typical peak shape cannot be generalised, a schematic representation of the recorded output is not shown in Fig. 2.8.

Sequential injection analysis relies on three cornerstone features: sequential aspiration, flow reversal, and controlled dispersion under reproducible timing.

The versatility, ruggedness and simplicity of sequential injection systems have been often recognised [76–78]. In fact, different analytical procedures could be carried out using the same manifold, commutation could be better exploited, and reagent addition was carried out only in the presence of the sample, thus reducing reagent consumption and waste generation.

Here, it is interesting to mention the stepwise injection system [79], a variant of sequential injection analysis (without a holding coil), which permits the accomplishment of all the steps involved in a specific application as if they were stationary. The related instrument is very robust and amenable to gas analysis.

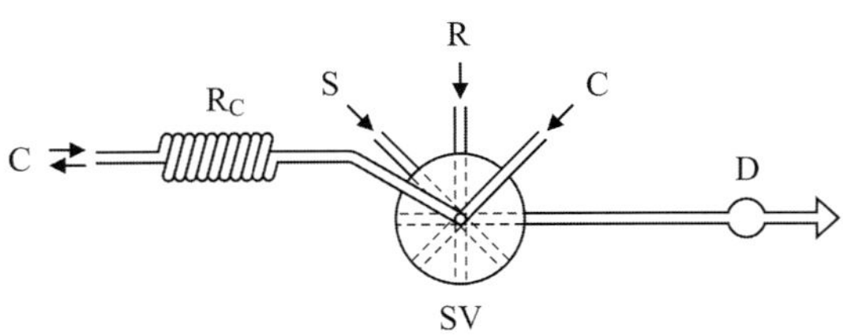

FIGURE 2.8 **Flow diagram of a simple sequential injection analyser.** S = sample; C = carrier/wash stream; R = reagent; R_C = holding coil; SV = selection valve; D = detector; arrow = site where pumping is applied.

Development of the sequential injection analyser, as well as intensive studies exploiting beads (including magnetic beads) [80–82], led to the appearance of flow analysers incorporating bead injection [83] and lab-on-valve [84] devices.

2.4.1. Bead Injection Analysis

The main characteristic of this system is its ability to efficiently manage functionalised beads [83,85]. The solutions involved are handled in the same way as in sequential injection analysis but the monitored species is normally adsorbed onto a solid phase, which helps to improve analytical sensitivity and/or selectivity.

A suspension of beads is introduced in the flow channel and the beads are trapped in a special cell. The analyte is concentrated in-line during passage of the sample through the cell by interacting with a specific reagent on the bead surface, thus forming a detectable species on the solid phase. Alternatively, an auxiliary reagent can be introduced into the system to perform this task [86]. The chemical species formed can be directly detected or otherwise eluted towards a detector. After each analytical cycle, the beads can be removed, allowing the cycle to be repeated with fresh beads.

The bead injection system was designed to operate in the sequential injection mode. It is possible to carry out bead injection analysis in a low-cost flow injection system with a unidirectional pump and without a computer, mainly for applications where samples and reagents are abundant and when there is no need for micro-volume control. To this end, it is necessary to use a flow cell able to achieve bead retention, accommodation of chemical reactions and detection [87,88].

The bead injection system was initially used for drug discovery [89], trace metal determinations [86] and *in-vivo* cellular metabolism evaluation [90–92]. Its potential for the determination of inorganic and organic chemical species, especially for agronomical, environmental, and pharmaceutical analysis, as well as nutrient monitoring, has been emphasised elsewhere [93–96].

2.4.2. Lab-on-valve

Sequential injection analysis with lab-on-valve was conceived by Ruzicka in 2000 [84] as a progression of the sequential injection system towards downscaling. All components are integrated in a single unit mounted on a multi-position valve. The detector compartment is reduced and placed very close to the sample inlet, thus minimising the required volumes of sample, reagent and wash solution.

The system also includes optical fibres coupled to the flow cell. This manifold architecture permits downscaling of the system using conventional pumps, detectors and other peripherals. The system typically operates at the microlitre scale but can be adapted to the nanolitre scale with an appropriate pumping device. Its versatility and robustness were initially demonstrated by a phosphate assay, an enzymatic activity assay of protease and a bio-ligand interaction assay of immunoglobulin G (IgG) based on its interaction with protein G immobilised on Sepharose beads [84]. The system was further applied to monitoring fermentation processes [97], evaluating environmental pollution [98], determining heavy metals in biological fluids by ICP-MS after in-line ion-exchange concentration [99], and separating anions in combination with capillary electrophoresis [100]. Recently, it was coupled with HPLC using a multi-syringe flow injection analyser for solid phase extraction prior to HPLC separation and detection [101]. More details of potential and applications of lab-on-valve flow systems can be found elsewhere [96,102–104].

2.5. MULTI-COMMUTED FLOW ANALYSIS

An important attribute of flow analysis started with the advent of computer facilities in the laboratory. Discretely operated devices in the manifold were exploited in advanced multi-commuted flow systems. These devices initially supported the main injector by performing specific tasks [105], especially those requiring external timing control and/or concentration-oriented feedback mechanisms. The concept became more established after a landmark article on multi-commutation [106].

Multi-commuted flow analysers are characterised by enhanced flexibility and easy operation and combine some of the attractive characteristics of both segmented and unsegmented flow analysis. In this way, segmented flow, flow injection and sequential injection analytical procedures are efficiently implemented in multi-commuted flow manifolds.

Multi-commuted flow systems can be classified in two modes, depending on the fluid delivery strategy involved. In the aspiration mode [106], several three-way valves are operated to select the solution to be aspirated towards the main channel at each step (Fig. 2.9). The system is designed in a straight (single line) configuration, requiring only one propulsion device, *e.g.*, one pump tube in a peristaltic pump or a single syringe. In the pumping mode [107], the solutions are pumped towards three-way valves that direct them either towards the analytical path or to

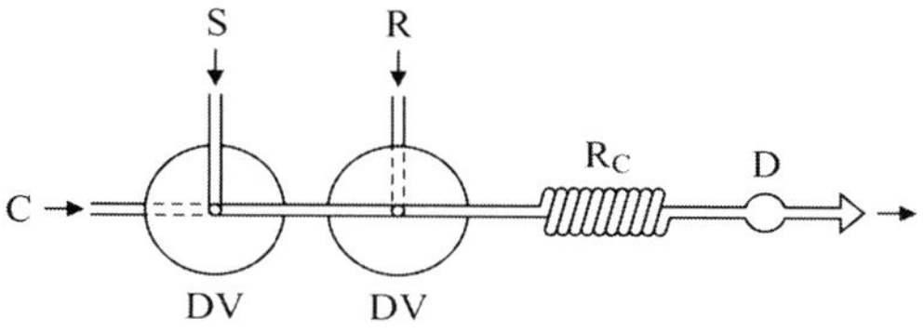

FIGURE 2.9 **Flow diagram of a simple multi-commuted analyser designed in the aspiration mode.** S = sample; C = carrier/wash stream; R = reagent; DV = directing valves; R_C = reactor; D = detector; arrows = sites where pumping is applied.

a recycling loop. Several pump tubes are then required and different solutions can be pumped simultaneously.

Regardless of the operating mode, and depending on the valve timing sequence, different sampling strategies relying on, *e.g.*, insertion of a sample plug into the carrier stream, exploitation of sample and reagent merging zones, and establishment of tandem flows [108] are easily implemented.

Multi-commuted flow analysis relies on three cornerstone features: discrete system operation, external timing and controlled dispersion.

Inside the analytical path, the steps inherent to a specific application take place in an analogous manner to conventional flow systems. A noteworthy feature of multi-commuted flow systems is the ease with which streams can be redirected and manifold components added or removed. Characteristics, attractive features, limitations and applications of multi-commuted systems have been comprehensively reviewed elsewhere [70,71].

2.5.1. Multi-syringe Flow Injection Analysis

Another alternative for designing multi-commuted flow systems operating in the pumping mode is to exploit syringes as fluid propulsion devices. This led to the proposal of multi-syringe flow injection analyser by Albertus *et al.* in 1999 [109] as an advanced means of managing multi-channel flow analysis.

Discretely operated multi-syringe piston pumps are utilised to accomplish fluidic delivery in a multi-channel flow manifold. Some drawbacks observed in other flow systems, *e.g.*, the maintenance required for peristaltic pumps and the variations in flow rates due to pump tube ageing, are overcome with this approach.

The system combines the advantageous characteristics of flow injection analysis, such as a high sample throughput and low detection limits, with those of sequential injection analysis such as robustness and versatility. It was initially exploited for industrial process control, with spectrophotometric detection, for the determination of highly acidic or alkaline protolytes using an ingenious titration procedure [109], free acidity and iron(III) in metallurgical solutions [110], and of nickel(II) and iron speciation [111], as well as for the determination of sulfur dioxide in wines [112], and arsenic in solid and water reference materials by hydride generation atomic fluorescence spectrometry [113]. Details of characteristics, attractive features, limitations and applications are given elsewhere [114].

2.5.2. Multi-pumping Flow Analysis

Laminar flows are inherent to most flow analysers, as already recognised in the 1970's [115], but their exploitation can lead to a pronounced broadening of the sample zone. This shortcoming can be circumvented by resorting to turbulent mixing, which can be achieved when membrane, solenoid or step-wise piston pumps which deliver pulsed flows [116–119] are used. Exploitation of pulsed flows led to the proposal of multi-pumping flow analysis by Lapa *et al.* in 1999 [120].

The multi-pumping flow system comprises several computer-controlled solenoid pumps strategically positioned in the manifold, leading to enhanced system versatility. The pumps provide a non-metallic, inert fluid path, recommended for dispensing high-purity or aggressive fluids [121]. Because discretely-operated devices and pulsed flows are involved, solution propulsion, improved mixing conditions, selection of sample and reagent aliquots, introduction of these aliquots into the analytical path, establishment of tandem streams, stopping the sample, and commuting facilities are efficiently attained, as demonstrated in different analytical applications with *e.g.*, spectrophotometric [122], fluorimetric [123] and chemiluminometric [124] detection.

The approach is very useful for *e.g.*, chemi- and bioluminescence detection when the monitoring of short-lived chemical species is done as close as possible to the detector. With pulsed flows, mixing conditions are no longer a weakness [125] because manifold simplicity and very fast and effective mixing of the sample with reagent(s) are characteristics of multi-pumping flow systems.

Another favourable aspect of pulsed flows is the exploitation of fluidised beds [126], a novel strategy for the utilisation of solid reagents (ion-exchangers, metallic particles, enzyme cartridges) in flow analysis. As the flow inside the fluidised bed undergoes abrupt variations, a diffuse yet reproducible geometry is attained, allowing the solid particles to be maintained in constant motion (floating, refluxing and circulating) inside a mini-column. Drawbacks inherent to solid-phase packed columns such as backpressure, preferential pathways and swelling are therefore minimised. A review of the potential and applications of multi-pumping flow systems is given elsewhere [127].

2.6. OTHER FLOW SYSTEMS

During the evolution of flow analysis, other specific systems for solving particular analytical problems have been developed, *e.g.*, the batch injection analyser [128] which was conceived simultaneously with the advent of the sequential injection analyser. It is very simple and low-cost, resembles the early flow injection analysers and exploits the fast injection of microlitre sample volumes toward a nearby detector that is in contact with a large-volume solution. This innovation was particularly suitable in conjunction with electrochemical sensors, where the reference electrode is usually immersed in the solution. High sample throughput, measurement reproducibility and manifold simplicity are attained, and pumps, valves and associated tubing are not required. The system can be manually operated, and sampling rates as high as 1800 h^{-1} can be attained using computer-operated pipettes [129]. Most batch injection analytical procedures involve electrochemical detection [130–133], but spectrophotometric [129] and fluorimetric detection [134] with fibre optics technology can also be used.

An ingenious system relying on a hanging drop was also proposed [135]. The sample and reagent solutions flow through a tube, resulting in a liquid drop where the analyte is monitored. The drop therefore acts as a windowless flow cell. This approach has been used mainly for gas analysis, involving solvent extraction for headspace sampling [136]. The drawback of a short optical path, inherent to a hanging liquid drop, was recently circumvented by using a simple flow-through type optical cell with open ends, so that a prolonged "liquid drop" was established [137]. The liquid drop acted as a reactor and optical cell for the spectrophotometric determination of formaldehyde in wastewaters. With this innovation, the advantages of a windowless optical cell provided by a liquid drop were retained, while the optical path was increased to over seven times of that of a traditional liquid drop. Recently, the area-to-volume ratio, and hence analytical sensitivity, was dramatically improved by designing a flow system relying on a soap bubble [138]. Details of the potential, limitations and applications of liquid drop flow cells can be found elsewhere [139].

In 1999, Honorato *et al.* conceived a flow system combining the favourable characteristic of both flow and batch analysis [140], that was called a *flow-batch system*. The main component in the manifold is a mini-chamber into which different solutions can be added or removed. As most of the steps related to a specific application, such as sample conditioning, reagent addition and detection, can be reproducibly carried out inside the chamber, it can be regarded as a mini-laboratory. The flow-batch system is amenable to self re-dimensioning by a concentration-oriented feedback mechanism, and has been successfully applied to individual sample conditioning [141,142], titrations [140,143], standard additions [144,145], water hardness screening [146], generation of concentration gradients [147], the in-line preparation of standard solutions [148], and capsule analyses [149]. It can also be operated according to the sequential injection analysis strategy, resembling a large lab-on-valve system. It should be stressed that the automated micro-batch analyser proposed earlier [150] has similar characteristics to the flow-batch analyser [148], but it cannot be considered as a typical flow-based analyser because a sample carrier stream is usually not used.

2.7. COMMUTATION IN FLOW ANALYSIS

Commutation refers to the act of modifying the manifold status [69], *e.g.*, substituting manifold components or changing the direction of a flowing stream by means of a commutator device, and its exploitation can enhance system versatility and hence system performance. There is a clear link between the developments in flow analysis and commutation [151].

Commutation was rudimentary in early air-segmented flow systems, as the only commuting part of the instrument was the sampling arm. Such systems, although robust, lacked versatility. Regarding flow injection analysis, syringes with or without needles were initially used to introduce selected sample aliquots into the flowing carrier stream [115]. Loop-based injection was a noteworthy improvement, as a more elaborate commutation was involved. This led to the development of rotary and sliding bar valves, the latter often referred to as single commutators. The potential of commutation in unsegmented flow became evident in 1980 when a flow injection system involving optional intercalation of a reducing mini-column into the manifold was proposed for the spectrophotometric determination of nitrate and nitrite in natural waters [152]. Further development demonstrated that the potential of unsegmented flow analysis could be greatly expanded by proper exploitation of commutation [71].

The multi-port selection valve of a sequential injection analyser can be regarded as an advanced commuting device, which selects one inlet port to be commuted to any one outlet port at a time. Also, the entire manifold can be accommodated in such a valve, leading to the lab-on-valve concept [84].

The presence of several discretely operated devices was the cornerstone feature of the modern exploitation of commutation. As a rule, multicommuted flow systems have high versatility and are suitable for implementing methods requiring external timing and / or concentration-oriented feedback mechanisms.

Proper exploitation of multi-commutation has led to the inception of different flow analysers such as the multi-pumping flow system, the multi-syringe flow injection system, the flow-batch system, the discontinuous flow system [153] and the all injection system [154].

In summary, the history of flow analysis is closely associated with the development and improvement of commuting devices.

2.8. FLOW PATTERN

A critical view of the development of flow analysis shows that efforts to improve mixing of the solutions have exploited flow regimes other than laminar flow.

In the first stopped-flow analysers, it was noted that turbulence was approached during the very sudden period of sample and reagent introduction (GO period) and that the reaction medium was stopped (STOP period) for a relatively long time interval [1,3]. Moreover, the authors reported the formation of vortices and their beneficial influence on mixing conditions. It should be stressed that the expression *turbulent flow* is not appropriate for characterising the flow regime during the GO period because the available time interval is too short to permit the establishment of a steady state situation [155].

In segmented flow analysis, the presence of successive drops of a second immiscible phase inside the flowing sample leads to the formation of vortices that define a circulating flow pattern between two successive solution plugs (see also Fig. 5.2). These vortices improve the mixing conditions and minimise sample broadening.

In flow injection analysis, the dominance of laminar flow was reported in 1977 when automated procedures for the potentiometric determination of potassium, sodium and nitrate in soil extracts and blood sera were proposed [47]. As broadening of the sample zone is more pronounced under a laminar flow regime, a high axial dispersion is more likely to occur in this mode of flow analysis. Although beneficial for some analytical applications, *e.g.*, those involving large sample dispersion [50] and/or exploiting concentration gradients [53], the effect is usually detrimental, as it can limit sample throughput. Consequently, several strategies have been exploited for minimising broadening, as discussed in section 3.1.1.4.

Here, it should be stressed that passing the flowing stream through platinum gauze can attain turbulent-like flows, and this greatly enhances mixing conditions, as emphasised by an improved chemiluminometric determination of formaldehyde [31]. The effect was also exploited in relation to discontinuous flow analysis, where the flowing sample and reagent solutions were mixed together after passing through a nozzle. In view of the geometry of this device and the high mean linear speed of the streams, turbulence was approached and excellent mixing conditions were reported [153]. Applicability of the approach was limited, however, probably because of the high pressure drop involved during the entire analytical cycle, thus necessitating a special pump.

Use of supercritical fluids as carrier streams was also proposed [156] as a means to improve mixing, but it was not commercialised because the system required special components and, from a practical point of view, there is only one chemical species (CO_2) suitable for this purpose. Hence system versatility and applicability were limited.

Gaseous carrier streams have been exploited in relation to gas-solid [157] and gas-liquid interfaces [158] but, in spite of the improvement of mixing conditions by exploiting fugacity, the number of applications to date is relatively low.

Beneficial aspects arising from the exploitation of pulsed flows [119] are increasingly being exploited. The flowing stream is established by successive insertions of fixed solution volumes (pulses) at a given frequency. The resulting stream then undergoes sudden successive stopping between each pulse, so that a steady laminar flow regime is not attained. As already mentioned, turbulence is approached during the very short GO periods, and this type of flow is strongly recommended in flow analysis [159].

2.9. INSTRUMENT CHARACTERISTICS

An important aspect to be considered in reviewing the historical development of flow analysis is instrument size, with the tendency being towards miniaturisation. This is confirmed by comparing the flow injection analysers in Figs 2.5 and 2.6 with modern flow analysers. Also, very low-volume flow systems with engraved micro-conduits [160] have been designed but not widely accepted, probably due to the lack of versatility and difficulty in maintaining the geometry for a long period.

Regarding flow systems with very low inner volumes, there has been intensive development of total analytical systems (μTAS) [161], as well as very low-volume flow injection analysers (μFIA) [160].

During the VIII International Conference on Flow Analysis [162,163] there was a consensus that extreme miniaturisation is not always desirable. In fact, the increase in system complexity and cost, as well as the deterioration of some analytical figures of merit, may not justify the benefits in portability, sample/reagent consumption, and possibility of *in vivo* analysis, inherent in the reduction of analyser dimensions.

Nowadays, different instruments exploiting different strategies and with different complexities are found in laboratories devoted to high throughput analysis. Novel systems with the combined characteristics of different modes of flow analysis are being proposed, *e.g.*, a segmented flow injection system for the determination of total cyanide in waters [164] and flow-batch systems for titrations [140].

There is a pronounced tendency towards coupling flow analysis with other analytical systems. Hyphenation has proved to be an excellent development, as the favourable characteristics of both sub-systems are better exploited [165].

2.10. TRENDS

Flow analysis has experienced amazing developments in recent decades and continues to evolve. The number of flow-based analytical procedures is increasing [166] and novel flow analysers with a high degree of automation are commercially available. Multi-commutation is an attractive feature and is also increasingly exploited [71].

In the past, some features, such as turbulent mixing and the presence of air bubbles, were considered to be a problem due to inherent technological challenges but are now useful in certain situations [159]. In view of the progress of flow analysis one can predict that future flow analysers will be very different from those in current use and emerging technologies may well allow old ideas to be harnessed in new analyser designs.

In the early stages of flow analysis, maximising sample throughput was the main objective and mechanisation of classical analytical procedures was usually the way this was achieved. It was soon realised, however, that flow analysers could be used to manipulate and process samples in ways that were not possible with batch analysers, and less emphasis was given to sample throughput. The impact of the phrase *"The instrument processes hundreds of samples per hour but I only have a few"* became less and less important for researchers and manufacturers. Nowadays, the dominance of novel flow-based methodologies with a relatively low sampling throughput but with special characteristics, *e.g.*, suitable for *in situ* and/or *in vivo* analysis [167], detecting and circumventing potential sources of inaccuracy [168], management of unstable solutions [169,170], addition of in-line formed crystals for assisting nucleation [171], slurry analysis [172], in-line sample preparation [173,174], accuracy assessment [175], and individual sample conditioning [141], is a reality.

The evolution of flow analysis has caused a paradigm shift in Analytical Chemistry. Modern innovations are more associated with the duality of chemical equilibria and kinetics than simply with chemical equilibria. Instrumentation has experienced continuous downsizing, and the availability of miniaturised systems generating lower volumes of waste is an attractive feature of flow analysis. The implementation of different analytical methods in the same flow manifold is common. The quantitative information provided by flow analysers is impressive, and flow injection analysis was thought to be a numbers game (maximising sample throughput) in the early 1980's [62]. This was a driving force towards the increasing number of bioanalytical applications and flow-based analytical procedures exploiting chemometrics. Additionally, the trend towards full automation, miniaturisation and *in situ* analysis is compatible with the attractive features of flow-based methodologies.

Regarding the evolution of Science and Technology, the unceasing search for new developments results in innovations and products with increasing potential and applicability, normally classified by different generations. As a rule, the first generation synthesises pre-existing concepts, innovations or products, the second generation replaces the first, the third generation replaces the second, and so on. As a rule, the most recent generation is the one of choice. This trend however is not the case in flow analysis. In fact, air-segmented flow analysers included favourable features of pre-existing conceptual, methodological and instrumental achievements; flow injection analysers emerged after two decades, sequential injection and multi-commuted flow analysers were then proposed as a consequence of technological advances, the availability of different commuting devices and the appearance of the microcomputer in the laboratory. A contradictory feature is that the appearance of a new mode of flow analyser has not resulted in the obsolescence of previously reported modes. This is the main reason why the development and manufacture of air-segmented flow and flow injection analysers continue to progress in spite of the appearance of other instruments with specialised analytical characteristics.

There are many variants of analytical flow systems, e.g., segmented flow analysis, flow injection analysis, sequential injection analysis, multi-syringe flow injection analysis, batch injection analysis, mono-segmented flow analysis, flow-batch analysis, multi-pumping flow analysis, all injection analysis and bead injection analysis, all of which have acronyms [176]. In view of the existence of several common features, however, all flow analysers can be broadly classified as either segmented or unsegmented, with the most common example of the later mode being the flow injection analyser.

References

[1] H. Hartridge, F.J.W. Roughton, A method of measuring the velocity of very rapid chemical reactions, Royal Soc. Proc. A104 (1923) 376.

[2] B. Chance, The accelerated flow method for rapid reactions. Part II. Design, construction, and tests. I. Apparatus construction, J. Franklin Inst. 229 (1940) 737.

[3] B. Chance, The stopped-flow method and chemical intermediates in enzyme reactions – a personal essay, Photosynth. Res. 80 (2004) 387.
[4] R.D. Simoni, R.L. Hill, M. Vaughan, H. Tabor, A classic instrument: the Beckman DU spectrophotometer and its inventor, Arnold O. Beckman, J. Biol. Chem. 278 (2003) 79.
[5] J.K. Foreman, P.B. Stockwell, Automatic Chemical Analysis, Ellis Horwood Limited, Chichester, 1975.
[6] G.D. Patterson Jr., Automatic Operations, Anal. Chem. 27 (1955) 574.
[7] V.A. Yarborough, J.F. Haskin, W.J. Lambdin, Temperature dependence of absorbance in ultraviolet spectra of organic molecules, Anal. Chem. 26 (1954) 1576.
[8] H.C. Marks, U.S. Patent 2,585,811 (1952).
[9] F.J. Krug, J. Mortatti, L.C.R. Pessenda, E.A.G. Zagatto, H. Bergamin-Filho, Flow injection spectrophotometric determination of boron in plant material with Azomethine-H, Anal. Chim. Acta 125 (1981) 29.
[10] D.B. Roodyn, Part I. Automated enzyme assays, in: T.S. Work, E. Work (Eds.), Laboratory Techniques in Biochemistry and Molecular Biology, vol. 2, North Holland, Amsterdam, 1970.
[11] C.J. Patton, personal communication, 1999.
[12] L.T. Skeggs, Persistence and prayer: From the artificial kidney to the autoanalyzer, Clin. Chem. 46 (2000) 1425.
[13] L.T. Skeggs Jr., Automatic method for colorimetric analysis, Am. J. Clin. Pathol. 28 (1957) 311. Tech. Bull. Registry Med. Technologists 27 (1957) 171.
[14] A.E. Teeri (Abstractor), Chem. Abstr. (1958) 1694i.
[15] W.E. van der Linden, Classification and definition of methods based on flowing media, Pure Appl. Chem. 66 (1994) 2493.
[16] J. Ruzicka, E.H. Hansen, Flow injection analyses. Part I. A new concept of fast continuous flow analyses, Anal. Chim. Acta 78 (1975) 145.
[17] J. Ruzicka, J.W.B. Stewart, Flow injection analysis. Part II. Ultrafast determination of phosphorus in plant material by continuous flow spectrophotometry, Anal. Chim. Acta 79 (1975) 79.
[18] J. Ruzicka, E.H. Hansen, H. Mosbaek, Flow injection analysis. Part IX. A new approach to continuous flow titrations, Anal. Chim. Acta 92 (1977) 235.
[19] H. Bergamin-Filho, J.X. Medeiros, B.F. Reis, E.A.G. Zagatto, Solvent extraction in continuous flow injection analysis. Determination of molybdenum in plant material, Anal. Chim. Acta 101 (1978) 9.
[20] L.T. Skeggs, New dimensions in medical diagnoses, Anal. Chem. 38 (1966) 31A.
[21] S. Coverly, Segmented flow analysis, in: P.J. Worsfold, A. Townshend, C.F. Poole (Eds.), Encyclopedia of Analytical Science, second ed., vol. 8 Elsevier, Oxford, 2005, p. 210.
[22] C.J. Patton, R. Crouch, Experimental comparison of flow-injection analysis and air-segmented continuous flow analysis, Anal. Chim. Acta 179 (1986) 189.
[23] D.A. Burns, Automated sample preparation, Anal. Chem. 53 (1981) 1403A.
[24] L. Snyder, J. Levine, R. Stoy, A. Conetta, Automated chemical analysis: update on continuous-flow approach, Anal. Chem. 48 (1976) 942A.
[25] Technicon Autoanalyser Bibliography 1957-67, Technicon Corp., Ardslay, N. York, 1968; Technicon Autoanalyser Bibliography 1967-75, Technicon Corp., Tarrytown, N. York, 1975.
[26] R.E. Thiers, J. Bryan, K. Oglesby, A multichannel continuous-flow analyzer, Clin. Chem. 12 (1966) 120.
[27] R.L. Habig, B.W. Schlein, L. Walters, R.E. Thiers, A bubble-gating flow cell for continuous-flow analysis, Clin. Chem. 15 (1969) 1045.
[28] W.H.C. Walker, Curve regeneration in continuous flow analysis, Clin. Chim. Acta 32 (1971) 305.

[29] P.T. Priestley, W.S. Sebborn, R.F.W. Selman, Continuous-flow enthalpimetry, Analyst 90 (1965) 589.
[30] W.R. McLean, G.E. Penketh, Automatic analysis with thermometric detection, Talanta 15 (1968) 1185.
[31] D. Slawinska, J. Slawinski, Chemiluminescent flow method for determination of formaldehyde, Anal. Chem. 47 (1975) 2101.
[32] W.J. Blaedel, G.P. Hicks, Continuous analysis by measurement of the rate of enzyme catalyzed reactions. Glucose determination, Anal. Chem. 34 (1962) 388.
[33] W.J. Blaedel, C.L. Olson, XI. Continuous analysis, in: S.Z. Lewin (Ed.), Topics in chemical instrumentation, J. Chem. Educ., vol. 40, 1963, p. A549.
[34] G. Nagy, Z. Feher, E. Pungor, Application of silicone rubber-based graphite electrodes for continuous flow measurements. 2. Voltammetric study of active substances injected into electrolyte streams, Anal. Chim. Acta 52 (1970) 47.
[35] H.U. Bergmeyer, A. Hagen, Ein neues Prinzip enzymatischer Analyse, Frezenius' Z, Anal. Chem. 261 (1972) 333.
[36] J.R. Sarbeck, P.A. St. John, J.D. Winefordner, Measurement of microsamples in atomic emission and atomic fluorescence flame spectrometry, Mikrochim. Acta 1 (1972) 55.
[37] V.R. White, J.M. Fitzgerald, Continuous determination of ascorbic acid by photobleaching of methylene blue, Anal. Chem. 44 (1972) 1267.
[38] E.W. Chlapowski, H.A. Mottola, Some comparative studies on data handling in variable-time kinetic determinations. Modification of manganese(II) catalysis with 1,10-phenanthroline and some analogs, Anal. Chim. Acta 76 (1975) 319.
[39] K.K. Stewart, G.R. Beecher, P.E. Hare, Rapid analysis of discrete samples: the use of nonsegmented, continuous flow, Anal. Biochem. 70 (1976) 167.
[40] R.S. Schifreen, D.A. Hanna, L.D. Bowers, P.W. Can, Analytical aspects of immobilized enzyme columns, Anal. Chem. 40 (1977) 1929.
[41] E.H. Hansen, Flow-injection analysis: leaving its teen-years and maturing. A personal reminiscence of its conception and early development, Anal. Chim. Acta 308 (1995) 3.
[42] E.H. Hansen, J. Ruzicka, FIA is already a routine tool in Brazil, Trends Anal. Chem. 2 (1983) 5.
[43] J.W.B. Stewart, J. Ruzicka, H. Bergamin-Filho, E.A.G. Zagatto, Flow injection analysis. Part III. Comparison of continuous flow spectrophotometry and potentiomety for the rapid determination of total nitrogen content in plant digests, Anal. Chim. Acta 81 (1976) 371.
[44] J. Ruzicka, J.W.B. Stewart, E.A.G. Zagatto, Flow injection analysis. Part IV. Stream sample splitting and its application to the continuous spectrophotometric determination of chloride in brackish waters, Anal. Chim. Acta 81 (1976) 387.
[45] J.W.B. Stewart, J. Ruzicka, Flow injection analysis. Part V. Simultaneous determination of nitrogen and phosphorus in acid digests of plant material with a single spectrophotometer, Anal. Chim. Acta 82 (1976) 137.
[46] J. Ruzicka, E.H. Hansen, Flow injection analysis. Part VI. The determination of phosphate and chloride in blood serum by dialysis and sample dilution, Anal. Chim. Acta 87 (1976) 353.
[47] J. Ruzicka, E.H. Hansen, E.A.G. Zagatto, Flow injection analysis. Part VII. Use of ion-selective electrodes for rapid analysis of soil extracts and blood serum. Determination of potassium, sodium and nitrate, Anal. Chim. Acta 88 (1977) 1.
[48] E.H. Hansen, J. Ruzicka, B. Rietz, Flow injection analysis. Part VIII. Determination of glucose in blood serum with glucose dehydrogenase, Anal. Chim. Acta 89 (1977) 241.
[49] J. Ruzicka, E.H. Hansen, Flow injection analysis. Part X. Theory, techniques and trends, Anal. Chim. Acta 99 (1978) 37.

[50] J. Ruzicka, E.H. Hansen, Flow Injection Analysis, J. Wiley & Sons, New York, 1981.
[51] M. Valcarcel, Analyst 130 (2005) 1121.
[52] K. Reichardt, P.L. Libardi, N.M.F. Meirelles, H.F.F. Ferreyra, E.A.G. Zagatto, E. Matsui, Extração e análise de nitratos em solução do solo, Rev. Bras. Ci. Solo 1 (1977) 130.
[53] D. Betteridge, B. Fields, Construction of pH gradients in flow injection analysis and their potential use for multielement analysis in a single sample bolus, Anal. Chem. 50 (1978) 654.
[54] M. Agudo, J. Marcos, A. Rios, M. Valcarcel, Analytical potential of flow gradients in unsegmented flow systems, Anal. Chim. Acta 239 (1990) 211.
[55] H. Bergamin-Filho, E.A.G. Zagatto, F.J. Krug, B.F. Reis, Merging zones in flow injection analysis. Part 1. Double proportional injector and reagent consumption, Anal. Chim. Acta 101 (1978) 17.
[56] J. Mindegaard, Flow multi-injection analysis - a system for the analysis of highly concentrated samples without prior dilution, Anal. Chim. Acta 104 (1979) 185.
[57] F.J. Krug, B.F. Reis, M.F. Gine, J.R. Ferreira, A.O. Jacintho, E.A.G. Zagatto, Zone trapping in flow injection analysis. Spectrophotometric determination of low levels of ammonium ion in natural waters, Anal. Chim. Acta 151 (1983) 39.
[58] B.F. Reis, E.A.G. Zagatto, P.B. Martelli, S.M.B. Brienza, Trial measurements in flow analysis, Analyst 118 (1993) 719.
[59] B.F. Reis, A.O. Jacintho, J. Mortatti, F.J. Krug, E.A.G. Zagatto, H. Bergamin- Filho, et al., Zone sampling processes in flow injection analysis, Anal. Chim. Acta 123 (1981) 221.
[60] J. Ruzicka, E.H. Hansen, Stopped-flow and merging zones. A new approach to enzymatic assay by flow injection analysis, Anal. Chim. Acta 106 (1979) 207.
[61] P.J. Worsfold, J. Ruzicka, E.H. Hansen, Rapid automated enzymatic method for the determination of alcohol in blood and beverages using flow injection analysis, Analyst 106 (1981) 1309.
[62] D. Betteridge, Analytical chemistry: the numbers game, Chem. Brit. 16 (1980) 646.
[63] C. Riley, B.F. Rocks, Flow-injection analysis - The end of the beginning? Segmented-flow analysis - The beginning of the end? J. Autom. Chem. 5 (1983) 1.
[64] C. Pasquini, W.A. Oliveira, Monosegmented system for continuous-flow analysis. Spectrophotometric determination of chromium(VI), ammonia, and phosphorus, Anal. Chem. 57 (1985) 2575.
[65] L.-C. Tian, X.-P. Sun, Y.-Y. Xu, Z.-L. Zhi, Segmental flow-injection analysis. Device and applications, Anal. Chim. Acta 238 (1990) 183.
[66] M.A. Evenson, G.P. Hicks, R.E. Thiers, Peak characteristics and computers in continuous flow analysis, Clin. Chem. 16 (1970) 606.
[67] R.E. Thiers, R.R. Cole, W.J. Kirsch, Kinetic parameters of continuous flow analysis, Clin. Chem. 13 (1967) 451.
[68] J. Slanina, F. Bakker, A. Bruyn-Hes, J.J. Möls, Computer-controlled multichannel continuous-flow analysis system applied to the measurement of nitrate, chloride and ammonium ions in small samples of rain water, Anal. Chim. Acta 113 (1980) 331.
[69] F.J. Krug, H. Bergamin-Filho, E.A.G. Zagatto, Commutation in flow injection analysis, Anal. Chim. Acta 179 (1986) 103.
[70] M.C. Icardo, J.V.G. Mateo, J.M. Calatayud, Multicommutation as a powerful new analytical tool, Trends Anal. Chem. 21 (2002) 366.
[71] M.A. Feres, P.R. Fortes, E.A.G. Zagatto, J.L.M. Santos, J.L.F.C. Lima, Multi-commutation in flow analysis: recent developments and applications, Anal. Chim. Acta 618 (2008) 1.
[72] S. Liu, P.K. Dasgupta, Sequential injection analysis in capillary format with an electroosmotic pump, Talanta 41 (1994) 1903.

[73] J. Ruzicka, G.D. Marshall, G.D. Christian, Variable flow rates and a sinusoidal flow pump for flow injection analysis, Anal. Chem. 62 (1990) 1861.
[74] T. Guebeli, G.D. Christian, J. Ruzicka, Fundamentals of sinusoidal flow sequential injection spectrophotometry, Anal. Chem. 63 (1991) 2407.
[75] J. Ruzicka, G.D. Marshall, Sequential injection: a new concept for chemical sensors, process analysis and laboratory assays, Anal. Chim. Acta 237 (1990) 329.
[76] G.D. Christian, Sequential injection analysis for electrochemical measurements and process analysis, Analyst 119 (1994) 2309.
[77] C.E. Lenehan, N.W. Barnett, S.W. Lewis, Sequential injection analysis, Analyst 127 (2002) 997.
[78] A. Economou, P.D. Tzanavaras, D.G. Themelis, Sequential-injection analysis: Principles, instrument construction, and demonstration by a simple experiment, J. Chem. Edu. 82 (2005) 1820.
[79] A.V. Bulatov, D.K. Goldvirt, L.N. Moskvin, A.L. Moskvin, E.A. Vaskova, Stepwise injection photometric determination of hydrogen sulphide in natural gas, J. Flow Injection Anal 23 (2006) 102.
[80] C.H. Pollema, J. Ruzicka, G.D. Christian, A. Lernmark, Sequential injection immunoassay utilizing immunomagnetic beads, Anal. Chem. 64 (1992) 1356.
[81] B.-I. Haukanes, C. Kvam, Application of magnetic beads in bioassays, Nat. Biotechnol. 11 (1993) 60.
[82] B. Willumsen, G.D. Christian, J. Ruzicka, Flow injection renewable surface immunoassay for real time monitoring of biospecific interactions, Anal. Chem. 69 (1997) 3482.
[83] J. Ruzicka, L. Scampavia, From flow injection to bead injection, Anal. Chem. 71 (1999) 257A.
[84] J. Ruzicka, Lab-on-valve: universal microflow analyzer based on sequential and bead injection, Analyst 125 (2000) 1053.
[85] K. R-Zhang, K. Hirakawa, M. Katayama, H. Nakajima, N. Soh, K. Nakano, T. Imato, Flow immunoassay based on sequential injection using microbeads, J. Flow Injection Anal 23 (2006) 117.
[86] C.C. Oliveira, E.A.G. Zagatto, J. Ruzicka, G.D. Christian, A bead injection system for copper determination, Anal. Lett. 33 (2000) 929.
[87] S.K. Hartwell, G.D. Christian, K. Grudpan, Bead injection with a simple flow-injection system: an economical alternative for trace analysis, Trends Anal. Chem. 23 (2004) 619.
[88] S.K. Hartwell, D. Somprayoon, P. Kongtawelert, S. Ongchai, O. Arppornchayanon, L. Ganranoo, et al., Online assay of bone specific alkaline phosphatase with a flow injection-bead injection system, Anal. Chim. Acta 600 (2007) 188.
[89] P.S. Hodder, J. Ruzicka, A flow injection renewable surface technique for cell-based drug discovery functional assays, Anal. Chem. 71 (1999) 1160.
[90] I. Lahdesmaki, L.D. Scampavia, C. Beeson, J. Ruzicka, Detection of oxygen consumption of cultured adherent cells by bead injection spectroscopy, Anal. Chem. 71 (1999) 5248.
[91] I. Lahdesmaki, C. Beeson, G.D. Christian, J. Ruzicka, Measurement of cellular stimulation through monitoring pH changes by bead injection fluorescence microscopy, Talanta 51 (2000) 497.
[92] P.S. Hodder, C. Beeson, J. Ruzicka, Equilibrium and kinetic measurements of muscarinic receptor antagonism on living cells using bead injection spectroscopy, Anal. Chem. 72 (2000) 3109.
[93] P. Solich, M. Polasek, J. Klimundova, J. Ruzicka, Sequential injection technique applied to pharmaceutical analysis, Trends Anal. Chem. 23 (2004) 116.

[94] M. Miro, E.H. Hansen, D. Buanuam, The potentials of the third generation of flow injection analysis for nutrient monitoring and fractionation analysis, Environ. Chem. 3 (2006) 26.

[95] M.J. Ruedas-Rama, A. Dominguez-Vidal, S. Radel, B. Lendl, Ultrasonic trapping of microparticles in suspension and reaction monitoring using Raman microspectroscopy, Anal. Chem. 79 (2007) 7853.

[96] M. Miro, S.K. Hartwell, J. Jakmunee, K. Grudpan, E.H. Hansen, Recent developments in automatic solid-phase extraction with renewable surfaces exploiting flow-based approaches, Trends Anal. Chem. 27 (2008) 749.

[97] C.-H. Wu, L. Scampavia, J. Ruzicka, B. Zamost, Micro sequential injection: fermentation monitoring of ammonia, glycerol, glucose, and free iron using the novel lab-on-valve system, Analyst 126 (2001) 291.

[98] C.-H. Wu, J. Ruzicka, Micro sequential injection: environmental monitoring of nitrogen and phosphate in water using a "Lab-on-Valve" system furnished with a microcolumn, Analyst 126 (2001) 1947.

[99] J.-H. Wang, E.H. Hansen, Interfacing sequential injection on-line preconcentration using a renewable micro-column incorporated in a 'lab-on-valve' system with direct injection nebulization inductively coupled plasma mass spectrometry, J. Anal. At. Spectrom. 16 (2001) 1349.

[100] C.-H. Wu, L. Scampavia, J. Ruzicka, Microsequential injection: anion separations using 'Lab-on-Valve' coupled with capillary electrophoresis, Analyst 127 (2002) 898.

[101] J.B. Quintana, M. Miro, J.M. Estela, V. Cerda, Automated on-line renewable solid-phase extraction-liquid chromatography exploiting multisyringe flow injection-bead injection lab-on-valve analysis, Anal. Chem. 78 (2006) 2832.

[102] X.-B. Long, M. Miro, E.H. Hansen, Universal approach for selective trace metal determinations via sequential injection-bead injection-Lab-on-Valve using renewable hydrophobic bead surfaces as reagent carriers, Anal. Chem. 77 (2005) 6032.

[103] J.-H. Wang, Lab-on-valve mesofluidic analytical system and its perspectives as a world-to-chip front end, Anal. Bioanal. Chem. 381 (2005) 809.

[104] M.D.L. Castro, J. Ruiz-Jimenez, J.A. Perez-Serradilla, Lab-on-valve: a useful tool in biochemical analysis, Trends Anal. Chem. 27 (2008) 118.

[105] B.F. Reis, M.F. Gine, F.J. Krug, H. Bergamin-Filho, Multipurpose flow-injection system. Part 1. Programmable dilutions and standard additions for plant digests analysis by inductively coupled plasma atomic emission spectrometry, J. Anal. At. Spectrom. 7 (1992) 865.

[106] B.F. Reis, M.F. Giné, E.A.G. Zagatto, J.L.F.C. Lima, R.A.S. Lapa, Multicommutation in flow analysis. Part 1. Binary sampling: concepts, instrumentation and spectrophotometric determination of iron in plant digests, Anal. Chim. Acta 293 (1994) 129.

[107] M.A. Feres, E.A.G. Zagatto, J.L.M. Santos, J.L.F.C. Lima, The concept of multicommutation in flow analysis, in: M. Trojanowicz (Ed.), Advances in Flow Analysis, Wiley VCH, Mainheim, 2008, p. 167 (Chapter 7).

[108] D.J. Malcome-Lawes, C. Pasquini, A novel approach to non-segmented flow analysis. Part 3. Nitrate, nitrite and ammonium in waters, J. Autom. Chem. 10 (1988) 192.

[109] F. Albertus, B. Horstkotte, A. Cladera, V. Cerda, A robust multisyringe system for process flow analysis. Part I. On-line dilution and single point titration of protolytes, Analyst 124 (1999) 1373.

[110] L. Albertus, A. Cladera, V. Cerda, A robust multisyringe system for process flow analysis. Part II. A multi-commuted injection system applied to the photometric determination of free acidity and iron(III) in metallurgical solutions, Analyst 125 (2000) 2364.

[111] F. Albertus, A. Cladera, E. Becerra, V. Cerda, A robust multi-syringe system for process flow analysis. Part 3. Time based injection applied to the spectrophotometric determination of nickel(II) and iron speciation, Analyst 126 (2001) 903.
[112] M.A. Segundo, A.O.S.S. Rangel, A. Cladera, V. Cerda, Multisyringe flow system: determination of sulfur dioxide in wines, Analyst 125 (2000) 1501.
[113] N.V. Semenova, L.O. Leal, R. Forteza, V. Cerda, Multisyringe flow-injection system for total inorganic arsenic determination by hydride generation-atomic fluorescence spectrometry, Anal. Chim. Acta 455 (2002) 277.
[114] M.A. Segundo, L.M. Magalhaes, Multisyringe flow injection analysis: State-of-the-art and perspectives, Anal. Sci. 22 (2006) 3.
[115] D. Betteridge, Flow injection analysis, Anal. Chem. 50 (1978) 832A.
[116] T. Korenaga, X.-J. Zhou, T. Moriwake, H. Muraki, T. Naito, S. Sanuki, Computer-controlled micropump suitable for precise microliter delivery and complete in-line mixing, Anal. Chem. 66 (1994) 73.
[117] X.D. Wang, T.J. Cardwell, R.W. Cattrall, G.E. Jenkins, Pulsed flow chemistry. A new approach to the generation of concentration profiles in flow analysis, Anal. Commun. 35 (1998) 97.
[118] E.B. van Akker, M. Bos, W.E. van der Linden, Continuous, pulsed and stopped flow in a μ-flow injection system (numerical vs experimental), Anal. Chim. Acta 378 (1999) 111.
[119] P.S. Francis, S.W. Lewis, K.F. Lim, K. Carlsson, B. Karlberg, Flow analysis based on a pulsed flow of solution: theory, instrumentation and applications, Talanta 58 (2002) 1029.
[120] R.A.S. Lapa, J.L.F.C. Lima, B.F. Reis, J.L.M. Santos, E.A.G. Zagatto, Multi-pumping in flow analysis: concepts, instrumentation, potentialities, Anal. Chim. Acta 466 (2002) 125.
[121] V. Cerda, C. Pons, Multicommutated flow techniques for developing analytical methods, Trends Anal. Chem. 25 (2006) 236.
[122] C. Pons, R. Forteza, V. Cerda, Multi-pumping flow system for the determination, solid-phase extraction and speciation analysis of iron, Anal. Chim. Acta 550 (2005) 33.
[123] P.C.A.G. Pinto, M.L.M.F.S. Saraiva, J.L.M. Santos, J.L.F.C. Lima, A pulsed sequential injection analysis flow system for the fluorimetric determination of indomethacin in pharmaceutical preparations, Anal. Chim. Acta 539 (2005) 173.
[124] F.R.P. Rocha, E. Rodenas-Torralba, B.F. Reis, A. Morales-Rubio, M. de La Guardia, A portable and low cost Equipment for flow injection chemiluminescence measurements, Talanta 67 (2005) 673.
[125] K.L. Marques, C.K. Pires, J.L.M. Santos, E.A.G. Zagatto, J.L.F.C. Lima, A multi-pumping flow system for chemiluminescent determination of ammonium in natural waters, Intern, J. Environ. Anal. Chem. 87 (2007) 77.
[126] M.F.T. Ribeiro, A.C.B. Dias, J.L.M. Santos, J.L.F.C. Lima, E.A.G. Zagatto, Fluidized beds in flow analysis: use with ion-exchange separation for spectrophotometric determination of zinc in plant digests, Anal. Bioanal. Chem. 384 (2006) 1019.
[127] J.L.M. Santos, M.F.T. Ribeiro, J.L.F.C. Lima, A.C.B. Dias, E.A.G. Zagatto, Multi pumping flow systems: the potential of simplicity, Anal. Chim. Acta 600 (2007) 21.
[128] J. Wang, Z. Taha, Batch injection analysis, Anal. Chem. 63 (1991) 1053.
[129] J. Wang, L. Angnes, Batch injection spectroscopy, Anal. Lett. 26 (1993) 2329.
[130] C.M.A. Brett, Electrochemical sensors for environmental monitoring. Strategy and examples, Pure Appl. Chem. 73 (2001) 1969.
[131] M. Trojanowicz, M. Szewczynska, M. Wcislo, Electroanalytical flow measurements: recent advances, Electroanal 15 (2003) 347.
[132] M.S.M. Quintino, L. Angnes, Batch injection analysis: an almost unexplored powerful tool, Electroanal 16 (2004) 513.

[133] F.R. Simoes, C.M.P. Vaz, C.M.A. Brett, Electroanalytical detection of the pesticide paraquat by Batch Injection Analysis, Anal. Lett. 40 (2007) 1800.
[134] J. Wang, G.D. Rayson, Z. Taha, Batch injection analysis using fiberoptic fluorimetric detection, Appl. Spectrosc. 46 (1992) 107.
[135] H. Liu, P.K. Dasgupta, A renewable liquid droplet as a sampler and windowless optical cell. An automated sensor for gaseous chlorine, Anal. Chem. 67 (1995) 4221.
[136] A. Tankeviciute, R. Kazlauskas, V. Vickackaite, Headspace extraction of alcohols into a single drop, Analyst 126 (2001) 1674.
[137] Z.-D. Zhou, G.-P. Zhong, Q. Tan, Y. Li, X.-D., On-line spectrophotometric system based on pseudo liquid drop and handheld CCD spectrometer for monitoring formaldehyde level in wastewater, Instrum. Sci. Technol. 33 (2005) 297.
[138] T. Kanyanee, W.L. Borst, J. Jakmunee, K. Grudpan, J.Z. Li, P.K. Dasgupta, Soap bubbles in analytical chemistry. Conductometric determination of sub-parts per million levels of sulfur dioxide with a soap bubble, Anal. Chem. 78 (2006) 2786.
[139] K.E. Miller, R.E. Synovec, Review of analytical measurements facilitated by drop formation technology, Talanta 51 (2000) 921.
[140] R.S. Honorato, M.C.U. Araujo, R.A.C. Lima, E.A.G. Zagatto, R.A.S. Lapa, J.L.F.C. Lima, A flow-batch titrator exploiting a one-dimensional optimisation algorithm for end point search, Anal. Chim. Acta 396 (1999) 91.
[141] J.M.T. Carneiro, R.S. Honorato, E.A.G. Zagatto, Individual sample conditioning in flow analysis. Determination of N-total in plant materials, Fresenius J. Anal. Chem. 368 (2000) 496.
[142] J.M.T. Carneiro, A.C.B. Dias, E.A.G. Zagatto, R.S. Honorato, Spectrophotometric catalytic determination of Fe(III) in estuarine waters using a flow-batch system, Anal. Chim. Acta 455 (2002) 327.
[143] E.V. Aquino, J.J.R. Roohwedder, C. Pasquini, A new approach to flow-batch titration. A monosegmented flow titrator with coulometric reagent generation and potentiometric or biamperometric detection, Anal. Bioanal. Chem. 386 (2006) 1921.
[144] L.F. Almeida, V.L. Martins, E.C. Silva, P.N.T. Moreira, M.C.U. Araujo, Implementation of an automatic standard addition method in a flow-batch system: application to copper determination in an alcoholic beverage by atomic absorption spectrometry, Anal. Chim. Acta 486 (2003) 143.
[145] J.E. Silva, F.A. Silva, M.F. Pimentel, R.S. Honorato, V.L. Silva, M.C.B.S.M. Montenegro, et al., A flow-batch internal standard procedure for iron determination in hydrated ethanol fuel by flame atomic absorption spectrometry, Talanta 70 (2006) 522.
[146] R.A.C. Lima, S.R.B. Santos, R.S. Costa, G.P.S. Marcone, R.S. Honorato, V.B. Nascimento, et al., Hardness screening of water using a flow-batch photometric system, Anal. Chim. Acta 518 (2004) 25.
[147] E.P. Medeiros, E.C.L. Nascimento, A.C.D. Medeiros, J.G. Veras-Neto, E.C. Silva, M.C.U. Araujo, Multicommutated generation of concentration gradients in a flow-batch system for metronidazole spectrophotometric determination in drugs, Anal. Chim. Acta 511 (2004) 113.
[148] L.F. Almeida, M.G.R. Vale, M.B. Dessuy, M.M. Silva, R.S. Lima, V.B. Santos, et al., A flow-batch analyzer with piston propulsion applied to automatic preparation of calibration solutions for Mn determination in mineral waters by ET AAS, Talanta 73 (2007) 906.
[149] J.M.T. Carneiro, A.C.B. Dias, E.A.G. Zagatto, J.L.M. Santos, J.L.F.C. Lima, An improved sampling approach in multi-pumping flow systems applied to the spectrophotometric determination of glucose and fructose in syrups, Anal. Chim. Acta 531 (2005) 279.
[150] J.A. Sweileh, P.K. Dasgupta, Applications of in situ detection with an automated microbatch analyzer, Anal. Chim. Acta 214 (1988) 107.

[151] E.A.G. Zagatto, B.F. Reis, C.C. Oliveira, R.P. Sartini, M.A.Z. Arruda, Evolution of the commutation concept associated with the development of flow analysis, Anal. Chim. Acta 400 (1999) 249.
[152] M.F. Gine, H. Bergamin-Filho, E.A.G. Zagatto, B.F. Reis, Simultaneous determination of nitrate and nitrite by flow injection analysis, Anal. Chim. Acta 114 (1980) 191.
[153] T.J. Cardwell, R.W. Cattrall, G.J. Cross, J.R. O'Cornell, J.D. Petty, G.R. Scollary, Determination of titratable acidity of wines and total acidity of vinegars by discontinuous flow analysis using photometric end-point detection, Analyst 116 (1991) 1051.
[154] H. Itabashi, H. Kawamoto, T. Kawashima, A novel flow injection technique: all injection analysis, Anal. Sci. 17 (2001) 229.
[155] A.C.B. Dias, J.L.M. Santos, J.L.F.C. Lima, C.M. Quintella, A.M.V. Lima, E.A.G. Zagatto, A critical comparison of analytical flow systems exploiting streamlined and pulsed flows, Anal. Bioanal. Chem. 388 (2007) 1303.
[156] S.R. Bysouth, J.F. Tyson, Supercritical carbon dioxide as a carrier for sample introduction in atomic absorption spectrometry, Anal. Chim. Acta 258 (1992) 55.
[157] S.M. Ramasamy, M.S.A. Jabbar, H.A. Motolla, Flow injection analysis based on two consecutive reactions at a gas-solid interface for determination of bromine and chlorine, Anal. Chem. 52 (1980) 2062.
[158] M.D.H. Silva, C. Pasquini, Bisegmented flow system for determination of low concentrations of gaseous constituents in gaseous samples, Anal. Chim. Acta 393 (1999) 121.
[159] P.R. Fortes, M.A. Feres, M.K. Sasaki, E.R. Alves, E.A.G. Zagatto, J.A.V. Prior, et al., Evidences of turbulent mixing in multi-pumping flow systems, Talanta 79 (2009) 978.
[160] J. Ruzicka, Flow injection analysis. From test tube to integrated microconduits, Anal. Chem. 55 (1983) 1040A.
[161] A. van den Berg, P. Bergveld, Lab-on-a-Chip: origen, highlights and future perspectives, Lab. Chip 6 (2006) 1266.
[162] A. van den Berg, T.S.J. Lammerink, Micro total analysis systems: microfluidic aspects, integration concept and applications, Top. Curr. Chem. 194 (1997) 21.
[163] G.D. Christian, Flow analysis and its role and importance in the analytical sciences, Anal. Chim. Acta 499 (2003) 5.
[164] L. Solujic, E.B. Milosavljevic, M.R. Straka, Total cyanide determination by a segmented flow injection-on-line UV digestion-amperometric method, Analyst 124 (1999) 1255.
[165] M. Trojanowicz (Ed.), Advances in Flow Analysis, Wiley-VCH, Mainheim, 2008.
[166] E. Rodenas-Torralba, A. Morales-Rubio, M. de la Guardia, Scientometric picture of the evolution of the literature of automation in spectroscopy and its current state, Spec. Lett. 39 (2006) 513.
[167] T. Yao, T. Yano, On-line microdialysis assay of L-lactate and pyruvate in vitro and in vivo by a flow-injection system with a dual enzyme electrode, Talanta 63 (2004) 771.
[168] E.A.G. Zagatto, F.R.P. Rocha, P.B. Martelli, B.F. Reis, Detecting and circumventing sources of inaccuracy in flow analysis, Pure Appl. Chem. 73 (2001) 45.
[169] R.C. Schothorst, J.M. Reijn, H. Poppe, G. den Boef, The application of strongly reducing agents in flow-injection analysis. Part 1. Chromium(II) and vanadium(II), Anal. Chim. Acta 145 (1983) 197.
[170] W.T. Suarez, H.J. Vieira, O. Fatibello-Filho, Generation and destruction of unstable reagent in flow injection system: determination of acetylcysteine in pharmaceutical formulations using bromine as reagent, J. Pharm. Biomed. Anal 37 (2005) 771.
[171] S.M.B. Brienza, R.P. Sartini, J.A. Gomes-Neto, E.A.G. Zagatto, Crystal seeding in flow-injection turbidimetry: determination of total sulfur in plants, Anal. Chim. Acta 308 (1995) 269.

[172] M. Hulsman, M. Bos, W.E. van der Linden, Automated injection of slurry samples in flow-injection analysis, Anal. Chim. Acta 324 (1996) 13.

[173] M. Burguera, J.L. Burguera, O.M. Alarcon, Flow-injection and microwave sample decomposition for determination of copper, zinc and iron in whole blood by atomic absorption spectrometry, Anal. Chim. Acta 179 (1986) 351.

[174] M. Miro, W. Frenzel, Automated membrane-based sampling and sample preparation exploiting flow-injection analysis, Trends Anal. Chem. 23 (2004) 624.

[175] C.C. Oliveira, R.P. Sartini, E.A.G. Zagatto, J.L.F.C. Lima, Flow analysis with accuracy assessment, Anal. Chim. Acta 350 (1997) 31.

[176] M. Polasek, Playful reflections on the use of abbreviations and acronyms in analytical flow methods, J. Flow Injection Anal 23 (2006) 81.

CHAPTER

3

Fundamentals

Such ebb and flow must ever be. **Wordsworth**

OUTLINE

This chapter focusses on important issues related to the flowing sample, such as the flow pattern, sample dispersion and interactions between flowing solutions, which are essential for system design, optimisation and operation. Strategies for visualisation of a flowing sample are also discussed, the commutation principle is highlighted, and guidelines for characterising a flow analyser and related methodologies are provided. The chapter concludes by emphasising the different strategies for obtaining the analytical results.

3.1. THE FLOWING SAMPLE

After the introduction of a selected sample volume into a flow system, it is transported through the analytical path, and subjected to several physical and chemical processes such as reagent addition, chemical reaction and dispersion. Understanding how these processes affect the flowing sample and the formation of the chemical species to be monitored is essential for optimising manifold design and, hence, analytical performance. To this end, it is important to consider the flow pattern and how to modify it by varying experimental parameters.

3.1.1. Flow Pattern

In this section, the flow pattern is discussed in terms of the flow regime, composition of the flowing stream and flow rate. Strategies for modifying the flow pattern are also presented.

3.1.1.1. Flow Regime

The flow regime of a solution flowing through straight, narrow-bore tubing can be classified as either laminar or turbulent. The former, also called streamlined flow, is steady and characterised by regular displacement of all fluid elements following parallel trajectories (Fig. 3.1, upper). It can be regarded as the relative motion of a set of concentric cylinders of fluid, the outer one being fixed at the tube wall and the others moving at increasing speeds as the centre of the tube is approached. Turbulent flow is on average steady, but with significant local fluctuations, as the velocity of each fluid element continuously undergoes changes in both magnitude and direction (Fig. 3.1. lower). Its main characteristic is the chaotic displacement of the fluid elements resulting from incessant variations in their speed and direction.

In 1883, Reynolds demonstrated that the transition from laminar to turbulent flow in a tube is associated with a dimensionless quantity called the Reynolds number, Re [1]:

$$\mathrm{Re} = 2\langle\Phi\rangle\, \mathrm{r}\, \rho/v \tag{3.1}$$

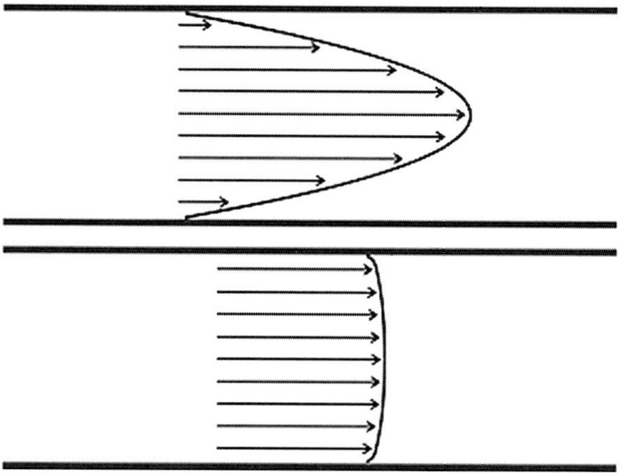

FIGURE 3.1 **Velocity distribution in laminar (upper) and turbulent (lower) flow.** The figure refers only to convective mass transport. Vectors indicate average linear velocities corresponding to the different radial distances from the tube axis.

where $\langle \Phi \rangle$ is the average linear velocity of flow, r is the tube radius, ρ is the mass density of the fluid and ν is the kinematic viscosity of the fluid. In order to be more readily applied to a flow system, Eq. 3.1 can be rewritten [2] as

$$Re = 2\rho Q/\pi r\nu \tag{3.2}$$

where Q is the volumetric flow rate. The parameters ρ, Q, r and ν are usually expressed in $g\,cm^{-3}$, $cm^3\,s^{-1}$, cm and poise ($g\,cm^{-1}\,s^{-1}$), respectively.

The value of Re depends on the experimental conditions; therefore, laminar flow can be changed to turbulent flow or *vice versa* by modifying some relevant experimental parameters. In practice, the transition between laminar and turbulent flow does not occur at a specific value of Re but over a range of values. In the classical Reynolds experiment involving a continuous increase in Re [3], it can be verified that for Re <2000, the flow of a fluid in a smooth tube is always laminar. Between 2000 and 4000 (the so-called transition region), there is a gradual change to turbulent flow. Generally, turbulence is fully established for Re >4000. Based on these values, and considering the flow rates (0.7–7 $mL\,min^{-1}$) and tubing dimensions (0.2–1.0 mm i.d.) of a typical flow analyser, one can conclude that laminar flow is the prevailing regime in flow analysis [4].

Under laminar flow conditions, Poiseuille flow must be considered [5]. This means that the linear velocity of the fluid element at the centre of the tube is about twice the average linear flow velocity $\langle \Phi \rangle$, whereas the velocity of fluid elements adjacent to the inner walls of the tubing approaches zero. For each fluid element, Eq. 3.3 expresses the linear velocity of the corresponding fluid line, Φ. This results in a typical parabolic profile of the axial linear velocity distribution (Fig. 3.1, upper):

$$\Phi = 2\langle\Phi\rangle \left(1 - r^2/a^2\right) \tag{3.3}$$

where r is the radial distance from the tube axis and a is the tube radius.

During turbulent flow, the linear velocity of each fluid element is the vector sum of their individual velocities. As a consequence of the chaotic displacement, the linear velocity tends to be the same for all the fluid elements and approaches the average linear flow velocity (Fig. 3.1, lower). In real situations, the linear velocities of fluid elements near the tubing walls are slightly decreased due to frictional energy losses. However, due to the chaotic movement, each fluid element only stays near the tubing wall for a very short time.

It is interesting to note that Ruzicka and Hansen initially suggested turbulence to explain the relatively low carryover between successive samples in their first article on flow injection analysis [6]. It did not take them long [7], however, to realise that laminar flow is inherent in flow injection (*i.e.*, unsegmented flow) analysis. The laminar mass transport in these systems was elegantly demonstrated by Silfwerbrand-Lindh *et al.* [8] using photographic equipment. Laminarity was also verified in the manifold regions near confluence points where there is a sudden increase in flow rate and turbulence is more likely to occur [9]. The laminar regime in flow injection analysis is illustrated in the photograph shown in Fig. 3.2.

Although most flow analysers operate under laminar flow conditions, there are specific situations where turbulent mixing is approached inside the manifold (see also 6.1.3.1).

3.1.1.2. *Composition of the Flowing Stream*

In the absence of a flowing sample, the chemical composition of the flowing stream is steady and defined in terms of the concentrations of the involved solutions (*e.g.*, carrier, reagent and diluent) and their corresponding volumetric fractions (Eq. 3.18). The composition of the main carrier stream can undergo sudden variations when confluent streams are added (Eq. 3.10).

In the presence of the sample, transient variations in chemical composition are observed, as concentration gradients are established along the sample zone. Although sometimes beneficial [11,12], these gradients are generally undesirable. They are more pronounced at the sample/carrier stream boundaries and are reduced as sample dispersion proceeds. An effective strategy for gradient reduction is to match

FIGURE 3.2 **A dispersing sample monitored by a microscope.** The photographs show a buffered dye solution (methylene blue) inserted into a buffered water carrier stream of a straight (single-line) flow injection system. From the top: sample arrival in the microscope field, the front portion of the sample, the most concentrated portion of the sample and part of the sample trailing edge. For experimental conditions, see Ref. [10]. *Source: Courtesy of T. Korenaga via F.J. Krug (1993).*

the matrices of the inserted sample and the carrier stream solutions (see 3.2.2).

Physically, the flowing stream can be either segmented or unsegmented, depending on whether a second phase (segmenter) is present or not (Figs 3.3a, 3.3b). The need for segmentation and the main characteristics of a segmented flow analyser are discussed in 2.2 and 5.1. In this context, two specific streams, namely tandem and mono-segmented, should be highlighted.

TANDEM STREAM

Tandem stream is a type of unsegmented flow that was originally implemented in a single-line system to permit random reagent access for multi-parametric determinations in natural waters [13]. This stream initially comprises several neighbouring plugs of different miscible solutions (in general sample and reagents), which are sequentially added as small aliquots into the flowing carrier stream (Fig. 3.3c). The delivery of very low volumes of these solutions in alternate plugs results in a series of parabolic interfaces between them, leading to improved mixing conditions, even for short lengths of straight tubing [14].

Tandem stream is especially attractive in relation to flow systems designed in the single-line configuration, such as the sequential injection

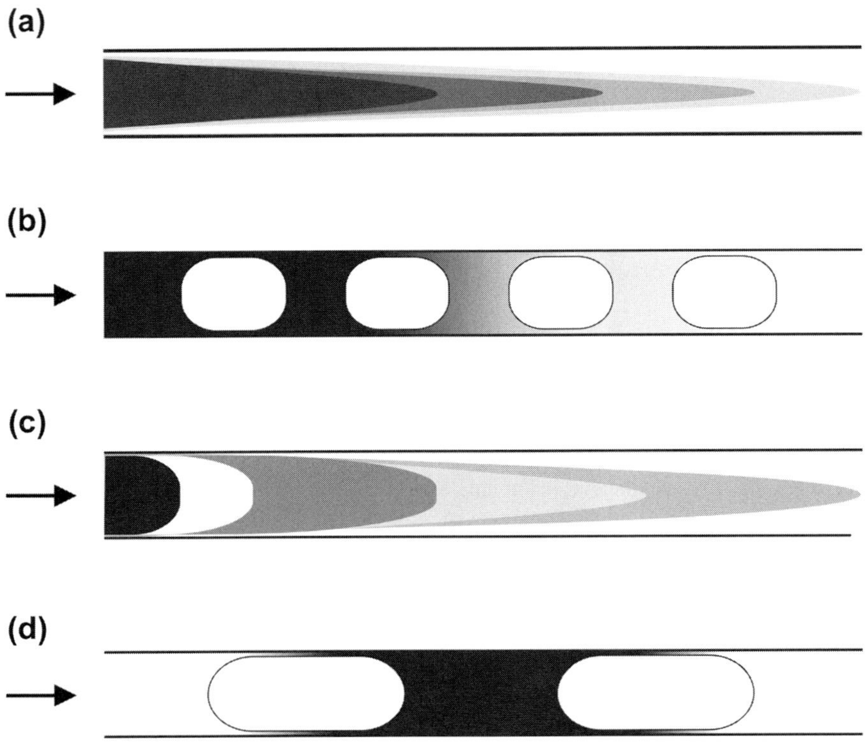

FIGURE 3.3 **Didactic representation of unsegmented (a), segmented (b), tandem (c) and mono-segmented (d) streams.** The figure refers to the front portion of the flowing sample, with exception of (d) which refers to the whole sample. Black = sample; white = carrier/wash stream; oval shapes = air phase; arrows = flow direction.

analyser, as several solutions can be handled inside a single analytical channel. It is somewhat surprising that this aspect has not yet been fully exploited.

The approach is also expeditious for controlling sample dilution, as demonstrated by Israel *et al.* [15] who exploited tandem streams for attaining high and variable degrees of sample dilution prior to ICP-OES and ICP-MS. Several plugs of sample and diluent solutions were successively aspirated, mixed together and directed towards the nebuliser of the detection unit. Although undulations in the detector output were reported, good mixing conditions were attained, leading to an almost steady-state situation. This innovation was named *tandem injection*.

Another possibility for establishing a tandem stream is to mix two different streams at a fast switching three-way valve. This innovation was referred to as *binary sampling* [16], and can also be implemented by using a reciprocating pump [14]. A similar approach involving insertion of sample/reagent plugs at a high frequency through nozzles was also proposed [17]. Turbulent mixing was attained, leading to improved mixing conditions and lower axial dispersion. The expressions *multi-insertion principle* [18] and *time-division multiplex technique* [19] have also been used in this context.

An attractive situation refers to the delivery of a number of sample/reagent plugs into the sampling loop of a flow-injection system [20]. Sample/reagent interactions start during the sampling step, thus increasing the mean available time for reaction development without affecting the sampling rate.

The tandem stream approach is also important in true flow-based titrations [17,21]. The number and length of carrier, titrant and titrand plugs can be efficiently modified in real time according to a concentration-orientated feed-back mechanism relying on an extrapolative algorithm for the determination of the titration endpoint. This approach was called *binary search*.

MONO-SEGMENTED STREAM

Mono-segmentation was conceived [22] in order to combine the favourable characteristics of both segmented and unsegmented flow analysis. The mono-segmented stream is established by inserting the selected sample volume between two air plugs into the carrier stream (Fig. 3.3d). Long sample residence times can then be attained without pronounced sample dispersion. The innovation is especially attractive for analytical applications involving relatively slow chemical reactions, especially if sensitivity is critical. Mono-segmentation requires instrumentation similar to ordinary flow injection systems, with advantages in sampling rate, sensitivity and sample/reagent consumption.

In a similar manner to segmented-flow analysers, the air plugs are normally removed before detection, and this can lead to sample loss, increased wash time and disturbance of the concentration profile along the sample zone. These drawbacks are minimised by exploiting the bubble-gating approach [23], but more complex and expensive instrumentation is required. In mono-segmented flow systems, the bubbles are usually removed through a gas permeation membrane [22], thus avoiding the above-mentioned shortcomings. Alternatively, only the central portion of the sample zone – without the gas phase – can be directed towards detection, and this can be efficiently accomplished by exploiting zone sampling [24] or multi-site detection [25].

The main drawback of mono-segmented flow systems is the difficulty of adding reagents by confluence but this can be circumvented by including microcomputer-controlled opto-switches to precisely locate the air bubbles in the analytical path [26].

3.1.1.3. *Temporal Variations in Flow Rates*

During the analytical cycle, flow rates may undergo variations over time. Depending on the kind of fluid-propelling device and the timescale of operation, different kinds of flow are dispensed, classified as follows.

CONSTANT FLOW

Most flow-based analytical procedures rely on constant flow rates (Fig. 3.4a). These have also been used in model systems designed for studying sample dispersion, in order to simplify the related mathematical models. Uniform solution delivery is accomplished by exploiting pumping devices relying on pistons such as automatic burettes or syringe pumps, peristaltic pumps with dampeners, or gravity as fluid propellers.

A strictly constant flow delivery cannot always be guaranteed, however, and this limitation becomes more pronounced for flow systems designed in the confluence configuration. Without pulsation, the confluent stream addition may become somewhat erratic [27]. In fact, the repetitive sudden variation in back pressure inherently ensures reproducible addition of the confluent stream. This aspect is inherent to pulsating flow.

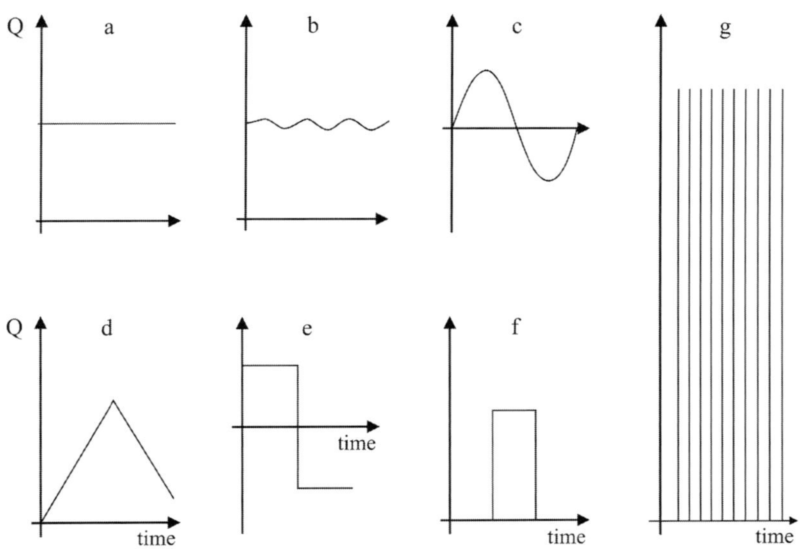

FIGURE 3.4 **Temporal variations in flow rates.** a = constant; b = pulsating; c = sinusoidal; d = linearly variable; e = reversed; f = intermittent; g = pulsed flows; *Q* = volumetric flow rate.

PULSATING FLOW

The peristaltic pump is by far the most common propulsion device in flow analysis, due to its low cost and easy incorporation into multi-channel flow systems. It delivers a pulsating flow (Fig. 3.4b) as a result of the vector sum of two effects. The first is the main action of the roller, which accounts for the main constant flow, whereas the second relates to the roller lift-off from the platen, which accounts for the slight sinusoidal fluctuations (ripple) on the main flow.

Pulsating flows are beneficial with regard to segmented flow analysis, as the bubble additions are synchronised with the flow pulsation. In order to enhance the pulsating nature of the confluent stream of air to be added, some peristaltic pumps (including the widely used Technicon AAII) incorporate a lifting bar to successively pinch the air delivery tubing.

Pulsating flows are also beneficial in unsegmented-flow systems, especially those designed in the confluence configuration (see above). There are, however, some situations where the slight ripple effect should be avoided, *e.g.*, when analyte separation/concentration relying on flat membranes is used, or when very low sample volumes are inserted into the main carrier stream. In the latter situation, less precise results are obtained because samples may be injected at slightly different carrier stream flow rates. In the extreme situation, one sample can be inserted when the carrier stream flow is at its maximum flow rate and the next is inserted at the minimum flow rate. As the carrier stream flow rate affects sample dispersion, and considering that dispersion is high for a low injected volume, measurement repeatability may be degraded. Different schemes have been proposed to reduce or circumvent this drawback, and an elegant and economical solution is simply to synchronise the instant of sample insertion with the pulsation of all involved streams by delivering them in synchronisation with roller lift-off [28].

SINUSOIDAL FLOW

This type of flow is characterised by variations in flow rate following a sinusoidal time dependence (Fig. 3.4c) and can be delivered by a cam-driven syringe (see also Figs 6.2 and 6.3). As a sinusoidal function is involved, the flowing stream undergoes backwards and forwards movements, without any sudden variation. Although not as flexible as a peristaltic pump, and requiring piston refilling that may impair the sampling rate, a sinusoidal flow pump presents several advantages such as the absence of pump tubing or check valves, an inert environment, no pulsation, simple construction, low initial and maintenance costs, amenability to computer control, adjustable flow rate, ability to withstand high back pressures and quiet operation [29]. A constant (or pulsating) flowing stream is usually used in flow systems exploiting sinusoidal flows in order to provide an averaged forward displacement. Although sinusoidal flows are very efficient for attaining reproducible sample dispersion, they are rarely used. However, they played an important role as a precursor of sequential injection analysis, which relies on reversed flow.

REVERSED FLOW

After flowing for a pre-selected time period, the stream flows in the opposite direction (Fig. 3.4d). This kind of flow is inherent to sequential injection analysis and its derivatives and can also be exploited for multiple monitoring of a single sample zone [30]. The sample is directed towards the flow cell and, after leaving it, the flow is reversed and the sample is monitored again. In this way, analytical procedures relying on a two-point rate assay can be efficiently accomplished [31]. Another possibility is the attainment of high and reproducible dispersion: the sample is inserted into the carrier stream and the established sample zone is directed backwards in order to discard most of the sample zone. The flow is then reversed, thus directing the remainder of the sample zone towards the main reactor and detector [32]. Moreover, successive and fast flow reversals can be accomplished, which is useful for controlling dispersion, to increase the sample residence time in the analytical path and to investigate reaction kinetics [33]. The iterative reversal of the flow direction is especially attractive in relation to liquid–liquid extraction [34], as the surface between the aqueous and organic phases (thin film on the inner tubing walls) is continuously renewed.

LINEARLY VARIABLE FLOW

The flow rate undergoes a linear increase (or decrease) as a function of time (Fig. 3.4e). These linear ramps were exploited in the 1980s to improve system washing and implement flow titrations [29]. This approach has been erroneously referred to as *flow gradient* but, in fact, a spatial function was not involved.

INTERMITTENT FLOW

The stream only flows for a short period within the total analytical cycle (Fig. 3.4f). This kind of flow is particularly attractive in the following situations:

- After achievement of the analytical signal, an intermittent washing stream at high flow rate is added near the injection port. Wash time is reduced without affecting mean sample residence time, thus improving sampling rate and/or reducing reagent consumption [35].

The approach has been exploited for analytical procedures relying on relatively slow chemical reactions.
- A confluent reagent stream is added to the main analytical channel only when the sample zone is passing through the confluence site. The merging zones configuration is simply implemented, promoting a significant reduction in reagent consumption [36].
- After addition of the required reagents to the flowing sample, all streams are stopped in order to increase the degree of development of the chemical reactions involved [37]. This innovation is useful for analytical procedures requiring long sample incubation periods. During the STOP period, sample dispersion almost ceases, as the convective component of dispersion (see Eq. 3.4) is zero. This approach, herein referred to as *sample stopping*, has been improperly called *stopped-flow*.

Sample stopping relying on intermittent streams can be implemented in association with the merging zones configuration in order to promote a pronounced reduction in reagent consumption, as demonstrated by Lim *et al.* who applied flow injection analysis for the first time in the field of immunoassay, to determine albumin by a homogeneous fluorescence energy transfer technique [38]. It should be stressed that, if a sample is halted in the flow cell of the detector, monitoring further progress of the reaction is straightforward [39].

PULSED FLOW

Pulsed flows are characterised by a number of sudden very short GO and relatively long STOP periods (Fig. 3.4g) and can be beneficial in view of the chaotic movement of fluid elements during pulse delivery (GO period) [17]. This flow was critically investigated by Owens *et al.*, who evaluated chemical reaction rates [40]. As the linear velocity is very high during this short period, turbulent mixing tends to be achieved, which is a positive factor for enhancing radial mass transport. On the other hand, dispersion tends to cease during the STOP periods between successive pulses. Consequently, the progressive sample spreading is low in comparison with laminar flow [41]. Pulsed flows are efficiently established using solenoid pumps. Other beneficial aspects inherent to these pumps are the good mixing conditions, improved commutating facilities, high radial mass transport, easy implementation of tandem streams, and the feasibility of acting as injection devices [42].

It should be stressed that the above-mentioned flows are generally established using specific pumping devices. An important aspect related to reversal and intermittent flows is that they can be implemented either by modifying the operation of the pumping device or, more efficiently, by resorting to commuting facilities [43].

3.1.1.4. *Alterations to the Flow Pattern*

Numerous attempts have been made to minimise the effects of the laminar flow regime in flow analysis in order to improve mixing conditions, limit zone broadening, and achieve long sample residence times inside the analytical path.

In this context, the addition of a second phase in segmented flow systems impairs the establishment of a typical streamlined flow, because the fluid elements cannot be moved at will. Vortices are formed, and this effect is beneficial for limiting carryover, washing the tube inner walls and improving mixing between the two phases [44]. Vortex formation is also inherent to most flow systems exploiting liquid—liquid extraction and/or mono-segmented streams (see also Chapter 8 and 3.1.1.2).

It should be stressed that this meaning of the term *turbulence* (vortex establishment) is quite common in everyday life, but is not strictly true, as vortices can also be established at very low Reynolds number.

The flow pattern is altered in the presence of solid particles, either originally present in the introduced sample plug (slurry analysis), formed as a reaction product (turbidimetric and nephelometric procedures), or added as solid suspensions [45]. The solid particles are located between neighbouring flow lines moving at different velocities; therefore, they may undergo rotation, resulting in pronounced modifications to the flow pattern. The effect was discussed by Harrow and Janata [46] who proposed a strategy to circumvent the influence of suspended matter in the analysis of slurry samples such as whole blood. A deeper investigation of this was further reported by Hulsman *et al.* [47].

The flow pattern is also modified when reactors other than straight open tubes are used. In coiled reactors, all fluid elements cannot be displaced on parallel trajectories, as the distances travelled are dependent on their relative positions. This results in split circulation of the fluid elements (Fig. 3.5), which is a consequence of the establishment of secondary flows [48]. The effect becomes more pronounced at higher flow rates. Its beneficial influence on mixing conditions, hence, on sample broadening and sampling rate, has often been emphasised [10,49]. An analogous but more pronounced effect is observed with knitted (or 3-D) reactors [50].

Packed bed reactors [51] incorporate beads, and the presence of these solid materials in the analytical path also has a beneficial influence on the flow pattern, reducing sample broadening and, hence, sample dispersion. However, hydrodynamic pressure tends to be increased, especially when small particles and/or particles with heterogeneous sizes are used. Single bead string reactors [5] were proposed to circumvent this drawback. These reactors utilise large (particle diameters about 70% or the tube inner

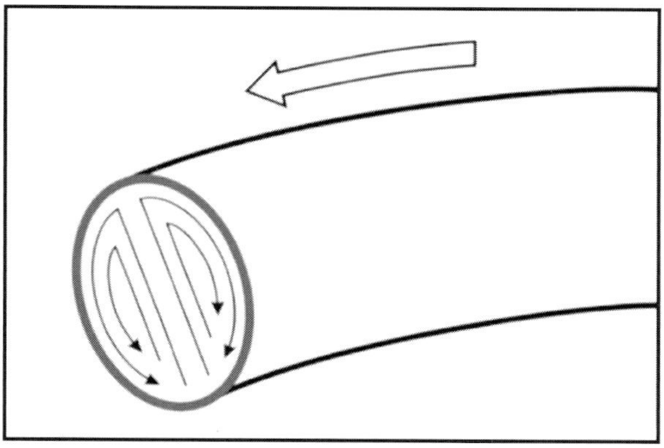

FIGURE 3.5 **Secondary flows induced by coiling a tubular reactor.** Empty arrow = main flow direction, other arrows = secondary flow movements.

diameter) and homogeneous particles and the hydrodynamic pressure is then no longer a limiting factor in the operation of the system and a helical movement of the stream is established. Similar results have been reported for solid reagents such as mini-columns and enzyme cartridges, as the channelling effects are analogous to those observed in high-performance liquid chromatography. A detailed investigation of the different reactor types affecting dispersion in single-line flow injection systems was given by Reijn *et al.* [52]. Beads are also integral to bead injection systems (see 2.4.1). As the beads are not rigidly packed, they are relatively free to move without altering the flow pattern in a pronounced manner.

Other artefacts in the analytical path, such as mixing chambers and phase separating devices, are additional sources of alteration to the flow pattern, and are discussed in 3.1.2.2. The flow pattern is also dependent on the characteristics of the flowing streams (see above), and on the addition of confluent streams, often required for implementing specific analytical procedures (see 3.1.2.3).

Last, but not least, the transient presence of the sample is an additional source of alteration to the flow pattern. Understanding how the sample interacts with the surrounding solution is therefore important, especially to assist manifold design. A common feature of all flow analysis techniques is that the sample undergoes a continuous dispersion process that leads to a continuous decrease in its concentration.

3.1.2. Sample Dispersion

Dispersion is a continuous process occurring during transport of the sample through the analytical path, due to the redistribution of material from one moving zone (sample zone) to another (carrier stream), caused by convective and diffusive mass transport. In other words, if sample and carrier solutions travelling in the same direction are placed in contact, one distributes into the other with a tendency towards obtaining a single homogeneous liquid. This phenomenon has been referred to as the dispersion process, a feature of crucial importance in flow analysis.

Dispersion is inherent to both segmented and unsegmented flow analysis, but is more important in the latter. The relative importance of the various factors influencing dispersion are different in unsegmented flow and segmented flow analysis, where the thin liquid film on the inner tubing wall plays an important role. The following sections only refer to unsegmented flow, and the issue of dispersion in segmented flow systems is addressed in 5.1.2.

3.1.2.1. *Dispersion Inside a Tubular Reactor*

The essence of unsegmented flow analysis is the controlled and, therefore, highly reproducible physical dispersion of an aqueous sample introduced into a continuously flowing carrier stream. For a better understanding of the process, it is advisable to initially consider laminar flow inside a straight open tubular reactor.

Sample introduction into a carrier stream generates a dispersing sample zone and each of its fluid elements is carried along at a velocity corresponding to its respective position in the stream. Convective mass transport is then a consequence of the parabolic distribution of the linear velocities of every fluid element (Fig. 3.1, upper). Diffusive mass transport is dependent mainly on the concentration differences between neighbouring fluid elements and their diffusion coefficients. Although diffusive mass transport occurs in an isotropic fashion, only its radial component is relevant as a factor influencing dispersion.

In this context, Griffiths reported in 1911 the interactions of an aqueous plug with a chemically inert carrier stream flowing through a narrow, straight tube [53]. He carried out the first experimental work demonstrating the essence of the dispersion process and concluded (without a mathematical treatment) that "a tracer injected into a water stream spreads out in a symmetrical manner about a plane in the cross section that moves with the mean flow velocity" [54]. He also pointed out the establishment of a fully developed laminar flow regime.

The theme was further investigated in the 1950s by Taylor [55,56], who considered Griffiths' results surprising. He injected a very small volume of a liquid tracer (solution with a dyestuff A) into a laminar flowing stream of water, and verified that the fluid elements near the centre of the tube moved with twice the mean velocity of the liquid tracer zone. He thus inferred that the water at the centre of the tube penetrated the tail end of the tracer zone and absorbed the mass of dyestuff A during passage through the zone; conversely, water near the tube walls penetrated the front portion of the tracer zone that moved at the centre of the tube. He also noted that tracer A spread symmetrically in relation to the plane moving at the mean flow velocity. Based on these observations, he applied the two-dimensional convective–diffusive equation (Eq. 3.4) to describe the concentration of dyestuff A in any fluid element as a function of its spatial and temporal coordinates:

$$D_m \left(\delta^2 C/\delta x^2 + \delta^2 C/\delta r^2 + 1/r\delta C/\delta r\right) = \Phi_0\left(1 - r^2/a^2\right)\delta C/\delta x + \delta C/\delta t \tag{3.4}$$

where D_m is the molecular diffusion coefficient, C is the concentration of dyestuff A in the specified fluid element, x is the distance along the tube, r is the radial distance from the tube axis, a is the tube radius, t is the time and Φ_0 is the maximum linear flow velocity at the centre of the tube.

Equation 3.4 is derived from the mass balance in a differential volume of the fluid, and considers the axial and radial concentration gradients as well as the linear velocities established in a laminar flow regime in the absence of chemical reactions. The first term ($\delta^2C/\delta x^2$) refers to axial diffusion and the other left-hand terms to radial diffusion. The expression Φ_0 $(1 - r^2/a^2)$ refers to the parabolic shape of the sample zone due to the different velocity of each fluid element (see also Fig. 3.1, upper). For fluid elements near the tube wall ($r \to a$), the corresponding linear velocity (Φ) tends to zero, whereas for the fluid elements in the centre of the tube ($r \cong 0$), the highest possible velocity ($\Phi = \Phi_0$) is attained. When radial mass transport is too fast, *e.g.*, in turbulent mixing [57], the expression Φ_0 $(1 - r^2/a^2)$ and the related convective–diffusive equation are simplified accordingly. Thus, it has been successfully applied to the plug flow model [58]. The expression $\delta C/\delta t$ refers to the temporal variation in concentration due solely to concentration gradients, and is the main contribution to the very low sample dispersion occurring in situations where the flow is stopped.

Equation 3.4 is presented in a differential manner, so numerical calculations are required and boundary conditions should be well defined in order to apply it to practical situations.

Taylor's approach was incomplete, as it only provided solutions for this equation in the two extreme situations (a,b) shown in Fig. 3.6, where different operating conditions are displayed as a function of dimensionless time ($\tau = D_m t/a^2$) and Peclet number ($pe = a\Phi_0/D_m$) [59,60]. Taylor assumed that molecular diffusion dominated in the first situation (a), whereas convection dominated in the second situation (b) due to minimal diffusive mass transport.

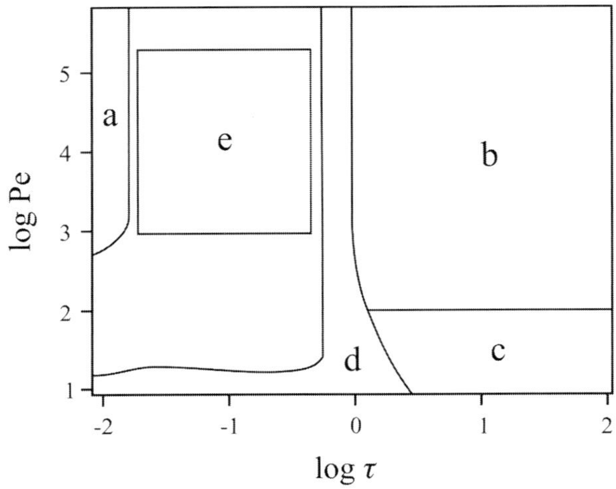

FIGURE 3.6 Regions of validity for various solutions of the diffusive–convective equation. Log *Pe* and log τ = logarithms of the Peclet number and the adimensional time. Regions a = Taylor a; b = Taylor b; c = Taylor-Aris; d = Ananthakrishnan–Gill–Barduhn approximations; e = region where most flow analysers operate. *Source: Figure adapted from Refs [59] and [60].*

The first Taylor's approximation was improved by Aris [61], who proposed a numerical solution where radial mixing by diffusion was substantially emphasised (Fig. 3.6c). Further, Ananthakrishnan *et al.* [62,63] provided solutions related to a region between the extremes of convective transport and molecular diffusion considered by Taylor (Fig. 3.6d). The authors obtained a complete description of the mechanisms controlling dispersion in a cylindrical tube. To this end, they proposed a numerical finite solution of Eq. 3.4, involving as the initial condition either a step change in the inlet concentration [62] or a plug stimulus [63]. Parameters were set in order to get dimensionless times and Peclet numbers matching the region specified in Fig. 3.6d.

The convective–diffusive model underwent further refinements, as discussed in 5.2.2. It permits a quantitative description of sample dispersion in unsegmented flow analysis and provides a good simulation to assist system design and method implementation. Other quantitative models have also been proposed for specific applications, and in this regard, the tanks-in-series model should be highlighted.

THE TANKS-IN-SERIES MODEL

This model is extensively used in chemical engineering [64], and has also been successfully applied to liquid chromatography, as it uses the concept of theoretical plates. The model assumes that an element of fluid flows through N ideally stirred tanks linked in a serial manner, experiencing successive equilibria in each compartment. If this condition is fulfilled, the mathematical operator $\mathbf{\Gamma}$ can be applied [49]:

$$\Gamma = \left[(t_r)_N\right]^{-1} \left[t/(t_r)_N\right]^{N-1} \left[(N-1)!\right]^{-1} \exp - \left[t/(t_r)_N\right] \qquad (3.5)$$

where $(t_r)_N$ is the mean residence time for an element of fluid in the N_{th} tank; t is the time, and N is the number of tanks.

The concentration C of tracer A in relation to its initial concentration, C_0, can be expressed as

$$C = \Gamma C_0$$

or:

$$C = C_0\left\{\left[(t_r)_N\right]^{-1} \left[t/(t_r)_N\right]^{N-1} \left[(N-1)!\right]^{-1} \exp - \left[t/(t_r)_N\right]\right\} \qquad (3.6)$$

Analysis of Eq. 3.6 reveals that the function $C = f(t)$ tends to be Gaussian and approaches the Taylor model when N increases. On the other hand, the validity of this equation is dubious for low N values, as the shape of the curve skews. This means that the model provides good results when applied to unsegmented-flow systems with long reactors but fails to describe sample dispersion in short reactors. This limitation of the model is not as relevant to efficient mixing devices such as the single bead string reactor [5,52].

There are situations where the entire system can be regarded as one well-stirred tank ($N = 1$), *e.g.*, when sample dispersion is large [2]. This also holds when a device with a large inner volume, such as a mixing chamber, is placed in the analytical path. The tanks-in-series model with ($N = 1$) is then a suitable tool for describing sample dispersion.

3.1.2.2. *Dispersion Inside a Mixing Chamber*

Mixing chambers have often been used in flow analysis, mainly to improve mixing conditions, attain a high degree of sample dispersion, provide exponential dilution, separate different immiscible phases and/or establish fluidised beads. They are essential for some specific tasks such as in-line sample preparation [65], reagent dissolution [66], analyte separation/concentration [67] and integrated reaction/detection [68]. Moreover, they are inherent to some modes of flow analysis, such as flow batch analysis.

IMPROVED MIXING CONDITIONS

Good mixing conditions should always be attained, and this aspect is particularly relevant in relation to *spectrophotometric* detection. The interaction between sample and reagent (dictated by the mixing conditions) should be reproducible because the analytical signal usually depends on both analyte and reagent concentrations. The spectrophotometric flow injection determination of calcium in a simple flow system able to analyse natural waters, plant digests and soil extracts after only minor modifications to the manifold illustrates this point [69]. Without a mixing chamber, the confluent addition of the colour-forming reagent stream (prepared in ethanol medium) resulted in a noisy baseline because a myriad of transient mirrors were randomly formed under poor mixing conditions, leading to the occurrence of a pronounced Schlieren effect [70]. This shortcoming was minimised by placing a mixing chamber in the analytical path, and the superior performance of the device in relation to an equivalent tubular coiled reactor was demonstrated.

Mixing conditions tend to be more critical when the flow system is designed in the single-line configuration, as a chemically inert carrier stream [71] cannot be exploited. With a mixing chamber, homogeneous concentration gradients along the flowing sample are established, resulting in the formation of relatively steady liquid lenses, another source of the Schlieren effect [70]. Reproducible measurements are then attained, as demonstrated in the catalytic determination of iodide in nutrition salts using a single-line sequential injection system [72]. The reaction medium required high acidity; therefore, pronounced concentration gradients were established along the sample zone. With a mixing chamber in the analytical path, net absorbance values between 0.08 and 0.8 were precisely quantified (r.s.d. $< 4\%$) on a very high (0.4 absorbance) yet reproducible blank value.

It should be stressed that vortices are established inside a mixing chamber and they may accelerate the chemical reactions involved, as demonstrated in relation to, *e.g.*, the determination of anisidine value in palm oil samples [73].

HIGH SAMPLE DISPERSION

A mixing chamber is a powerful way to attain a high degree of sample dispersion. This can be demonstrated by considering a very small (ideally infinitesimal) sample aliquot introduced as a plug directly into an ideal well-stirred mixing chamber. Considering that immediate homogenisation with the solution takes place inside the chamber, the concentration of tracer A in this aliquot, C_S, is suddenly reduced to C_0 according to [74]:

$$C_0 = C_S V_S / V_{MC} \tag{3.7}$$

where V_S is the sample inserted volume and V_{MC} is the inner volume of the mixing chamber.

Very low C/C_0 ratios are efficiently attained because the chamber and sample volumes are usually not restricted. The multi-syringe flow system for single-point titrations [75] can be used to demonstrate this. Precise dilutions were accomplished in two steps, first by splitting the sample zone and then by diluting it inside a large mixing chamber. Dispersion coefficients [2] as high as 1150 were efficiently attained. The superior efficiency of a mixing chamber for sample dispersion relative to a mixing coil of identical inner volume was also demonstrated by Garn *et al.* [76] who proposed a strategy relying on two sequentially placed mixing chambers. Dispersion coefficients as high as 1.0×10^6 and excellent measurement repeatability were reported.

EXPONENTIAL DILUTION

After insertion of the sample volume into the mixing chamber, the established concentration of tracer A inside it (C_0 – Eq. 3.7) undergoes a continuous reduction as a function of time. The time-dependent concentration, C, is determined by assuming that the instant reduction in concentration, $-dC/dt$, is proportional to the actual concentration [74], or

$$-\mathrm{d}C/\mathrm{d}t = k\,C \tag{3.8}$$

where k is the Q/V_{MC} and Q is the flow rate through the mixing chamber.

Equation 3.8 is converted to

$$-\mathrm{d}C/C = k\,\mathrm{d}t$$

or

$$\ln C/C_0 = -k\,t$$

where t is the time interval elapsed after solution introduction. So

$$C = C_0 \exp(-k\,t)$$

or

$$C = C_0 \exp(-Q\,t/V_{\mathrm{MC}}) \tag{3.9}$$

Equation 3.9 is within the scope of the tanks-in-series model with $N = 1$ [39]. Consequently, the recorded peak (Fig. 3.7) is very skewed and tailing effects are evident. In an ideal flow system with a well-stirred mixing chamber, the C_0 value is proportional to the recorded peak height, whereas the $C(t)$ values (peak shape) reflect the sample broadening effect, dependent on the Q/V_{MC} relationship. As the concentration of the tracer A decreases exponentially with time, the sample zone leaving the chamber is characterised by a long trailing edge. This broadening effect may impair sample throughput.

The exponential reduction of the time-dependent concentration of tracer A at the chamber outlet (Eq. 3.9) has been exploited to, *e.g.*, expand the dynamic concentration range [77] and implement flow titrations [78] (see also 7.3.2.1 and Chapter 8).

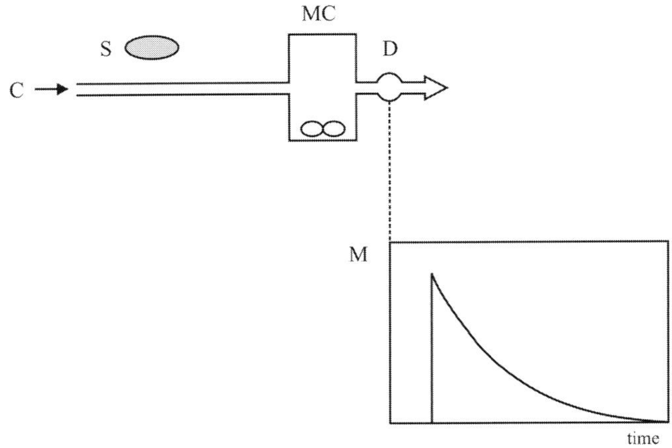

FIGURE 3.7 A model flow system with a mixing chamber. S = sample; C = carrier/washing stream; MC = stirred mixing chamber; D = ideal detector; M = measurement proportional to concentration. The recorder output reflects the ideal situation described by Eqs 3.7 and 3.9. For further details, see text.

It should be stressed, however, that measurements performed on the trailing edge of a dispersing sample tend to be less precise than at the peak maximum, as elegantly demonstrated by Gisin *et al.* [79]. They evaluated the measurement precision associated with the different fluid elements leaving a mixing chamber or a straight capillary tube, and noted that precision was ultimately limited by the *N* value. Increasing this value allowed more Gaussian-shape peaks to be recorded, improving measurement precision. Poorer measurement precision in regions of the flowing sample with pronounced analyte concentration gradients was also noted in relation to a flow-injection system exploiting zone sampling and designed to permit precise monitoring of every fluid element of a dispersing zone [80]. Different sample plugs were simultaneously introduced into the same carrier stream and overlapping occurred. A complex concentration–time function was then seen at the detector. The best precision was found in regions of the flowing sample associated with maxima and minima in the concentration–time function. Regarding measurements performed in the sample portions with more pronounced concentration gradients, those performed at the sample front were less precise than at the trailing edge, emphasising the beneficial aspect of mean residence time on the mixing conditions.

Equations 3.7 and 3.9 can be applied to flow analysis with the following caveats:

- perfect mixing is not possible from a practical point of view;
- dispersion always occurs before the sample reaches the mixing chamber, and is also inherent to the injection process [81,82];

- instant introduction of an infinitesimal aliquot of sample into the chamber is not feasible;
- additional dispersion occurs between mixing chamber and detector, as a connecting line between these devices is always present;
- the detector flow cell is also a chamber-like element; and
- the detector is not ideal.

These aspects also influence zone broadening and are the main reason why the $C(t)$ function tends to be a C-curve [83]. Hence, Eqs 3.7 and 3.9 have been expanded to describe flow systems with mixing chambers, as in the models of Pardue *et al.* [84–89] and Pungor *et al.* [90]. For specific applications, these models have been further refined by Tyson *et al.* [91–95].

ESTABLISHMENT OF FLUIDISED BEADS

Solid reagents such as slightly soluble chemicals, ion-exchange resins, immobilised reagents and enzymes, functionalised beads, and biological materials have often been exploited to expand the potential of flow analysis. In general, the solid reagents are packed inside a cylindrical chamber and chemical reactions take place at the solid/liquid interfaces. Drawbacks inherent to solid-phase devices such as increased back pressure, preferential pathways and swelling effects, as well as some limitations inherent to immobilised reagents – especially the unavailability of some sites for reaction – are minimised by taking advantage of fluidised beads [96], which involves the establishment of a diffuse and reproducible geometry where the solid particles are maintained in constant movement (floating, refluxing and circulating) inside the chamber.

Early applications were related to industrial processes and focussed on the sampling of metal ions in natural waters [96], the determination of phenylalanine using in-line ion-exchange [97], modelling of heavy metal adsorption [98], use of pulsed flows for nickel removal from acidic solutions [99], investigation of properties inherent to fluidised beds and transport reactors [100], and treatment of radioactive wastes [101]. The characteristics, limitations and applications of fluidised beads are provided elsewhere [102–104].

In flow analysis, the reproducible pattern of the distribution of solid particles inside a mixing chamber can be attained by vertically admitting the main stream at the chamber bottom. The steady reflux of the moving particles establishes the fluidised bed, which is more efficiently accomplished with pulsed flows. The feasibility of the approach was demonstrated in the spectrophotometric determination of zinc in plant digests using a multi-pumping flow system [105]. Dowex 1-X8 anionic exchanger beads were kept freely circulating inside a mini-chamber, and zinc chlorocomplexes were retained on the moving particles; after elution, derivatisation with the Zincon reagent was performed, yielding the coloured species to be monitored.

MANIFOLD COMPONENTS BEHAVING AS MIXING CHAMBERS

Manifold components used for accomplishing specific steps such as analyte separation/concentration, in-line sample preparation and reagent dissolution can have relatively large inner volumes, thus behaving as mixing chambers.

The flow-through detector may also behave as a chamber-like component, as already recognised in the early development of flow injection analysis [106]. In fact, with too large a detector, sample dilution and broadening become more pronounced (Eqs 3.7 and 3.9), impairing sensitivity and sample throughput, thus increasing reagent consumption and waste generation.

Regarding segmented flow analysis, a relatively large de-bubbler is usually required and this device may act as a mixing chamber, thus impairing system performance. This limitation is typical of older instruments. The bubble-gating approach [23] can be exploited to circumvent this drawback, but it has not been universally implemented.

In summary, the versatility of a mixing chamber, the multiple tasks it can be used for and the distinct role it can have in the flow manifold are the driving forces for more advanced methodologies using mixing chambers, not only as a component of a flow system but also as a complete small-scale laboratory. Based on an active chamber module, flow batch [107] and lab-on-valve [108] systems can be included in this category, as they can perform diverse and relatively complex functions, as discussed in 2.6 and 5.3.3.

3.1.2.3. Dispersion by Confluent Streams

Sample dispersion is altered when the flowing sample merges with a confluent stream [27]. At the vicinity of the confluence site, convective mass transport is strongly altered by the sudden change in concentrations caused by the convergence of the confluent and sample carrier streams. The laminar flow regime tends to be maintained; therefore, interactions between the sample carrier stream and the confluent stream are mainly dictated by radial diffusion.

A quantitative description of the influence of a confluent stream on sample dispersion is given below. For simplicity, radial mixing is considered as a very fast process, thus the slices perpendicular to the tubing walls can be considered as the fluid elements. Under ideal mixing conditions, the concentration C_1 of a chemical species in a fluid slice inside a flowing solution (F_1) immediately before a confluence point (Fig. 3.8) is modified to C_1' according to [27,59]:

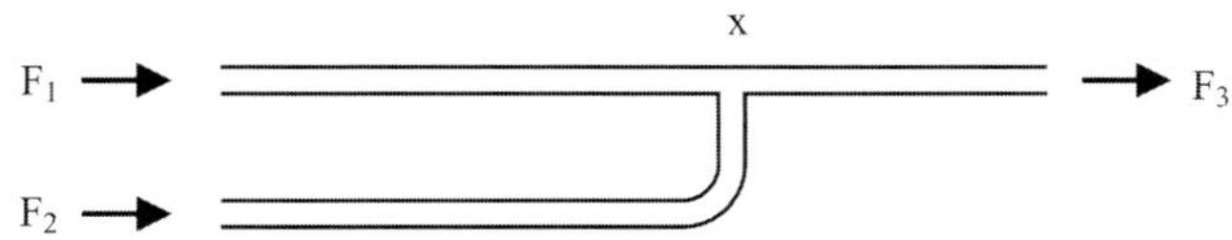

FIGURE 3.8 **Representation of a confluent stream addition.** F_1, F_2 = inlet convergent streams; F_3 = outlet stream; x = confluence site.

$$C_1' = C_1[Q_1/(Q_1 + Q_2)] = C_1(Q_1/Q_3) \quad (3.10)$$

where Q_1, Q_2 and Q_3 are the flow rates of F_1 and F_2 inlet convergent streams and of F_3 the outlet stream; $Q_3 = Q_1 + Q_2$.

Conversely, the steady concentration C_2 of confluent stream F_2 is reduced to C_2':

$$C_2' = C_2[Q_2/(Q_1 + Q_2)] = C_1(Q_2/Q_3) \quad (3.11)$$

A deeper examination of Eqs 3.10 and 3.11 suggests that the result of adding a confluent stream is analogous to a dilution. This is perhaps the main reason why terms like *sample dilution* and *dilution-like process* are often found in the scientific literature.

A practical result of Eq. 3.11 is that the reagent concentration in a confluent stream should be modified when its flow rate is changed (see also 3.1.2.4).

Accordingly, the axial length (mean thickness) of the slice, x_1, is enlarged to x'_1 after passing the confluence point [27,59]:

$$x_1' = x_1[(Q_1 + Q_2)/Q_1] = x_1(Q_3/Q_1) \quad (3.12)$$

The passage of a moving slice through a confluence point results in a decrease in its concentration and an increase in its mean thickness. Ideal flow-through detectors placed immediately before and after the confluence point would register peaks with widths at half the peak height equal to σ and σ', respectively. Without dispersion, Eq. 3.12 can be converted to [27]:

$$\sigma_1' = \sigma_1[(Q_1 + Q_2)/Q_1] = \sigma_1(Q_3/Q_1) \quad (3.13)$$

The continuous sample dispersion process occurring while it passes through the confluence point should also be taken into account. However, expansion of Eqs 3.12 and 3.13 to consider all the slices along the entire flowing sample, as well as those in the confluent stream that interact with the sample, has not yet been reported.

Lowering the sample concentration and hence increasing its mean axial length is more evident in flow systems with low sample dispersion such as segmented and mono-segmented flow systems. However, the recorded peak width is practically unaffected by confluent stream addition, as the increase in sample mean axial length is exactly compensated by the increase in overall flow rate. This is the main reason why modifying flow rates in segmented-flow analysers only slightly affects the recorded peak width (expressed in volume). In flow systems with large sample axial dispersion, these effects are masked by the sample dispersion occurring downstream (see also 5.2.3.2).

3.1.2.4. *Practical Indices for Expressing Sample Dispersion*

Several parameters have been used to gather information about sample dispersion in flow analysis: peak variance [109], time of appearance of the analytical signal, also known as *baseline-to-baseline time* [110], number of tanks in the tanks-in-series model [111], the Peclet number in the axially dispersed plug flow model [112], the Peclet number and the mean residence time in the diffusive–convective equation [113].

To design and optimise flow-based procedures, however, the analytical chemist needs a practical index that indicates the extent of the dispersion process and thus the concentrations of sample, carrier and reagent(s) along the sample zone. This index may also provide an approximation of the improvement in analytical sensitivity that could, in principle, be attained simply by reducing sample dispersion.

In segmented flow analysis, axial sample dispersion is not pronounced, being influenced mainly by the characteristics of the thin liquid film established at the tubing inner wall and by the number of segments per sample (see 5.1.2). Reduction in sample concentration is therefore strongly dependent on the addition of confluent streams. Hence, the flow rates of the sample/wash and confluent streams are the main parameters determining sample dispersion, and application of Eqs 3.10 and 3.13 can then make this practical index readily available.

In flow injection analysis, the first index proposed for this purpose was the *dilution factor, D* [7], further defined as *"the ratio of concentrations before and after the dispersion process has taken place in the element of fluid that yields the analytical readout"* [114]. This index has also been called *dispersion number, dispersion D, dispersion value, dilution ratio* and *dispersion coefficient* [115], and the latter term has been generally accepted. Its reciprocal was called the *dispersion factor* [116]. The dispersion coefficient also holds for flow injection systems with reagent injection into the flowing sample [117].

In 1978, dispersion in flow injection analysis was classified as large ($D > 10$), medium ($10 > D > 3$) and limited ($D < 3$), and was solely dependent on the dispersion coefficient [39]. This classification is valid only for single-line flow systems.

Expansion of the dispersion coefficient concept led to the D_g value that analogously provided an approximate indication of the extent of dispersion in specific portions of a dispersing sample [114]. Furthermore, the *standard dispersion* (analogous to D_g) was proposed in relation to the standard addition method [118]. In flow systems with liquid–liquid extraction, D_{seg}, D_{unseg} and D_{sep} were used [119] to specify the extent of dispersion occurring inside the aqueous/organic segmented stream, organic stream and phase-separating device, respectively.

All of the above-mentioned indices, however, indicate only the extent of dispersion of a given chemical species, usually the analyte, in the considered fluid element, without emphasising its volumetric composition. They provide scant information about the global composition of a given fluid element, the influence of a confluence stream addition or the saturation index [27]. Although useful in describing single-line flow systems, they are not easily applied to sequential injection analysis or to more elaborate flow systems. Moreover, they are restricted to the instant of detection, and are undefined in the absence of a sample. Extending the well-known concept of *volumetric fraction* to flow analysis [115] circumvents these limitations.

THE VOLUMETRIC FRACTION

The volumetric fraction, X, is a practical index that encompasses the dispersion coefficient. It includes all the above-mentioned indices and holds for the various modes of segmented and unsegmented-flow systems, as well as for batch analysis. This general index expresses the relative contribution of a solution to any given fluid element located anywhere in the manifold at any time. In this book, the notation for the volumetric fraction [115] is

$$X_{\mathrm{s,t,k}}$$

where s is the solution considered (sample, carrier and confluent stream), t is the temporal coordinate (time interval elapsed after sample introduction) and k is the spatial coordinate.

Since averaging effects in *spectrophotometric* measurements are unavoidable, a mean volumetric fraction corresponding to the average of the volumetric fractions within the slice considered is more relevant. For simplicity, however, the term *volumetric fraction* and its notation X have herein the meaning of both volumetric fraction and averaged volumetric fraction.

When s, t and k are not specified, they are assumed to relate to the sample solution, the instant corresponding to peak maximum and the flow cell, respectively. The single notation X then corresponds to the reciprocal of D, the dispersion coefficient [114]. When the temporal coordinate does not refer to the peak maximum, but the sample and flow cell are defined, X is the reciprocal of the D_g value [118]. The spatial coordinate refers to a specific location in the manifold. The volumetric fraction is fully defined by also specifying the solution considered. When s corresponds to a standard solution to be added to the sample, and t and k are not specified, the volumetric fraction corresponds to the reciprocal of the *standard dispersion* [118].

Information about the actual concentration of a given chemical species in the considered fluid element, C_i, can then be gathered by applying Eq. 3.14:

$$C_i = C_{0i} X_{s,t,k} \tag{3.14}$$

where C_{0i} is the initial concentration of this chemical species in the considered solution (sample, carrier and reagent).

Volumetric fractions can be steady or time dependent. A steady X holds in situations where the sample is not present, and is calculated as

$$X_{s,k} = Q_{s,k}/\Sigma Q_k \tag{3.15}$$

where $Q_{s,k}$ is the flow rate of solution s through k and $\sum Q_k$ is the total flow rate through k.

The summation of the steady volumetric fractions of all the involved solutions is equal to unity.

The time-dependent volumetric fraction can be evaluated by inserting a dyestuff A and considering the resulting concentration–time function (see below). It is calculated as

$$X_{s,t,k} = C_{s,t,k}/C_0 \tag{3.16}$$

where $C_{s,t,k}$ is the concentration of dyestuff A at time t inside a specific manifold site k and C_0 is the original concentration of dyestuff A.

All the volumetric fractions related to a given fluid element can be displayed as a pie-chart (Fig. 3.9) that consists of several sectors with areas proportional to the related X values. The pie-chart is also specified in terms of spatial and temporal coordinates. Sectors associated with

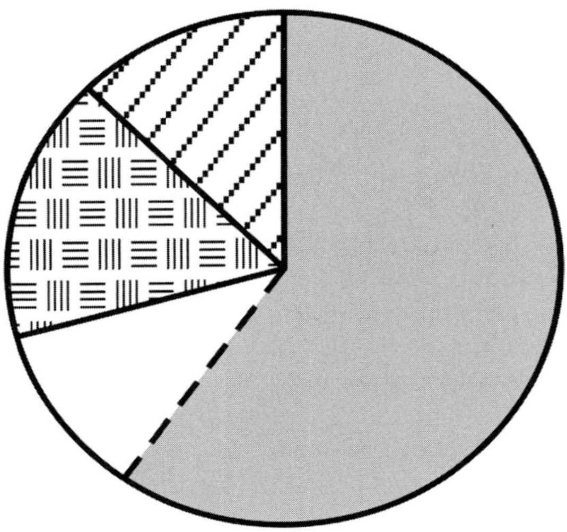

FIGURE 3.9 Volumetric fraction pie-chart. The figure refers to a flow system in which the sample (grey) is inserted into a chemically inert carrier stream (white) and two confluent reagent solutions (hatched, squared) are added downstream. Full and dashed lines = steady and transient borders, respectively.

solutions inserted into the flow system and with the related carrier streams should be adjacent to each other in order to provide information about the saturation index, as discussed in 5.2.3.2. The boundary between these sectors is time dependent, and their total area is constant. The other boundaries are steady and dependent only on flow rates (Eq. 3.15).

The concept of volumetric fraction is not a final answer to the problem of specifying the extent of dispersion in flow analysis. There are situations where its application is cumbersome, especially in flow systems exploiting solute focussing, chemical species with rather different diffusion coefficients, and/or low response time detectors.

Moreover, the presence of an artefact in the analytical path for analyte separation/concentration is not included in the definition of X. Also, the sample, reagent and carrier solutions are considered as a whole, so that differences in the diffusion coefficients of the different chemical species are not considered. The occurrence of chemical reactions altering the sample dispersion [113] is not considered.

Regarding detector response time, the experimental determination of X involves measurements carried out under both transient and steady-state conditions (see below). For faster detector response times, Eq. 3.16 is not readily applicable, as the influence of this parameter needs to be compensated.

EXPERIMENTAL DETERMINATION

The volumetric fraction can be experimentally determined using the dye approach [114]. To this end, a buffered coloured solution of a dyestuff A and a blank solution (same buffer constituents) are used. As the matrix composition of the coloured and blank solutions is the same, Schlieren effects [69] do not manifest themselves and variations in the molar absorptivity coefficient of dyestuff A due to pH variations are avoided.

Initially, the dye solution is used to mimic the sample, whereas the blank solution is used to mimic the carrier and confluent streams. Then the dye solution is used successively as the carrier and confluent streams, and the other solutions are replaced by the blank solution. In this way, measurements related to the sample, carrier and confluent streams are performed on different fluid elements at any time after sample introduction. The volumetric composition of any fluid element is then determined by applying Eq. 3.16.

A particular situation is the sample volumetric fraction at the detector at the instant of analytical signal acquisition. Providing that linearity between measurement and dye concentration holds, Eq. 3.16 can be simplified to

$$X = M/M_0 = C/C_0 \tag{3.17}$$

where M, M_0 are the measurements related to peak maximum (transient) and to initial dyestuff concentration (steady state) and C, C_0 is the related dyestuff concentrations.

Bromocresol green (BCG) and sodium tetraborate have often been used as dyestuff and buffer, respectively, in view of the stability of the resulting solutions, their nonadherence on the inner tube wall for most tubing materials, and the linear relationship between absorbance and dye concentration. Normally, 0.01 mol L^{-1} $Na_2B_4O_7$ is used and measurements are carried out at 512 nm with a 10 mm optical path flow-through cuvette. Under these conditions, 4.0×10^{-5} mol L^{-1} BCG (28 mg L^{-1}) corresponds to an absorbance of 1.0.

PRACTICAL SITUATIONS

In order to illustrate the relevance of X, some specific situations are discussed below.

DIFFERENT FLOW SYSTEMS WITH THE SAME **X** Although they have the same X (actually 0.25) the single-line and confluence flow systems in Fig. 3.10 behave differently. Increasing the sample volume in the single-line system increases X_S but decreases X_R or, in other words, causes a variation in the sample-to-reagent volumetric ratio. Conversely, increasing the sample volume in the confluence flow system increases X_S and decreases X_C, whilst keeping the amount of the added confluent reagent solution constant. Consequently, improved sensitivity by using larger inserted sample volumes is better attained in the confluence flow system.

PASSING A CONFLUENCE POINT The X pie-chart permits a rapid evaluation of the effects of a confluent stream addition. Analysis of

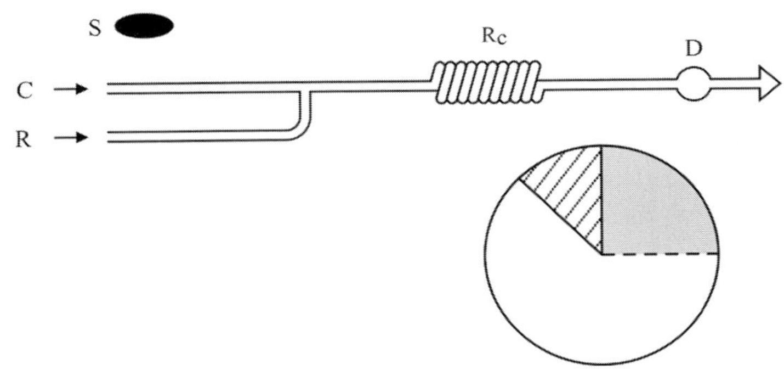

FIGURE 3.10 **Typical single-line (upper) and confluence (lower) flow systems with corresponding volumetric fractions.** S = sample [black (grey after dispersion)]; R = reagent (hatched); C = chemically inert carrier stream (white). RC = reaction coil; D = detector. Full and dashed lines = steady and transient borders respectively. For further details, see text.

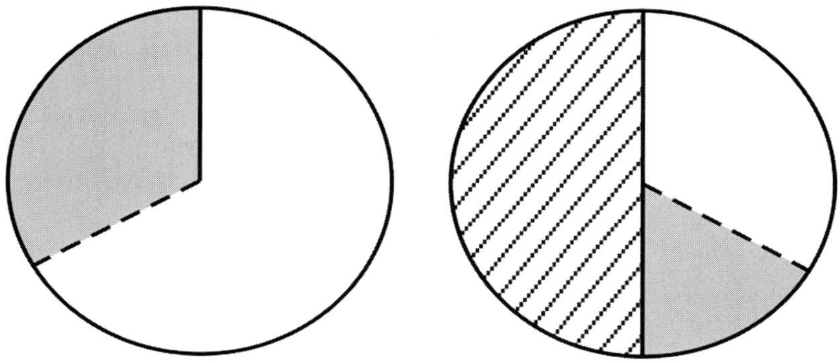

FIGURE 3.11 **Passing a confluence point.** Volumetric pie charts showing the situations before (left) and after (right) a 1:1 (v/v) confluence. Symbols as in Fig. 3.10.

Fig. 3.11 reveals that, after a 1:1 (v/v) confluence, the X values related to the sample and carrier solutions are halved and the volumetric fraction of the added stream is 0.5. The X value for the carrier stream is reduced, leading to an increase in the saturation index [27]. This confirms that the injected volume should be re-defined after any modification of the confluent stream flow rate.

FLOW SYSTEM WITH INJECTION OF REAGENT In flow injection analysis, it is possible to inject the reagent into a continuously flowing sample stream. The resulting system has been referred to as *reversal flow injection analysis* [117], and is often exploited for process monitoring. Better sensitivity relative to ordinary flow injection systems has been claimed [2]. However, if sensitivity is dependent mainly on the sample dispersion, the reversal flow-injection system could give a sensitivity better than, worse then or even equal to that of the analogous flow-injection system, depending on whether the sample volumetric fraction is higher then, lower than or equal to 0.5. Therefore, there is no reason to keep the concept of reversal flow injection analysis. With merging zones, both systems are equivalent.

SYSTEM SCALING When adapting a given flow-based procedure to routine analytical use in the laboratory, the available pumping device may not be able to provide exactly the required flow rates. The reagent concentrations should be then modified in such a way as to guarantee the same concentrations inside the reactors as in the original procedure. For scaling purposes, Eq. 3.18 (derived from Eq. 3.14) is applicable:

$$C_{\mathrm{t,k}} = C_o \, X_{\mathrm{s,t,k}} \tag{3.18}$$

where $C_{t,k}$ is the reagent concentration of the solution inside k (usually a reactor) at time t and C_o is the reagent concentration of the undispersed solution s. For reagent solutions added by confluence, it is not necessary to consider t, and Eq. 3.12 is readily applicable. This emphasises the usefulness of the concept of the steady volumetric fraction.

Some practical worked examples relating to exploitation of the volumetric fraction concept are given below.

WORKED EXAMPLES

Worked Example #1

The flow-injection system in Fig. 2.7 was designed for the spectrophotometric analysis of plant digests with a mean acidity of 0.10 mol L^{-1} $HClO_4$. The specific procedure should be carried out at pH 4.7. Determine the concentration of sodium acetate in the confluent buffer/masking/chromogenic reagent stream R, by taking into account that the pH is not affected by the presence of perchlorate anions, chromogenic reagent or masking agents.

Answer: The acidity of the sample carrier stream should be kept at 0.10 mol L^{-1} in order to avoid the establishment of undesirable pH gradients. Its contribution inside the main reactor can be calculated (Eq. 3.14) as 0.08 mol L^{-1}. The pH to be adjusted matches the pKa value of the acetic acid acetate buffer system; therefore, equimolar concentrations of acetate ion and acetic acid should be established inside the main reactor. Considering that the reaction between acetate and perchloric acid is stoichiometric, the acetate contribution from the confluent stream should be double or 0.16 mol L^{-1}. This corresponds (Eq. 3.14) to 0.80 mol L^{-1} in the reagent reservoir. It should be noted that this result does not depend on the inserted sample volume because a confluent flow system is used.

Worked Example #2

In the air-segmented analytical procedure outlined in Fig. 2.4, suggested flow rates for the sample/wash, air and reagent streams are 1.6, 0.20 and 0.40 mL min^{-1}, respectively. The available fluid propulsion device provides flow rates of 1.8, 0.20 and 0.20 mL min^{-1}. Determine the correction factor for adapting the reagent concentrations. Do not consider any kinetic aspect or any deficiency in mixing conditions as relevant.

Answer: For segmented flow analysis, the amount of added air is not considered in determining X. The sample undergoes 20 or 10% (v/v) dilution when passing through the confluence point depending on whether the reference system or the laboratory system is considered. By taking into account the X values, one can calculate that the sample and reagent solutions should undergo a 1:8/9 (v/v) dilution and a two-fold concentration for proper adaptation of the procedure to the laboratory conditions. It is not good practice to make a slight manual sample dilution prior to its introduction into the flow system; therefore, it is advisable to maintain the sample dilution and to increase the reagent concentration by a factor of ($2 \times 9/8$).

3.1.3. Visualisation of the Dispersing Sample Zone

In flow analysis with *spectrophotometric* detection, transient modifications of the monitored light during passage of the flowing sample through the detector are recorded as a peak. Each fluid element of the flowing sample corresponds to a different handling time; therefore, the entire recorded peak refers to an infinite number of development stages of the flowing sample, each one associated with a specific residence time and dispersion.

In the absence of chemical reactions (with a coloured analyte), the recorded peak reflects the temporal variation of the analyte concentration due solely to the dispersion process. Conversely, the peak shape may provide useful information on reaction kinetics in situations involving chemical reactions. Successive measurements performed on a flowing sample can therefore be exploited, as illustrated by the spectrophotometric determination of vanadium and iron in steel alloys [120], which relies on a novel strategy for implementing differential kinetic analysis.

The concept of *temporal variations in concentration* at the flow-through detector explains why pronounced skewed peaks are often observed in flow analysis, especially with loop-based sample introduction. Taylor assumed that dispersion is symmetric in relation to an observer located at the dispersing zone [55,56], but in practice the recorded peaks are usually characterised by a rise time much shorter than the fall time (see also Fig. 1.3e). This skew effect is explained by the fact that the front and trailing portions of the flowing sample, which relate to the rise time and the fall time, respectively, have different residence times in the manifold and are therefore subjected to different extents of dispersion.

The shape of the recorded peak does not match the true axial distribution of the analyte along the flowing sample. Therefore, expressions such as *peak profile*, *analyte distribution* and *concentration gradients* (typically associated with spatial variations) should be avoided.

Knowledge of the true spatial distribution of the dispersing sample is, however, necessary for a better understanding of sample interactions with neighbouring solutions. Different strategies have been suggested for achieving images of the spatial distribution (concentration gradient) of a tracer along a sample zone flowing through narrow-bore tubing.

In this regard, computer-simulated images of the sample zone were independently presented in 1986 by several research teams who proceeded with the numerical solution of Eq. 3.4. Sample spatial distributions and/or analyte concentration/time functions under laminar flow conditions were predicted by Vanderslice *et al.* [121] and Wada *et al.* [122]. Results were in good agreement with those obtained with a model single-line flow-injection system into which a dye solution was inserted. Wada *et al.* [122] also demonstrated the relevance of chemical reactions in altering sample dispersion, thus confirming earlier findings [113], whereas Korenaga *et al.* [123] considered a hypothetical situation without diffusion, demonstrating the relevance of this process to sample dispersion. It is important to emphasise that Vanderslice *et al.* initially devised the underlying mathematical strategy in 1981 [110].

The tanks-in-series model, as well as other more elaborate models, has been also exploited for this purpose [91,124,125].

For ordinary [126] and micro-conduit [127] flow injection analysers, an ingenious strategy involving a slide projector permitted a coloured flowing sample to be displayed on a large screen. The true spatial distribution of concentrations of tracer A along the sample zone was also obtained using a microscope to photograph the flowing sample [10] (Fig. 3.12). High-quality photographs allowing a quantitative estimation of the concentration gradients along a flowing sample were obtained (see also Fig. 3.2).

Images of the flowing sample can also be obtained by exploiting the random walk stochastic model, as demonstrated by Betteridge *et al.* [128]. The model considers the sample as a discrete number of individual fluid elements; each one being allowed to move under laminar flow conditions for a given time interval and then take a random step. The process is repeated a number of times. It is interesting to recall that Einstein exploited the random walk model of diffusion to explain Brownian motion, demonstrating that a group of suspended chemical species taking a series of random steps finally reaches a Gaussian distribution around the origin. He also stressed that the number of steps and the mean size of each step determine the spread.

For a fluid element experiencing laminar flow, as in a flow system, the total displacement per step is considered a combination of the effects of the random walk and convective mass transport. In contrast to other models applied to flow analysis, the model deals with individual fluid elements and not assemblages. Consequently, it is conceptually simpler than those based on a series of imaginary tanks and/or requiring the solution of a set of differential equations. The model is therefore a powerful tool for simulating the effects of, *e.g.*, sample size, timing, reactor geometry and chemical kinetics, as well as the competing effects of reaction rate and physical dispersion in flow analysis.

FIGURE 3.12 **A dispersing sample monitored by a microscope at different available times for reaction development.** An acidic dye solution (methyl orange in 0.001 mol L^{-1} HCl) is inserted into an alkaline water carrier stream of a single-line flow-injection system. For experimental conditions, see Ref. [10]. *Source: Courtesy of T. Korenaga via F.J. Krug (1993).*

When applied to a flow-injection system designed in a single-line configuration [128], the model yielded reliable results. However, a pronounced statistical noise was noted because only 500–5000 fluid elements were considered in view of the relatively low capacity of the computer (Apple microcomputer fitted with an accelerator card using a 3.5-MHz 6502C microprocessor). Simulations of 1500 fluid elements over 60 simulated seconds took about 7 h to complete. The approach was further applied to a flow-injection system with merging zones [129] and proved to be suitable for studying the influence of the main parameters involved in system design. Furthermore, the random walk model was exploited to broaden the basis for introducing the concept of sinusoidal flow for sample propulsion [29]. It is, however, surprising that in spite of improvements in computer technology, this model has not been further investigated.

Images of the real sample distribution related to a given mean residence time, and the concentration/time function related to a given manifold site were obtained [130] by taking advantage of the zone sampling process [131]. The sample was inserted into its carrier stream and, after a pre-defined time interval, t_s, an aliquot of the dispersing sample was selected, re-sampled and inserted into a second carrier stream for further processing (see also 7.2). The time-dependent concentration

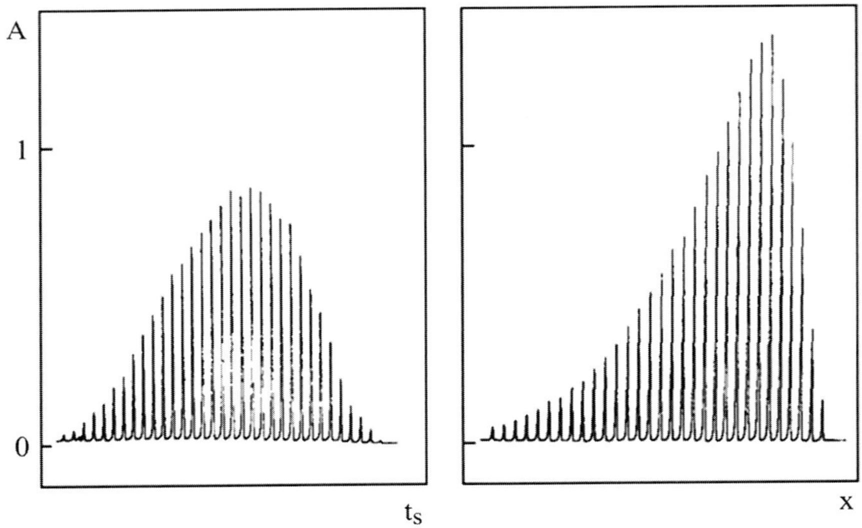

FIGURE 3.13 **True axial sample distribution (left) and time-dependent concentration (right) obtained by scanning the resampling time, t_s, or the reactor length, X.** For experimental conditions, see Ref. [130]. *Reprinted from Anal. Chim. Acta 193 (1987) 309, E.A.G. Zagatto, O. Bahia-Filho, H. Bergamin-Filho, Recording the real sample distribution and concentration time functions in flow injection analysis, with permission from Elsevier.*

inside the re-sampling port was then attained by scanning the t_S value, whereas the true axial sample distribution inside the main reactor was attained by fixing the t_S value and proceeding with a scan of reactor length (Fig. 3.13).

In 2004, three-dimensional images of flowing samples were obtained by laser-induced fluorescence [132]. The laser spot was 0.031 mm^2 and the tubular flow cell (5.0 cm length, 0.8 mm i.d.) was made from borosilicate glass. A step motor was programmed to permit the laser focus to stepwise scan the flow cell in the radial direction. Twelve different measurements were performed along the cell radius, enough for proper visualisation of an entire slice of the flowing sample. The fluorescent species Rhodamine B was used to mimic the sample and monoethylene glycol was used as the background. In view of the large data set involved, highly resolved images were obtained. The approach was further applied to a multi-pumping flow system, demonstrating the superior mixing conditions of pulsed flows and the tendency towards turbulence [41]. The laser-induced fluorescence technique, however, has not yet been exploited to display the true spatial analyte distribution.

Recently, webcam images of the flowing samples were obtained [133] in order to emphasise the main differences between sample dispersion in laminar or pulsed flows (Fig. 3.14).

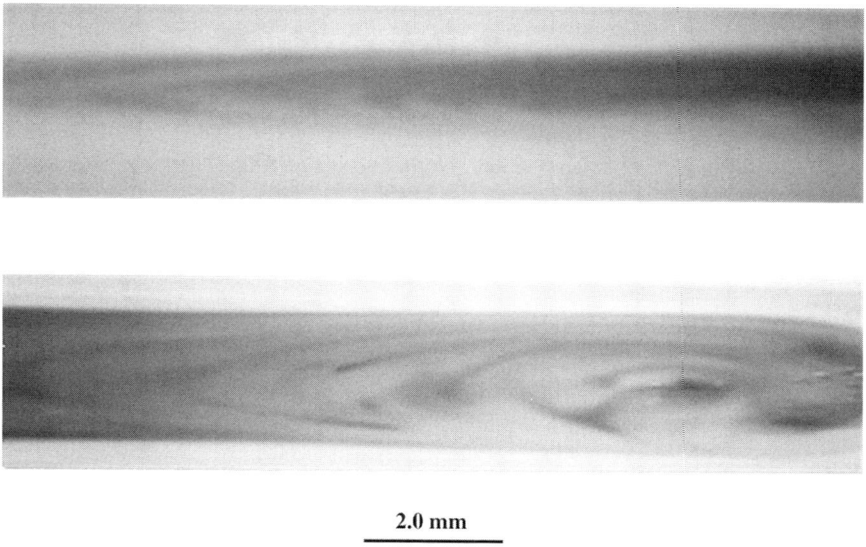

FIGURE 3.14 **Webcam images of a flowing sample under laminar (upper) and pulsed (lower) flow.** Tube walls not shown; sample inlet at the right hand side; flow from right to left. For details, see Ref. [133]. *Reprinted from Talanta 79 (2009) 978, P.R. Fortes, M.A. Feres, M.K. Sasaki, E.R. Alves, E.A.G. Zagatto, J.A.V. Prior, J.L.M. Santos, J.L.F.C. Lima, Evidences of turbulence mixing in multi-pumping flow systems, with permission from Elsevier.*

3.2. SYSTEM CONFIGURATIONS

The different modalities of flow analysers can be designed in single-line or confluence configurations (see also Fig. 3.10).

In single-line (also called straight or single channel) flow systems, the required reactants are present in the sample carrier stream and are added to the sample zone as a consequence of dispersion. The configuration is associated with the inception of flow injection analysis and is characteristic of the sequential injection analyser. Flow injection systems comprising two or more streams that converge to form the main carrier stream into which the sample is inserted [134] are also considered as single-line flow systems.

Conversely, confluence flow injection systems rely on sample insertion into a chemically inert carrier stream and the required reagents are added by confluence. The configuration is characteristic of the segmented flow analyser. The carrier (or background) stream is a solution similar to the sample but without the chemical species under determination. Distilled water, soil extracting solution, ethanol and synthetic seawater are examples of chemically inert carrier streams for the analysis of natural waters, soil extracts, spirits and seawater, respectively.

3.2.1. Single-line Flow Systems

This very simple system involves only two solutions, the sample and the carrier stream; the sum of their volumetric fractions is then always unity. As a corollary, the reagent concentration is reduced in the presence of the sample (Fig. 3.15), and maximum depletion of reagent concentration is associated with the most concentrated region of the sample zone. This aspect has not often been highlighted in the literature as it can be circumvented by increasing the reagent concentration, but this also increases reagent consumption and waste generation.

There are situations where the reagent may not reach the central portion of the dispersing sample. Two reaction zones are then established at the front and trailing edges of the dispersing sample and coalesce as a consequence of dispersion, leading to a distortion in the recorded peak. In extreme situations, when a large sample plug is inserted and low sample dispersion is involved, double peaks are recorded, the effect being critically examined by Tyson [92]. The two maxima associated with a double peak reflect the regions where the sample-to-reagent volumetric fractions are optimised in relation to the monitored analytical signal. This effect has been exploited mainly for expanding the dynamic concentration range, as at least two linear ranges are easily obtained [135]. Moreover, it is useful for differential kinetic analysis, as two analytical signals corresponding to different sample handling times are available. As an illustration, the flow injection spectrophotometric determination of cobalt and nickel exploiting formation of two reactive zones inside the flowing sample is highlighted [136].

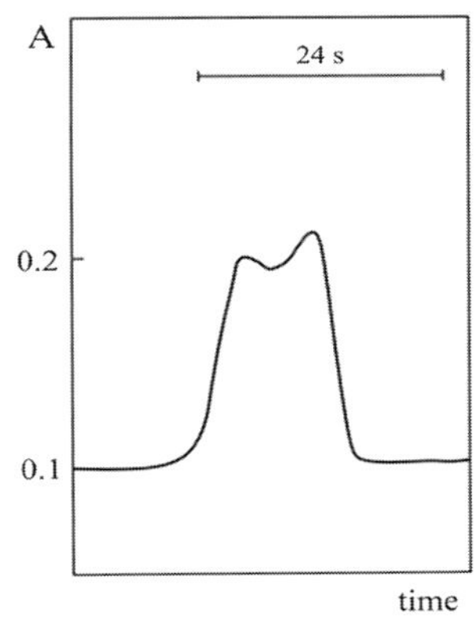

FIGURE 3.15 Peak distortion caused by reagent depletion at the centre of the sample zone. Figure refers to a single-line flow system into which a 5.00 mg L^{-1} chloride solution is inserted into a coloured reagent carrier stream. For experimental details, see Ref. [71].

Because the reagent concentration is depleted in the presence of the dispersed sample, various factors should be taken into account when designing flow systems in the straight configuration:

- The reagent consumption is, in principle, higher than in other system configurations because of the need to compensate for reagent depletion in the region of the sample zone yielding the analytical signal.
- A given time interval is needed for the reagent to reach the region of the sample zone yielding the analytical signal. The mean available time for reaction development is then lower than the mean sample residence time in the flow manifold. This may limit sensitivity, especially in analytical procedures based on relatively slow chemical reactions or with detection techniques based on reaction rate measurements such as chemiluminescence and bioluminescence.
- The reagent carrier stream may absorb (or fluoresce) at the selected wavelengths. Since the baseline decreases during passage of the processed sample through the detector, and the net analytical signal is a vector sum of baseline plus sample signals, inverted (often called negative) peaks can be observed for samples with lower analyte concentrations.
- The injected sample volume cannot be increased indefinitely.
- The configuration is restricted to single reagent procedures. This shortcoming is overcome in some analytical procedures by incorporating, *e.g.*, reagent cartridges or mini-columns, and/or exploiting light as a reagent (photochemical procedures).
- The Schlieren effect tends to be pronounced. It should be noted that undesirable concentration gradients leading to the appearance of the Schlieren effect (see also 4.2) are dependent on differences in the original composition of the sample and carrier solutions and that the effects are more pronounced in single-line systems.
- Inaccurate results may be obtained in poorly designed systems. As the analytical signal refers to a region with optimised sample-to-reagent volumetric fractions, the presence of an interfering species that consumes the reagent forces the system to select another region where the reagent is more concentrated; in this region, the sample volumetric fraction is lower, leading to inaccurate results. This limitation was reported in relation to sequential injection analysis [137] and its influence in other modes of flow analysers has not yet been investigated.

The single-line configuration is therefore recommended when high sample dispersion is required, sample/reagent interactions are not limiting, simplicity in system design is desirable and/or reagent addition is not required. This configuration is particularly attractive when analytical procedures relying on gradient exploitation are used, because the establishment of concentration gradients along the sample zone is inherent to this system configuration.

An alternative single-line system involves reagent insertion into a sample stream [116]. Although it has been called a *reversed flow-injection system*, it is actually a reagent injection system. It has often been applied to process monitoring, where the sample stream into which the reagent is injected is taken directly from the monitored process. This alternative configuration is characterised by a similar sensitivity to that inherent in single-line systems, as the maximum analytical signal corresponds to the region with optimised sample-to-reagent volumetric fractions (see also 3.1.2.4).

3.2.2. Confluence Flow Systems

In this configuration, the sample is inserted into a chemically inert carrier stream and the reagents are added by confluence (Fig. 3.10, lower). Although a dilution-like effect occurs at the confluence sites, better sensitivity in relation to single-line systems is usually attained. Confluence systems are therefore potentially *high sensitivity systems* [114].

It should be emphasised that radial diffusion is the main mechanism yielding analyte/reactant interactions, both in single-line and confluence configurations. In the former, parallel axial fluid lines are established as a consequence of the parabolic distribution of the fluid element linear velocities, whereas in the latter, parallelism of the fluid lines is attained immediately after the confluence site. The confluence thus predisposes the system to sample/reagent interactions.

As the reagent concentration is constant along the sample zone, some important aspects should be considered in designing confluence flow injection systems.

- Reagent consumption is, in principle, less than in the single-line configuration because there is no depletion in reagent concentration at the central portion of the flowing sample.
- The reagent is added to all portions of the sample zone at the confluence point; consequently, no time interval is required for the reagent to reach the central portion of the sample zone. The confluence configuration is therefore attractive for analytical procedures based on relatively slow chemical reactions. Moreover, the configuration is also useful in relation to very fast processes such as those inherent to chemi- and bio-luminescence reactions, where the confluent reagent is usually added very close to the detector.
- The baseline does not undergo depletion even if the reagent carrier stream absorbs (or fluoresces) at the selected wavelengths. Inverted peaks are therefore not usually observed in confluence systems.

- The injected sample volume can be increased indefinitely, and this aspect is relevant when limited sample dispersion is required and/or when the detector requires a relatively large time interval for proper quantification of the sample, *e.g.*, for fluorimetric and chemiluminometric assays [138].
- Several reagent solutions can be added at different confluence points.
- The Schlieren effect tends to be minimal because exploitation of a chemically inert carrier minimises any matrix differences between the inserted sample and carrier solutions. It should be noted, however, that in the analysis of sample batches with a highly variable matrix composition, a chemically inert carrier stream cannot be prepared.

A steady concentration of the reactant is not observed when the confluent reagent stream is not properly added. Pronounced differences in matrix composition (*e.g.*, colour and suspended matter), flow rates and viscosity of the carrier and confluent streams may result in a pulsed addition of the confluent stream. The effect is random, but if the fluid-propelling device is a peristaltic pump, it is characterised by a typical frequency, dictated by the rotation speed of the peristaltic pump. A pulsating flow is then established, leading to undulations in the recorded peak. The effect is reduced if the involved streams converge with similar mean linear velocities. As the effect is characterised by a constant frequency, the modulated signal (ripple) is easily filtered out.

3.3. THE DETECTOR RESPONSE

The analytical results are obtained by evaluating the detector response obtained after handling the samples and standard solutions. They generally rely on transient peaks recorded during passage of the flowing zones through the detector.

Most *spectrophotometric* flow-based analytical procedures involve single analyte determination or multi-parametric determinations carried out in multi-channel analysers, with an independent analytical channel for each analyte. Consequently, analytical results are obtained after converting all of the recorded peaks to single values [139], usually referred to as the analytical signals.

Either flat or bell-shaped peaks are normally recorded in flow analysis; therefore, these types of peaks are dealt with separately.

3.3.1. Flat Peaks

These peaks are inherent to segmented-flow analysers and to certain unsegmented-flow systems using large sample volumes. A plateau is observed in the central portion of the peak (Figs 2.4 and 5.5), thus the only parameters useful for measurement are the height of this plateau or, in some instances, the integrated peak area. Some favourable features intrinsic to flow systems yielding flat peaks are:

- Measurements are performed on a steady detector response, leading to enhanced ruggedness.
- Averaged analytical signals can be estimated by considering several successive measurements, and this may improve the precision.
- When peak height is measured, accuracy tends to be unaffected by variations in sample volume.
- The flat portion of the recorded peak can be inspected for diagnostic purposes in real time and this is an additional guarantee of reliability.
- Fast transient measurement disturbances associated with, *e.g.*, the presence of mini-air bubbles inside the flowing sample are less pronounced relative to flow analysers relying on bell-shaped peaks. The related transient signals can easily be inspected and corrected.

As a plateau is involved, sample throughput is lower and the required volumes of sample and reagent solutions are larger in relation to flow systems yielding bell-shaped recorded peaks.

3.3.2. Bell-Shaped Peaks

These peaks are inherent to flow injection analysers and other flow systems characterised by large sample dispersion. They tend towards a Gaussian shape, although skewed peaks are often reported.

Peak height, peak area and peak width have often been selected as the analytical signals. Moreover, measurements related to the rise and/or fall regions of the peaks can be also exploited.

3.3.2.1. Peak Height

The height of the recorded peak is by far the most used parameter for converting the entire recorded peak into a single analytical signal. This can be explained by recalling that measurements performed on the most concentrated portion of the flowing sample are characterised by higher signal-to-noise ratio in relation to other regions of the sample zone. As concentration gradients in this region are less pronounced, measurements at the top of the peak tend to be more precise. The recorded peak height is determined as the difference in measurements related to the peak maximum and the baseline and is therefore obtained easily and quickly.

In practice, the baseline is not restored during handling of successive samples, leading to a lack of baseline measurements. This aspect becomes more pronounced for analytical procedures with higher sample carryover. In order to estimate the baseline value related to the peak maximum, a blank solution is processed from time to time, and the related signals are taken into account when obtaining the baseline value at the sample peak maximum (Fig. 1.5). With this strategy, baseline drift is easily compensated.

An alternative approach is to consider the differences in measurements related to the instants of signal appearance and peak maximum recording. Although largely exploited in laboratories devoted to large-scale analysis, the approach is seldom emphasised in the scientific literature.

Another possibility is to select a lower sampling rate, in order to permit quantitative baseline restoration between the monitoring of successive samples. This procedure cannot, however, always be recommended in view of its deleterious influence on sample throughput and hence on reagent consumption.

Another strategy is to consider only the measurement related to the peak maximum as the analytical signal. Although often exploited, the strategy is potentially susceptible to baseline drift affecting measurement reliability.

It is interesting to note that the peak shape is altered for detectors characterised by a large flow cell volume and/or time constant. In this situation, the peak height resembles the peak area.

3.3.2.2. *Peak Area*

Estimation of the peak area for attaining the analytical signal can be used in situations where an integrated measurement is required. This approach has the favourable characteristics inherent to flat peak recording (see above), but to a lesser extent because a plateau is not usually reached.

Peak area evaluation is particularly useful when a large time interval is needed to obtain the analytical signal, *e.g.*, for analytical procedures relying on chemiluminometric measurements [140] and for detectors with a large time constant.

The recorded area, however, cannot be determined quickly because the integration limits are not well defined, as the baseline is approached in an asymptotic manner. This drawback can be minimised by using triangulation procedures [139].

There are also some limitations in the use of peak area as the measurement basis. As measurements other than those related to the top of the peak are used, the signal-to-noise ratio for peak area measurements is relatively poor compared with peak height measurements. Moreover, no overlap between successive flowing samples is permitted because the entire recorded peak is used for obtaining the analytical signal. Consequently, the sampling rate is greatly reduced.

3.3.2.3. *Peak Width*

Peak width at half-maximum height is indicative of sample dispersion and has been often used for determining diffusion coefficients (see 5.2.3.1). Peak width near the baseline, often called *peak appearance time* [110] is used in flow systems relying on time-based measurements, with the aim of a wider dynamic range or implementation of pseudo flow titrations (see Chapter 8).

3.3.3. Gathering the Calibration Model

Analytical signal(s) are related to analyte concentration(s) via a calibration model. In flow analysis, the single one-dimensional model relying on the analytical curve is by far the most common. For other models, as well as for concentration estimation by extrapolative strategies, *e.g.*, the standard addition method, see Chapter 8.

Normally, a linear regression of several data points is estimated, a given point including the analyte concentration and the corresponding analytical signal. In some situations, *e.g.*, a wide concentration range, quadratic functions or even more complex functions are used.

As a rule, the working standard solutions are run prior to and after the samples, and results are estimated based on the two analytical curves. If the difference in the coefficients of the related analytical equations is higher than a pre-set value, the entire process should be repeated or the equation coefficients should be recalculated in order to compensate for the effect. A useful strategy is to consider a timely linear variation in the equation coefficient [141], as illustrated in the working example #3.

Finally, it is good practice to run the initial calibration twice and then run the samples only if the two analytical equations are identical according to a certain pre-selected condition. This ensures that more steady conditions are attained before running the first sample.

Worked example #3. In a routine laboratory devoted to large-scale analysis, 40 samples are run between the successive standard sets. Estimate the analyte concentration for sample # 10, providing that: measurement for this sample is 0.6328 A and the initial and final analytical equations are:

$$A = 0.0028(A) + 0.3496(A\,\mathrm{L\,mg^{-1}})C(\mathrm{mg\,L^{-1}}) \text{ and}$$

$$A = 0.0020(A) + 0.3503(A\,\mathrm{L\,mg^{-1}})C(\mathrm{mg\,L^{-1}}).$$

where A is the absorbance and C is the analyte concentration in $\mathrm{mg\,L^{-1}}$.

Answer: Considering a timely linear variation in the coefficients of the analytical equations, the linear (a_0) and angular (a_1) coefficients for sample # 10 are determined as:

$$a_0 = [(40 - 10) \times 0.0028 + 10 \times 0.0020]/40 = 0.0026\,A$$

$$a_1 = [(40 - 10) \times 0.3496 + 10 \times 0.3503]/40 = 0.3498\,A\,L\,mg^{-1}$$

Analyte concentration in sample # 10 is then:

$$C = 0.6328/0.3498 - 0.0026 = 1.8064\,\mathrm{mg\,L^{-1}}.$$

References

[1] W.J. Duncan, A.S. Thom, A.D. Young, Mechanics of Fluids, second ed., Edward Arnold, London, 1970.

[2] J. Ruzicka, E.H. Hansen, Flow Injection Analysis, second ed., Wiley-Interscience, New York, 1988.

[3] J.T. Davies, Turbulence Phenomena. An Introduction to the Eddy Transfer of Momentum, Mass and Heat, Particularly at Interfaces, Academic Press, New York, 1972.

[4] D. Betteridge, Flow injection analysis, Anal. Chem. 50 (1978) 832A.

[5] J.M. Reijn, W.E. van der Linden, H. Poppe, Dispersion in open tubes and tubes packed with large glass beads. The single bead string reactor, Anal. Chim. Acta 123 (1981) 229.

[6] J. Ruzicka, E.H. Hansen, Flow injection analyses. Part I. A new concept of fast continuous flow analysis, Anal. Chim. Acta 78 (1975) 145.

[7] J. Ruzicka, E.H. Hansen, E.A.G. Zagatto, Flow injection analysis. Part VII. Use of ion-selective electrodes for rapid analysis of soil extracts and blood serum. Determination of potassium, sodium and nitrate, Anal. Chim. Acta 88 (1977) 1.

[8] C. Silfwerbrand-Lindh, L. Nord, L.G. Danielsson, F. Ingman, The analysis of aqueous solutions with ethanol-soluble reagents in a flow injection system. Spectrophotometric determination of uranium, Anal. Chim. Acta 160 (1984) 11.

[9] A. Bertsch, S. Heimgartner, P. Cousseau, P. Renaud, Static micromixers based on large-scale industrial mixer geometry, Lab Chip 1 (2001) 56.

[10] T. Korenaga, F.H. Shen, H. Yoshida, T. Takahashi, K.K. Stewart, Study of sample dispersion in flow injection analysis, Anal. Chim. Acta 214 (1988) 97.

[11] D. Betteridge, B. Fields, Construction of pH gradients in flow injection analysis and their potential use for multielement analysis in a single sample bolus, Anal. Chem. 50 (1978) 654.

[12] A.E. Davies, A.G. Fogg, Flow injection sandwich technique applied to indirect visible spectrophotometric determinations. Bromimetric and iodimetric determination of sulphite, Anal. Chim. Acta 318 (1995) 55.

[13] D.J. Malcome-Lawes, C. Pasquini, A novel approach to non-segmented flow analysis. Part 3. Sequential determination of ammonium, nitrite and nitrate ions, J. Autom. Chem. 10 (1988) 192.

[14] T. Korenaga, X.-L. Zhou, T. Moriwake, H. Muraki, T. Naito, S. Sanuki, Computer-controlled micropump suitable for precise microliter delivery and complete in-line mixing, Anal. Chem. 66 (1994) 73.

[15] Y. Israel, A. Lasztity, R.M. Barnes, Online, steady-state concentrations for inductively coupled plasma atomic emission and mass-spectrometry achieved by tandem injection and merging-stream flow injection, Analyst 114 (1989) 1259.

[16] B.F. Reis, M.F. Gine, E.A.G. Zagatto, J.L.F.C. Lima, R.A. Lapa, Multicommutation in flow analysis. Part 1. Binary sampling: concepts, instrumentation and spectrophotometric determination of iron in plant digests, Anal. Chim. Acta 293 (1994) 129.
[17] X.-D. Wang, T.J. Cardwell, R.W. Cattrall, G.E. Jenkins, Pulsed flow chemistry. A new approach to the generation of concentration profiles in flow analysis, Anal. Commun. 35 (1998) 97.
[18] J.M. Calatayud, J.V.G. Mateo, V. David, Multi-insertion of small controlled volumes of solutions in a flow assembly for determination of nitrate (photoreduction) and nitrite with proflavin sulphate, Analyst 123 (1998) 429.
[19] X.-D. Wang, T.J. Cardwell, R.W. Cattrall, R.P. Dyson, G.E. Jenkins, Time-division multiplex technique for producing concentration profiles in flow analysis, Anal. Chim. Acta 368 (1998) 105.
[20] P.B. Martelli, B.F. Reis, M. Korn, I.A. Rufini, The use of ion exchange resin for reagent immobilization and concentration in flow systems. Determination of nickel in steel alloys and iron speciation in waters, J. Braz. Chem. Soc. 8 (1997) 479.
[21] M. Korn, L.F.B.P. Gouveia, E. Oliveira, B.F. Reis, Binary search in flow titration employing photometric end-point detection, Anal. Chim. Acta 313 (1995) 177.
[22] C. Pasquini, W.A. Oliveira, Monosegmented system for continuous-flow analysis. Spectrophotometric determination of chromium(VI), ammonia, and phosphorus, Anal. Chem. 57 (1985) 2575.
[23] R.L. Habig, B.W. Schlein, L. Walters, R.E. Thiers, A bubble-gating flow cell for continuous-flow analysis, Clin. Chem. 15 (1969) 1045.
[24] C. Pasquini, Mechanical removal of the central sample zone to avoid air bubbles in monosegmented continuous-flow analysis, Anal. Chem. 58 (1986) 2346.
[25] A.R.A. Nogueira, S.M.B. Brienza, E.A.G. Zagatto, J.L.F.C. Lima, A.N. Araujo, Multi-site detection in flow analysis. Part 2. Monosegmented systems with relocating detectors for the spectrophotometric determination of boron in plants, Anal. Chim. Acta 276 (1993) 121.
[26] I.M. Raimundo Jr., C. Pasquini, Automated monosegmented flow analyser. Determination of glucose, creatinine and urea, Analyst 122 (1997) 1039.
[27] E.A.G. Zagatto, B.F. Reis, M. Martinelli, F.J. Krug, H. Bergamin-Filho, M.F. Gine, Confluent streams in flow injection analysis, Anal. Chim. Acta 198 (1987) 153.
[28] I. Lopez Garcia, P. Vifias, N. Campillo, M. Hernandez Cordoba, Use of submicroliter-volume samples for extending the dynamic range of flow injection flame atomic absorption spectrometry, Anal. Chim. Acta 308 (1995) 85.
[29] J. Ruzicka, G.D. Marshall, G.D. Christian, Variable flow rates and a sinusoidal flow pump for flow injection analysis, Anal. Chem. 62 (1990) 1861.
[30] M. Romero-Saldana, A. Rios, M. Valcarcel, Continuous sample monitoring by flow reversal methodology, Fresenius J. Anal. Chem. 342 (1992) 547.
[31] J. Toei, Determination of glucose in clinical samples by flow reversal flow injection analysis, Analyst 113 (1988) 475.
[32] G.D. Clark, J. Ruzicka, G.D. Christian, Split zone flow injection analysis: an approach to automated dilutions, Anal. Chem. 61 (1989) 1773.
[33] D. Betteridge, P.B. Oates, A.P. Wade, Control of dispersion and variation of reaction coil length in flow injection analyzers by flow reversals, Anal. Chem. 59 (1987) 1236.
[34] F. Canete, A. Rios, M.D. Luque de Castro, M. Valcarcel, Kinetics of ion-pair extraction in continuous flow systems, Anal. Chim. Acta 224 (1989) 169.
[35] E.A.G. Zagatto, A.O. Jacintho, J. Mortatti, H. Bergamin-Filho, An improved flow injection determination of nitrite in waters by using intermittent flows, Anal. Chim. Acta 120 (1980) 399.
[36] J. Ruzicka, E.H. Hansen, Flow injection analysis. Principles, applications and trends, Anal. Chim. Acta 114 (1980) 19.

[37] J. Ruzicka, E.H. Hansen, Stopped-flow and merging zones. A new approach to enzymatic assay by flow injection analysis, Anal. Chim. Acta 106 (1979) 207.

[38] C.S. Lim, J.N. Miller, J.W. Bridges, Automation of an energy-transfer immunoassay by using stopped-flow injection analysis with merging zones, Anal. Chim. Acta 114 (1980) 183.

[39] J. Ruzicka, E.H. Hansen, Flow injection analysis. Part X. theory, techniques and trends, Anal. Chim. Acta 99 (1978) 37.

[40] G.D. Owens, R.W. Taylor, T.Y. Ridley, D.W. Margerum, Pulsed-flow instruments for measurements of fast reactions in solution, Anal. Chem. 52 (1980) 130.

[41] A.C.B. Dias, J.L.M. Santos, J.L.F.C. Lima, C.M. Quintella, A.M.V. Lima, E.A.G. Zagatto, A critical comparison of analytical flow systems exploiting streamlined and pulsed flows, Anal. Bioanal. Chem. 388 (2007) 1303.

[42] R.A.S. Lapa, J.L.F.C. Lima, B.F. Reis, J.L.M. Santos, E.A.G. Zagatto, Multi-pumping in flow analysis: concepts, instrumentation, potentialities, Anal. Chim. Acta 466 (2002) 125.

[43] F.R.P. Rocha, B.F. Reis, E.A.G. Zagatto, J.L.F.C. Lima, R.A.S. Lapa, J.L.M. Santos, Multicommutation in flow analysis: concepts, applications and trends, Anal. Chim. Acta 468 (2002) 119.

[44] E.A.G. Zagatto, P.J. Worsfold, Flow analysis: overview, in: P.J. Worsfold, A. Townshend, C.F. Poole (Eds.), Encyclopedia of Analytical Science, second ed., Elsevier, Oxford, 2005. v.3, p. 24.

[45] S.M.B. Brienza, R.P. Sartini, J.A. Gomes-Neto, E.A.G. Zagatto, Crystal seeding in flow injection turbidimetry: determination of total sulfur in plants, Anal. Chim. Acta 308 (1995) 269.

[46] J. Harrow, J. Janata, Heterogeneous samples in flow injection systems. Part 1. Whole blood, Anal. Chim. Acta 174 (1985) 115.

[47] M. Hulsman, M. Bos, W.E. van der Linden, Dispersion behaviour of solid particles in flow injection analysis, Anal. Chim. Acta 346 (1997) 351.

[48] R. Tijssen, Axial dispersion and flow phenomena in helically coiled tubular reactors for flow analysis and chromatography, Anal. Chim. Acta 114 (1980) 71.

[49] M. Valcarcel, M.D. Luque de Castro, Flow Injection Analysis. Principles and Applications, Ellis Horwood, Chichester, 1987.

[50] U.D. Neue, H. Engelhardt, Reaction detector with three-dimensional coiled open tubes in HPLC, Chromatographia 15 (1982) 403.

[51] L. Gorton, L. Ögren, Flow injection analysis for glucose and urea with enzyme reactors and on-line dialysis, Anal. Chim. Acta 130 (1981) 45.

[52] J.M. Reijn, H. Poppe, W.E. van der Linden, A possible approach to the optimization of flow injection analysis, Anal. Chim. Acta 145 (1983) 59.

[53] A. Grifftihs, On the movement of a coloured index along a capillary tube, and its application to the measurement of the circulation of water in a closed circuit, Proc. Phys. Soc. London 23 (1911) 190.

[54] R.J. Nunge, W.N. Gill, Mechanisms affecting dispersion and miscible displacement, in: R.J. Nunge (Ed.), Flow Through Porous Media, Am. Chem. Soc, Washington, 1970, p. 179.

[55] G.I. Taylor, Dispersion of soluble matter in solvent flowing slowly through a tube, Proc. Roy. Soc. London, Series A (Math. Phys. Sci.) A219 (1953) 186.

[56] G.I. Taylor, Diffusion and mass transport in tubes, Proc. Phys. Soc. London 67 (1954) 857.

[57] P.S. Francis, S.W. Lewis, K.F. Lim, K. Carlsson, B. Karlberg, Flow analysis based on a pulsed flow of solution: theory, instrumentation and applications, Talanta 58 (2002) 1029.

[58] F.A. Inon, M.B. Tudino, Theoretical Aspects of Flow Analysis, in: M. Trojanowicz (Ed.), Advances in Flow Analysis, Wiley VCH, Mainheim, 2008, p. 3. Chapter 1.
[59] R.D. Hull, R.E. Malick, J.G. Dorsey, Dispersion phenomena in flow injection systems, Anal. Chim. Acta 267 (1992) 1.
[60] G. Horvai, E. Pungor, H.A. Mottola, Theoretical backgrounds of flow analysis, CRC Crit. Rev. Anal. Chem. 17 (1987) 231.
[61] R. Aris, On the dispersion of a solute in a fluid flowing through a tube, Proc. Roy. Soc. A235 (1956) 67.
[62] V. Ananthakrishnan, W.N. Gill, A.J. Barduhn, Laminar dispersion in capillaries. I. Mathematical analysis, AICHE J 11 (1965) 1063.
[63] W.N. Gill, V. Ananthakrishnan, Laminar dispersion in capillaries. 4. Slug stimulus, AICHE J 13 (1967) 801.
[64] J.K. Foreman, P.B. Stockwell, Automatic Chemical Analysis, Ellis Horwood, Chichester, 1975.
[65] M. Burguera, J.L. Burguera, O.M. Alarcon, Flow injection and microwave-oven sample decomposition for determination of copper, zinc and iron in whole blood by atomic absorption spectrometry, Anal. Chim. Acta 179 (1986) 351.
[66] G.D. Marshall, D. Wolcott, D. Olson, Zone fluidics in flow analysis: potentialities and applications, Anal. Chim. Acta 499 (2003) 29.
[67] H. Bergamin-Filho, J.X. Medeiros, B.F. Reis, E.A.G. Zagatto, Solvent extraction in continuous flow injection analysis. Determination of molybdenum in plant material, Anal. Chim. Acta 101 (1978) 9.
[68] F. Lazaro, M.D. Luque de Castro, M. Valcarcel, Integrated reaction/spectrophotometric detection in unsegmented flow systems, Anal. Chim. Acta 214 (1988) 217.
[69] A.O. Jacintho, E.A.G. Zagatto, B.F. Reis, L.C.R. Pessenda, F.J. Krug, Merging zones in flow injection analysis. Part 6. Determination of calcium in natural waters, soil and plant materials with glyoxal bis(2-hydroxyanil), Anal. Chim. Acta 130 (1981).
[70] A.C.B. Dias, E.P. Borges, E.A.G. Zagatto, P.J. Worsfold, A critical examination of the components of the Schlieren effect in flow analysis, Talanta 68 (2006) 1076.
[71] H. Bergamin-Filho, B.F. Reis, E.A.G. Zagatto, A new device for improving sensitivity and stabilization in flow injection analysis, Anal. Chim. Acta 97 (1978) 427.
[72] A.N. Araujo, J.L.F.C. Lima, M.L.M.F.S. Saraiva, R.P. Sartini, E.A.G. Zagatto, A SIA system with mixing chamber for handling high concentrated solutions: spectrophotometric catalytic determination of iodide in nutrition salts, J. Flow Injection Anal 14 (1997) 151.
[73] B. Saad, W.T. Wai, B.P. Lim, M.I. Saleh, Flow injection determination of anasidine value in palm oil samples using a triiodide potentiometric detector, Anal. Chim. Acta 591 (2007) 248.
[74] E.A.G. Zagatto, J.M.T. Carneiro, S. Vicente, P.R. Fortes, J.L.M. Santos, J.L.F.C. Lima, Mixing chambers in flow analysis: a review, J. Anal. Chem.York. (Russia) 64 (2009) 540.
[75] F. Albertus, B. Horstkotte, A. Cladera, V. Cerda, A robust multisyringe system for process flow analysis. Part I. On line dilution and single point titration of protolytes, Analyst 124 (1999) 1373.
[76] M.B. Garn, M. Gisin, H. Gross, P. King, W. Schmidt, C. Thommen, Extensive flow injection dilution for in-line sample pre-treatment. Comparison between single-stage and dual-stage modules, Anal. Chim. Acta 207 (1988) 225.
[77] K.K. Stewart, A.G. Rosenfeld, Exponential dilution chambers for scale expansion in flow injection analysis, Anal. Chem. 54 (1982) 2368.
[78] J. Ruzicka, E.H. Hansen, H. Mosbaek, Flow injection analysis. Part IX. A new approach to continuous flow titrations, Anal. Chim. Acta 92 (1977) 235.

[79] M. Gisin, C. Thommen, K.F. Mansfield, Hydrodynamically limited precision of gradient techniques in flow injection analysis, Anal. Chim. Acta 179 (1986) 149.
[80] E.A.G. Zagatto, M.F. Gine, E.A.N. Fernandes, B.F. Reis, F.J. Krug, Sequential injections in flow systems as an alternative to gradient exploitation, Anal. Chim. Acta 173 (1985) 289.
[81] A.U. Ramsing, J. Ruzicka, E.H. Hansen, The principles and theory of high-speed titrations by flow injection analysis, Anal. Chim. Acta 129 (1981) 1.
[82] R.D. Hull, R.E. Malick, J.G. Dorsey, Dispersion phenomena in flow injection systems, Anal. Chim. Acta 267 (1992) 1.
[83] O. Levenspiel, Chemical Reaction Engineering, second ed., John Wiley & Sons, New York, 1972.
[84] H.L. Pardue, B. Fields, Kinetic treatment of unsegmented flow systems. Part 1. Subjective and semi-quantitative evaluations of flow injection systems with gradient chamber, Anal. Chim. Acta 124 (1981) 39.
[85] H.L. Pardue, B. Fields, Kinetic treatment of unsegmented flow systems. Part 2. Detailed treatment of flow injection systems with gradient chamber, Anal. Chim. Acta 124 (1981) 65.
[86] H.L. Pardue, P. Jager, Kinetic treatment of unsegmented flow systems. Part 3. Flow injection system with gradient chamber evaluated with a linearly responding detector, Anal. Chim. Acta 179 (1986) 169.
[87] P. Jager, H.L. Pardue, Kinetic treatment of unsegmented flow systems. Part 4. Equations for a system with gradient chamber corrected to account for detectors with finite sensitivities, Anal. Chim. Acta 187 (1986) 343.
[88] M.J. James, H.L. Pardue, Experimental evaluation of theoretical response equations for an unsegmented flow system with a well-stirred mixing chamber, Anal. Chim. Acta 270 (1992) 195.
[89] J.M. Jordan, S.H. Hoke, H.L. Pardue, Evaluation of different data-processing options for a flow system with a well-stirred mixing chamber, Anal. Chim. Acta 272 (1993) 115.
[90] E. Pungor, Z. Feher, G. Nagy, K. Toth, G. Horvai, M. Gratzl, Injection techniques in dynamic flow-through analysis with electroanalytical sensors, Anal. Chim. Acta 109 (1979) 1.
[91] J.F. Tyson, Peak width and reagent dispersion in flow injection analysis, Anal. Chim. Acta 179 (1986) 131.
[92] J.F. Tyson, Analytical information from doublet peaks in flow injection analysis. Part 1. Basic equation and applications to flow injection titrations, Analyst 112 (1987) 523.
[93] D.C. Stone, J.F. Tyson, Models for dispersion in flow injection. Part 2. Two tanks in parallel model, Analyst 114 (1989) 1453.
[94] R.T. Echols, J.F. Tyson, Comparison of mixing devices for flow injection determination based on doublet peak formation, Anal. Chim. Acta 286 (1994) 169.
[95] R.T. Echols, J.F. Tyson, Determination of rate constants by a double-line flow injection method incorporating a well-stirred tank reactor, Talanta 41 (1994) 1775.
[96] J.W. Hofstraat, J.A. Tielrooij, H. Compaan, W.H. Mulder, Fluidized-bed solid- phase extraction. A novel approach to time-integrated sampling of trace metals in surface water, Environ. Sci. Technol 25 (1991) 1722.
[97] J.H. Koh, N.H.L. Wang, P.C. Wankat, Ion-exchange of phenylalanine in fluidized/ expanded beds, Ind. Eng. Chem. Res. 34 (1995) 2700.
[98] P. Menoud, L. Cavin, A. Renken, Modelling of heavy metals adsorption to a chelating resin in a fluidized bed reactor, Chem. Eng. Process 37 (1998) 89.
[99] H. Tokuyama, S. Maeda, K. Takahashi, Development of a novel moving bed with liquid-pulse and experimental analysis of nickel removal from acidic solution, Sep. Purif. Technol. 38 (2004) 139.

[100] M.A. Hastaoglu, A simple sampling method for the reaction history in fluidized or transport reactors, Ind. Eng. Chem. Res. 43 (2004) 7975.
[101] X.-G. Feng, S. Jing, Q.-L. Wu, J. Chen, C. Song, The hydrodynamic behaviour of the liquid-solid circulating fluidized bed ion exchange system for caesium removal, Powder Technol. 134 (2003) 235.
[102] L.S. Fan, Summary paper on fluidization and transport phenomena, Powder Technol. 88 (1996) 245.
[103] C.L. Briens, S. Barghi, Defluidized zones in liquid-solid fluidized beds, Powder Technol. 114 (2001) 186.
[104] Y. Zheng, J.-X. Zhu, N.S. Marwaha, A.S. Bassi, Radial solids flow structure in a liquid-solids circulating fluidized bed, Chem. Eng. J. 88 (2002) 141.
[105] M.F.T. Ribeiro, A.C.B. Dias, J.L.M. Santos, J.L.F.C. Lima, E.A.G. Zagatto, Fluidized beds in flow analysis: use with ion-exchange separation for spectrophotometric determination of zinc in plant digests, Anal. Bioanal. Chem. 384 (2006) 1019.
[106] J.M. Reijn, W.E. van der Linden, H. Poppe, Some theoretical aspects of flow injection analysis, Anal. Chim. Acta 114 (1980) 105.
[107] R.S. Honorato, M.C.U. Araujo, R.A.C. Lima, E.A.G. Zagatto, R.A.S. Lapa, J.L.F.C. Lima, A flow-batch titrator exploiting a one-dimensional optimisation algorithm for end point search, Anal. Chim. Acta 396 (1999) 91.
[108] J. Ruzicka, Lab-on-valve: universal microflow analyzer based on sequential and bead injection, Analyst 125 (2000) 1053.
[109] D. Robinson, J.E. Anderson, J.-L. Lin, Measurement of diffusion coefficients of some indoles and ascorbic acid by flow injection analysis, J. Phys. Chem. 94 (1990) 1003.
[110] J.T. Vanderslice, K.K. Stewart, A.G. Rosenfeld, D.J. Higgs, Laminar dispersion in flow injection analysis, Talanta 28 (1981) 11.
[111] J.M. Hungerford, G.D. Christian, Chemical kinetics with reagent dispersion in single-line flow injection systems, Anal. Chim. Acta 200 (1987) 1.
[112] S.D. Kolev, E. Pungor, Mathematical modelling of single-line flow injection analysis systems without chemical reaction, Anal. Chem. 60 (1988) 1700.
[113] C.C. Painton, H.A. Mottola, Kinetics in continuous flow sample processing. Chemical contributions to dispersion in flow injection techniques, Anal. Chim. Acta 158 (1984) 67.
[114] J. Ruzicka, E.H. Hansen, Flow Injection Analysis, J. Wiley & Sons, New York, 1981.
[115] E.A.G. Zagatto, B.F. Reis, H. Bergamin-Filho, The concept of volumetric fraction in flow injection analysis, Anal. Chim. Acta 226 (1989) 129.
[116] B.F. Reis, H. Bergamin-Filho, E.A.G. Zagatto, F.J. Krug, Merging zones in flow injection analysis. Part 3. Spectrophotometric determination of aluminium in plant and soil materials with sequential addition of pulsed reagents, Anal. Chim. Acta 107 (1979) 309.
[117] K.S. Johnson, R.L. Petty, Determination of phosphate in sea water by flow injection analysis with injection of reagent, Anal. Chem. 54 (1982) 1185.
[118] J.F. Tyson, J.M.H. Appleton, A.B. Idris, Flow injection sample introduction methods for atomic absorption spectrometry, Analyst 108 (1983) 153.
[119] K. Bäckström, L.G. Danielsson, L. Nord, Dispersion in phase separators for flow injection extraction systems, Anal. Chim. Acta 187 (1986) 255.
[120] P.R. Fortes, S.R.P. Meneses, E.A.G. Zagatto, A novel flow-based strategy for implementing differential kinetic analysis, Anal. Chim. Acta 572 (2006) 316.
[121] J.T. Vanderslice, A.G. Rosenfeld, G.H. Beecher, Laminar-flow bolus shapes in flow injection analysis, Anal. Chim. Acta 179 (1986) 119.
[122] H. Wada, S. Hiraoka, A. Yuchi, G. Nakagawa, Sample dispersion with chemical reaction in a flow injection system, Anal. Chim. Acta 179 (1986) 181.
[123] T. Korenaga, H. Yoshida, Y. Yokota, S. Kaseno, T. Takahashi, Simulation of FIA peak without molecular diffusion in laminar capillary flow, J. Flow Injection Anal. 3 (1986) 91.

[124] D.F. Leclerc, P.A. Bloxham, E.C. Toren Jr., Axial dispersion in coiled tubular reactors, Anal. Chim. Acta 184 (1986) 173.

[125] S.D. Kolev, E. Pungor, Description of an axially-dispersed plug flow model for the flow pattern in elements of fluid systems, Anal. Chim. Acta 185 (1986) 315.

[126] J.F. Tyson, A simple lecture demonstration of flow injection analysis, Fresenius J. Anal. Chem. 329 (1988) 675.

[127] I.D. McKelvie, T.J. Cardwell, R.W. Cattral, A microconduit flow-injection analysis demonstration using a 35-mm slide projector, J. Chem. Edu. 67 (1990) 262.

[128] D. Betteridge, C.Z. Marczewski, A.P. Wade, A random walk simulation of flow injection analysis, Anal. Chim. Acta 165 (1984) 227.

[129] C.D. Crowe, H.W. Levin, D. Betteridge, A.P. Wade, A random-walk simulation of flow injection systems with merging zones, Anal. Chim. Acta 194 (1987) 49.

[130] E.A.G. Zagatto, O. Bahia-Filho, H. Bergamin-Filho, Recording the real sample distribution and concentration time functions in flow injection analysis, Anal. Chim. Acta 193 (1987) 309.

[131] B.F. Reis, A.O. Jacintho, J. Mortatti, F.J. Krug, E.A.G. Zagatto, H. Bergamin-Filho, L.C.R. Pessenda, Zone sampling processes in flow injection analysis, Anal. Chim. Acta 123 (1981) 221.

[132] C.M. Quintella, Y.N. Watanabe, A.M.V. Lima, M. Korn, M. Embirucu, A.P.S. Musse, Evaluation of transversal and longitudinal dispersion in a flow injection system by exploiting laser induced fluorescence: influence of flow-cell positioning, Anal. Chim. Acta 523 (2004) 293.

[133] P.R. Fortes, M.A. Feres, M.K. Sasaki, E.R. Alves, E.A.G. Zagatto, J.A.V. Prior, J.L.M. Santos, J.L.F.C. Lima, Evidences of turbulence mixing in multi-pumping flow systems, Talanta 79 (2009) 978.

[134] J. Ruzicka, J.W.B. Stewart, Flow injection analysis. Part II. Ultrafast determination of phosphorus in plant material by continuous flow spectrophotometry, Anal. Chim. Acta 79 (1975) 79.

[135] M. Martinelli, H. Bergamin-Filho, M.A.Z. Arruda, E.A.G. Zagatto, A new approach for wide-range flow injection spectrophotometry: determination of cobalt in livestock mineral supplements, Quim. Anal 8 (1989) 153.

[136] A. Femandez, M.D. Luque de Castro, M. Valcarcel, Formation of two reaction zones in flow injection systems for kinetic determinations of cobalt and nickel, Anal. Chim. Acta 193 (1987) 107.

[137] E.A.G. Zagatto, F.R.P. Rocha, P.B. Martelli, B.F. Reis, Detecting and circumventing sources of inaccuracy in flow analysis, Pure Appl. Chem. 73 (2001) 45.

[138] A. Townshend, J.A.M. Pulgarin, M.T.A. Pardo, Flow injection-chemiluminescence determination of propranolol in pharmaceutical preparations, Anal. Chim. Acta 488 (2003) 81.

[139] R. Tryzell, B. Karlberg, Comparison of various peak evaluation techniques for limited and extended dynamic ranges in flow injection systems, Anal. Chim. Acta 364 (1998) 97.

[140] J.A.M. Pulgarin, A.A. Molina, P.F. López, Automatic chemiluminescence-based determination of carbaryl in various types of matrices, Talanta 68 (2006) 586.

[141] J.W.B. Stewart, J. Ruzicka, H. Bergamin-Filho, E.A.G. Zagatto, Flow injection analysis. Part III. Comparison of continuous flow spectrophotometry and potentiometry for the rapid determination of total nitrogen content in plant digests, Anal. Chim. Acta 81 (1976) 371.

CHAPTER

4

Interaction of Radiation with the Flowing Sample

Images of broken light which dance before me like a million eyes
Well you can radiate everything you are
Yes, you can indicate everything you see. **The Beatles**

OUTLINE

The attenuation of a beam of UV or visible radiation by the flowing sample relies on the same processes that occur in batch analysis with a cuvette and is also true for the radiation emitted by excited chemical species. The significance of each process may however be different because the flowing sample is usually monitored under dynamic conditions inside a flow cell. Complete mixing of the sample with the carrier/wash and reagent streams is usually not achieved; hence, the presence of concentration gradients and sharp liquid interfaces is almost unavoidable. As a consequence, mixing boundaries within different fluid elements can give rise to scattering, refraction and reflection of the radiation. The geometry of the flow-through detector and the optical performance of the instrument also play important roles in this context.

This chapter deals with fundamental aspects related to *spectrophotometric* monitoring of the flowing sample. Emphasis is given to the effects arising from partial mixing, *e.g.*, the Schlieren effect, and to special monitoring strategies, *e.g.*, dual-wavelength spectrophotometry and solid-phase spectrophotometry, for enhancing sensitivity. Approaches used to increase the actual optical path length and to perform multi-parametric determinations are discussed in Chapters 6 and 8, respectively. This chapter concludes with a brief discussion of the presence of immiscible phases in the flowing sample.

4.1. FUNDAMENTALS

The different interactions of radiation with the flowing sample in relation to the different *spectrophotometric* techniques are presented below. Basic aspects are dealt with in relation to a static cuvette, whereas those aspects more related to flow analysis are discussed in relation to a flow-through cuvette.

4.1.1. UV–Visible Spectrophotometry

UV–Visible spectrophotometry relies on the wavelength dependent absorption of electromagnetic radiation by the processed sample. The

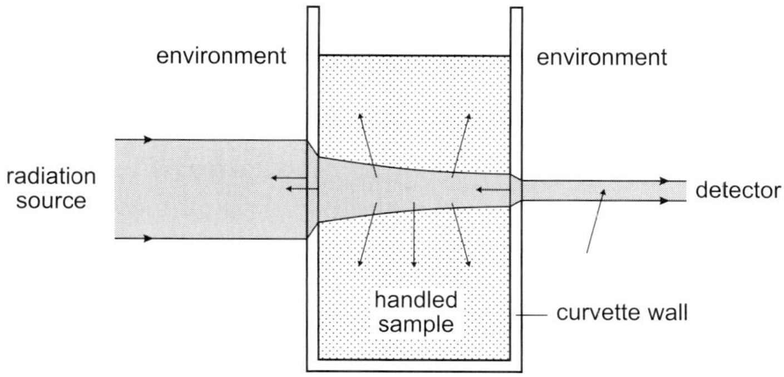

FIGURE 4.1 **A beam of radiation traversing an ordinary cuvette.** Radiation power is represented by the beam width; large arrows = incident and emergent beams; small arrows = main processes involved [absorption by the light absorbing species inside the sample, reflection and refraction at interfaces (including liquid interfaces), scattering by small particles, stray radiation]. *Source: Adapted from Ref. [2].*

wavelength range is typically between *ca* 200 and 900 nm, and can be limited by the solvent, as common solvents such as water, ethanol, hexane and cyclohexane absorb significantly below 190, 210, 195 and 210 nm, respectively [1]. As a rule, absorption is associated with electronic transitions and there is a relationship between the proportion of radiation absorbed and the concentration of the absorbing species, and hence the analyte concentration in the sample, as described below.

When a monochromatic beam of radiation traverses a cuvette containing the processed sample (Fig. 4.1) and some of the radiation is absorbed (ideally, this is the only process occurring), the power of the incident light beam (Pi) is reduced to Pe, the power of the emergent beam; the ratio Pe/Pi is known as the transmittance (T). The absorbance (A) is defined as

$$A = -\log T = \log Pi/Pe \tag{4.1}$$

which ideally is proportional to the concentration of the absorbing species.

4.1.1.1. *Losses of Radiation*

In real situations, absorption of radiation by the analyte (or related chemical species) is not the only process that attenuates the incident beam because radiation losses by reflection and refraction at the gas/solid, gas/liquid, solid/liquid and liquid/liquid interfaces, absorption by the cuvette walls and by the sample matrix and scattering inside the sample (Fig. 4.1) can also occur. The contribution of these secondary processes should be kept as small as possible because they may impact on the measurement signal. Moreover, absorption of radiation by species other than the analyte (or related chemical species) should be avoided.

RADIATION LOSSES AT INTERFACES

Losses of radiation by reflection or refraction at the environment/cuvette/sample interfaces are a consequence of the establishment of mirrors and lenses. A mirror is just an interface between two transparent media with different refractive indices, whereas different interface shapes can give rise to the formation of lenses. As the influence of these artefacts is strongly dependent on the collimation of the incident beam, the proportion of lost radiation is minimised by properly designing the shape of the cuvette and using a spectrophotometer with good optical characteristics [2].

Radiation losses are reduced with a well collimated radiation beam, as orthogonality between the incident beam and any interfaces results in less pronounced refraction. Regarding flow analysis, transient liquid/liquid interfaces that are oblique to the incident beam can be established inside the flowing sample and this is related to occurrence of the Schlieren effect [3]. This effect is more pronounced in spectrophotometers with inferior optical components and design.

In this regard, flow-through cuvettes with parallel windows are preferred, although tubular flow cells with concave/convex windows have also been used, especially when the signal-to-noise ratio is not critical and simplicity of system design is required [4]. The attenuation of radiation can be observed with the naked eye by removing the flow cell from the spectrophotometer and looking through it towards the light. A clear or hazy image is seen depending on whether there is a parallel window or a tubular flow cell, respectively.

RADIATION LOSSES AT THE CUVETTE WALLS

Electromagnetic radiation can be absorbed by the cuvette walls and the effect is minimised by properly selecting the material used for manufacturing the transparent parts of the cuvette. Table 4.1 specifies the materials most frequently used for cuvettes. Depending on the selected material, different wavelength ranges are available and a practical

TABLE 4.1 Wavelength Ranges Compatible with Some Cuvette Materials

Material	Wavelength Range/nm
Fused silica or quartz	180–2500
Borosilicate glass	380–2000
Poly(methyl methacrylate)	380–800

example is cuvettes made from ordinary glass that cannot be used in the UV spectral region. Moreover, accumulation of undesirable material such as solid particles, solvent drops or gas micro-bubbles on the cuvette inner walls, and condensation of vapour on the outside walls, must be avoided.

RADIATION LOSSES INSIDE THE SAMPLE

Loss of radiation by absorption inside the flowing sample zone is usually due to the presence of interfering radiation-absorbing chemical species in the sample matrix. In this context, potential interfering chemical species should be previously separated or effectively masked. Moreover, strategies for sample clean up should be applied if the sample is inherently coloured. Unreacted (excess) reagents can also be a limiting factor in the signal-to-noise ratio, especially when coloured or UV absorbing reagents are used. In flow systems where the reagent is added by stream confluence, the absorption of radiation by the reagents generally manifests itself as an elevated baseline. Under this condition, the reagent concentration must be limited.

Loss of radiation by reflection and refraction inside the sample is mainly due to the Schlieren effect, which can manifest itself as a consequence of the establishment of lenses and mirrors between neighbouring fluid elements with different refractive indices (see also 4.2). These losses are more pronounced in flow analysis [3], as perfect homogeneity of the monitored sample is not attained in practice. Losses are minimised by avoiding the presence of discontinuities along the optical beam, i.e., by efficient mixing of the sample with the carrier/reagent streams. In this regard, exploitation of a chemically inert carrier/wash stream is beneficial [5].

Loss of radiation by scattering inside the sample can also occur; Tyndall and Rayleigh effects are often relevant in this regard (see also 4.1.2). In flow analysis, scattering is more evident in relation to cytometric analysis [6], the determination of large molecules [7] or when colloidal species (particles or other immiscible phases) are present in the monitored sample.

COMPENSATION OF RADIATION LOSSES

Attenuation of radiation due to all processes other than absorption by the analyte (or related chemical species) is compensated by measuring the *Pe* values of the processed sample (P) and of a reference or blank (P_0) solution. Absorbance is then expressed as

$$A = \log P_0/P \tag{4.2}$$

An alternative procedure for compensation is to adjust the instrument settings in order to get 0% T when the radiation beam is blocked and

100% T for the blank. For differential spectrophotometry, the lowest standard is used instead the blank [8].

In flow analysis, the baseline is established by the carrier and reagent streams. In situations where the sample matrix contributes significantly to attenuation of the incident beam, a chemically inert sample carrier stream (with composition as close as possible to that of the sample) is recommended, and 100% *T* should correspond to the baseline.

4.1.1.2. The Lambert–Beer Law

Lambert's (or Bouguer's) and Beer's laws both relate to the absorption of radiant energy by the absorbing medium. The former states that each layer of equal thickness of the medium absorbs an equal fraction of the energy traversing it. The latter states that the absorptive capacity of an absorbing substance is directly proportional to its concentration in a solution [9]. The equation ideally relating absorbance to the concentration of the radiation-absorbing species is therefore expressed as

$$A = \log P_0/P = \mathrm{a\,b\,C} \qquad (4.3)$$

where A is the absorbance; a is the absorptivity coefficient[1]; b is the optical path length; and C is the concentration of the radiation-absorbing species inside the cuvette.

Equation 4.3 is commonly known as the Lambert–Beer law. In flow analysis, C refers to the transient averaged concentration of the radiation-absorbing species inside the illuminated volume of the flow-through cuvette.

When the concentration of the absorbing species is expressed in mol L^{-1} and the optical path length in cm, the absorptivity coefficient is called the molar absorptivity coefficient (ε), with units of L mol^{-1} cm^{-1}, and the absorbance is expressed as

$$\mathrm{A} = \varepsilon\,\mathrm{b\,C} \qquad (4.4)$$

The ε value is dependent on the temperature and refractive index of the solution involved [10], but these variations are usually negligible for concentrations lower than 0.01 mol L^{-1} [11]. The b value is normally constant and typically 1 cm. Strategies for increasing the actual optical path length are discussed in Chapter 6.

Analysis of Eqs. 4.3 and 4.4 reveals a linear relationship between absorbance and the averaged concentration of the radiation-absorbing species in the considered fluid element. This latter parameter is proportional to the analyte concentration in the sample (C_0). If the fluid element is that associated with the analytical signal and the Lambert–Beer law

[1]The expression absorptivity (and related ones) is often used, but IUPAC recommends absorptivity coefficient.

TABLE 4.2 Molar Absorptivity Coefficients for Compounds Formed During the Classical Determinations of Some Metal Ions. λ = Wavelength of Maximum Absorption. For Experimental Conditions, see Ref. [13]

Element	Reagent	λ (nm)	ε (L mol^{-1} cm^{-1})
Al	Eriochrome cyanine R	535	6.5×10^4
Ca	Glyoxal bis(2-hydroxyanil)	516	1.8×10^4
Co	Nitroso R salt	415	3.5×10^4
Fe	1,10-Phenanthroline	512	1.1×10^4
Mg	Eriochrome Black T	525	1.8×10^4
Mn	Formaldoxime	455	1.12×10^4
Ni	Dimethylglyoxime	360	3.4×10^3
Th	Arsenazo(III)	655	1.15×10^5

holds, the slope of the resulting linear function relating A and C_0 (proportional to ε) is the analytical sensitivity [12]. Table 4.2 shows ε values for some commonly measured species.

In flow analysis, the averaged concentration of the absorbing species in the fluid element yielding the analytical signal can also be determined by applying Eq. 4.4 and the analyte concentration in the sample is

$$C_0 = k' \, A \, \varepsilon^{-1} \, b^{-1} \tag{4.5}$$

where k′ is the experimentally defined factor that includes the sample dispersion.

As dispersion is directly related to sensitivity, the extent of sample dispersion is of the utmost importance in the design of a flow system.

4.1.1.3. *Practical Aspects of the Lambert–Beer Law*

Equation 4.4 is valid under ideal situations where the concentration of the radiation-absorbing chemical species is not a critical factor, the radiation is monochromatic, the sample is homogeneous, chemical deviations are not present, temperature is maintained, there is no stray radiation and the detector is ideal. In addition, other aspects may prejudice the linearity of the analytical curve. Fundamental issues relating to deviations from the Lambert–Beer law are presented below.

CONCENTRATION OF THE RADIATION-ABSORBING SPECIES

This concentration should not exceed a given threshold, generally 0.001–0.01 mol L^{-1} [1,2,14]. For higher concentrations, the average distance between the absorbing species is reduced and, consequently, each particle can affect the charge distribution of neighbouring particles. This may result in negative deviations from linearity of the absorbance/concentration function, and hence the analytical response curve, at higher concentrations. This phenomenon can also manifest itself at lower concentrations of the radiation-absorbing species if other chemical species are present in the medium at higher concentrations and/or when increasing the concentration modifies the refractive index of the medium and hence the absorptivity coefficient [1].

This phenomenon can become more pronounced in flow analysis, especially if good mixing conditions are not attained. The presence of fluid elements with transient concentrations higher than the maximum averaged concentration (central portion of the flowing sample) can lead to variations in the overall light beam attenuation due to variations in the charge distribution of neighbouring species. This effect has not been extensively reported but can certainly be avoided by proper design of the reaction manifold and the flow-through cuvette.

MONOCHROMATICITY OF THE RADIATION

This can be a limiting factor in the linearity of the absorbance/concentration function. A perfectly monochromatic beam is not practically achievable (unless laser technology is exploited), because it is impossible to isolate it from the spectrum of the continuum light source and therefore the energy of the incident beam has a finite spectral bandwidth. This can result in a negative deviation from the Lambert–Beer law at higher concentrations of the radiation-absorbing chemical species. The mathematical treatment describing this limitation is given elsewhere [14].

Moreover, increasing spectral width of the incident light beam can result in an averaged ε value lower than the theoretical one, as the ε value is wavelength dependent. In this situation, sensitivity is degraded.

Whenever possible, absorbance measurements should be performed in the spectral region of maximum absorption inside which ε is practically constant. Slight variations in wavelength do not lead to a significant error in the analytical results.

SAMPLE HOMOGENEITY

Lack of homogeneity of the sample may also be a source of deviation from the Lambert–Beer law in view of the increased possibility of radiation losses at the liquid/liquid interfaces, as well as the presence of gas and/or liquid mini-bubbles and suspended matter (see also Fig. 4.1). In flow analysis, effects due to discontinuities along the monitored sample zone tend to be more severe; therefore the Schlieren effect can become more pronounced, the possibility of evolution of gaseous mini-bubbles is higher and the conditions for the eventual settling of suspended matter are less favourable.

CHEMICAL DEVIATIONS

Linearity of the absorbance/concentration function can also be impaired by some reactions, *e.g.*, dissociation and association reactions, or interaction of the solvent with the monitored chemical species. A classical example is potassium dichromate ($K_2Cr_2O_7$) solutions, which contain $Cr_2O_4^{2-}$, $HCrO_4^-$ and $Cr_2O_7^{2-}$ ions and other Cr(VI) species [15]. As these ions have different absorption spectra and their relative proportions depend on the total chromium concentration and pH of the solution, a perfectly linear relationship between absorbance and Cr(VI) concentration is not attainable. Moreover, a simple dilution may alter the species distribution. For spectrophotometric measurements, pH buffering is therefore required.

TEMPERATURE

This parameter should always be constant, as the absorptivity coefficient is generally temperature dependent [10]. In addition, variations in this parameter can lead to changes in the chemical equilibria, refractive indices and dielectric constants of the monitored solutions, which may limit the applicability of the Lambert–Beer law. A more constant temperature from sample to sample is generally more easily attainable in flow analysis than in batch analysis.

STRAY RADIATION

External radiation may reach the detector, leading to an enhancement of the transmitted light that can introduce systematic errors in the measured absorbance. This extraneous radiation is referred to as stray radiation (or diffuse light) and reaches the detector as a consequence of scattering, reflection and refraction effects on the mirrors, lenses, filters, windows and other optical components. It may differ in wavelength from the principal radiation and may not pass through the monitored sample. In flow analysis, the power of the stray radiation is probably higher than in batch analysis due to the establishment of transient optical artefacts inside the flowing sample [3].

In the presence of stray radiation, the absorbance is altered to A′ as follows [16]:

$$\mathrm{A}' = -\log(P_0 + P_S)/(P + P_S) \neq \log P_0/P = \mathrm{A} \tag{4.6}$$

where P_S is the power of the stray radiation, which can be *a priori* experimentally determined.

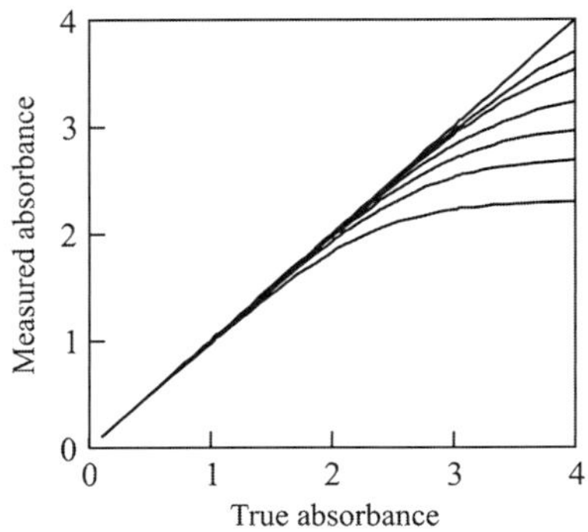

FIGURE 4.2 **Influence of stray radiation on the measured absorbance.** From the top, the different lines refer to stray radiation powers corresponding to 0.000, 0.003, 0.01, 0.03, 0.1, 0.3 and 1.0% that of the incident radiation beam (Eq. 4.6).

One can see from Eq. 4.6 that stray light can cause deviations from the Lambert–Beer law under low transmittance conditions, i.e., high absorbances (Fig. 4.2). This is the main reason why the expansion of the absorbance range is limited by the optical performance of the spectrophotometer. In fact, Eq. 4.6 may not be applicable for higher absorbance values, as P_S might be much higher than P.

The classical sources of deviations from the Lambert–Beer law listed above are generally of minor concern. However, there are other situations where the linearity of the analytical curve is also affected.

Strictly linear analytical response curves are not observed in some procedures where radiation is absorbed by the reagent(s). Increasing the analyte concentration reduces the concentration of the unreacted reagent; hence, a decrease in absorbance is the basis for measurement. The determination of sucrose in molasses involving reaction with the alkaline hexacyanoferrate(III) reagent [17] is a good example. It was carried out in a segmented flow analyser that also enabled in-line dialysis and sucrose hydrolysis: as the reagent was the absorbing species, its original concentration gave a baseline reading of about 1.0 absorbance. Increasing the sucrose concentration in the sample decreased the baseline but a strictly linear absorbance decrease with analyte concentration was not attained. This is inherent to most analytical procedures exploiting absorbance decreases. In fact, in the hypothetically extreme situation of analyte concentration approaching infinity, the absorbance of the reagent system would tend to zero, but this value would not be attained due to the chemical equilibria involved.

Non-linearity of the analytical response curve is also observed when the reagent concentration is not high enough but cannot be increased at will due to its inherent colour and/or solubility. The determinations of aluminium and boron relying on the highly coloured Eriochrome cyanine

R [13] and the slightly soluble Azomethine-H [18] reagents, respectively, provide good examples of this phenomenon. Although both methods were characterised by enhanced analytical figures of merit, strictly linear analytical response curves were not observed due to insufficient reagent concentrations. This is of major concern in flow analysis and hence the flow system should be designed to provide good sample/reagent interaction (mixing).

Another non-ideal situation can occur when the product formed during the analytical procedure influences the main reaction rate, known as an autocatalytic effect. This affects the linearity of the analytical response curve [19].

4.1.1.4. *Special Strategies*

During the development of UV–Visible spectrophotometry, different strategies have been proposed to improve the performance of the spectrophotometer and/or figures of merit for the analytical procedures involved. Some of those that are relevant to flow analysis are considered below.

DUAL-WAVELENGTH SPECTROPHOTOMETRY

Dual-wavelength spectrophotometry, DWS, was conceived as a means of compensating for variations in the power source and hence in the incident radiation beam [20]. The approach uses a single radiation beam with at least two wavelength bands traversing the sample. The analytical wavelength usually corresponds to the wavelength of maximum absorption of the monitored chemical species and the reference wavelength is selected outside of the absorbance profile of this chemical species and accounts for non-specific radiation losses. The power fractions of the emergent beam associated with the different wavelengths are quantified and the corresponding absorbances subtracted from each other (Fig. 4.3). The absorbance difference, ΔA (or the transmittance ratio) is calculated as

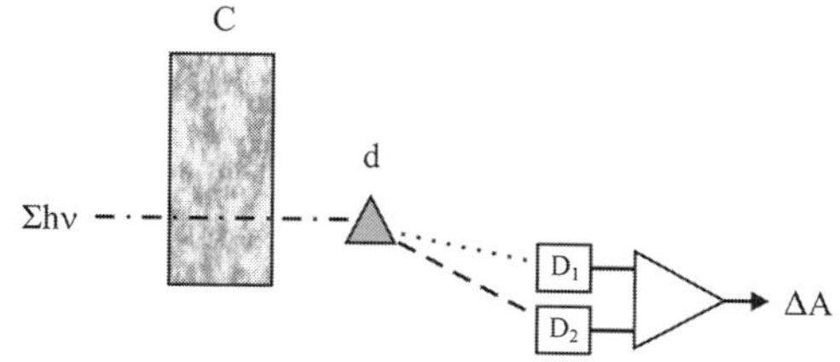

FIGURE 4.3 **Didactic representation of a dual-wavelength spectrophotometer.** C = cuvette with the handled sample; $\sum h\nu$ = polychromatic incident radiation beam [entire source emitted spectrum (diode array spectrophotometer) or two wavelength bands (optical fibre technology)]; d = optical dispersing element; D_1, D_2 = detectors; ΔA = real-time corrected absorbance. For details, see text.

$$\Delta A = T_{\lambda 1}/T_{\lambda 2} = A_{\lambda 1} - A_{\lambda 2} = (\log P_0/P_{\lambda 1}) - (\log P_0/P_{\lambda 2}) = \log(P_{\lambda 2}/P_{\lambda 1}) \quad (4.7)$$

where λ_1 and λ_2 are the analytical and reference wavelengths; ΔA is the real-time corrected absorbance (analytical signal); and other symbols as in Eq. 4.1.

The last term on the right (Eq. 4.7) allows one to conclude that the analytical signal is not dependent on P_0, the power of the incident radiation beam. In practice, this means that fluctuations in the power of the radiation source are compensated.

Measurements at two wavelengths were first carried out in relation to chemical kinetics, as the experiments usually required fast monitoring of the ratio of two concentrations. DWS exploitation was efficient for accomplishing this task [20] as the wavelengths could be selected in order to correspond to the absorbance profiles of the different chemical species involved.

Alternatively, the isosbestic point, the wavelength at which two chemical species present in the cuvette (usually one reactant and one reaction product) have the same molar absorptivity, could be selected as the reference wavelength because the related transmittance is not affected by the development of the chemical reactions. Fluctuations in the source radiation could then be efficiently compensated.

It was soon realised that DWS was also useful for the analysis of turbid samples. Multiple scattering occurs in the presence of suspended matter, leading to an apparent increase in the optical path length, as well as an attenuation of the incident radiation beam, both processes leading to a positive deviation in the absorbance. As scattering tends to be wavelength-independent, there is non-specific attenuation of the radiation, and DWS is an appropriate tool to circumvent this [21].

Regarding exploitation of the isosbestic point in DWS for the analysis of turbid samples, a good example is the determination of hydrogen peroxide in wood pulp bleaching streams relying on peroxide association with the molybdate ion [22]. The influence of dispersed lignin in the pulp bleaching stream, a critical spectral interferent in the analytical procedure, was quantitatively circumvented.

DWS was initially implemented by letting two monochromatic beams simultaneously or sequentially (using a chopper) traverse the cuvette; the emergent beam was dispersed, reaching two separate detectors. Real-time subtraction of the wavelength-independent noise was then achieved. This development led to the appearance of double-beam instruments relying on separate beams traversing the analytical and reference cuvettes [23]. More recently, optical fibre technology and diode arrays have also been used [24].

In flow analysis, DWS has been used since the 1970s, mainly with segmented flow analysis, and an ingenious strategy for differentiating and counting particles in whole blood [25] is a good example. The emergent radiation beam was quantified at different wavelengths, allowing the evaluation of both the radiation scattered by solid particles and the radiation absorbed by the different coloured cell types. The system handled 60 samples per hour and was able to measure up to 10,000 cells per sample. More recently, a variable-wavelength double-beam spectrophotometric detector designed specifically for segmented flow analysis permitted the rapid and simultaneous determination of metals using DWS [26]. DWS is now an option in standard segmented flow analysers.

DWS is also useful for suppressing solvent peaks in liquid chromatography [27] and for compensating the Schlieren effect in flow analysis [28], and this latter potential is discussed in 4.2.3.5. Optical fibre technology [29,30] and rapid spectral scanning using photodiode arrays [31,32] are also useful for this application. This innovation is also important for minimising optical propagation losses and optical source fluctuations in LED-based instruments [33,34].

SOLID-PHASE SPECTROPHOTOMETRY

Solid-phase spectrophotometry, SPS, relies on the measurement of the power of the emergent beam after attenuation by the solid particles onto which the analyte (or related chemical species) has been sorbed [35,36]. To this end, support materials such as ion-exchangers, cellulose membranes, filters and polyurethane foams have all been used. Both analyte concentration and measurement are then carried out in the presence of the solid support. SPS is an elegant way to improve the sensitivity and selectivity of *spectrophotometric* methods [37]. In this context, the landmark work on SPS carried out in the mid-1970s [38] should be highlighted. Chromium(VI), iron(II), copper(II) and cobalt(II) were determined after reacting with diphenylcarbazide, 1,10-phenanthroline, Zincon and thiocyanate, respectively, which were previously sorbed onto an ion-exchange resin. The radiation-absorbing complexes were directly measured in the resin phase and a tenfold improvement in sensitivity was obtained.

The power of the emergent radiation beam is quantified after the interaction with solid particles; therefore, both radiation absorption and scattering by the particles contribute to attenuating the radiation beam. Radiation scattering is the main drawback, as it can lead to a high blank value and deterioration in the measurement repeatability. This effect can be minimised by reducing the number of solid particles, by making the solid-phase as transparent as possible and/or by exploiting DWS. In this way, the blank value can be compensated (Eq. 4.2) and the signal-to-noise ratio, and hence the detection limit, improved. These strategies are fundamental for flow systems where the reaction usually takes place inside the flow-through detector [39–41] and/or renewable sensors [42–44] are exploited. In this regard, there is a tendency for SPS exploitation in bead injection analysis, as emphasised in 2.4.1. Further discussion of the simplicity, versatility and sensitivity of SPS can be found elsewhere [45].

4.1.2. Turbidimetry

Turbidimetry relies on the attenuation of a radiation beam as a result of the scattering of electromagnetic radiation by suspended material formed during sample processing. No excited state is involved. Analogously to spectrophotometry, measurement of the power of the transmitted radiation allows the turbidimetric determination of the analyte.

Some particles and molecules have the ability to scatter UV–Visible electromagnetic radiation, and this phenomenon can be exploited in turbidimetric analytical procedures. When the particles are characterised by a radius less than approximately one-tenth of the radiation wavelength, Rayleigh scattering (which radiates in all directions) dominates and the intensity of the effect is proportional to the fourth power of the frequency of the incident radiation [9]. Increasing the particle size decreases the probability of occurrence of the Rayleigh effect and forward scatter dominates; scattering intensity is then no longer proportional to the fourth power of the frequency. For particles with a radius only slightly larger than the radiation wavelength, scattering intensity is practically independent of frequency. This is why colloidal suspensions of colourless substances such as barium sulphate are white.

The Tyndall effect plays an important role in this context [21] and can be explained by recalling that under diffuse light, colloidal suspensions seem to be perfectly clear and appear to contain no solids. The presence of suspended matter can be detected by shining the beam of a flashlight into the solution; as particles of colloidal dimensions scatter visible radiation, the path of the beam through the colloidal suspension can easily be seen [1].

As these effects manifest themselves simultaneously, scattering intensity is wavelength dependent. If the particles do not absorb the incident radiation, as happens with, *e.g.*, AgCl crystals, the power of scattered radiation is higher for shorter wavelengths, leading to improved sensitivity. Consequently, the UV spectral region is in principle preferred for turbidimetric analytical procedures, but the wavelength cannot be decreased at will, as the spectral limits for applying turbidimetry are also dependent on the solvent and sample characteristics. Selective scattering occurs in the presence of particles able to absorb some of the incident radiation, and hence more effectively scatter at specific wavelengths, as in *e.g.*, the products of the Nessler reaction [46].

Analogously to spectrophotometry, a cuvette containing the processed sample, traversed by a collimated radiation beam (Fig. 4.1), is utilised in turbidimetry. If some of the radiation is scattered, the power of the incident light beam (*Pi*) is decreased to *Pe*, the power of the emergent beam, and Eq. 4.1 holds, providing that turbidance is considered instead of absorbance.

4.1.2.1. *Losses of Radiation*

Radiation losses by reflection at the environment/cuvette/sample interfaces and inside the sample (Schlieren effect) and by absorption on the cuvette walls and inside the processed sample have been discussed in the previous section and should be kept as low as possible.

Attenuation due to all processes other than scattering by the monitored suspension should be compensated by measuring the *Pe* values related to the processed sample and to another system that acts as a reference. Turbidance (S) is then expressed as

$$S = \log P_0/P = a' \, b \, C \tag{4.8}$$

where a′ is the experimentally defined coefficient; b is the optical path length; and C is the concentration of the radiation scattering species in the cuvette, which is ideally proportional to the analyte concentration in the sample.

4.1.2.2. *Relationship between Turbidance and Analyte Concentration*

Analysis of Eqs. 4.3 and 4.8 allows one to conclude that the Lambert–Beer law can be applied to turbidimetry, providing that the number of particles involved is proportional to the analyte concentration.

When the analyte concentration is expressed in mol L^{-1} and the optical path length in cm, then, the a′ coefficient is the turbidity coefficient (τ), with units of L mol^{-1} cm^{-1}. So

$$S = \tau \, b \, C \tag{4.9}$$

In flow turbidimetry, the transient averaged concentration of the particles inside the illuminated portion of the flow-through cuvette is ideally proportional to the analyte concentration in the sample (C_0) and the mathematical relationship involved is analogous to Eq. 4.5:

$$C_0 = k' \, S \, \tau^{-1} \, b^{-1} \tag{4.10}$$

where k′ is the experimentally determined coefficient.

The k′ value is closely related to the parameters involved in the design of the flow system. It should be stressed again that the system should be carefully dimensioned in view of the close relationship between sample dispersion and k′, and hence analytical sensitivity.

4.1.2.3. *Practical Aspects*

Application of Eqs. 4.9 and 4.10 presupposes an ideal situation where the concentration and size distribution of the particles in the monitored suspension are not limiting factors, scattering is the only process attenuating the incident beam, the radiation is monochromatic, sample discontinuities are not present, chemical deviations do not manifest themselves, rate of turbidity formation is not critical, temperature is maintained, stray radiation is not detectable and the detector is ideal. These aspects are discussed in detail below.

CHARACTERISTICS OF THE PARTICLES

The concentration and particle size distribution of the monitored suspension can affect the linearity of the analytical response curve. Radiation scattering is proportional to the number of suspended particles only when there is perfect uniformity in particle size and the influence of particle concentration is negligible.

If the number of suspended particles is too high, each particle can influence the scattering (or diffraction [21]) of neighbouring particles. There is also a tendency for particles to aggregate and this effect is more evident at higher particle concentrations. In situations of pronounced particle aggregation, the increase in the number of particles with concentration is less than expected and the total particle surface area is no longer proportional to particle concentration. As a consequence, the linearity of the function relating turbidance and particle concentration (Eqs. 4.9 and 4.10) deteriorates. This effect becomes more pronounced in extreme situations where the large particles are not maintained in suspension and settle out. The addition of colloid protectors and/or control of the ionic strength can minimise these effects [47–49].

Bending of the analytical response curve at higher concentrations is due to aggregation of the slightly soluble crystals formed. Curvature at lower concentrations is related to insufficient time for turbidity formation and/or to unfavourable relative supersaturation conditions. Consequently, a sigmoidal analytical curve is observed for some flow-based turbidimetric procedures, and the determination of homatropine methylbromide in pharmaceutical formulations using silicotungstic acid as precipitant [50] is a good example. The linear dynamic range was narrow (8.1×10^{-5} mol L^{-1} – 2.2×10^{-4} mol L^{-1}) and the detection limit (5.0×10^{-6} mol L^{-1}) was far below the linear portion of the analytical response curve.

In this context, addition of surfactants or polymers such as gelatins, alcohols and gums can avoid particle aggregation and sorption on the inner walls of the flow system and improve the uniformity of particle size [51]. Its concentration should be higher than the critical micelle concentration but too high a concentration should be avoided. Knowledge of the nature of the particles formed is usually needed to select the appropriate surfactant. The charge on the particles is dependent on the reagent, which is normally present in excess. For example, if sulphate ion is the analyte and barium ion is the reagent the particles tend to be positively charged, but if barium is the analyte and sulphate is the reagent the particles tend to be negatively charged. For positively charged particles, it is usually necessary to add a negatively charged surfactant in order to promote particle stabilisation, thus avoiding excessive growth. On the other hand, if the particles formed are negatively charged or neutral, it is usually more convenient to add a positively charged or neutral surfactant. This effect was observed in a study of colloid stability in supra-molecular assemblies of the synthetic cationic lipid dioctadecyldimethylammonium bromide on carboxymethyl cellulose [52], where the effect of the lipid on particle size and zeta potential were evident.

MONOCHROMATICITY OF RADIATION

The influence of monochromaticity of the incident radiation beam in turbidimetry is not as critical as in spectrophotometry because scattering is involved. Linearity of the analytical response curve is generally not influenced by the spectral bandwidth of the incident beam in a significant manner because the turbidance versus wavelength function is characterised by broader bands. Consequently, relatively large wavelength bandwidths (typically 10–100 nm) can be used. This favours the exploitation of LED-based technology [53].

SAMPLE UNIFORMITY

The solutions involved should be perfectly mixed, and this aspect is more relevant in flow turbidimetry because discontinuities along the monitored sample zone tend to be more pronounced. Proper design of the manifold is therefore crucial for avoiding the formation of undesirable interfaces, minimising the evolution of gaseous mini-bubbles and avoiding settling of suspended matter.

CHEMICAL DEVIATIONS (INCLUDING CO-PRECIPITATION)

There are several situations where chemical deviations can affect turbidimetric analytical procedures, some of which are outlined below:

- Reaction of the analyte with interferent species to form a slightly soluble product with a different stoichiometry or solubility. If the stoichiometry is higher than the original product, turbidity is lower than expected, as more ions are required to form the same amount of precipitate. Conversely, if the stoichiometry is lower than expected, a positive deviation is observed.

- Formation of a concomitant precipitate at a different rate during particle formation.
- Interaction of the precipitate with other chemical species present in the solution, which can increase or decrease its solubility [54].
- Co-precipitation during crystal growth. Some foreign ions that frequently form soluble compounds can be carried down from the solution, contaminating the desired precipitate. The effect is more important in turbidimetric methods because the particles tend to be very small (1–100 μm), presenting a large surface area for absorbing ions from solution.
- Water occlusion and inclusion, adsorption, isomorphous replacement and post-precipitation processes that may contribute to co-precipitation.
  - Occlusion is not very important in flow turbidimetry because, although water trapped in the interior of the crystal may contain some interferents, precipitate growth is limited and the precipitate is not dried as in gravimetric analysis.
  - Inclusion is associated with trapping of ions of similar charge and size to the analytes and reagent, which causes isomorphous inclusion in the precipitate and tends to occur in colloidal precipitates due to the large surface area.
  - Post-precipitation is the slow precipitation of other ions present in the sample that react with the precipitant reagents. The process is not important in flow turbidimetry as normally, there is not sufficient time for post-precipitation to occur.
  - Isomorphous replacement occurs when different ions yield slightly soluble crystals in a geometric form similar to the analyte. The interferent ion substitutes the analyte in the crystalline structure, forming a mix of precipitates and consequently causes error in the turbidimetric analysis.
  - Adsorption is a more serious interference affecting the formed precipitate. As the solution contains a high concentration of the precipitating ions, the solid absorbs these ions, becoming charged on its surface. The electrostatic attraction causes other ions present in the precipitating solution to be absorbed on the solid surface. An example is the precipitation of sulphate in an excess of barium ions: the barium sulphate crystals formed are positively charged and can absorb negative ions such as chloride and nitrate.

RATE OF TURBIDITY FORMATION

There are situations where measurements depend on the kinetics of turbidity formation and Ca^{2+} interaction with F^{-} is a good example.

Quantitative precipitation may be observed after a few seconds, and can be confirmed by potentiometry, but turbidity is sometimes observed only after several weeks [9]. Therefore, the timing of turbidimetric procedures is generally a very important aspect of system design [55]. In this context, the rate of turbidity formation can be selected as the measurement basis for kinetic procedures, as elegantly demonstrated in the flow injection determination of immunoglobulin G in serum [56].

A clear distinction between crystallisation and precipitation is not always possible from a practical point of view [57]; hence, it is more convenient to consider precipitation as a very fast crystallisation process. Crystallisation is a result of the combined effects of nucleation, nucleus growth and secondary processes inside the suspension such as agglomeration, ageing and re-crystallisation. Depending on the reaction conditions, the above processes can occur together or sequentially during the crystallisation period.

Nucleation kinetics, crystal growth and ageing influence the size, purity and amount of the precipitate formed. These characteristics depend on precipitation parameters such as temperature, precipitate solubility, reagent concentration, speed and the manner of reagent addition, as well as the presence of other electrolytes and/or non-electrolytes [57]. In order to obtain the ideal particle size and homogeneity for monitoring purposes, a large number of stabilised small particles is preferred to a smaller number of larger aggregates that can precipitate out of solution. It is therefore necessary to control these parameters and relative supersaturation is important in this context. The relative supersaturation (s_r) is

$$s_r = (C_t - C_s)/C_s \tag{4.11}$$

where C_t is the total concentration of the species to be precipitated and C_s is the solubility of microscopic crystals of the precipitate.

Analysis of Eq. 4.11 shows that if precipitation is carried out under conditions of high solubility, with low reagent and analyte concentrations, the resulting precipitate contains fewer but larger particles because there is enough time for particle growth; the particles are then particularly suitable for gravimetric analysis. On the other hand, if precipitation occurs under high relative supersaturation conditions, the number of particles formed is increased and particle size reduced because there is not enough time for particle growth due to the rapid neutralisation of the charged particles formed by the excess reagent. This situation is favourable for turbidimetry.

The relationship between relative supersaturation and rate of crystal growth (v) is given by [9,58]

$$v = k\,s_r = k(C_t - C_s)/C_s \tag{4.12}$$

where k is the proportionality constant.

Nucleation and particle growth are the two main mechanisms influencing the rate of crystal growth and the size of the particles. For high relative supersaturation, formation of nuclei dominates and the precipitate is composed of a large number of small particles; particle growth predominates in situations of low relative supersaturation, and the resulting precipitate is composed of a much smaller number of well-defined, large particles. Therefore, the precipitate tends to be colloidal when the relative supersaturation is high and forms a crystalline precipitate when the relative supersaturation is low [59]. Supersaturation is therefore the key variable in the precipitation process. In flow analysis, addition of a merging stream with the same chemical species as the analyte is an efficient way to modify the relative supersaturation conditions [55].

The presence of other chemical species can also influence the nucleation kinetics, crystal growth and aggregation characteristics, resulting in a modified crystal morphology due to complexation of some precipitant species and/or obstruction of some active growth sites [57,60].

In short, precipitates can consist of several polymorphic, amorphic and/or hybrid phases but in some applications, only a specific form is acceptable [52]; it is therefore necessary to control the precipitation conditions in order to guarantee the formation of the desirable product. In addition, the repeatability of particle characteristics should be maintained from one determination to another [57] and this is efficiently achieved in flow analysis.

STRAY RADIATION

Analogously to spectrophotometry, stray radiation can alter the measured turbidance (Eq. 4.6), the limitation becoming more severe at lower transmitted power. The presence of suspended matter in the processed sample is inherent to turbidimetry and leads to an amplification of stray light due to scattering effects. This effect has not been systematically investigated in flow analysis.

4.1.2.4. Special Strategies

The analytical response associated with flow turbidimetry is generally susceptible to the same limitations as in flow spectrophotometry, with some variations. Turbidimetric methods tend to generate analytical response curves with a sigmoidal profile. At low analyte concentrations, it is difficult to get a concentration-dependent turbidance due to limitations associated with the solubility product and rate of turbidity formation. This effect is sometimes circumvented by adding a confluent stream consisting of a solution with the chemical species under investigation [55] or a preformed suspension (crystal seeding) [61] in order to improve the relative supersaturation conditions. On the other hand, negative deviations from linearity are observed at high analyte concentrations. This is due to several processes but, most importantly, the formation of larger particles, and can be minimised by adding a suitable colloid protector.

4.1.2.5. *Final Remarks*

In the early development of turbidimetry, scattering of light was not reproducible from one sample to another because it was difficult to control the processing of the sample. In 1927, this led Wells to voice his scepticism of the reliability of results obtained from instruments relying on the measurement of scattered radiation [62]. In view of this, turbidity measurements were not accepted as standard methods at that time. Once the limitations of such optical methods were understood, however, their real potential could be exploited. Nowadays, a reproducible suspension is easily attained, particularly in flow analysis. As a consequence, several turbidimetric procedures have been successfully implemented and Wells' statement is no longer valid.

4.1.3. Nephelometry

Nephelometry is a variant of turbidimetry in which the power of the scattered radiation is measured at an angle of $<180^\circ$ to the source (normally 90°). Forward angles give the best potential sensitivity for larger scattering species but it is not easy to differentiate scattered light from the incident beam. Nephelometry is applied to lower concentrations than turbidity because the dark background is zero compared with full illumination in turbidimetry, a fact that was first reported in 1921 [63].

4.1.4. Fluorimetry and Phosphorimetry

Fluorimetry and phosphorimetry are analytical techniques based on the absorption of electromagnetic radiation by the processed sample, which contains or generates radiation-absorbing molecules in excited states; when the excited molecules return to the ground state, radiation is emitted, i.e., the excited molecules luminesce.

Energy transitions in fluorimetry do not involve a change in electron spin, whereas in phosphorimetry they do. For this reason, the emission process in phosphorimetry is much slower than in fluorimetry. The lifetime of an excited singlet state in fluorimetry is usually 10^{-9}–10^{-6} s, whereas the lifetime of the excited molecule in phosphorimetry is in the 10^{-4}–100 s range, because in the latter situation the singlet excited state must be converted to a triplet state by intersystem crossing, which involves a change in electron spin [64,65]. As a singlet/triplet transition is involved, the rate of the process, and hence the probability of phosphorescence occurring, are decreased due to greater non-radiative energy

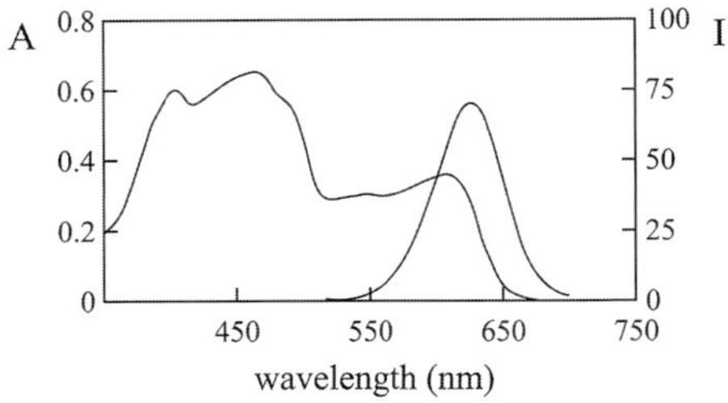

FIGURE 4.4 Typical absorption (left) and emission (right) spectra in fluorimetry. A = absorbance; I = relative emission intensity; instrument = model LS Perkin Elmer luminescence spectrometer; slits = 6 nm; fluorescent material = CdTe quantum dots synthesised according to Ref. [66] (aqueous medium; 1:6 v/v water dilution before measurements); wavelength for maximum absorption = 467 nm; wavelength for maximum emission = 632 nm.

losses. Phosphorescence is therefore less compatible with flow systems and this section focuses more on the combination of flow analysis with fluorescence detection.

In both techniques, the spectrum of the emitted radiation is characterised by lower energy than the excitation spectrum (Fig. 4.4), as the excitation process usually requires not only energy for a molecular electronic transition but also an increase in vibrational energy.

The quantitative relationship between fluorescence power and analyte concentration can be derived from the number of molecules in the excited state and the radiant power absorbed by the processed sample:

$$P_f \alpha \varnothing_f (P_i - P_e) \tag{4.13}$$

where P_f is the fluorescence radiant power, $ø_f$ is the fluorescence efficiency, P_i is the power of incident light beam, P_e is the radiant power emerging from the sample and hence $(P_i - P_e)$ is the radiant power absorbed by the sample.

The fluorescence efficiency (also called the *quantum yield of fluorescence*) is the fraction of the absorbed photons emitted during fluorescence (k_f) relative to competing deactivation processes (k_d):

$$\varnothing_f = k_f/(k_f + k_d) \tag{4.14}$$

As the Lambert–Beer law holds:

$$P_f = \varnothing_f P_i (1 - e^{-\varepsilon b C}) \tag{4.15}$$

In fluorimetry and phosphorimetry, Eq. 4.15 can be expanded to a more general relationship [67]:

$$P_f = \varnothing_f P_i \varepsilon b C \left[1 - \varepsilon b C/2! + (\varepsilon b C)^2/3! + \cdots + (\varepsilon b C)^n/(n+1)!\right] \tag{4.16}$$

When ε b C < 0.05, a condition inherent to diluted processed samples, Eq. 4.16 can be simplified to:

$$P_f = \varnothing_f P_i \varepsilon \mathrm{b} \mathrm{C} \quad (4.17)$$

As the fluorescence radiation is emitted in all directions, only a fraction of it is detected in real applications. In addition, the detection efficiency is not 100% because of quenching effects, dead time aspects and spectral bandwidths [64]. The flowing sample also undergoes dispersion while being transported through the analytical path of the flow analyser. After considering these aspects, the detector response F can be related to the analyte concentration by:

$$F = \mathrm{k}'' \varnothing_f P_i \varepsilon \mathrm{b} \mathrm{C} \quad (4.18)$$

where k″ is the experimentally defined factor including reaction quantum yield, sample dispersion, geometry of detection and detector efficiency.

In order to optimise the manifold design of the flow system, the lifetimes of the excited states of the molecules should be considered, because the processed sample is in motion. In the situation of too high a flow rate and too small a flow cell, a fraction of the excited molecules could exit the flow cell without emitting light. On the other hand, if the flow rate is too low, a significant fraction of the molecules may emit radiation before reaching the flow cell. Both situations can lead to a decrease in the analytical sensitivity. This feature can be considered as a "time window" in flow analysis and the effect is more pronounced in phosphorimetric methods where light emission is slow relative to fluorimetric methods [65]. This is also true for chemiluminescence and bioluminescence, as discussed in the next section.

It is interesting to note that the possibility of using two sites in the manifold for excitation and emission, as well as the possibility of inducing light emission in order to speed up the process, has not been investigated in connection with flow analysis.

Regarding flow cell geometry, the arrangement of the illuminating and viewing angle should be either frontal (37°) for solid samples, highly concentrated samples and/or samples presenting suspended matter, or orthogonal (90°) for dilute and clear liquid samples [64].

4.1.5. Chemiluminescence and Bioluminescence

Chemiluminescence (CL) is the emission of light, usually in the visible or near infrared spectral region, as a result of an excited electronic molecular state, formed in a chemical reaction, returning to the ground state. Bioluminescence (BL) is the same emission process but involves a biochemical process such as an enzyme-mediated reaction for excitation. Chemiluminescence is a relatively rare phenomenon because in most chemical reactions, excited molecules lose the excitation energy by non-radiative processes, i.e., heat.

For a reaction to produce chemiluminescence, an excited state molecule must be produced during the course of the reaction. The observed emission stems from the ejection of a photon from this excited state species. Three essential features are associated with these reactions:

- The reaction must be exothermic in order to generate sufficient energy for formation of the electronically excited state (for emission in the visible region, the minimum energy requirement is 180 kJ mol^{-1});
- A pathway must exist for the formation of the electronically excited state;
- The excited state must be capable of deactivation by the emission of radiation or energy transfer to a fluorophore.

The equations relating analytical signal and analyte concentration in the sample are similar to fluorimetry, but there is no incident power beam. This reduces the instrumentation required to a flow system for handling solutions, a flow cell and a detector. The mathematical relationship between power of the emitted radiation (thus detector response) and analyte concentration is

$$F = P_{CL} \approx k''' \Phi_{ex} \Phi_{em} \Delta C/\Delta t \tag{4.19}$$

where F is the detector response, P_{CL} is the the chemiluminescence emission power, Φ_{ex} and Φ_{em} are the excitation and emission quantum yields; and k''' is the experimentally defined factor (see Eq. 4.18).

Flow analysis has been used to investigate the fundamental chemistry of chemiluminescence and bioluminescence reactions, to optimise post-column chemiluminescence reaction conditions for liquid chromatographic detection and to quantify analytes in relatively simple or synthetic matrices [68]. In recent years, there has been a pronounced increase in the application of these methods to the analysis of real sample matrices [69]. This has usually been achieved by a combination of efficient in-line sample treatment, *e.g.*, use of solid-phase reagents for concentrating selected analytes and/or for removing the sample matrix and exploitation of more inherently selective reactions [70,71].

Most flow-based chemiluminescence procedures are based on variants of the luminol reaction and procedures exploiting the oxidising potential of, *e.g.*, permanganate or cerium(IV). There are also several references to 1,10-phenanthroline, gallic acid/pyrogallol, lucigenin and peroxyoxalate [69,72–74]. Other reagents that have been used include ruthenium(II), periodate, TCNQ (tetracyanoquinodimethane), fluorescein and quercetin. The relatively small number of applications involving peroxyoxalate is in marked contrast to the situation in liquid chromatography. This is because the majority of flow-based chemiluminescence applications involve aqueous matrices, whereas the peroxyoxalate reaction works best in non-aqueous (reversed phase) media.

4.2. THE SCHLIEREN EFFECT

In 1864, Toepler introduced the term *schliere* as "an inequality of the optical refraction of an otherwise homogeneous mass" [75]. The phenomenon is now known as the Schlieren effect [76] and in English-language documents, terms such as *streak* and *stria* have also been used.

In flow analysis with *spectrophotometric* detection, the Schlieren effect is generally the ultimate limitation on the signal-to-noise ratio and hence on the detection limit. Furthermore, it can lead to systematic measurement deviations, thus degrading accuracy.

It is interesting to note that, although not reported, a high blank related to the Schlieren effect was noted in the very early development of flow injection analysis, as shown in Fig. 4.5, which depicts the chart recorder output from a routine run for total nitrogen determination in Kjeldahl digests [77]. The highly concentrated digests were injected into a concentrated alkaline phenol sample carrier stream, yielding a pronounced blank value that impaired the linearity of the analytical response curve and the signal-to-noise ratio. Moreover, measurement reproducibility was poor compared with other analytical procedures carried out at that time with similar instrumentation [78,79].

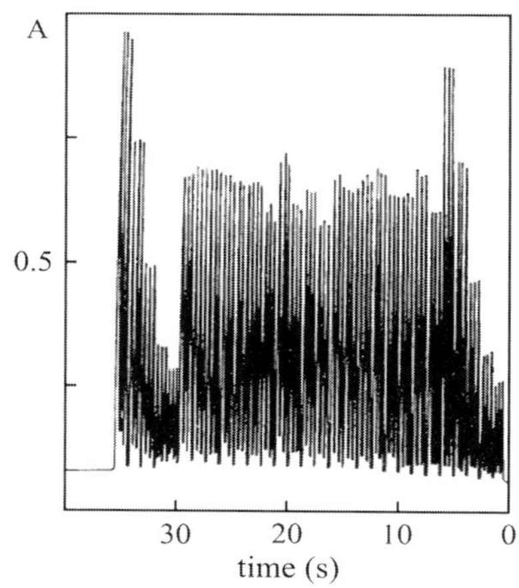

FIGURE 4.5 **Recorder tracing of a routine run with a high blank signal.** From right, analytical signals in triplicate correspond to six standard solutions (0.5 – 1.0 – 2.0 – 4.0 – 6.0 % m/m N, dry basis), the digests of sample leaves, and the standard solutions again. *Adapted from Anal. Chim. Acta 81 (1976) 371, J.W.B. Stewart, J. Ruzicka, H. Bergamin-Filho, E.A.G. Zagatto, Flow Injection Analysis. Part III. Comparison of continuous flow spectrophotometry and potentiometry for the rapid determination of total nitrogen content in plant digests, with permission from Elsevier (Ref. [77]).*

4.2.1. Physical Principles

An interface established between two homogeneous media with η_1 and η_2 refractive indices can be regarded as a mirror. If a radiation beam impinges on this interface, it can be either reflected or refracted (Fig. 4.6). The reflecting angle of the emergent beam is the same as that of the incident beam, whereas the refracting angle of the emergent beam is calculated according to Snell's law (also known as Descarte's law):

$$\eta_1 \sin \theta_1 = \eta_2 \sin \theta_2 \qquad (4.20)$$

where θ_1 and θ_2 are the angles of incidence and refraction.

In general, different radiation wavelengths correspond to slightly different indices of refraction. The main consequence of this aberration is that the refraction angle is wavelength dependent [80]. This aspect is very important for minimising the Schlieren effect by dual-wavelength spectrophotometry (see also 4.2.3.5).

Refraction of radiation can also occur in fluidic regions with refractive index gradients because optical artefacts analogous to fluidic lenses are established. Regardless of the process (reflection or refraction) and the optical artefact (mirror or lens) involved, the deflection of radiation inside a fluidic environment can result in external distortions, characterised by the Schlieren effect.

In *spectrophotometry,* this effect alters the power of the emergent radiation beam (Fig. 4.1), because some radiation is deflected. Consequently, a systematic deviation in the measurement is observed.

The Schlieren effect in flow analysis has often been referred to as a *refractive index effect*, but both reflection and refraction are involved (Fig. 4.6). Indeed, it is a phenomenon more related to light reflection.

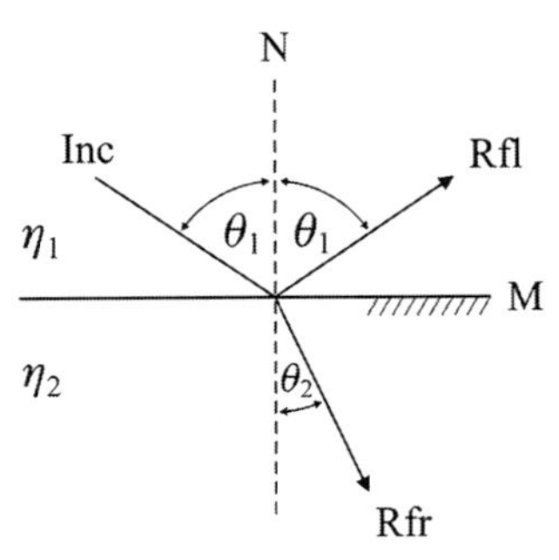

FIGURE 4.6 **Deflection of a beam of radiation at an interface.** Inc, Rfl and Rfr = incident, reflected and refracted radiation beams; M = mirror interface; N = plane normal to the interface; θ_1 and θ_2 = angles of incidence and refraction; η_1 and η_2 = refractive indices.

4.2.2. Occurrence

In the 17th century, some optical properties of the atmosphere and of aqueous solutions related to the shadowgraph and Schlieren effects were investigated by Robert Hooke, the father of the optics of inhomogeneous media. He noted the deflection of a light beam passing through a fluidic medium and proposed the classical experiments demonstrating that both temperature and concentration non-homogeneity can alter light propagation [75] (Fig. 4.7).

In 1942, a critical review, including historical examples of the shadowgraph and Schlieren effects, was published [81]. The natural phenomena involved and the design of novel strategies for proper visualisation of the Schlieren effect were discussed in 2001 [82]. The authors emphasised examples presented in the 19th century such as sunlight shadowgraphs of blast waves from a dynamite explosion and image distortions due to convection inside irregularly heated media.

Shadows are routinely observed in shadowgraph imaging (Fig. 4.8), the effects becoming more pronounced when the radiation from the light

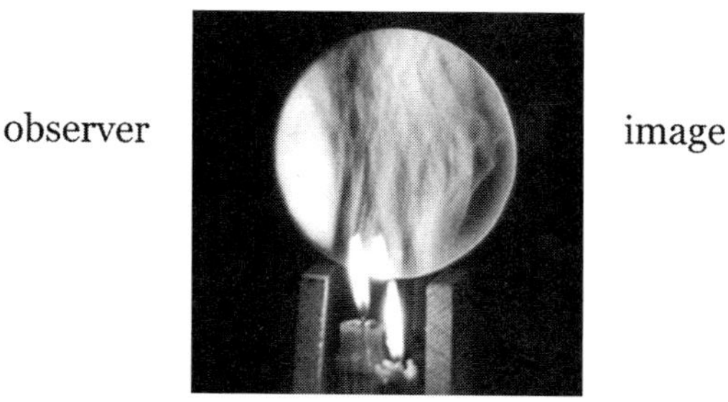

FIGURE 4.7 Experimental design for Hooke's classical experiments.

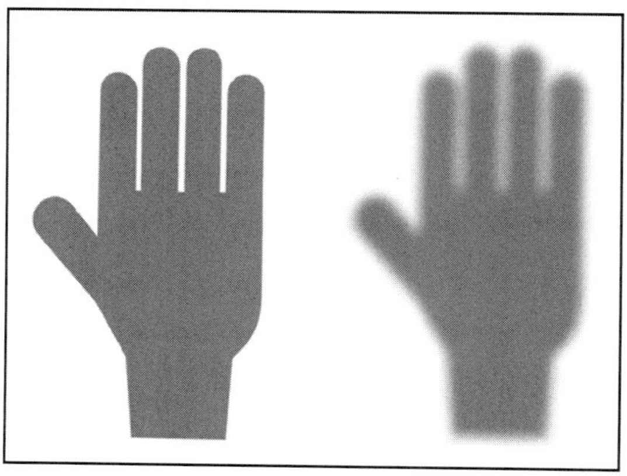

FIGURE 4.8 Shadows of a human hand illuminated with a well (left) or badly (right) collimated radiation source.

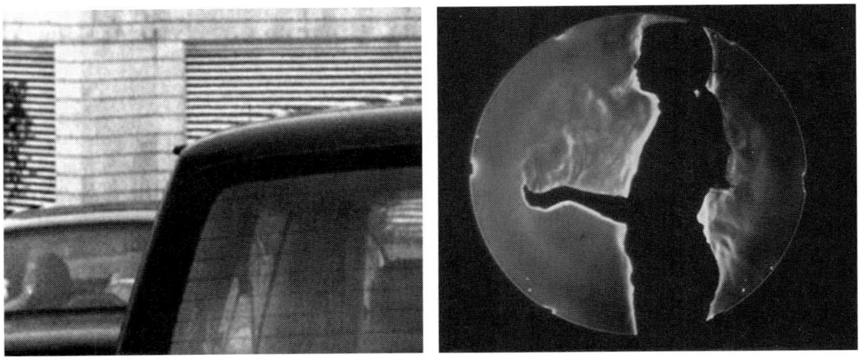

FIGURE 4.9 **Observations of the Schlieren effect in everyday life.** *Source: Courtesy of G.S. Settles and reproduced with permission from "Schlieren and shadowgraph techniques" by G.S. Settles, ISBN 978-3-540-661559, Springer (2001), see Ref. [82].*

source is not well collimated. This is an important aspect to be considered when designing detectors and/or in compensating the Schlieren effect in flow analysis.

Conversely, Schlieren imaging requires specific optical arrangements because the transparent medium under investigation should be placed between the light source and the observer (Fig. 4.7). This optical arrangement is commonly observed and a practical example is the appearance of low-definition moving images above the surface of asphalt highways. If the air temperature immediately above the road is higher than the ambient air temperature, differences in the refractive indices of the poorly mixed air masses may give rise to the Schlieren effect, and this is the reason why the driving path looks like a wet layer above the road.

Other image distortions commonly observed in daily life are shown in Fig. 4.9, from which one can infer that the effect is governed by the combined influence of two components. One component (hereafter referred as to A) is associated with the deflection of radiation at regions with fuzzy moving fluidic interfaces, whereas the other, (B), is associated with regions that have relatively steady refractive index gradients. Depending on the dominant component, relatively stable (Fig. 4.9, left) or fuzzy (Fig. 4.9, right) images are observed. This phenomenon was recently demonstrated in the context of flow analysis [3].

The Schlieren effect is also observed in liquid media. A pedagogical demonstration involves the addition of a concentrated sucrose solution to a beaker containing water. Immediately after addition, the aqueous medium becomes turbid (Fig. 4.10, left). Transparency is restored by letting the solution stand for a few minutes (Fig. 4.10, centre). The medium may become turbid again if the solution is gently stirred again

FIGURE 4.10 **Batchwise observation of the Schlieren effect.** Photographs show a beaker of water to which a concentrated sucrose solution is added. From left to right: the situation during sucrose addition, after about 5 min standing time, and after gentle shaking.

(Fig. 4.10, right). It is interesting to note that the horizon line in Fig. 4.10 becomes less visible in situations of pronounced Schlieren effect, because most of the incident light is moved in other directions.

The solution turbidity is associated with component A of the Schlieren effect, which is caused by the reflection of radiation at the interfaces between the poorly mixed neighbouring fluid elements of water and sucrose solution. The restoration of transparency during standing is due to diffusion processes that reduce the solution discontinuities at the mixing boundaries, the phenomenon being mainly related to short-distance mixing. Consequently, component A vanishes and component B becomes dominant, but its influence is not evident, as the related modifications in the image (mainly size) are not easily seen. The reappearance of turbidity after stirring is associated with the relocation of fluid elements with different refractive indices in a neighbouring situation (long-distance mixing) or, in other words, to the re-establishment of liquid mirrors.

In *spectrophotometry*, the Schlieren effect can alter the analytical results because it influences the propagation of the radiation beams involved (Fig. 4.1). Collimation of the light beam plays an important role in this respect, as an effect analogous to a shadowgraph is observed when a spectrophotometer with poor optical properties is used [83].

Regarding the influence of the Schlieren effect on analytical procedures, the effect of differences between the temperature of the ambient environment and the sample on the absorbance was originally reported in relation to stopped-flow procedures [84]. The systematic deviation in absorbance was proportional to the temperature difference and was also dependent on the optical properties of the instrument. The effect also manifested itself when identical solutions at different temperatures were mixed [85]: a linear dependence of the measurement on the temperature difference and the formation of liquid lenses was reported. Consequently, the Schlieren effect can be exploited in enthalpimetric analytical procedures relying on optical beam deflections induced by the heat of reaction [86].

The Schlieren effect was also observed in older liquid chromatographs incorporating heated columns because injected samples eluted from a heated column and reached the flow-through detector which was at a lower temperature [87], and relatively high and different analyte concentrations were involved. The refractive index gradients established along the flowing sample stream were therefore a consequence of gradual variations in temperature and/or concentration. The resulting Schlieren effect was reproducible, as it reflected a relatively steady situation (dominance of component B). Moreover, this effect has often been exploited as a basis for measurement (Schlieren optics) [88–91].

4.2.3. The Schlieren Effect in Flow Analysis

In flow analysis, the Schlieren effect tends to be more pronounced [28] because a perfectly mixed flowing sample is not practically achievable, and an analyte concentration gradient is always present. Undesirable concentration gradients and/or discontinuities along the monitored sample can give rise to the formation of relatively steady liquid lenses as well as a myriad of randomly distributed transient mirrors. These optical artefacts lead to fluctuations in the emergent radiation that can alter the measured signal.

The influence of the Schlieren effect on the analytical output depends on the optical performance of the spectrophotometer [83], the number of mirror interfaces and the intensity and direction of the concentration gradients. The latter parameters are closely related to the concentrations of the solutions involved, the mixing conditions and the kinetics of solution inter-mixing and hence the transport number of the related chemical species, as discussed below.

The intensity of the effect is higher at the leading and trailing edges of the flowing sample (Fig. 4.11), where the concentration gradients are usually higher. This means that more pronounced measurement noise is associated with the rising and falling portions of the recorded peak (dominance of component A) or with a smooth peak followed by an inverted one (dominance of component B).

The Schlieren effect degrades the signal-to-noise ratio, and hence the detection limit, in flow-based *spectrophotometric* analytical procedures and the accuracy because the related noise is superimposed on the analytical signal.

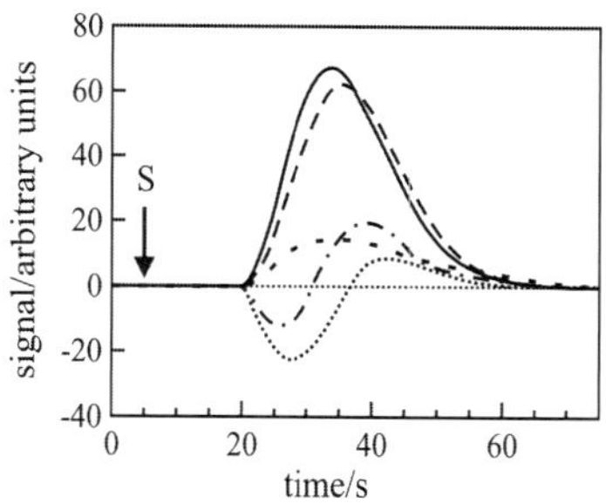

FIGURE 4.11 **Recorder traces showing the influence of the Schlieren effect caused by a 35‰ salinity matrix on the signals for different P concentrations under good mixing conditions. •••••• 0 mg L^{-1} P, 35 ‰; ▪▪▪▪▪▪▪ 50 mg L^{-1} P, 0 ‰; ▬ ▪ ▬ 50 mg L^{-1} P, 35 ‰; ▬▬▬ 200 mg L^{-1} P, 0 ‰; ▬ ▬ 200 mg L^{-1} P, 35 ‰. For details, see Ref. 92.** *Adapted from Anal. Chim. Acta 351 (1997) 265, I.D. McKelvie, D.M.W. Peat, G.P. Matthews, P.J. Worsfold, Elimination of the Schlieren effect in the determination of reactive phosphorus in estuarine waters by flow-injection analysis, with permission from Elsevier.*

The Schlieren effect has been investigated mainly in relation to UV–Visible spectrophotometry [3,28,92,93], but it is also important in fluorimetry. Analytical procedures involving chemi- or bioluminescence detection are less affected due to the geometry of the flow cell and the integrating nature of the light collection process. The influence of the Schlieren effect on turbidimetric and nephelometric detection is less pronounced because mixing tends to be more efficient due to the presence of solid particles in the flowing sample.

Little attention has been given to the influence of these transient liquid mirrors and lenses on the amount of stray light reaching the detector. Investigation of this process is strongly recommended, as the Schlieren effect can lead to an increase in detected radiation, hence degrading the linearity of the analytical response curve [13].

4.2.3.1. *Historical Aspects*

The Schlieren effect in flow injection analysis was originally reported in 1977 in relation to the turbidimetric determination of sulphate in natural waters and plant digests: "a mixing boundary (Schlieren pattern) could be seen very clearly when the flow-through cuvette was removed from the spectrophotometer and observed visually after injecting a blank or a sample" [94]. This effect was also noted in the following year, when an LED-based device was used as a detector in a flow injection system [95]. The authors removed the flow cell from its housing and verified with the naked eye that, during passage of the flowing sample, the image seen through the flow cell was initially amplified and then reduced. This resulted in the recording of a reproducible "positive" peak followed by an inverted one (or *vice versa*), with heights proportional to the concentration of the injected solution [96]. During the 1980s, many similar reports on this effect appeared in the literature *e.g.*, [97,98].

From these historical reports, it is also possible to conclude that the Schlieren effect in flow analysis is a consequence of the combined effects of two components, namely deflection of the beam of radiation by transient (A) and/or relatively steady (B) optical artefacts [3].

4.2.3.2. *The Two Components of the Schlieren Effect*

The transient formation of sharp liquid interfaces between neighbouring fluid elements with different refractive indices prevails under poor mixing conditions, whereas relatively steady concentration gradients dominate under good mixing conditions [3]. Consequently, dominance of component A or B is strongly influenced by the mixing conditions. For didactic purposes, therefore, the effect is herein discussed by considering flow systems characterised by poor or good mixing conditions, respectively.

POOR MIXING CONDITIONS

Under poor mixing conditions (dominance of component A), there are sharp concentration discontinuities instead of a true gradient function along the flowing sample. A myriad of transient liquid mirrors is established within the poorly mixed fluid elements with different concentrations, and hence refractive indices. The incident beam then undergoes attenuation, mainly by reflection, leading to a decrease in the power of transmitted radiation and therefore to an increase in, *e.g.*, absorbance.

As mirrors are involved, wavelength-independent reflective phenomena are more likely to occur. Compensation of this effect can efficiently be accomplished by dual-wavelength spectrophotometry (see 4.2.3.5).

In view of the random formation and fading of mirrors, the amount of light reaching the detector is erratic. Consequently, the analytical signal becomes noisy, affecting the measurement repeatability and signal-to-noise ratio.

The intensity of the effect depends on the composition of the solutions involved, becoming more pronounced for more concentrated solutions and/or when solution inter-mixing is slower. A linear relationship between the concentration of the Schlieren-generating species and the intensity of the effect is not observed [3].

The effect becomes less pronounced by improving the mixing conditions. In fact, as mixing improves, a continuous extinction of the transient mirrors is achieved and relatively steady concentration gradients established.

GOOD MIXING CONDITIONS

Under good mixing conditions (dominance of component B), the Schlieren effect is due to the concentration gradients formed. As the flow analyser usually operates under laminar flow conditions, isolines of different concentrations, and hence refractive indices acting as lenses, are established along the flowing sample. The liquid lenses either focus light onto the detector or scatter it during passage of the flowing sample through the flow cell. Consequently, the baseline can be increased or decreased and this is the reason why inverted peaks are often reported [99]. It is interesting to note that negative absorbance values (i.e., >100% transmittance) have also been reported. This effect is observed in spectrophotometers with poor collimation, where radiation focussing by the liquid lenses may re-direct part of the radiation towards the detector, in an way analogous to shadowgraphy (see also Fig. 4.8).

Under good mixing conditions, the concentration gradients along the flowing sample are true continuous functions because the concentrations of successive neighbouring fluid elements are continuously varying. Consequently, a relatively steady situation is established, yielding a reproducible Schlieren signal (see also Fig. 4.11). Measurement precision is then affected to a lesser extent and the intensity of the Schlieren signal is proportional to the concentration of the species involved. In this situation, the Schlieren signal should not be considered as noise but as a blank. The above comments are summarised in Table 4.3.

TABLE 4.3 Features Related to the Components of the Schlieren Effect

Prevailing Component	**A**	**B**
Mixing conditions*	Poor	Good
Optical element	Myriad of mirrors	Liquid lenses
Stability	Transient	Relatively steady
Concentration dependence	Partial	Yes
Geometry	Fuzzy	Organised
Repeatability	Low	High
Characteristic	Noise	Blank
Compensation by DWS**	Quantitative	Partial

* *including the transport number of the chemical species involved;*
** *dual-wavelength spectrophotometry.*
Source: Adapted from Ref. [3].

It is interesting to note that:

- Different samples with the same concentration of the Schlieren-generating species can yield different Schlieren signals, depending on the transport number of the species involved. The differences in the magnitude of the Schlieren effect are clearly illustrated in Fig. 4.12. Sodium chloride, sodium hydroxide and ethanol are rapidly dissolved in water, whereas potassium sodium tartrate, sodium citrate and sucrose are only homogenised with water after several minutes. The difference in dissolution abilities explains the pronounced differences between the recorder tracings labelled b, d and f and those labelled c, e and g. It should be stressed that for ethanol (tracing f), Schlieren component A has almost vanished, and the dominance of component B results in a small smooth inverted peak followed by a normal one [see also Ref. [100]].
- Different samples with the same refractive index may yield different Schlieren signals. All solutions in Fig. 4.13 are characterised by the same refractive index ($\eta = 1.4398$), but not the same transport number. As discussed above, different recorder tracings are then obtained.

4.2.3.3. *Applications*

The Schlieren effect can be exploited for analytical purposes, especially when concentration gradients are present [101]. To achieve this, the flow system is designed with good mixing conditions in order to allow Schlieren component B to prevail. Non-specific light scattering is then exploited for the determination of a major analyte.

The determination of glycol in aviation de-icing/anti-icing fluids [102] provides a good illustration. A large sample plug was inserted into an unsegmented water stream and two inverted peaks relating to the leading

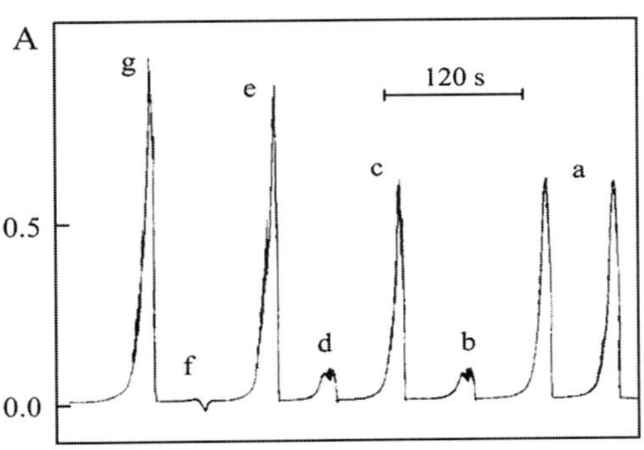

FIGURE 4.12 Schlieren signals recorded for different iso-concentrated solutions. The figure refers to a single line manifold; tubing = 0.8 mm i.d. polyethylene, carrier stream = 0.01 mol L^{-1} $Na_2B_4O_7$; flow rate = 2.5 mL min^{-1}; sampling loop = 50 cm; coiled reactor length = 20 cm; wavelength = 617 nm; b, c, d, e, f, g = 1.00 mol L^{-1} NaCl, potassium sodium tartrate, NaOH, sodium citrate, ethanol and sucrose respectively. For comparative purposes, a dye solution (also in tetraborate) is also shown (a). *Reprinted from Anal. Chim. Acta 234 (1990) 153, E.A.G. Zagatto, M.A.Z. Arruda, A.O. Jacintho, I.L. Mattos, Compensation of the Schlieren effect in flow-injection analysis by using dual-wavelength spectrophotometry, with permission from Elsevier (Ref. [28]).*

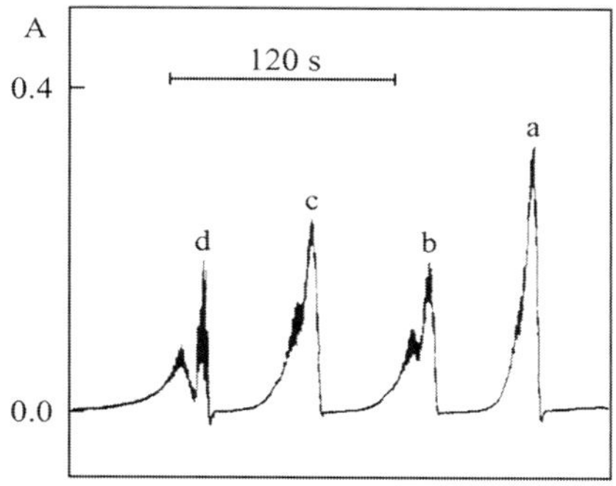

FIGURE 4.13 **Schlieren signals recorded for different solutions with the same refractive index.** Carrier stream = water; coiled reactor length = 100 cm; a. b, c, d = 2.0 mol L^{-1} HCl, 11.2% (m/v) sucrose, 14% (m/v) glycerol, and 24.13% (m/v) ethanol respectively. Other conditions are as in Fig. 4.13. *Reprinted from Anal. Chim. Acta 234 (1990) 153, E.A.G. Zagatto, M.A.Z. Arruda, A.O. Jacintho, I.L. Mattos, Compensation of the Schlieren effect in flow-injection analysis by using dual-wavelength spectrophotometry, with permission from Elsevier (Ref. [28]).*

and trailing edges of the flowing sample were recorded. The heights of these peaks were proportional to the analyte concentration in the sample, regardless of whether positive, negative or both peaks were used as the basis for measurement.

This strategy was also exploited for the determination of alcoholic strength in beverages [103]. As the proposed flow injection system included an LED-based detector, the Schlieren signal was increased due to the poor collimation of the incident radiation beam. About 120 samples were run per hour, and precise results (r.s.d. = 3.4%) were obtained.

Potential alcoholic beverage adulteration was identified by exploiting a similar strategy [104]. The partial least squares statistical algorithm was applied to a number of measurements along the entire recorded analytical signal, which allowed effective discrimination between authentic and adulterated beverages. The detection of concentration gradients by exploiting Schlieren optics is discussed in detail elsewhere [101].

4.2.3.4. *Circumstances for Occurrence*

There are specific circumstances leading to the occurrence of the Schlieren effect, some of which are highlighted below.

DIFFERENCES BETWEEN SAMPLE AND CARRIER SOLUTIONS

The composition of the sample aliquot introduced into the flow manifold is always different from that of the carrier stream. Interactions between these solutions can give rise to the formation of the optical artefacts that cause the Schlieren effect. This becomes more severe when:

- Concentrations of the sample matrix are very high and not matched with the carrier stream;
- Sample lots with high variability in the sample matrix (*e.g.*, estuarine waters) are assayed;
- The flow system is designed in the straight configuration. The carrier stream is then the reagent solution, which is usually very different from the sample;
- Chemical species with low transport number are involved, impairing diffusional transport between neighbouring fluid elements;
- The temperature of the inserted sample aliquot is different from that of the carrier stream.

PULSED SAMPLE INLET INTO THE FLOW CELL

If the flowing sample reaches the flow cell in a pulsed manner, the Schlieren effect can manifest itself as an undulation of the monitored signal. This undulation is superimposed on the analytical signal and its frequency is determined by the flow pulsation of the flowing stream [105].

This phenomenon is often observed and is caused by the intermittent addition of small sample pulses into the illuminated region of the flow cell. As the sample concentration in the pulse tends to be different from the mean concentration inside the illuminated region, optical artefacts are formed along the radiation beam, which affects the measurements in a regular fashion. This leads to sinusoidal absorbance fluctuations superimposed on the main recorded signal (Fig. 4.13). As a constant frequency is involved, the related undulation can easily and efficiently be filtered out (see *e.g.*, Ref. [4]). The signal-to-noise ratio, however, is adversely affected.

This effect is more pronounced in sample regions associated with the rise and fall of the peak. In fact, at the top of the peak, there is very little difference between the instantaneous sample concentration (sample pulse) and the mean concentration inside the illuminated sample portion. The same is true on the baseline. This effect has not been reported in relation to segmented flow analysis, probably because the analytical signal is associated with the flat region of the recorded peak (Fig. 2.5).

PULSED ADDITION OF MERGING STREAMS

Undulations on the recorded signal are also observed in flow systems with merging streams, especially when pronounced differences in concentrations and poor mixing conditions between the convergent streams are involved. Liquid optical artefacts are then established immediately after the confluence junction, leading to undulations in both the baseline and the recorded peak, which are a consequence of the pulsed addition of the confluent stream (see also 3.2.2).

EXPLOITATION OF INTERMITTENT STREAMS

Intermittent stream addition is often exploited to achieve merging zones [106], to speed up the flushing time in a flow analyser [107] or to implement multi-analyte determinations [108]. At the confluence point, the sample/carrier stream concentrations undergo sudden variations caused by the addition of the merging stream (Eq. 3.10). Consequently, the Schlieren effect can manifest itself as a spurious transient signal, usually observed before sample arrival in the detector flow cell. This effect is more likely to occur in highly versatile flow systems such as multi-commuted flow systems [109].

As a rule, the recorded peak height is not affected. If it is, the system should be modified in order to provide a temporal separation between the flowing sample and the Schlieren zone formed by the addition/removal of the intermittent stream.

ADDITION/REMOVAL OF MANIFOLD COMPONENTS

Different components such as reactors, detectors and mini-columns can be added to or removed from the manifold in order to improve system performance but the strategy can lead to occurrence of the Schlieren effect. Reactor interchange has been used to attain two different analytical signals under two different sample handling conditions, thus allowing the implementation of differential kinetics. The feasibility of this approach was demonstrated in relation to the spectrophotometric determination of cobalt and nickel in steels [110]. The method involved complexation with citrate followed by time-dependent dissociation and reaction with 4-(2-pyridylazoresorcinol).

Multi-site detection involves the displacement of the detection unit from one manifold site to one or more other sites (see 7.6.2). As different solutions are inherent to the different sites involved, occurrence of the Schlieren effect is unavoidable. This approach is particularly attractive for improving sample throughput, implementing sequential determinations, re-conditioning the detector and/or exploiting differential kinetics. The potential applications of this approach have been reviewed elsewhere [111].

Mini-columns of solid reagents, *e.g.*, ion-exchangers or immobilised enzymes, can be inserted into the main analytical channel in a similar manner to the sampling loop and the direct shipboard determination of manganese in deep sea water samples involving in-line analyte separation/concentration [112] is a good example of this. The method relied on the catalytic effect of Mn^{2+} on the oxidation of N,N-diethylaniline by periodate in a neutral pH medium. During the concentration step, the sample flowed through the resin mini-column at a constant flow rate and Mn^{2+} was retained. Variation of the concentration (loading) time from 10 s to 10 min allowed the determination of manganese in the 10 ng L^{-1}–20 μg L^{-1} concentration range. The mini-column was inserted into the eluent carrier stream, which displaced the retained analyte towards the detector. As two rather different solutions were interchanged, the Schlieren effect manifested itself. This is probably the main reason why there are fewer flow-based *spectrophotometric* analytical applications exploiting solid-phase extraction compared with other detection techniques that are not affected by the Schlieren effect.

4.2.3.5. Minimising the Schlieren Effect

As a rule, occurrence of the Schlieren effect degrades analytical figures of merit such as accuracy and detection limit. Furthermore, interaction of the Schlieren-generating chemical species with the tubing inner wall can impair the sampling rate, as the adsorption and release processes involved may increase the wash time.

Efforts have been made to minimise the Schlieren effect, its influence on the analytical signal or both. Most of the strategies for circumventing the effect rely on improved system design or the subtraction of signals monitored at different wavelengths.

IMPROVEMENT OF SYSTEM DESIGN

A useful and somewhat obvious expedient for Schlieren minimisation is to design the flow system to reduce undesirable concentration gradients and/or discontinuities along the monitored sample zone. To this end, the use of less concentrated reagents, sample/carrier matrix matching, improved mixing conditions, exploitation of special monitoring strategies and/or enhanced flow cell design are all possible strategies.

The formation of liquid optical artefacts along the flowing sample is minimised by using less concentrated solutions, preferably involving chemical species with a high transport number. This approach was used when solid-phase extraction was implemented in a flow injection manifold for the spectrophotometric determination of ammonium in natural waters [113]. As pronounced concentration gradients are inherent to solid-phase extraction, a relatively dilute (0.5 mol L^{-1} NaOH) solution was selected as the eluent.

Sample/carrier matrix matching can be efficiently accomplished using a chemically inert carrier stream (or wash solution) with a bulk composition as close as possible to that of the injectate. With the sample composition matching that of the carrier solution, concentration gradients and/or discontinuities are minimised. In this way, for the analysis of, *e.g.*, soil extracts or plant digests, the corresponding chemically inert carrier stream should ideally be the extractant solution or the digestion blank, respectively. The system should then be designed in the confluence configuration where the required reagents are added by confluence (see also 3.2.2).

Another possibility is to use a solution with a refractive index similar to that of the injectate as the carrier stream, as demonstrated in the flow injection determination of reactive phosphorus in estuarine waters [92]. The Schlieren effect was reduced by injecting the acidic molybdophosphate reagent into a carrier stream of sodium chloride solution of similar refractive index and letting the established reagent zone merge with the sample (the salinity of which may vary widely from sample to sample) and a reducing agent. Despite the occurrence of sizeable blank signals, reactive phosphorus was successfully measured in samples with salinities ranging from 0 to 34, and the detection limit was estimated as 6 $\mu g\ L^{-1}$ P.

Matrix matching however is not always possible. Examples of situations where this strategy cannot be exploited are analyses of sample lots with high and variable salinity (estuarine waters) or acidity (plant digests) and sample injection into a concentrated reagent stream. Specific strategies for Schlieren compensation are then needed. In this regard, an ingenious approach was proposed for reactive silicate in sea water [97]. The sample stream was split and reagents were added in defined proportions to both branches, thus providing two streams with similar salinity; a sample/reagent aliquot was thereafter sampled from one stream and inserted into the other one, thus minimising the establishment of concentration gradients. The Schlieren effect was reduced but sample replacement was cumbersome and impaired the sampling frequency. Another possibility is to establish similar but opposite concentration gradients inside two symmetrical merging streams and letting them converge. At the confluence point, the gradients are cancelled out by summation of their vectors. This approach was applied to the sequential determination of metals [114]: the analytes were retained as metal chlorocomplexes on an anion-exchange resin mini-column and sequentially eluted by exploiting a pronounced chloride concentration gradient and the main stream merged with a solution with a similar but opposite concentration gradient that was established by inserting water into a concentrated chloride solution.

Regarding improvement of the mixing conditions for Schlieren compensation, it is well known that the Schlieren effect becomes less pronounced as mixing conditions are improved. Under very good mixing conditions however, component B is still present and complete extinction of the effect is therefore not achieved (see also next section). Another aspect related to mixing conditions is the addition of confluent streams. A carrier-to-confluent flow rate ratio that is too high should be avoided as it implies that the concentration of the confluent stream is too high and this creates mixing difficulties, with the consequent appearance of a sinusoidal signal on the main analytical signal. Carrier-to-confluent flow rate ratio should therefore be selected as a compromise between sample dilution at the confluence and baseline noise.

With respect to special monitoring strategies for minimising the Schlieren effect, quantifying only the central portion of the flowing sample, where the concentration gradients are less pronounced, has been proposed [115]. In spite of reducing the Schlieren effect, this strategy impaired sample throughput because a very large inserted sample volume was required. The strategy is particularly attractive in segmented flow analysis, where the signal associated with the central portion of the flowing sample is used for measurement. This is perhaps the main reason why losses of radiation caused by scattering effects have seldom been highlighted in relation to segmented flow analysis.

Another possibility to minimise the Schlieren effect is to stop the sample inside the detector. During the STOP period, the effect is slowly reduced due to diffusion processes, which minimise Schlieren component A. One can confirm from Fig. 4.14 that dispersion practically ceases during the STOP period. In fact, if one imagines that the recorder trace relating to this period is deleted, the right and left hand recorder traces would be similar. However, this possibility has not yet been exploited for real sample matrices.

With respect to the design of the flow-through detector, the influence of flow cell geometry on the Schlieren effect for cylindrical and square capillaries, sheath flow cells and 'U' and 'Z'-pattern flow cells has been investigated by several researchers in relation to liquid chromatography [116], and a theoretical model describing the influence of the Schlieren effect on absorbance measurements has been proposed [83].

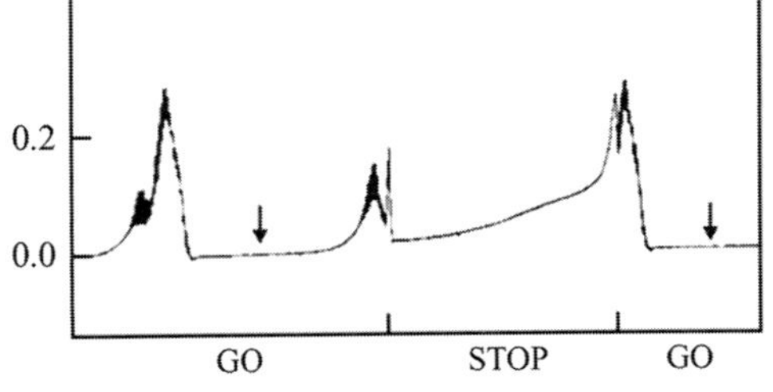

FIGURE 4.14 Schlieren effect fading during the time that the sample is stopped. The recorder trace refers to insertions of 1.0 mol L^{-1} potassium sodium tartrate (upper) or sucrose (lower) into a single line flow injection system. Arrows = times of injection. *Reprinted from Anal. Chim. Acta 234 (1990) 153, E.A.G. Zagatto, M.A.Z. Arruda, A.O. Jacintho, I.L. Mattos, Compensation of the Schlieren effect in flow-injection analysis by using dual-wavelength spectrophotometry, with permission from Elsevier (Ref. [28]).*

The use of a multi-reflection flow cell is another strategy for reducing the Schlieren effect [117]. The incident radiation beam is directed into a silver-coated capillary through a sidewall aperture and emerges through a similar aperture after undergoing several reflections. The Schlieren effect is reduced by the radiation path being perpendicular to the iso-concentration lines along the flowing sample. This approach also leads to a sensitivity enhancement because the optical path length is increased. Similar results have been reported for a miniaturised multi-reflection flow cell [118] that could also be used in microfluidic analytical systems (see 6.3.1 for further discussion).

SUBTRACTION OF MONITORED SIGNALS

The influence of the Schlieren effect can also be minimised by subtracting the corresponding signal from the analytical signal or, in other words, by attaining the net absorbance. This strategy is not needed when several standard solutions are run to derive the analytical response curve because blank subtraction is inherent to this procedure.

Signal subtraction is more efficient when Schlieren component B prevails and the catalytic determination of iodide in table salts [119], highlighted in 3.1.2.2, illustrates this aspect. After sample passage through a mixing chamber, concentration discontinuities were eliminated and the influence of Schlieren component A became insignificant. Net absorbance values were precisely quantified on top of a very high (0.4 absorbance) yet reproducible blank value. Precise results (r.s.d. $<4\%$) were then obtained.

When component A manifests itself in a pronounced manner, the subtraction approach cannot easily be applied due to increased noise. Moreover, the noise is more intense at the rising and falling portions of the recorded signal whereas the analytical signal is usually measured at the top of the peak. In this situation, peak area measurements are recommended.

Peak smoothing is also an option, as demonstrated in the spectrophotometric determination of nitrate and nitrite [93]. The signals recorded for the blank and the sample were processed by applying a simple algorithm, allowing the blank signal to be subtracted. Analytical repeatability, however, was impaired due to the error propagation effects involved.

The drawbacks inherent to signal subtraction can be circumvented by exploiting real-time subtraction, and this strategy is the basis of dual-wavelength spectrophotometry (see 4.1.1.4). The beneficial influence of DWS in flow injection analysis was demonstrated using the dye approach where buffered coloured solutions mimicked the sample and sucrose was added as the Schlieren-generating species [28]. The Schlieren effect was almost quantitatively suppressed (Fig. 4.15 and Table 4.4).

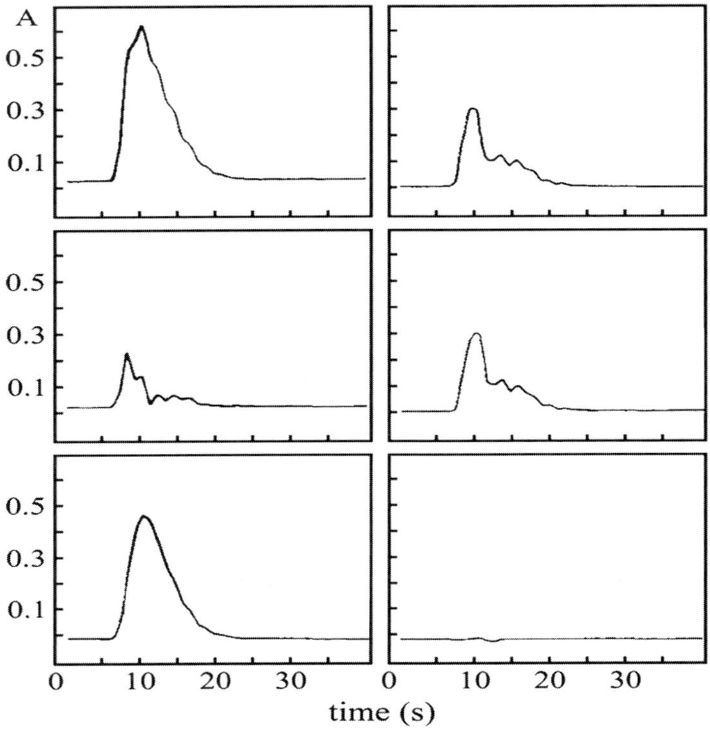

FIGURE 4.15 **Schlieren compensation by dual-wavelength spectrophotometry.** Top, middle and bottom sections refer to the analytical (512 nm), reference (800 nm) and net (ΔA) absorbance values respectively. The right and left columns refer to BCG + sucrose and pure sucrose solutions respectively. Other conditions are as in Table 4.4. *Reprinted from Anal. Chim. Acta 234 (1990) 153, E.A.G. Zagatto, M.A.Z. Arruda, A.O. Jacintho, I.L. Mattos, Compensation of the Schlieren effect in flow-injection analysis by using dual-wavelength spectrophotometry, with permission from Elsevier (Ref. [28]).*

The determination of zinc in plants involving solid-phase extraction was the first application of this strategy to real samples in flow injection analysis [98]. The pronounced Schlieren noise arising from the insertion of the ion-exchange resin mini-column into the eluent carrier stream was successfully minimised by DWS.

Recently, it was demonstrated that the Schlieren effect can be more effectively circumvented when component B prevails [3]. As the related spectrum approaches an ideal white spectrum, the intensity of the effect is not wavelength dependent and hence quantitative suppression by DWS can be attained. Paradoxically, it has often been emphasised that the Schlieren effect becomes less pronounced as mixing conditions are improved. However, under good mixing conditions, component A of the Schlieren effect is dominant. Hence, the DWS efficiency is somewhat

TABLE 4.4 Minimisation of the Schlieren Effect by Dual-Wavelength Spectrophotometry. Data Refer to Analytical Signals, in Arbitrary Units. Dye Solutions [Buffered Bromocresol Green (BCG) Solutions] Inserted into the Buffer Carrier Stream of a Single line Flow Injection Manifold. For Experimental Details, see Ref. [28].

	Sucrose Addition		No sucrose Addition
BCG, mg L^{-1}	Measured	Corrected	Measured
5.00	27099	7698	7649
10.00	31133	14757	15113
20.00	46725	29750	29245
25.00	54071	36981	36984
50.00	92301	72831	72183

reduced because the refraction processes involved are wavelength dependent. For DWS exploitation, therefore, a compromise between mixing conditions and spectral characteristics should always be considered.

Although very effective for compensating the influence of the Schlieren effect, Eq. 4.7 holds only when the intensity of the Schlieren effect is not wavelength dependent, otherwise, correction factors should be added. As the transient mirrors established between fluid elements of different refractive indices are not ideal and the incident light is also partially refracted, the refraction angle is strongly wavelength dependent (Eq. 4.15). Hence, the use of Eq. 4.7 for Schlieren compensation may be subject to restrictions. Moreover, it cannot be directly applied for Schlieren compensation in flow systems with turbidimetric or nephelometric detection.

Finally, Schlieren effect minimisation is inherent to most strategies already proposed for multivariate analyses relying on spectral exploitation. In fact, the calibration models are usually built after data decorrelation. As the Schlieren noise superimposed on every data point leads to an enhancement of the data correlation matrix, procedures for data decorrelation implicitly lead to Schlieren minimisation. This aspect has been demonstrated in relation to the partial least squares algorithm [120].

4.3. PRESENCE OF IMMISCIBLE PHASES

The sample zone may contain immiscible elements (*e.g.*, air, gas, solvent bubbles and suspended particulate matter) in close contact with each other, which may impair analyte detection. The most important

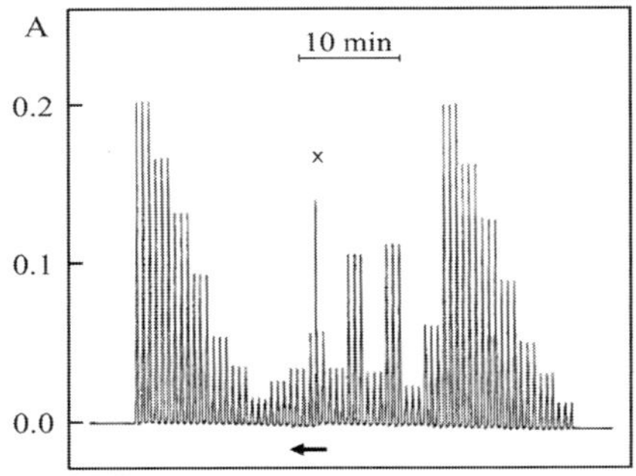

FIGURE 4.16 Recorder tracing of a routine run for the spectrophotometric flow injection determination of ammonium. From right to left, the analytical signals in triplicate correspond to seven standard solutions (0, 50, 100, 200, 300, 400 and 500 µg L^{-1} $N\text{-}NH_4$), nine samples and the standard solutions again. The peak marked x is caused by a spurious air bubble formed inside the flow system, easily distinguished by the very abrupt change in absorbance. *Reprinted from Anal. Chim. Acta 151 (1983) 39, F.J. Krug, B.F. Reis, M.F. Giné, J.R. Ferreira, A.O. Jacintho, E.A.G. Zagatto, Zone trapping in flow injection analysis, Spectrophotometric determination of low levels of ammonium ion in natural waters, with permission from Elsevier (Ref. [121]).*

example is the monitoring of a flowing sample containing air bubbles: as large and relatively steady mirror interfaces are established in the bulk sample, total light reflection may occur, giving a high transient absorbance.

When small drops of immiscible fluids are present in the sample, they can accumulate inside the detector and a continuous baseline drift is then observed. The effect is evident in Fig. 4.16, which refers to the large-scale determination of ammonium ions in natural waters [121]. The chemical reactions take place at about 40°C, and the generation and transport of micro-bubbles leads to an increasingly noisy baseline. When a large air bubble is formed, the micro-bubbles coalesce, resulting in the anomalous signal marked with an X that indicates a large bubble being expelled from the optical path, after which the baseline is restored. A Schlieren effect caused by temperature (thus refractive index) differences is also observed. This conclusion can be made because the blank values are much higher than expected, arising from the insertion of a warm solution that was trapped in a thermostated water bath into the main carrier stream.

The presence of air bubbles in some modes of flow analysis, *e.g.*, segmented, mono-segmented and multi-commuted flow analysis, is beneficial, but some limitations should be mentioned [122]:

- Due to air compressibility, the segmented flow can become a pulsating flow;
- A homogeneous bubble pattern is not easily achieved because it is difficult to control the size of the bubbles;

- In the presence of air bubbles, the pressure drop and flow rate undergo variations, depending on the tubing materials used for building up the manifold;
- The air bubbles may act as electric insulators, increasing the electrostatic charge, which affects electrochemical sensors;
- The efficiency of dialysis or gas diffusion may be hindered, due to the lessening of the effective available surface between the phases involved;
- The main flowing stream cannot be efficiently managed; therefore analytical procedures requiring, e.g., sample stopping, stream splitting or flow reversal, cannot be implemented.

In unsegmented flow systems, the presence of air bubbles is usually a serious drawback as both sample dispersion and detection can be affected. In spite of its relevance, the presence of spurious gas bubbles in the manifold is rarely highlighted in scientific articles on unsegmented flow analysis [123].

Situations leading to the appearance of air bubbles in the manifold include the following:

- The use of bad connectors for stream additions. This is more relevant in flow systems relying on aspiration of the involved solutions;
- Inlet of ambient air into the sampling probe during its movement between the different sample cups containing the solutions (samples and reagents) to be aliquoted;
- The presence of dissolved gases in the involved solutions, which may be released during the coalescence of solution due to their solubility. This aspect becomes more severe in concentrated solutions and water-miscible organic solvents and/or when the system is operated under negative pressure (see also 5.4). The process used to obtain distilled-deionised water in the laboratory for reagent preparation can also have an effect;
- The use of recently warmed solutions. As thermal equilibrium may not have been reached, the evolution of gas bubbles may occur. Conversely, the use of relatively cold solutions can lead to the formation of gas bubbles due to unavoidable heating;
- Heating of the reaction medium, which is required in some analytical procedures relying on chemical reactions (or specific physico-chemical processes such as in-line distillation) that are better accomplished at higher than ambient temperatures. Raising the temperature lessens the solubility of the gas in water, making liberation of bubbles easier;
- Gas permeation through the tubing walls. This becomes more pronounced for thinner tubing and situations involving negative hydrodynamic pressure and
- Formation of gaseous chemical species during sample handling.

Some approaches for preventing the presence of gas bubbles in the analytical path of unsegmented flow systems are the design of specially made flow-through cuvettes, increasing the hydrodynamic pressure of the system and solution degassing [124].

An increase in back pressure has been accomplished by inserting a restrictor coil – usually made from long and thin tubing – after the detection unit. In this way, high temperatures (up to 140°C) have been exploited without liberation of gas bubbles [125]. Analogously, gas bubbles were avoided in the spectrophotometric determination of sulphate in natural waters by using a 10 m long, 0.5 mm i.d. restrictor coil [126]. The authors reported that the small tubing diameter used increased the possibility of system blockage by the precipitate formed and/or the suspended matter inherent to the collected samples. Moreover, the fluid propulsion device should be high performance due to the high hydrodynamic pressure involved.

Degassing involves chemical, physical and/or electrolytic procedures, as well as the use of gas permeation units [127,128]. Addition of nitrogen and the use of reduced pressure permit a faster and more efficient degassing. These procedures are not however permanent, as there is always a tendency for the solutions to re-equilibrate with the environment. To date, no single efficient and completely safe approach for solution degassing in flow analysis has been reported.

When some dyestuffs adhere to the tubing and/or cuvette inner walls, baseline drift can occur, as illustrated by the determination of chromium using diphenylcarbazide as the colour-forming reagent [129]. If polyethylene tubing is used, baseline drift increases slowly, as a consequence of reagent release from the inner walls after in-line preconcentration. Moreover, the analytical signal increases slightly as the effective reagent concentration increases with time. This drawback can be circumvented by using other materials for reactor construction and/or by adding the reagent by merging zones. In this latter situation, the reagent carrier stream should be a solution in which the reagent is rapidly dissolved. Finally, the presence of suspended particulate matter can also cause a pronounced matrix effect [130]. The influence of solid particles on the analytical signal can usually be overcome by properly filtering the samples and/or by applying the standard addition method (see 8.6.3).

References

[1] D.A. Skoog, F.J. Holler, T.A. Nieman, Principles of Instrumental Analysis, fifth ed., Harcourt Brace College Publishers, London, 1998.

[2] H.A. Strobel, Chemical Instrumentation: a Systematic Approach, second ed., Addison-Wesley Pub. Co., London, 1973.

[3] A.C.B. Dias, E.P. Borges, E.A.G. Zagatto, P.J. Worsfold, A critical examination of the components of the Schlieren effect in flow analysis, Talanta 68 (2006) 1076.
[4] R.S. Khandpur, Biomedical Instrumentation: Technology and Applications, McGraw Hill, Columbus OH, 2005.
[5] H. Bergamin-Filho, B.F. Reis, E.A.G. Zagatto, A new device for improving sensitivity and stabilization in flow injection analysis, Anal. Chim. Acta 97 (1978) 427.
[6] L.W.M.M. Terstappen, R.A. Mickaels, R. Dost, M.R. Loken, Increasing light scattering resolution facilitates multidimensional flow cytometric analysis, Citometry 11 (1990) 506.
[7] E. Vidal, M.E. Palomeque, A.G. Lista, B.S.F. Band, Flow injection analysis: Rayleigh light scattering technique for total protein determination, Anal. Bioanal. Chem. 376 (2003) 38.
[8] L. Sommer, Analytical absorption spectrophotometry in the visible and ultraviolet. The principles, in: E. Pungor, W. Simon, H. Malissa, J. Inczedy (Eds.), Studies in Analytical Chemistry, vol. 8, Elsevier, Amsterdam, 1989.
[9] I.M. Kolthoff, E.B. Sandell, E.J. Meehan, S. Bruckenstein (see also Encyclopaedia Britannica, 1999), in: Quantitative Chemical Analysis, fourth ed., The MacMillan Co., London, 1969.
[10] V.A. Yarborough, J.F. Haskin, W.J. Lambdin, Temperature dependence of absorbance in ultraviolet spectra of organic molecules, Anal. Chem. 26 (1954) 1576.
[11] G. Kortum, M. Seiler, Die kritische Auswahl colorimetrischer, spectralphotometrischer und spektrographischer Methoden zur Absorptionsmessung, Angew. Chem. 52 (1939) 687.
[12] H.L. Pardue, The inseparable triad: analytical sensitivity, measurement uncertainty, and quantitative resolution, Clin. Chem. 43 (1997) 1831.
[13] Z. Marczenko, Spectrophotometric Determination of Elements, Ellis Horwood, Chichester, 1976.
[14] M.L.S.S. Gonçalves, Métodos Instrumentais para Análise de Soluções, fourth ed., Fundação Calouste Gulbenkian, Lisbon, 2001.
[15] M.M. Sena, I.S. Scarminio, K.E. Collins, C.H. Collins, Speciation of aqueous chromium(VI) solutions with the aid of Q-mode factor analysis followed by oblique projection, Talanta 53 (2000) 453.
[16] H.H. Willard, L.L. Merritt Jr., J.A. Dean, F.A. Settle Jr., in: Instrumental Methods of Analysis, sixth ed., D. van Nostrand Co., New York, 1981.
[17] H. Janshekar, J.R. Mor, Automated determination of sugar in molasses, Anal. Chim. Acta 90 (1977) 201.
[18] R. Capelle, Microdosage colorimetrique du bore en milieu aqueux, au moyen de reactifs a groupment azoique ou imine derives des acides H et K, Anal. Chim. Acta 24 (1961) 555.
[19] M. Mesquita, A.O. Jacintho, E.A.G. Zagatto, R.F. Antonio, Autocatalysis in the spectrophotometric flow-injection determination of manganese as permanganate. Soil, plant and rock analysis, J. Braz. Chem. Soc. 1 (1990) 28.
[20] B. Chance, Rapid and sensitive spectrophotometry. III. A double beam apparatus, Rev. Sci. Instrum. 22 (1951) 634.
[21] G.W. Ewing, Instrumental Methods of Chemical Analysis, McGraw-Hill, Inc., Columbus OH, 1969.
[22] X.S. Chai, Q.X. Hou, Q. Luo, J.Y. Zhu, Rapid determination of peroxide in the wood pulp bleaching streams by a dual-wavelength spectroscopic method, Anal. Chim. Acta 507 (2004) 281.
[23] B. Chance, D. Mayer, N. Graham, V. Legallais, Compensation of light source noise in a sensitive dual wavelength spectrophotometer, Rev. Sci. Instrum. 41 (1970) 111.

[24] Y. Talmi, Spectrophotometry and spectrofluorometry with the self-scanned photodiode array, Appl. Spectrosc. 36 (1982) 1.
[25] H.P. Mansberg, A.M. Saunders, W. Groner, Hemalog-D white cell differential system, J. Histochem. Cytochem. 22 (1974) 711.
[26] Y.A. Zolotov, E.I. Morosanova, S.V. Shalovannaya, S.S. Dyukarev, Continuous-flow analysis: simultaneous spectrophotometric determinations of metals, Anal. Chim. Acta 308 (1995) 386.
[27] L.R. Snyder, J.J. Kirkland, in: Introduction to Modern Liquid Chromatography, second ed., Wiley, New York, 1979.
[28] E.A.G. Zagatto, M.A.Z. Arruda, A.O. Jacintho, I.L. Mattos, Compensation of the Schlieren effect in flow-injection analysis by using dual-wavelength spectrophotometry, Anal. Chim. Acta 234 (1990) 153.
[29] Q. Song, G.-X. Zhang, Z.-R. Qiu, A.-P. Zhang, G. Hao, Optical signal processing in dual-wavelength optoeletronic drop analyzer, J. Mater. Sci-Mater. El 17 (2006) 287.
[30] Y. Suzuki, H. Hori, M. Twatsuki, T. Yamane, A four-wavelength channel absorbance detector with a light emitting diode-fibre optics assembly for simplifying the flow-injection analysis system, Anal. Sci. 19 (2003) 1025.
[31] I.Y. Eom, P.K. Dasgupta, Frequency-selective absorbance detection: refractive index and turbidity compensation with dual-wavelength measurement, Talanta 69 (2006) 906.
[32] M. Seel, E. Wildermuth, W. Zinth, A multichannel detection system for application in ultra-fast spectroscopy, Meas. Sci. Technol. 8 (1997) 449.
[33] G. Murtaza, J.M. Senior, Methods for providing stable optical signals in dual wavelength referenced LED based sensors, IEEE Photonic Tech. Lett. 6 (1994) 1020.
[34] B. Horstkotte, E. Werner, A.K. Seresht, G. Cornelissen, O. Elsholz, V. Cerda, R. Luttmann, Sequential injection analyzer for glycerol monitoring in yeast cultivation medium, Talanta 71 (2007) 941.
[35] G.D. Brykina, D.Y. Marchenko, O.A. Shpigun, Solid-phase spectrophotometry, J. Anal. Chem. 50 (1995) 440.
[36] P.D. Tzanavaras, D.G. Themelis, Review of recent applications of flow injection spectrophotometry to pharmaceutical analysis, Anal. Chim. Acta 588 (2007) 1.
[37] R. Lobinski, Z. Marczenko, Recent advances in ultraviolet-visible spectrophotometry, Crit. Rev. Anal. Chem. 23 (1992) 55.
[38] K. Yoshimura, H. Waki, S. Ohashi, Ion-exchanger colorimetry. 1. micro-determination of chromium, iron, copper and cobalt in water, Talanta 23 (1976) 449.
[39] K. Yoshimura, H. Waki, Ion-exchanger phase absorptiometry for trace analysis, Talanta 32 (1985) 345.
[40] M. Miro, W. Frenzel, Flow-through sorptive preconcentration with direct optosensing at surfaces for trace-ion analysis, Trends Anal. Chem. 23 (2004) 11.
[41] L.S.G. Teixeira, E.S. Leao, A.F. Dantas, H.L.C. Pinheiro, A.C.S. Costa, J.B. Andrade, Determination of formaldehyde in Brazilian alcohol fuels by flow-injection solid phase spectrophotometry, Talanta 64 (2004) 711.
[42] J. Ruzicka, C.H. Pollema, K.M. Scudder, Jet ring cell: a tool for flow injection spectroscopy and microscopy on a renewable solid support, Anal. Chem. 65 (1993) 3566.
[43] O. Egorov, J. Ruzicka, Flow-injection renewable fibre optic sensor system. Principle and validation on spectrophotometry of chromium (VI), Analyst 120 (1995) 1959.
[44] J. Ruzicka, L. Scampavia, From flow injection to bead injection, Anal. Chem. 71 (1999) 257A.
[45] G. Ni, C.-M. Xiong, H.-D. Xu, J.-Z. Gao, L. Yuan, The progress of solid-phase spectrophotometry, Spectrosc. Spect. Anal. 22 (2002) 301.
[46] G.L. Miller, E.E. Miller, Determination of nitrogen in biological materials. improved Kjeldahl-Nessler method, Anal. Chem. 20 (1948) 481.

[47] C.F. Rodríguez, S.A. Morales, F.F. Rodriguez, Cessation of growth in crystals, J. Crystal Growth 208 (2000) 651.
[48] A. Millan, O. Sohnel, F. Grases, The influence of crystal morphology on the kinetics of growth of calcium oxalate monohydrate, J. Crystal Growth 179 (1997) 231.
[49] E. Ballesteros, M. Gallego, M. Valcarcel, Sequential determination of D- and L-glutamic acid by continuous fractional crystallization, Anal. Chem. 68 (1996) 322.
[50] L.S. Canaes, O.D. Leite, O. Fatibello-Filho, Flow-injection turbidimetric determination of homatropine methylbromide in pharmaceutical formulations using silicotungstic acid as precipitant reagent, Talanta 69 (2006) 239.
[51] C.C. West, J.H. Harwell, Surfactants and surface remediation, Environ. Sci. Technol. 26 (1992) 2324.
[52] F.M. Correia, D.F.S. Petri, A.M. Carmona-Ribeiro, Colloid stability of lipid/polyelectrolyte decorated latex, Langmuir 20 (2004) 9535.
[53] E. Rodenas-Torralba, A. Morales-Rubio, A.F. Lavorante, B.F. Reis, M. de la Guardia, Micropumping multicommutation turbidimetric analysis of waters, Talanta 73 (2007) 742.
[54] G.D. Christian, in: Analytical Chemistry, sixth ed., John Wiley & Sons, Inc., New Jersey, USA, 2004.
[55] F.J. Krug, E.A.G. Zagatto, B.F. Reis, O. Bahia-Filho, A.O. Jacintho, S.S. Jørgensen, Turbidimetric determination of sulphate in plant digests and natural waters by flow injection analysis with alternating streams, Anal. Chim. Acta 145 (1983) 179.
[56] P.J. Worsfold, A. Hughes, D.J. Mowthorpe, Determination of human serum immunoglobulin G using flow injection analysis with rate turbidimetric detection, Analyst 110 (1985) 1303.
[57] O. Söhnel, J. Garside, Precipitation: Basic Principles and Industrial Applications, Buttheworth-Heinemann, Oxford, 1992.
[58] M.L. Salutsky, Precipitates: their formation, properties and purity, in: I.M. Kolthoff, P.J. Elving, E.B. Sandell (Eds.), Treatise on Analytical Chemistry, Interscience Pub., New York, 1959.
[59] A.G. Walton, The Formation and Properties of Precipitates, Interscience Pub., New York, 1967.
[60] A. Costa-Bauza, Estudios sobre Nucleación, Crescimiento y Agregatión Cristalina. Aplicación al Estudio de los Mecanismos de la Urolitogénesis y Aplicaciones Analiticas, Prensa Universitaria, Palma de Mallorca, Spain, 1992.
[61] S.M.B. Brienza, R.P. Sartini, J.A. Gomes-Neto, E.A.G. Zagatto, Crystal seeding in flow-injection turbidimetry: determination of total sulfur in plants, Anal. Chim. Acta 308 (1995) 269.
[62] P.V. Wells, The present status of turbidity measurements, Chem. Rev. 3 (1927) 331.
[63] W. Denis, On the substitution of turbidimetry for nephelometry in certain biochemical methods of analysis, J. Biol. Chem. 47 (1921) 27.
[64] M.E. Diaz-Garcia, R. Badia-Laino, Fluorescence: overview, in: P.J. Worsfold, A. Townshend, C.F. Poole (Eds.),, Encyclopedia of Analytical Science, second ed., vol. 3, Elsevier, Oxford, 2005, p. 97.
[65] A. Sanz-Medel, J.M. Costa-Fernandez, Phosphorescence: principles and instrumentation, in: P.J. Worsfold, A. Townshend, C.F. Poole (Eds.), Encyclopedia of Analytical Science, second ed., vol. 7, Elsevier, Oxford, 2005, p. 149.
[66] L. Zou, Z.-Y. Gu, N. Zhang, Y.-L. Zhang, Z. Fang, W.-U. Zhu, X.-H. Zhong, Ultrafast synthesis of highly luminescent green- to near infrared emitting CdTe nanocrystals in aqueous phase, J. Mater. Chem. 18 (2008) 2807.
[67] J.N. Miller, Fluorescence: quantitative analysis, in: P.J. Worsfold, A. Townshend, C.F. Poole (Eds.), Encyclopedia of Analytical Science, second ed., vol. 3, Elsevier, Oxford, 2005, p. 163.

[68] M.G. Sanders, K.N. Andrew, P.J. Worsfold, Trends in chemiluminescence detection for liquid separations, Anal. Commun. 34 (1997). 13H.

[69] P. Fletcher, K.N. Andrew, A.C. Calokerinos, S. Forbes, P.J. Worsfold, Analytical applications of flow injection with chemiluminescence detection – a review, Luminescence 16 (2001) 1.

[70] J.L. Lopez-Paz, M. Catala-Icardo, B. Anton-Garrido, Determination of diquat by flow injection chemiluminescence, Anal. Bioanal. Chem. 394 (2009) 1073.

[71] E.J. Llorent-Martínez, P. Ortega-Barrales, M.L. Fernandez de Cordova, A. Ruiz-Medina, Development of an automated chemiluminescence flow-through sensor for the determination of 5-aminosalicylic acid in pharmaceuticals: a comparative study between sequential and multicommutated flow techniques, Anal. Bioanal. Chem. 394 (2009) 845.

[72] K. Robards, P.J. Worsfold, Analytical applications of liquid-phase chemiluminescence – a review, Anal. Chim. Acta 266 (1992) 147.

[73] S.W. Lewis, D. Price, P.J. Worsfold, Flow injection assays with chemiluminescence and bioluminescence detection – a review, J. Biolum. Chemilum. 8 (1993) 183.

[74] A.R. Bowie, M.G. Sanders, P.J. Worsfold, Analytical applications of liquid phase chemiluminescence reactions – a review, J. Biolum. Chemilum 11 (1996) 61.

[75] J. Rienitz, Optical inhomogeneities: schlieren and shadowgraphic method in the seventeenth and eighteenth centuries, Endeavour 21 (1997) 77.

[76] J. Rienitz, Schlieren experiments 300 years ago, Nature 254 (1975) 293.

[77] J.W.B. Stewart, J. Ruzicka, H. Bergamin-Filho, E.A.G. Zagatto, Flow injection analysis. Part III. Comparison of continuous flow spectrophotometry and potentiometry for the rapid determination of total nitrogen content in plant digests, Anal. Chim. Acta 81 (1976) 371.

[78] J. Ruzicka, E.H. Hansen, Flow injection analyses. Part I. A new concept of fast continuous flow analyses, Anal. Chim. Acta 78 (1975) 145.

[79] J. Ruzicka, J.W.B. Stewart, Flow injection analysis. Part II. Ultrafast determination of phosphorus in plant material by continuous flow spectrophotometry, Anal. Chim. Acta 79 (1975) 79.

[80] G.R. Fowles, Introduction to Modern Optics, Dover Pub., Mineola, NY, 1989.

[81] H. Schardin, Die Schlierenverfahren und ihre Anwendungen, Ergeb. Exakten Naturwissenschaften 20 (1942) 303.

[82] G.S. Settles, Schlieren and Shadowgraph Techniques, Springer, Weinheim, 2001.

[83] C.E. Evans, J.G. Shabushnig, V.L. McGuffin, Experimental and theoretical model of refractive index artifacts in absorbance detection, J. Chromatogr. 459 (1988) 119.

[84] P.K. Chattopadhyay, J.F. Coetzee, Effective temperature in stopped-flow measurements: a potential source of error with certain instruments, Anal. Chem. 44 (1972) 2117.

[85] M.L. Miller, G. Gordon, Refractive index anomalies in stopped-flow measurements, Anal. Chem. 48 (1976) 778.

[86] X.-Z. Wu, H. Shindoh, M. Yamada, E. Kobayashi, T. Hobo, Development of a novel detector for flow-injection analysis based on optical beam deflection induced by reaction heat, Anal. Sci. 10 (1994) 203.

[87] J.E. Stewart, Refractive index gradients in stopped-flow and temperature-jump kinetics and liquid chromatography, Anal. Chem. 53 (1981) 1125.

[88] J. Pawliszyn, Properties and applications of the concentration gradient sensor to detection of flowing samples, Anal. Chem. 58 (1986) 3207.

[89] J. Pawliszyn, J.Q. Wu, Moving boundary capillary electrophoresis with concentration gradient detection, J. Chromatogr. 559 (1991) 111.

[90] A. Hanning, J. Roeraade, A spectroscopic refractometer for temperature-independent refractive index detection, Anal. Chem. 69 (1997) 1496.

[91] C. Bergins, M. Nowak, M. Urban, The flow of a dilute cationic surfactant solution past a circular cylinder, Exp. Fluids 30 (2001) 410.

[92] I.D. McKelvie, D.M.W. Peat, G.P. Matthews, P.J. Worsfold, Elimination of the Schlieren effect in the determination of reactive phosphorus in estuarine waters by flow-injection analysis, Anal. Chim. Acta 351 (1997) 265.

[93] V. David, J.M. Calatayud, J.V.G. Mateo, Signal processing algorithm for Schlieren effect correction in flow analysis, Anal. Lett. 34 (2001) 1553.

[94] F.J. Krug, H. Bergamin-Filho, E.A.G. Zagatto, S.S. Jørgensen, Rapid determination of sulphate in natural waters and plant digests by continuous flow injection turbidimetry, Analyst 102 (1977) 503.

[95] D. Betteridge, E.L. Dagless, B. Fields, N.F. Graves, A highly sensitive flow-through phototransducer for unsegmented continuous-flow analysis demonstrating high-speed spectrophotometry at the parts per 10^9 level and a new method of refractometric determinations, Analyst 103 (1978) 897.

[96] K.S. Johnson, R.L. Petty, J. Thomsen, Flow-injection analysis for seawater micronutrients, Adv. Chem. Ser. 209 (1984) 7.

[97] J. Thomsen, K.S. Johnson, R.L. Petty, Determination of reactive silicate in seawater by flow injection analysis, Anal. Chem. 55 (1983) 2378.

[98] J.R. Ferreira, E.A.G. Zagatto, M.A.Z. Arruda, S.M.B. Brienza, Determination of zinc in plants by flow injection spectrophotometry with ion-exchange separation, Analyst 115 (1990) 779.

[99] W.E. Bauer, A.P. Wade, S.R. Crouch, Peak readout device for transient signals from flow injection analysis, Anal. Chem. 60 (1988) 287.

[100] C. Thommen, A. Fromageat, P. Obergfell, H.M. Widmer, Application of a capillary flow cell to sophisticated flow-injection systems, Anal. Chim. Acta 234 (1990) 141.

[101] J. Pawliszyn, Concentration gradient detection based on Schlieren optics, Spectrochim. Acta Rev. 13 (1990) 311.

[102] A. Wijk, B. Karlberg, Determination of glycol in aircraft ground deicing/anti-icing fluids using flow injection with refractive index detection, Talanta 41 (1994) 395.

[103] S.R.B. Santos, M.C.U. Araujo, R.A. Barbosa, An automated FIA system to determine alcoholic grade in beverages based on Schlieren effect measurements using an LED-photocolorimeter, Analyst 127 (2002) 324.

[104] R.S. Costa, S.R.B. Santos, L.F. Almeida, E.C.L. Nascimento, M.J.C. Pontes, R.A.C. Lima, S.S. Simoes, M.C.U. Araujo, A novel strategy to verification of adulteration in alcoholic beverages based on Schlieren effect measurements and chemometric techniques, Microchem. J. 78 (2004) 27.

[105] E.A.G. Zagatto, B.F. Reis, M. Martinelli, F.J. Krug, H. Bergamin-Filho, M.F. Gine, Confluent streams in flow injection analysis, Anal. Chim. Acta 198 (1987) 153.

[106] J. Ruzicka, E.H. Hansen, Flow injection analysis. Principles, applications and trends, Anal. Chim. Acta 114 (1980) 19.

[107] E.A.G. Zagatto, A.O. Jacintho, J. Mortatti, H. Bergamin-Filho, An improved flow injection determination of nitrite in waters by using intermittent flows, Anal. Chim. Acta 120 (1980) 399.

[108] D.J. Malcome-Lawes, C. Pasquini, A novel approach to non-segmented flow analysis. Part 2. A prototype high-performance analyser, J. Autom. Chem. 10 (1988) 192.

[109] B.F. Reis, M.F. Gine, E.A.G. Zagatto, J.L.F.C. Lima, R.A.S. Lapa, Multicommutation in flow analysis. Part 1. Binary sampling: concepts, instrumentation and spectrophotometric determination of iron in plant digests, Anal. Chim. Acta 293 (1994) 129.

[110] M.A.Z. Arruda, E.A.G. Zagatto, N. Maniasso, Kinetic determination of cobalt and nickel by flow-injection spectrophotometry, Anal. Chim. Acta 283 (1993) 476.

[111] V. Grassi, E.A.G. Zagatto, J.L.F.C. Lima, Flow-injection systems with multi-site detection, Trends Anal. Chem. 24 (2005) 880.

[112] I.Y. Kolotyrkina, L.K. Shpigun, Y.A. Zolotov, G.I. Tsysin, Shipboard flow injection method for the determination of manganese in sea-water using in-valve preconcentration and catalytic spectrophotometric detection, Analyst 116 (1991) 707.
[113] H. Bergamin-Filho, B.F. Reis, A.O. Jacintho, E.A.G. Zagatto, Ion exchange in flow injection analysis. Determination of ammonium ions at the μg/l level in natural waters with pulsed Nessler reagent, Anal. Chim. Acta 117 (1980) 81.
[114] N. Maniasso, E.A.G. Zagatto, R.E. Santelli, A new approach for compensating concentration gradients in flow analysis, Anal. Chim. Acta 331 (1996) 17.
[115] F.R.P. Rocha, J.A. Nobrega, Overcoming the Schlieren effect in flow injection spectrophotometry by introduction of large sample volumes. Determination of chloride in the electrolyte of lead-acid batteries, J. Braz. Chem. Soc. 8 (1997) 625.
[116] M. Miro, W. Frenzel, Flow injection analysis: detection techniques, in: P.J. Worsfold, A. Townshend, C.F. Poole (Eds.), Encyclopedia of Analytical Science, second ed., vol. 3, Elsevier, Oxford, 2005, p. 48.
[117] P.S. Ellis, A.J. Lyddy-Meaney, P.J. Worsfold, I.D. McKelvie, Multi-reflection photometric flow cell for use in flow injection analysis of estuarine waters, Anal. Chim. Acta 499 (2003) 81.
[118] S.K. Mishra, P.K. Dasgupta, Capillary scale light emitting diode based multi-reflection absorbance detector, Anal. Chim. Acta 605 (2007) 166.
[119] A.N. Araújo, J.L.F.C. Lima, M.L.M.F.S. Saraiva, R.P. Sartini, E.A.G. Zagatto, A SIA system with mixing chamber for handling high concentrated solutions: Spectrophotometric catalytic determination of iodide in nutrition salts, J. Flow Injection Anal. 14 (1997) 151.
[120] R.L. Carneiro, J.W.B. Braga, R.J. Poppi, R. Tauler, Multivariate curve resolution of pH gradient flow injection mixture analysis with correction of the Schlieren effect, Analyst 133 (2008) 774.
[121] F.J. Krug, B.F. Reis, M.F. Giné, J.R. Ferreira, A.O. Jacintho, E.A.G. Zagatto, Zone trapping in flow injection analysis. Spectrophotometric determination of low levels of ammonium ion in natural waters, Anal. Chim. Acta 151 (1983) 39.
[122] J. Ruzicka, E.H. Hansen, Flow Injection Analysis, second ed., Wiley-Interscience, NY, 1988.
[123] W. Frenzel, The Role of FIA/SIA in Solving Analytical Problems in Environmental Research: A Critical Stock-taking. 10th International Conference on Flow Analysis, Book of Abstracts, SPQ: Porto, (2006) p. 295.
[124] S.-H. Fan, S.-L. Wang, X.-F. Fan, F.-R. Wang, Z.-L. Fang, Studies on sequential injection spectrophotometric system using a reflective flow cell for process analysis, Spectrosc. Spectral Anal. 24 (2004) 368.
[125] Y. Hirai, N. Yoza, S. Ohashi, Flow injection analysis of inorganic polyphosphates, Anal. Chim. Acta 115 (1980) 269.
[126] O. Kondo, H. Miyata, K. Toei, Determination of sulfate in river water by flow injection analysis, Anal. Chim. Acta 134 (1982) 353.
[127] J.J. Pedrotti, L. Angnes, I.G.H. Gutz, A fast, highly efficient, continuous degassing device and its application to oxygen removal in flow-injection analysis with amperometric detection, Anal. Chim. Acta 298 (1994) 393.
[128] J.J. Pedrotti, S. Lima, N. Coichev, I.G.R. Gutz, Overcoming oxygen quenching in fluorescence spectrometry with a highly efficient in-line degassing device interfaced with a flow cell, Anal. Chim. Acta 422 (2000) 131.
[129] J.C. Andrade, J.C. Rocha, C. Pasquini, N. Baccan, The effect of on-line complex-formation kinetics on the FIA signal. The spectrophotometric determination of Cr(VI), Analyst 108 (1983) 621.
[130] J. Harrow, J. Janata, Heterogeneous samples in flow injection systems. Part 1. Whole blood, Anal. Chim. Acta 174 (1985) 115.

CHAPTER

5

Flow Analysers

Plus ça change, plus c'est la même chose. **Jean-Baptiste Alphonse Karr**

OUTLINE

It is difficult to classify the different modes of flow analysers because most of them have many common features and therefore this chapter presents the main types in order of their historical development. A critical overview of the principles of commutation is also presented. The chapter concludes with guidelines for describing any flow-based analytical procedure, with an emphasis on the analytical figures of merit.

5.1. THE SEGMENTED FLOW ANALYSER

The segmented flow analyser, based on Skeggs original proposal [1], was the first automatic analyser dedicated to serial assays. The approach utilises air-segmented flowing streams and, in principle, permits the successful implementation of any manual protocol for wet chemical assays.

5.1.1. Characteristics

In segmented flow analysis [2,3], the sample and the carrier/wash solutions are sequentially aspirated from an auto-sampler towards the analytical path (reaction manifold), and the main stream undergoes segmentation after merging with a confluent gaseous stream (*e.g.*, air and nitrogen). The required reagents are added downstream (Fig. 2.5); whenever required, a confluent diluting solution is added upstream of the reagent additions. If the analytical procedure involves a separation step (*e.g.*, dialysis, liquid–liquid extraction, filtration and distillation), the analyte is collected by an already segmented acceptor stream (Fig. 5.1).

In the sampling step, small volumes of air are introduced into the sampling tube during its displacement between the different cups of the

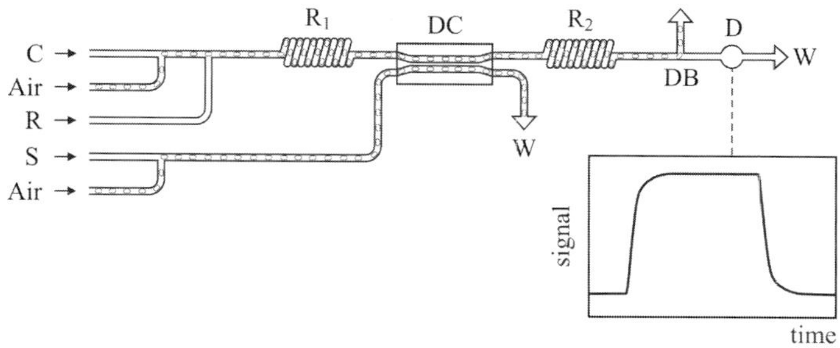

FIGURE 5.1 **Diagram of a simple segmented flow system incorporating dialysis.** S = sample/carrier stream; Air = air; R = reagent (optional); C = carrier stream; DC = dialysis chamber; R_i = reaction coils; DB = de-bubbler; D = detector; W = waste. For details, see text.

auto-sampler. This feature is beneficial to system performance because it minimises mixing between samples and hence carryover effects are reduced.

The addition of a confluent gaseous stream to the reaction manifold promotes segmentation of the sample/wash stream, and the resulting discrete liquid segments are transported towards the detector. While moving through the manifold, each of these segments is subjected to carefully controlled conditions specific to a particular assay, such as incubation temperature, reaction time and reagent concentrations, in order to generate the chemical species to be monitored.

The presence of an immiscible gas phase in the analytical path results in the bubble pattern shown in Fig. 5.2. Consequently, establishment of

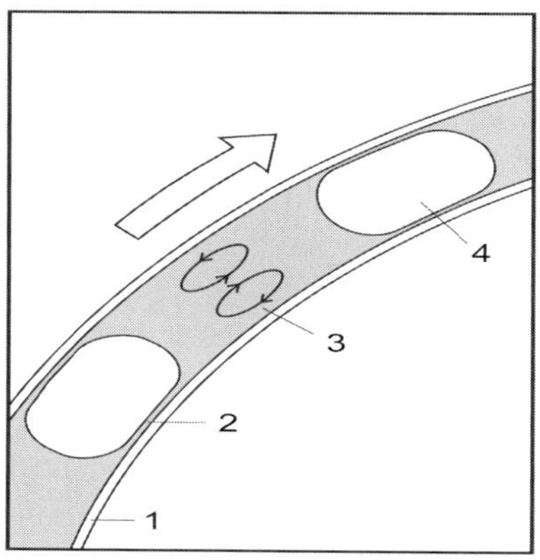

FIGURE 5.2 **Flow pattern inherent to segmented flow analysis.** 1 = tube wall of part of a coiled reactor; 2 = thin liquid layer adhering to it; 3 = vortices inside the liquid phase; 4 = air bubble; outer large arrow = overall displacement of the segmented flow. *Adapted from P.J. Worsfold, A. Townshend, C.F. Poole (Eds), Encyclopedia of Analytical Science, 2nd Edn, Oxford, 2005, v.3, p.24, E.A.G. Zagatto, P.J. Worsfold, Flow Analysis: Overview, with permission from Elsevier (Ref. [4]).*

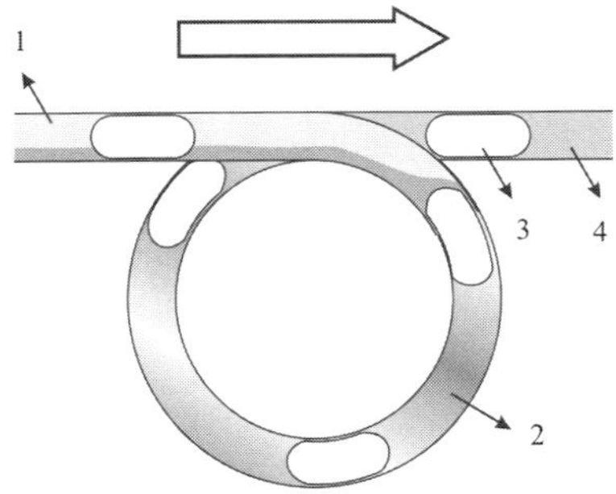

FIGURE 5.3 **Mixing of miscible fluid elements caused by density differences.** 1 = two miscible solutions with different densities; 2 = more dense solution falls downwards; 3 = air bubble; 4 = mixed segment; large arrow = flow direction. For didactic purposes, the influence of vortex establishment (Fig. 5.2) is not shown.

the typical parabolic velocity profile of a streamlined flow regime (Fig. 3.1) is impaired, and vortices are formed inside the individual fluid segments that behave as individual well-stirred tanks. Due to the presence of secondary (rotational) flow, the mixing conditions are improved.

Any difference in the densities of the solutions involved will also contribute to improved mixing, especially near the confluence points (Fig. 5.3). Broadening of the flowing sample is then constrained, decreasing axial dispersion and thereby minimising overlap between successive samples, i.e., reducing carryover. This phenomenon is beneficial for improving the sampling rate and minimising reagent consumption. The presence of gas bubbles inside the analytical path may however lead to some instability in the system, due to temperature effects, compressibility of the gas and the possibility of coalescence with other bubbles.

The gaseous phase is usually removed at the outlet stream of a debubbler positioned immediately before the detection unit; in some systems, the flow cell itself has the facility for bubble removal via a third port. Another possibility is to take advantage of the bubble gating approach [2,5], which involves switching off the emergent radiation beam output when a bubble is present in the flow cell (Fig. 5.4). This allows the bubbles to pass through the flow cell, thereby simplifying the manifold architecture, reducing sample dispersion and avoiding the need to pull the liquid stream through the flow cell, which could lead to sample losses.

Passage of the sample through the detector causes a variation in the monitored radiation, and the related transient signal is recorded as a peak which is ideally proportional to the analyte concentration in the sample. Figs 2.5 and 5.5 show typical recorder tracings from a segmented flow analyser, where a tendency towards a plateau is observed as a consequence of the low sample axial dispersion involved. Important parameters related to the recorded peak shape are the lag phase, peak width at

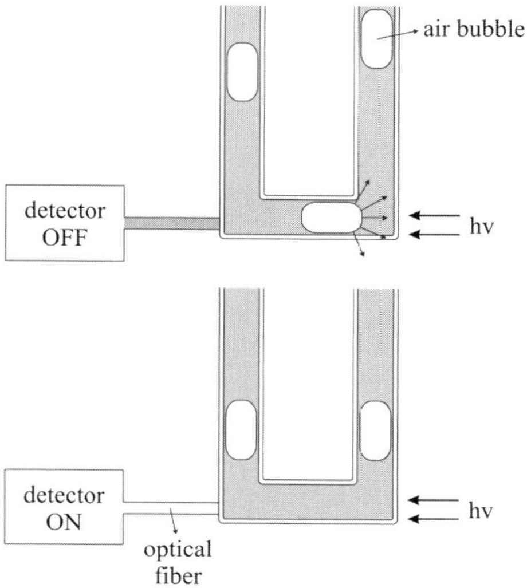

FIGURE 5.4 **The bubble gating approach.** When an air bubble is in the optical path, the detector is OFF (upper figure). Without any bubble in the optical path, the incident beam (hυ) passes through the flow cell and the detector is ON (lower figure). Detector ON/OFF is triggered by means of an auxiliary radiation beam positioned near the inlet tube of the flow-through cuvette (not shown). For details, see Ref. [5].

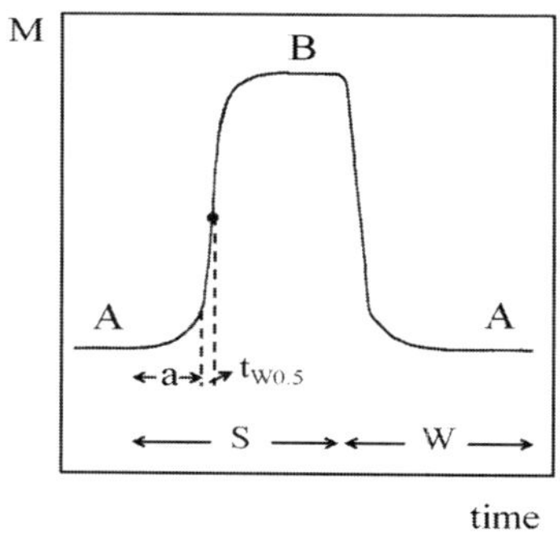

FIGURE 5.5 **A typical recorder output related to segmented flow analysis.** A and B = steady states related to the plateau and baseline; S and W = sampling arm in the sample and blank solution cups; a = lag phase; $t_{w0.5}$ = half wash time. *Adapted from Anal. Chim. Acta 78 (1975) 145, J. Ruzicka, E.H. Hansen, Flow Injection Analyses. Part I. A new concept of fast continuous flow analyses, with permission from Elsevier (Ref. [6]).*

half maximum height and wash time, which provide information on the chemical kinetics involved, the time interval available for measurement and sampling rate, respectively. Advantages and disadvantages inherent in the shape of the recorded peak are discussed in 3.3.1.

The half wash time corresponds to the time interval required to achieve a reduction of 50% in the sample volumetric fraction, X. Provided the rise and fall profiles of the recorded peak are equal to each other, the $t_{w0.5}$ value can be considered as the time interval for attaining $X = 0.5$. As the tanks-in-series model can only be approximately applied in this situation, it is good practice to select a total wash time (baseline to plateau or vice versa) of about 4 times $t_{w0.5}$, i.e., enough to include the lag phase [6]. The inverse of this time period is the *sampling rate* or *sampling frequency*[1], expressed in h^{-1}.

Several samples are simultaneously handled in the analytical path; the number should be as high as possible at any one time in order to optimise the sampling frequency for a given mean sample residence time and an acceptable degree of carryover [7]. However, this number cannot be increased at will due to tailing effects, which are a consequence of sample broadening, and manifest themselves mainly at the sample edges (Fig. 5.5).

Peak tailing confirms that axial dispersion also occurs in segmented flow; it results from the mass transference between successive segments through the thin film of solution established on the tubing inner wall (Fig. 5.2). Consequently, efforts have been made to study sample dispersion and inter-sample contamination in segmented flow analysis [7], which are influenced by, *e.g.*, the number of air bubbles, distance between the bubbles, flow rates and tubing diameter. Understanding the causes of dispersion is important for designing an effective segmented flow system.

5.1.2. Sample Dispersion

The addition of confluent streams is the main process that contributes to dispersion in segmented flow analysis. The extent of this process can be determined from Eqs. 3.10 and 3.11, which are presented here again for didactic purposes:

$$C_1' = C_1[Q_1/(Q_1 + Q_2)] = C_1(Q_1/Q_3) \quad (3.10)$$

$$C_2' = C_2[Q_2/(Q_1 + Q_2)] = C_2(Q_2/Q_3) \quad (3.11)$$

where C_1 and C_2 are the concentrations of F_1 and F_2 inlet convergent streams; C_1' and C_2' are the concentrations of F_3, the outlet stream (see Fig. 3.8); and Q_1, Q_2 and Q_3 are the flow rates of streams F_1, F_2 and F_3.

Equations. 3.10 and 3.11 provide quantitative information on the dilution of the sample and reagent concentrations at a confluent site, as influenced by the flow rates of the convergent streams involved. In applying these equations, the volumetric flow rate of the gaseous phase should not be considered. Their application in relation to flow systems relying on

[1]Terms such as *analytical rate, sample throughput, measurement frequency* and other derived expressions, often found in the literature, are not considered in this monograph in order to avoid misunderstanding with the concepts of sample handling, sample analysis and measurement characteristics, respectively.

liquid–liquid extraction is not straightforward, as the volumes of the aqueous and organic phases are modified by mutual solubility effects.

However, these equations only provide an estimate of the sample and reagent concentrations in the reaction medium or, in other words, the volumetric fraction pie chart (see 3.1.2.4). This information is of practical relevance, especially for assisting in the design of novel flow-based analytical procedures. For a more detailed description of dispersion in segmented flow analysis, however, its axial component should also be taken into consideration.

In this regard, Thiers and co-workers [8–12] and Snyder and Adler [13,14] provided a quantitative description of the dispersion process by considering a model system without aqueous confluent streams, into which a sample solution (dyestuff A) was introduced.

There is some leakage between successive segments through the liquid film established on the inner tube wall (Fig. 5.2). This film is characterised by an average thickness, thus a constant volume, which collects a defined amount of dyestuff A from the previous segment, and transfers it to the next one.

Some of the dyestuff A initially present in any segment (Fig. 5.6) is therefore transferred to the following segments and this process can be regarded as axial dispersion. The change in dyestuff A concentration is more pronounced in the segments at the leading and trailing edges of the entire flowing sample because they are more likely to interact with the carrier/wash solution. On the other hand, the axial dispersion is usually negligible in the segments at the central portion of the entire flowing sample. This explains the typical plateau inherent to the recorded peak (Figs 2.5 and 5.5).

For deriving a quantitative description of axial dispersion in segmented flow analysis, the above-named researchers assumed the following boundary conditions:

- the sample segments established immediately after segmentation are characterised by the same concentrations as the sample;
- the concentration of a flowing plug between two air bubbles is uniform or, in other words, perfect mixing is attained inside the segments;

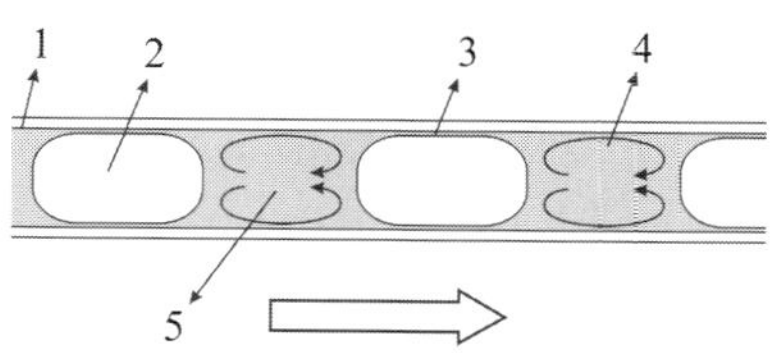

FIGURE 5.6 **The segments established in a segmented flow analyser.** 1 = tube wall; 2 = air bubble; 3 = thin liquid film on the tube wall; 4 and 5 = liquid segments with indication of the vortices (rounded arrows); large arrow = overall flow movement.

- the film adhering to the inner wall of the tubing has the same dyestuff concentration as the previous segment; and
- diffusion inside the film is negligible.

Under these conditions, the concentration of dyestuff A in the ith segment (C_i) is

$$C_i = C_0 \left(e^{-q} q^i \right) i! \qquad (5.1)$$

where C_0 is the initial concentration of dyestuff A in 0th segment (before dispersion); q is the liquid film/liquid segment volumetric ratio, expressed as

$$q = V_f / V_s = (4d_f \, L)/(L_s \, d) \qquad (5.2)$$

where V_f is the volume of the liquid film; V_s is the volume of the liquid segment; d_f is the average thickness of the liquid film, L is the length of the analytical path (tube length); L_s is the length of the liquid segment; and d is the inner diameter of the tube.

Equation 5.1, which dictates both the peak maximum and peak shape, is the Poisson distribution function and can be found in standard mathematical texts. It also enables the segments with the maximum dyestuff concentration to be identified as a function of q [13]. For large q values, the Poisson function tends to be Gaussian with the variance (σ^2) approaching q. It should be stressed that q and σ are dimensionless, and expressed as numbers of segments (*e.g.*, $q \approx 12.5$, $\sigma \approx 12.5^{1/2} \approx 3.5$). If a series of sample segments is considered, dispersion of the successive plugs occurs around the edge of the step function ($i = 0$) to yield a sigmoidal sample concentration profile which is given by the sum of Eq. 5.1 for each initial sample segment. This results in a cumulative Poisson function analogous to Eq. 5.1 for a single initial segment.

Axial dispersion is influenced the analytical path length, plug length, tubing diameter and average thickness of the thin film on the tubing inner wall.

In the 1970s, an equation that allows estimation of d_f values when the contact angle between the liquid and the tube walls is zero, leading to low d_f values, was proposed [15]:

$$d_f = 0.67 \, \pi \, d(\mu v/\gamma)^{2/3} \qquad (5.3)$$

where μ is the linear velocity of the segmented stream; v is the liquid viscosity; and γ is the liquid surface tension.

Although experimental measurements of the film thickness are in agreement with Eq. 5.3, this equation cannot be directly used to predict dispersion in segmented flow analysis. Combination of Eqs. 5.1 and 5.3, however, gives a complete description of an ideal segmented flow analyser:

$$q = \left[0.67\,\pi L\, d^2 (\mu\eta/\gamma)^{2/3}\right] / V_s \tag{5.4}$$

Application of Eq. 5.4 to experimental data [13] led to values for linear coefficients of around 0.5, which is lower than 0.67, the theoretical value. The difference between the theoretical and experimental values can be explained in terms of the contact angle, which is greater than zero in the aqueous surfactant solutions used in the experiments. The decrease in liquid tension tends towards lower linear coefficients. For most practical situations, the following coefficients can be assumed: 0.5 for aqueous surfactant solutions, 0.35 for most organic liquids and 0.6 for water without surfactant [13]. These numbers demonstrate the need for surfactant addition in most analytical procedures implemented in segmented flow systems, and Brij-35 (1.0 ml L^{-1} polyoxyethylene lauryl ether[2] solution) is by far the mostly commonly used surfactant [16].

It should be stressed that the ideal model considers that dispersion is defined by a single parameter q (Eq. 5.4) which is almost unaffected by variations in coiling diameter and sample matrix. Under situations of high axial dispersion, the centre of the sample distribution is delayed, and parameter q becomes different from σ^2, the variance related to the distribution of the dyestuff concentration. This means that the linear velocity of the most concentrated portion of the flowing sample becomes lower than the overall flow velocity. In this situation, the ideal model fails as the dispersion increases and σ^2 is larger than q:

$$\sigma^2 = H\,q \tag{5.5}$$

where H is the empirical correction factor.

The H value approaches unity under conditions of very fast mixing. In this situation, σ^2 is equal to q and the ideal model holds. For most practical situations, however, H is larger than unity, and the model should be expanded. A detailed description of axial dispersion in segmented flow analysis is given in Ref. [14].

5.1.3. Controlling Sample Dispersion

The key to controlling sample dispersion in air-segmented flow analysis is the addition of confluent streams. After the confluence point, an increase in the length of each plug (Eq. 3.12) leads to an increase in the length of the entire sample zone. This does not result in a broader recorded peak, however, because the sample linear speed is

[2]Non-ionic surfactant, $(C_2H_4O)_{23}C_{12}H_{25}O_H$, M = 1199.57.

increased, and the sample passes through the flow cell in a similar time (Fig. 5.7).

When a high dispersion is required, higher flow rates should be selected for the confluent stream(s). In order to avoid an excessive increase in the length of the sample zone, removal of part of the processed sample is sometimes recommended. This can be accomplished by increasing the outlet flow rate of the de-bubbler or by adding an additional T-joint in the analytical path to permit continuous purging of some of the sample to waste.

When high sensitivity is critical and/or the sample integrity needs to be preserved, the flow rates of any confluent streams should be as low as possible. With a high sample-to-reagent flow rate ratio, dilution of the sample concentration during its passage through the confluent point becomes less pronounced (Eq. 3.10). However, the lower the reagent flow rate, the higher the required reagent concentration and this may limit how low the reagent flow rate can be due to solubility effects.

Another relevant parameter affecting sample dispersion is the sample volume. In fact, during sample passage through the analytical path, the total leakage of fluid around the bubbles is considerable and, at the end of the tube, the first sample plugs are diluted by leakage from the carrier/wash stream [17]. If there are sufficient plugs along each sample, those at the central portion of the sample may remain relatively undiluted. Therefore, the aspirated sample volume should be sufficiently large to achieve a plateau region at the centre of the flowing sample during its passage through the detector. Selection of too large a volume should be avoided in order to prevent a pronounced drop in the sampling rate.

Another parameter influencing axial sample dispersion is the number of sample plugs for a given sample volume, which is defined by the number of bubbles along the sample zone or, in other words, the

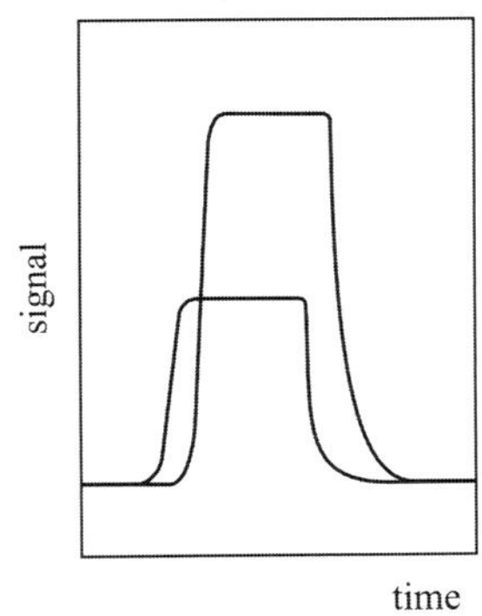

FIGURE 5.7 Influence of a 1:1 (v/v) confluence on the recorded peak. Recorded peaks obtained using a dye as the sample with (lower peak) or without (higher peak) the addition of a confluent stream. The lower sample residence time associated with the lower peak reflects an increase in the overall flow rate due to the confluence stream addition.

frequency of bubble introduction. The size of the air bubbles does not play a significant role [7]. Using typical manifold dimensions, it was demonstrated (see Fig. 5.8) that six air bubbles, five water plugs of 5.3 μL and a total load time of 22 s were enough attain a sampling rate of about 120 h^{-1} with a carryover approaching zero. Fig. 5.8 is a general overview of these results, showing the combined influence of the number of bubbles and plug length on the carryover for a sampling rate of 120 h^{-1}.

With the development of multichannel biochemical analysers [2], it was possible to carry out 12 separate assays on a single sample, which was later increased to 18 and 20 parameters on each sample at rates of 150 h^{-1}, leading to an analytical frequency comparable with modern flow analysers.

The feasibility of exploiting segmentation by adding a confluent stream of an organic immiscible solvent to the flowing sample was demonstrated by Frei and co-workers [18]. The organic plugs minimised sample broadening and also acted as the reagent. Moreover, drawbacks related to thermal contraction/expansion inherent to air-segmentation were avoided. In spite of these favourable features, the approach has seldom been exploited, probably due to the occurrence of pronounced Schlieren effects and the use of organic solvents [19].

Another development was to coat the tubing inner wall with a chemically inert liquid species, which was originally perfluorocarbon [20]. This reduces the thickness of the liquid film on the tubing inner wall, minimises the mass transference between successive liquid segments and reduces the frictional forces at the solid–liquid interface. Consequently, carryover is also reduced. The integrity of the air bubbles is better

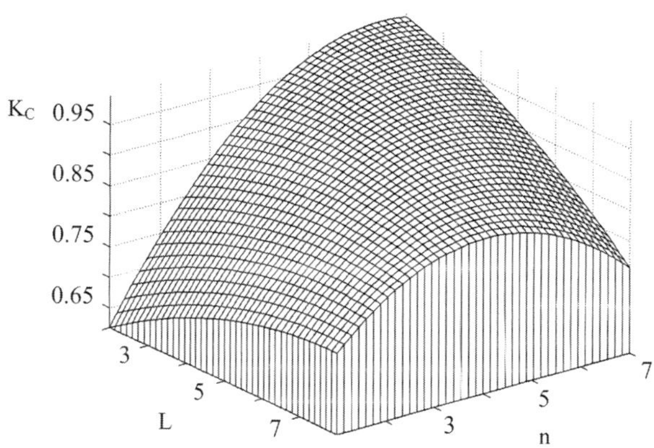

FIGURE 5.8 **Interdependence of carryover, number of bubbles and plug length in a typical segmented flow analyser.** K_C = carryover coefficient for a sampling rate of 120 h^{-1}; n = number of bubbles inside the flowing sample; L = plug length, in s (1 s = 2.7 μL). For experimental conditions see Ref. [7]. *Adapted from Anal. Chim. Acta 408 (2000) 21, G. Liu, B.J. van Wie, D. Leatzow, B. Weyrauch, T. Tiffany, Experimental design modeling of carryover to optimize air-segmented continuous flow analysis, with permission from Elsevier.*

preserved, even at high flow rates, allowing a sampling rate of 720 h^{-1}. This approach has often been exploited in microfluidic devices [21].

Nowadays, the parallel development of the different modes of unsegmented flow analysis has led to innovations in segmented flow systems, *e.g.*, by exploiting both sample insertion into unsegmented carrier stream and air-segmentation, as demonstrated by the potentiometric determination of fluoride in waters [22].

When a long sample residence time is required, *e.g.*, for slow reactions, the option is normally a segmented flow system, whereas unsegmented flow systems, such as the flow injection analyser, are generally preferred when enhanced versatility is required.

5.2. THE FLOW INJECTION ANALYSER

The most outstanding characteristics of the flow injection analyser are the beauty of simplicity and its worldwide acceptance as an analytical tool [23,24].

5.2.1. Characteristics

In flow injection systems [6,25], the aqueous sample is rapidly inserted into a continuously flowing, unsegmented carrier stream, forming a reproducible sample zone that undergoes continuous dispersion while being transported through the analytical path (see also Fig. 2.7). The sample zone participates in all steps inherent in the specific analytical application before reaching the flow-through detector, where it is continuously monitored. The processed sample is then directed to a waste receptacle. The key characteristics of a flow injection analyser are sample injection, controlled dispersion and reproducible timing [26].

As segmentation is not involved, the main flowing stream behaves as an uncompressible homogeneous liquid column, allowing processes such as stream splitting, stream re-directing, sample stopping, flow reversal and zone sampling to be efficiently accomplished (see Chapter 7). The main features of flow injection analysis were critically discussed by Fang who defined it as "a non-chromatographic flow analysis technique for quantitative analysis, performed by reproducibly manipulating sample and reagent zones in a flow stream under thermodynamically non-equilibrated conditions" [27].

5.2.2. Sample Dispersion

Axial dispersion is the main process that governs the continuous mixing of the sample with the surrounding solution in a flow injection system. If the system is designed in the confluence configuration, the addition of the confluent stream also influences this process (see also 5.2.3.2).

The dispersion process can be better understood by considering a model single line system into which a solution of dye A is inserted into an inert carrier stream. At the time of insertion (time zero), the plug of this solution is ideally a perfect cylinder and the associated concentration/time function is hypothetical and rectangular in shape (Fig. 5.9a).

As the sample plug starts being pushed forward, axial diffusion is the main component of the dispersion process, due to the high concentration gradients at the sample/carrier stream interface. The hypothetical peak shape associated with the flowing sample is shown in Fig. 5.9b, which corresponds to the first theoretical Taylor solution [28,29] for the diffusive–convective equation (Eq. 3.4). Situations associated with Fig. 5.9a,b never occur in practice in flow injection analysis.

For longer residence times, the axial convection inherent to the laminar flow regime starts to contribute more significantly to the dispersion process, especially when higher carrier stream flow rates are used, and a tendency of the sample zone towards a skewed Gaussian distribution due to the combined influences of convection and diffusion manifests itself in a more pronounced manner.

The convection process establishes pronounced radial concentration gradients; hence, radial molecular diffusion contributes to lessening the dispersion process. Dye A undergoes radial transport towards the inner wall of the tubing at the front of the sample zone and towards the centre of the flowing stream at the rear of the sample zone, leading to the recorded peak shape shown in Fig. 5.9c. This situation is typical of ordinary flow injection systems and corresponds to the time domain between the two extremes of the theoretical Taylor solutions for the diffusive–convective equation, and was investigated by Gill *et al.* [30–33]. They provided a solution by assuming that the diffusive length is proportional to the

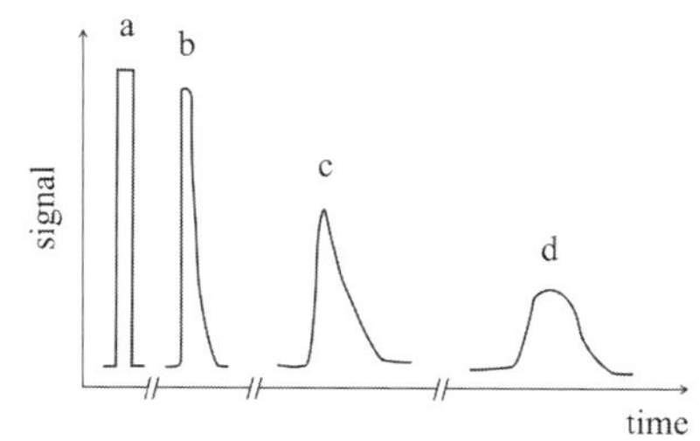

FIGURE 5.9 **Recorded peaks associated with different sample residence times.** Peaks are hypothetical and refer to instants of sample injection; (a) with a very short residence time, (b) with a typical sample residence time in an ordinary flow injection system, and (c) with a very long residence time (d). The peaks only give an estimate of sample dispersion because they correspond to temporal variations and not to concentration/time functions.

square root of time and the convective length is proportional to the first power of time.

Finally, another Taylor solution for Eq. 3.4 is valid for very large residence times (Fig. 3.6). In this situation (Fig. 5.9d), the contribution of molecular diffusion to dispersion is neglected in view of the dominance of convection [34].

From the above it is clear that the diffusive–convective equation does not provide solutions for the region in which most unsegmented flow analysers operate (Fig. 3.6). This led Vanderslice *et al.* [35] to apply the methods of Gill *et al.* [30–33] and Bate *et al.* [36,37] to obtain solutions related to other regions.

Equation 3.4, expressed in terms of Peclet number ($Pe = a\Phi/D_m$), reduced distance ($e = D_m x/a^2\Phi_0$) and reduced time ($\tau = D_m\, t/a^2$), was numerically integrated for all values of $\tau < 0.8$ that corresponded to the region of interest for unsegmented flow analysis. The authors concluded that for $\tau < 0.01$ dispersion is governed by convection, whereas diffusion dominates for $\tau > 0.1$. Therefore, for $\tau < 0.01$, the usual sharp rise and exponential decay associated with convective mass transport are observed, and skewed Gaussian curves are observed for $\tau > 0.1$, as dispersion is controlled mainly by diffusion. For $0.01 \leq \tau \leq 0.1$, convection and diffusion are both important, resulting in an unusual distribution (Fig. 5.10).

The convective–diffusive model has been continuously expanded and applied to specific situations. The influences of reactor coiling and packing a reactor with beads were investigated in the early 1980s [38,39]. Moreover, Eq. 3.4 was expanded in order to evaluate the travel time and baseline-to-baseline time [40]: the predicted results were in good agreement with experimental values but some correction factors were needed.

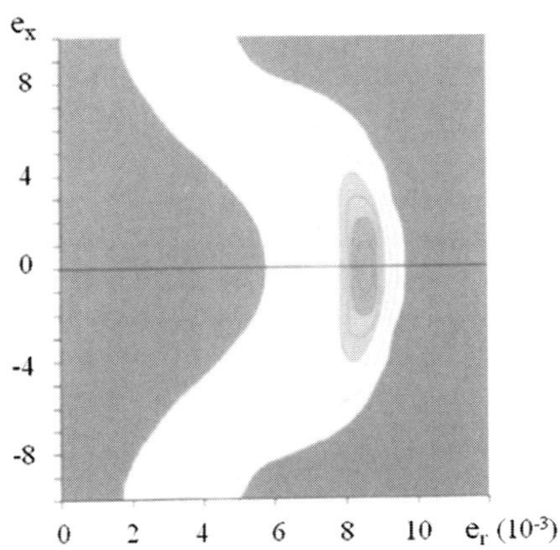

FIGURE 5.10 Bolus shape for τ = 0.10. e_r = reduced radial distance; e_x = reduced axial distance. *Adapted from Anal. Chim. Acta 179 (1986) 119, J.T. Vanderslice, A.G. Rosenfeld, G.H. Beecher, Laminar-flow bolus shapes in flow injection analysis, with permission from Elsevier (Ref. [35]).*

It was further realised that these factors were influenced by tubing length and diameter and by flow rate. In this way, more precise expressions for evaluating the travel and baseline-to-baseline times were derived [41]. The approach was further modified [42] in order to be applicable to larger injected volumes.

Other contributions have involved system modelling for specific situations such as a flow-based viscosimeter [43] and the development of novel strategies for system calibration [44]. The influence of tubular reactor shapes (coiled, 3-D, twisted and wave-shaped) has also been investigated [45–49] and comprehensively reviewed [50]. In this regard, contributions by Kolev and Pungor [51–54] are also relevant.

Nowadays, there is a satisfactory description of physical dispersion in single-channel flow injection systems in which artefacts do not play a pronounced role in the process. Numerous attempts have been made to extend the model to situations involving chemical reactions. In general, the strategy is to consider modifications to the concentrations of the reactant species and D_m. These parameters refer initially only to the main reactants but, as the chemical reactions proceed, D_m of the formed products (and their instant concentrations) should be also taken into account [55]. For a complete description of flow systems comprising several confluent streams, Eqs. 3.4 and 3.10–3.12 should be combined.

5.2.3. Controlling Sample Dispersion

Sample dispersion in flow injection systems is driven by the combined effects of several parameters that, for didactical purposes, are herein classified into two groups depending on whether they are more related to the characteristics of the reaction medium or to the system geometry.

5.2.3.1. Dispersion Parameters – Reaction Medium

Parameters included in this group, *e.g.*, temperature, viscosity, diffusion coefficient, composition of the sample and reagent solutions, occurrence of chemical reactions and the presence of solid particles, are usually less susceptible to variations, because in most applications they cannot be modified at will.

TEMPERATURE

Temperature influences both molecular diffusion and viscosity, and hence the distribution of velocities of the different fluid lines (see also Fig. 3.1). Consequently, both convective and diffusive mass transport are, in principle, affected. An increase in temperature (and the concomitant decrease in viscosity) promotes radial mixing, thus reducing sample broadening and increasing the recorded peak height.

The influence of temperature on sample dispersion is, however, not very important. This can be verified by applying the radial and axial convective–diffusive model (*cf.* Eq. 3.4) or the axial dispersion plug flow model [51], which is a simplification of the former. This was demonstrated in a single line flow injection system with a 100-cm coiled tubular reactor (0.8 mm i.d., 2.0-cm winding diameter), into which a 30-μL sample aliquot was injected [56]. Increasing the temperature from 26 to 70°C continuously increased the sample volumetric fraction up to 21.1%, the effect being more evident at lower flow rates. The experimental data were in agreement with those predicted by the random walk simulation model [57]. Similar findings were reported in relation to flame atomic absorption spectrometry: "the shape of the transient signal can be influenced only to a small extent by chemical or physical properties of the sample such as temperature and viscosity" [58]. Similarly, reduced dispersion with increasing temperature was observed in the determination of non-volatile nitrosamines using liquid chromatography, post-column redox reactions and voltammetric detection [59]. When the reactor temperature was increased from 25 to 70°C, a 16% increase in sample volumetric fraction was observed; other beneficial effects, such as a 30% decrease in the variance of the recorded peak and attainment of a more Gaussian peak, were also reported.

From a practical point of view, temperature control is generally not required in analytical applications, unless the physico-chemical processes involved are temperature dependent. A common practice is therefore to operate the flow injection system at ambient temperature, but a temperature-controlled laboratory environment is desirable.

Above ambient temperatures are generally required to speed up the physico-chemical processes involved, *e.g.*, for enzymatic assays. Temperatures should not be too high however, in order to prevent the formation of gas bubbles in the analytical path, which would affect sample dispersion and impair detection. This drawback can be minimised by increasing the hydrodynamic pressure, *e.g.*, by adding a back-pressure coil, which is usually a large thin coil placed after the detection unit (see also 4.3).

SAMPLE VISCOSITY

The influence of this parameter on sample dispersion was recognised in the 1970s when a novel strategy for measuring viscosity was proposed [60]. Glycerol solutions (viscosity standards) were inserted into a single line flow injection system, and a coloured solution acted as the carrier stream. Thus, the monitored absorbance transiently decreased during passage of the glycerol zone through the detector and an inverted peak was recorded. As viscosity inhibits sample dispersion, lower dispersion was observed for higher glycerol concentrations. The recorded peak heights were therefore proportional to the viscosity of the injected solution. This strategy was applied to the design of viscometers [43,61,62] but only for highly viscous samples.

From an analytical point of view, there are very few specific applications involving highly viscous samples, *e.g.*, whole blood, molasses or syrups, where this aspect needs to be carefully considered. In general, manual sample dilution prior to insertion into the flow manifold is carried out for such sample matrices and surfactant addition to the carrier stream may also be beneficial.

For blood analysis, the use of matrix-matched standards is an efficient strategy to compensate for the influence of viscosity, as emphasised by Rocks and co-workers, who determined zinc and copper in blood serum by flow injection flame atomic absorption spectrometry [63]. Another possibility is to exploit the standard addition method, since different viscosities can also cause a matrix effect. This approach was demonstrated by Harrow and Janata in the potentiometric evaluation of pH in blood serum [64], where the effects of sample viscosity and the presence of solid particles were successfully compensated.

Viscosity can also affect the analyte separation/concentration step, *e.g.*, with in-line dialysis [65]. The authors showed that variations in viscosity and/or pressure effects impaired the operation of a flat dialysis membrane.

As a rule, the influence of viscosity on sample dispersion is not pronounced. In fact, viscosity differences between sample and carrier solutions seldom impair accuracy. Moreover, analyses of highly viscous injectates are seldom reported.

DIFFUSION COEFFICIENT

In a given environment, different chemical species will have different diffusion coefficients. This parameter (D_m – Eq. 3.4) plays an important role in radial mass transport, thereby influencing sample dispersion [66]. Lower sample dispersion is associated with higher D_m values. Generally, this parameter is not of major concern when designing flow injection systems, as the same D_m value applies to any specific chemical species (analyte) in both the sample and standard solutions.

The influence of D_m is more relevant when the "analyte" comprises several similar chemical species, *e.g.*, triazines, pesticides or sugars. The determination of total reducing sugars (invert sucrose plus reducing sugars [67]) in sugar-cane juices exploiting carbohydrate alkaline degradation [68] is a good example. As the D_m values for glucose and fructose were slightly different, efforts were made to provide reaction conditions that exactly compensated for these differences. Hence, the heights of the peaks recorded for these carbohydrates were the same and highly accurate results were obtained.

Here, it is relevant to recall that dispersion in open tubular reactors is a consequence of the combined effects of convection and diffusion. D_m can then be determined as a function of the time required for the flowing sample to pass through the detector. To this end, the solution with the chemical species under investigation is inserted into the flow manifold and monitored downstream; either the peak width at half maximum or the baseline-to-baseline time can be selected as the basis for measurement. This strategy was initially exploited in relation to gas chromatography and further applied to liquid chromatography for the D_m evaluation of alkylbenzenes in chloroform [69]. It can be implemented in flow injection analysis, as pointed out by Vanderslice *et al.* [40], who derived an empirical equation for describing dispersion. This equation involved several parameters, including the baseline-to-baseline time and D_m with the following conditions applied:

- the flow should be laminar;
- the flow pattern should not be disturbed by the injection process;
- the sample zone should flow through a long, straight tube of circular cross section; and
- monitoring should be done in a plane perpendicular to the direction of flow.

Alternatively, coiled reactors have been used. The coiling effect which causes the establishment of secondary flows was compensated by modifying Eq. 3.4 according to the Taylor–Aris and Golay treatments. The flow injection determination of D_m values of different proteins [70] is a good example. Using the modified equation, D_m values were in agreement with literature values and a linear relationship between D_m and molecular mass was reported. Recently, an ingenious H-shaped device with two chip-based spectrophotometric sensors at its ends was designed for both D_m and molar mass determinations [71]. Inside this device, the chemical species under investigation diffused from the flowing sample towards an acceptor stream after merging under laminar flow at the device inlet. A chemometric approach relying on time and wavelength scans allowed successful application to a variety of analytes, including sodium azide, benzoic acid, amino acids, peptides and proteins.

For practical applications, the strategy was initially exploited by Gerhardt and Adams who determined the diffusion coefficients of several chemical species of neuro-biological interest [72]. They reported the influence of the environment on the diffusion of $[Fe(CN)_6]^{3-}$ and $[Fe(CN)_6]^{4-}$ into 1.0–2.0 mol L^{-1} KCl and 1.0 mol L^{-1} NaOH solutions; D_m values within the $4.7 \times 10^{-6} - 7.6 \times 10^{-6}\, cm^2\, s^{-1}$ range were determined. Moreover, they determined the D_m value for the thallium(I) ion (Tl^+) in 1.0 mol L^{-1} KCl as $1.49 \times 10^{-5}\, cm^2\, s^{-1}$, which is in agreement with literature data. Other applications are summarised in Table 5.1. This

TABLE 5.1 Flow Injection Determination of Diffusion Coefficients

Chemical Species	Environment	D_m value or range, 10^{-6} $cm^2\ s^{-1}$	Detection Technique	Comments	Ref.
Biogenic amine neurotransmitter-related compounds	0.1 mol L^{-1} phosphate buffer, pH 7.4	4.1–6.6	Spectrophotometry	Baseline-to-baseline time as the measurement basis	[72]
Dopamine, epinephrine, norepinephrine, ascorbic acid	0.1 mol L^{-1} phosphate buffer, pH 7.4.	6.0, 4.4, 6.0, 6.8	Amperometry; voltammetry	Syringe pump; straight tubular reactor; peak width at half height measurements	[73]
Several electroactive substances	0.1–1.0 mol L^{-1} KCl	6.27–17.5	Amperometry	Limiting current proportional to the two-thirds power of D_m.	[74]
Iron(II) complexes with ligands containing =N–C=C–N=	Acetonitrile-water solutions	3.25–6.13	Spectrophotometry; amperometry	Gravity flow; non-linear relationship between D_m and molecular mass.	[75]
Proteins	0.08 mol L^{-1} Na_2SO_4, pH 7	0.26–1.05	Spectrophotometry	10-m coiled reactor (Teflon tubing, 0.5 mm i.d., 1 cm winding diameter) immersed into a 20°C thermostatic water bath.	[70]
PAN, 1-(2′-pyridylazo)-2-naphthol	40–100% water–ethanol	2.43–8.31	LED-based photometry	Straight tubular reactor inside a thermostatic jacket; comparison with NMR	[76]
Monodispersed and polydispersed polymers	Dilute Na_2SO_4 solutions	0.2–3.0	Not stated	Linear relationship between D_m and molecular mass within 1000 and 90000 g mol^{-1}	[77]
Indoles, ascorbic acid	0.1 mol L^{-1} phosphate buffer, pH 7.4	4.61–6.75	Differential refractive index detection	D_m little affected by the analyte concentration	[78]

strategy has been used more in relation to large molecules and, in recent years, capillary electrophoresis has emerged as a more suitable approach.

COMPOSITION OF THE SAMPLE AND REAGENT SOLUTIONS

Characteristics of the sample and reagent solutions can also affect sample dispersion and special attention should be given to highly viscous samples, the presence of gaseous micro-bubbles (common in, *e.g.*, acidified river water samples) and very concentrated reagent solutions. In spite of the importance of these factors, they have not been widely investigated.

Another relevant aspect is the presence of solid particles in the injectate (sample). Under laminar flow conditions, these particles undergo rotation, altering the convective mass transport and hence sample dispersion. This can be regarded as a matrix effect and the standard addition method may successfully circumvent it [64].

Finally, the influence of chemical reactions on sample dispersion should be stressed again. After reaction, the reagents are converted into another species with a different D_m value in the specified medium, thus influencing the dispersion [79].

OTHERS

Specific sample handling procedures, *e.g.*, in-line microwave irradiation, in-line sonication and rate of analyte transport from a donor towards an acceptor stream, can influence sample dispersion but literature information on these factors is sparse.

5.2.3.2. Dispersion Parameters – System Geometry

Parameters included in this group, such as sample injected volume, injection mode, addition of confluent streams, flow rates, characteristics of the analytical path and manifold architecture, are generally more likely to be modified. Consequently, they play a relevant role in system design. These parameters are herein discussed in terms of an ideal flow injection analyser using the dye approach [26].

SAMPLE VOLUME

The injected sample volume is by far the most important parameter affecting dispersion in flow injection analysis, as it has a pronounced influence on the height, width and area of the recorded peak (Fig. 5.11).

There is always a linear relationship between sample injected volume and the area of the recorded peak. Regarding the use of peak height or peak width, three main scenarios can be considered, each one associated with a different influence of the injected sample volume on the recorded peak shape.

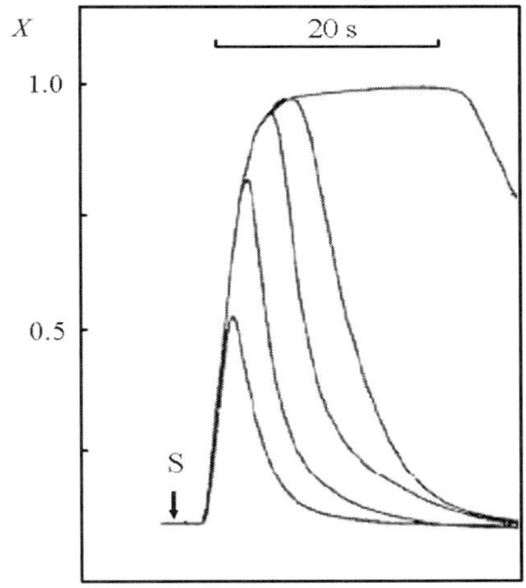

FIGURE 5.11 Recorded peaks for different sample injected volumes. X = sample volumetric fraction; S = injection instant. The recording tracings correspond to loop-based injections of 59, 108, 206, 403 and 795 μL into a single line flow injection system. *Adapted from Anal. Chim. Acta 99 (1978) 37, J. Ruzicka, E.H. Hansen, Flow injection analysis. Part X. Theory, techniques and trends, with permission from Elsevier (Ref. [80]).*

- First scenario: very small sample volumes, thus large sample dispersion. Increasing the sample injected volume (V_S) results in a proportional increase in the height of the recorded peak (h), whereas the peak width (σ) tends to be unaffected (Fig. 5.12).
- Second scenario: intermediate volumes, thus medium sample dispersion, situations where a typical flow injection system operates. Increasing V_S increases both h and σ, the former being more pronounced for lower V_S values (Figs 5.11 and 5.12).
- Third scenario: large volumes, thus limited sample dispersion. Increasing V_S results in a proportional increase in σ, whereas h tends to be unaffected. A tendency towards a flat peak recording is noted (Fig. 5.11).

In summary, increasing V_S initially increases the sample volumetric fraction (X) and this increase becomes less pronounced for larger sample volumes or, in other words, the system tends to reach saturation, a situation where further increase in V_S no longer alters h in a significant manner. For very small injected volumes, there is an almost linear X vs V_S relationship, and exploitation of this aspect allows analytical calibration graphs to be obtained using only one standard solution [81].

The relationship between V_S and X, adapted from the tanks-in-series model with N = 1 [80], can be approximately expressed as

$$X = C/C_0 = 1 - \exp(-0.693 V_s/V_{0.5}) = 1 - 2^{-k} \qquad (5.6)$$

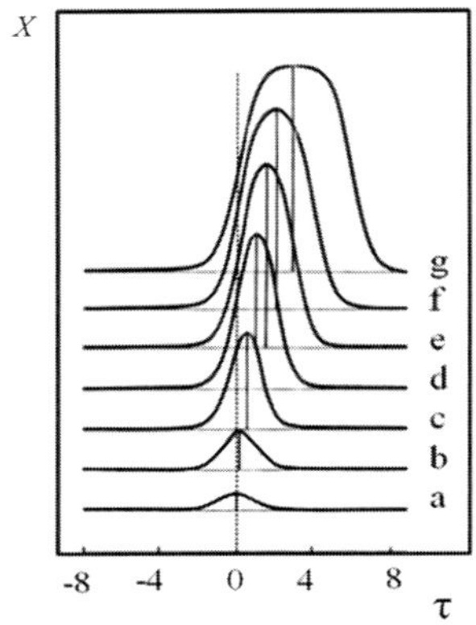

FIGURE 5.12 **Recorded peak shapes corresponding to the different sample volumes (expressed in semi-volumes) injected into a single line flow injection system.** X = volumetric fraction, τ = dimensionless time; a, b, c, d, e, f, g = 0.2, 0.5, 1.0, 2.5, 3.7, 5.0, 7.5 $S_{0.5}$. For details, see Ref. [82]. *Adapted with permission from "R.S. Schifreen, D.A. Hanna, L.D. Bowers, P.W. Carr, Analytical aspects of immobilized enzyme columns, Anal. Chem. 49 (1977) 1929". Copyright 1977, American Chemical Society.*

where X is the sample volumetric fraction; C_0 and C are the sample concentrations before and after dispersion; V_S is the sample injected volume, $V_{0.5}$ is the sample semi-volume (corresponding to $X = 0.5$ [26]), k = V_S expressed in semi-volumes.

One can calculate from Eq. 5.6 that $X = 0.5$, 0.75, 0.875, 0.9375 and 0.96875 for k = 1, 2, 3, 4 or 5, respectively. These data (and Fig. 5.12) strongly suggest that it is good practice to set a V_S value of 3–4 $V_{0.5}$ in designing flow injection analytical procedures where sensitivity is critical and h constitutes the measurement basis. In fact, higher V_S values would promote only a slight increase in h (or X), but sampling rate would severely deteriorate in view of the increased peak width and washing time.

The theoretically expected recorded peak shapes corresponding to the different k values are shown in Fig. 5.12, which also emphasises the main aspects related to the above-mentioned three scenarios.

From this figure, one can also see the delayed attainment of peak maximum for higher injected volumes. This confirms that the analytical path is increased when the sampling loop is inserted into the sample carrier stream [83].

Another interesting aspect is that when V_S tends to zero, σ tends to a constant value different from zero (Fig. 5.12) or, in other words, there is a lower limit value (σ^*) for the volume of the sample zone at the detection unit [84].

The existence of σ^* explains the tendency towards a linear V_S–X relationship for very low V_S values. In fact, an increase in the sample aliquot results in a linear increase in concentration only if the final volume is maintained.

For a better understanding of the influence of V_S on X, a saturation index, I_S, was proposed [84] as

$$I_S = (\sigma - \sigma^*)/\sigma \qquad (5.7)$$

where σ is the actual recorded peak width and σ^* is the limit value for the recorded peak width.

Analysis of Eq. 5.7 reveals that I_S tends to zero for very low V_S values, emphasising that the system is not saturated and tends towards a linear $V_S{-}X$ relationship. On the other hand, σ becomes much higher than σ^* for large V_S values. In this situation, I_S approaches unity and the system becomes saturated; hence, there is no point in increasing V_S to improve sensitivity in methods relying on peak height measurements. Any sensitivity enhancement would be too small to compensate for the deterioration in sampling rate.

INJECTION MODE

Another parameter affecting sample dispersion in flow injection analysis is the way that the sample is inserted. The influence of V_S on the shape of the recorded signal is, however, relatively independent of the mode of sample insertion, and this is perhaps the main reason why this aspect is rarely reported.

Most flow injection analysers exploit either loop-based or time-based injections (see 6.2.2), and the main difference between these injection modes is the shape of the initial sample plug. A critical comparison of these strategies is given elsewhere [85]. This aspect is of minor concern as a parameter affecting sample dispersion in flow injection analysis.

DIMENSIONS OF THE ANALYTICAL PATH

The length of the analytical path plays an important role in the extent of sample dispersion in flow injection analysis. Increasing this length decreases the recorded peak height and increases sample broadening, with these effects being more evident for lower path lengths (Fig. 5.13).

Analysis of Fig. 5.13 confirms the relationship [80]:

$$h = kT^{0.5} = k'L^{0.5} \qquad (5.8)$$

where h is the recorded peak height; k and k′ are the experimentally defined coefficients; T is the mean sample residence time; and L is the analytical path length.

As h is proportional to the square root of T or L, there is an asymptotic tendency for dispersion to level off at higher travelled distances. In this situation, it is not good practice to promote a further increase in L to

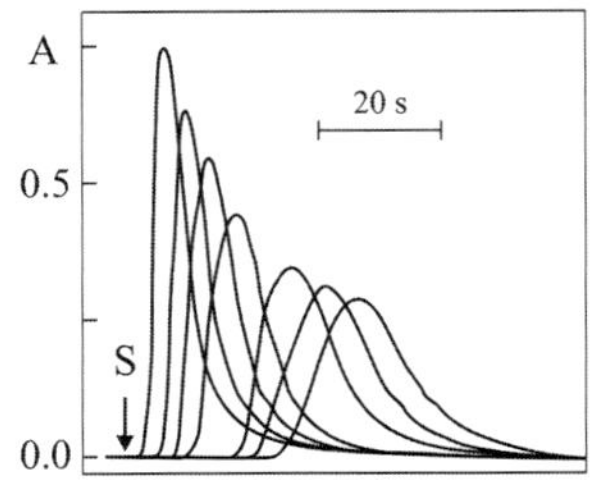

FIGURE 5.13 **Recorded peaks for different analytical path lengths.** A = absorbance; S = injection instant. The recording tracings correspond to loop-based injections of 50 μL of a dye solution into a flow injection system with a straight tube (25, 75, 125, 175, 250, 300 and 350 cm) acting as the main reactor. *Adapted from Anal. Chim. Acta 99 (1978) 37, J. Ruzicka, E.H. Hansen, Flow injection analysis. Part X. Theory, techniques and trends, with permission from Elsevier (Ref. [80]).*

increase the degree of dispersion, as the increase would be too low to compensate the pronounced drop in sampling rate and increased reagent consumption. As a rule, the length of the analytical path should be kept as short as possible yet long enough to guarantee suitable mixing conditions.

The L value can be reduced by using more efficient reactors such as tightly coiled, knitted or packed bed configurations (see 6.2.3). In analytical procedures requiring longer sample residence times, L should not be increased; the total flow rate should be decreased instead.

Here, it is important to emphasise that the analytical path includes part of the detector inner volume [83]. This volume should be kept as small as possible so as not to act as a chamber-like element.

TUBING INNER DIAMETER

The inner diameter of the manifold tubing plays an important role in sample dispersion. Increasing this parameter decreases *X* (thus the analytical signal) in a pronounced way (Fig. 5.14).

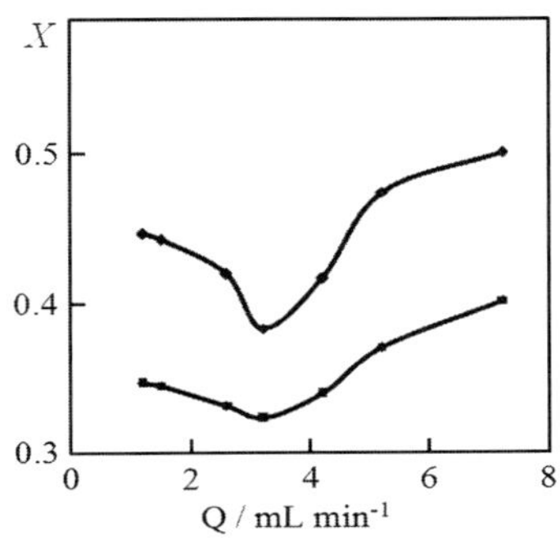

FIGURE 5.14 **Influence of tubing inner diameter on sample dispersion.** *X* = sample volumetric fraction; Q = volumetric flow rate; upper and lower curves = 0.50 and 0.75 mm tubing diameter. Figure adapted from Ref. [86].

The tubing diameter should be kept as narrow as possible but it cannot be reduced indefinitely. In fact, for too narrow an internal diameter, the system hydrodynamic pressure would be too high and the probability of system blocking by small suspended particles inside the sample zone would increase. Therefore, inner diameters within the range 0.5–1.0 mm are most commonly used.

FLOW RATE

Variations in flow rate (Q) can affect sample dispersion in flow injection systems, as it influences both the parabolic distribution of linear flow velocities (Fig. 3.1) and the time available for sample diffusion. Therefore, two different trends have been reported in relation to the *X vs* Q function (see also Fig. 5.15). For lower flow rates, *X* decreases with decreasing flow rate [80], whereas an inverse effect is observed for higher flow rates [41,87]. Consequently, there is a single point in the *X vs* Q function, Q* (Fig. 5.14), where $\Delta_X/\Delta Q$ is zero. The Q* value is independent of V_S, L and tubing internal diameter, and is dependent only on the diffusion coefficient of the substance (D_m – Eq. 3.4) and this opens up the possibility of a novel way to estimate diffusion coefficients [86]. The two different trends relating to the influence of Q on sample dispersion under laminar flow conditions were recently confirmed [88].

A deeper examination of Fig. 5.14 allows one to infer that the influence of flow rate on sample dispersion is not so pronounced. In fact, when Q is

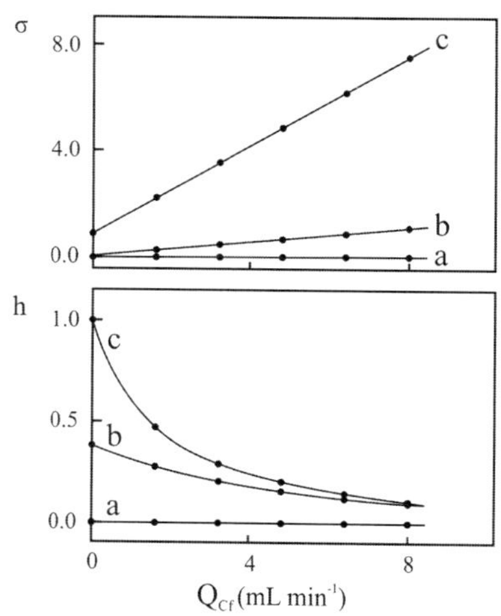

FIGURE 5.15 **Influence of the confluence stream flow rate Q_{Cf}.** h and s = recorded peak height (A) and width (mL); carrier stream flow rate = 1.6 mL min^{-1}; a, b, c = 5, 50 and 250-cm loop lengths (*ca* 25, 250 and 1250 μL injected volumes); other experimental conditions as in Ref. [84]. *Adapted from Anal. Chim. Acta 198 (1987) 153, E.A.G. Zagatto, B.F. Reis, M. Martinelli, F.J. Krug, H. Bergamin-Filho, M.F. Gine, Confluent streams in flow injection analysis, with permission from Elsevier.*

varied from 1.0 to 7.0 mL min^{-1} (typical range in ordinary flow injection systems with 0.75 i.d. tubing diameter), the sample volumetric fraction varies from 0.35 at 1.0 mL min^{-1} to 0.40 at 7.3 mL min^{-1}, reaching a minimum value of 0.32 at 3.2 mL min^{-1}. This corroborates the general recommendation that, for increasing the mean available time for reaction development, it is a good practice to reduce the total flow rate.

It should be noted that this trend is only valid for laminar flow regimes. With pulsed flows, Q is determined by the pulse volume and pulse frequency, and modifications to Q are easily accomplished by altering the pulse frequency or, in other words, increasing the STOP period between pulses. As dispersion practically ceases during the STOP period, variations in Q do not practically affect sample dispersion in flow injection systems [88].

CONFLUENT STREAM ADDITIONS

Better system ruggedness and analytical sensitivity are attained in flow injection systems designed in the confluence configuration. Reagent addition by confluence is therefore an important parameter affecting sample dispersion.

According to Eqs. 3.10 and 3.13, the inlet of a confluent stream promotes a decrease in the recorded peak height, h, and an increase in sample width, σ (Fig. 5.15), but these effects are only evident for systems with limited sample dispersion. In this situation, larger sample volumes are used, the saturation index approaches unity, and sample dispersion occurring after the confluence site is very low [84]. Consequently, modifications in h and σ caused by confluence addition cannot be compensated downstream, and manifest themselves in the shape of the recorded peaks (see also Fig. 5.16, right).

Conversely, the flow system can be designed with a small injected volume, with the aim of achieving large sample dispersion. The saturation index approaches zero and the effects of a confluence stream addition immediately after the injection port can be masked. In fact, the decrease in analyte concentration along the sample zone is compensated by the reduction in the dispersion occurring downstream, whereas the increase in length (expressed in monitoring time) is masked by the concomitant increase in total flow rate caused by the confluent stream addition. In the extreme situation, this compensation can be quantitative and a paradoxical situation is established where the influence of a confluent addition does not alter the recorded peak shape (Figs 5.15.a and 5.16, left). This paradox can be illustrated by considering a 1:5 (v/v) confluence addition immediately after the injection port: the concentrations of the fluid elements along sample zone are reduced to 20% during their passage through the confluence site and their volumes undergo a 5-fold increase. These effects are evident (Fig. 5.16, right) for large sample volumes. On the other hand, the σ^* limit value has to be reached in the

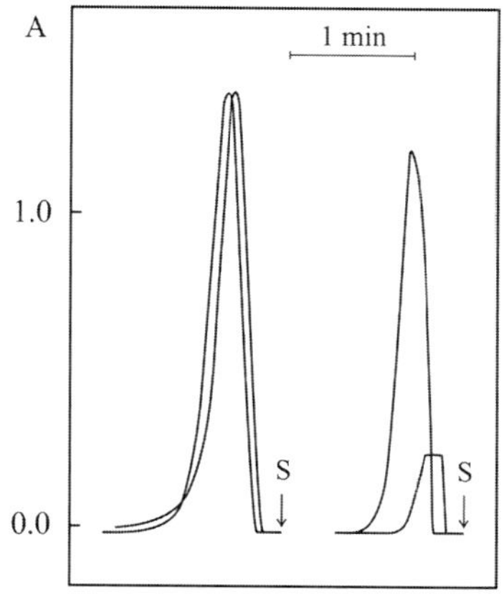

FIGURE 5.16 **Influence of confluent stream addition on the recorded peak shape.** Recorder tracings refer to the injections of 25 μL of a 150 mg L^{-1} bromocresol green, BCG, solution (left) or 250 μL of a 15 mg L^{-1} BCG (right) into a 1.6 mL min^{-1} carrier stream of a single line flow injection system to which an 8.0 mL min^{-1} confluent stream is optionally added. S = point of injection, A = absorbance. The first peak to reach the detector refers to the confluence addition. Other experimental conditions as in Ref. [84]. *Adapted from Anal. Chim. Acta 198 (1987) 153, E.A.G. Zagatto, B.F. Reis, M. Martinelli, F.J. Krug, H. Bergamin-Filho, M.F. Gine, Confluent streams in flow injection analysis, with permission from Elsevier.*

situation of large sample dispersion, so that dispersion downstream is reduced, exactly compensating the effects caused by the confluence stream addition.

It should be stressed that these effects refer only to the sample volumetric fraction, and not to the analytical signal. In fact, without confluent stream addition, the sample is dispersed in the carrier stream, whereas with it, the sample is dispersed mainly in the reagent. In this way, a pronounced enhancement in sensitivity is achieved by adopting the confluence configuration (see also 3.2.2).

Therefore, the flow rate of the confluent reagent stream should be as low as possible for sensitivity enhancement. However, it cannot be reduced indefinitely, as it will be limited by solubility effects and/or the difficulties in adding a very concentrated stream at a very low flow rate (see also 3.2.2).

Conversely, if large sample dispersion is required, a high confluent stream flow rate should be selected, and the reagent should be diluted accordingly. Reagent addition should take place far from the injection port. In this way, pronounced sample dispersion into the carrier stream occurs before sample zone enlargement at the confluence site, so the downstream compensation becomes negligible. The site of reagent stream addition is therefore a relevant parameter affecting sample dispersion in flow injection analysis [84].

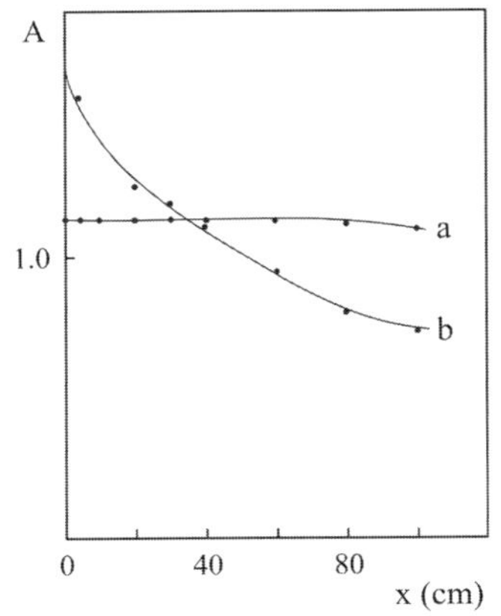

FIGURE 5.17 **Influence of the site of confluent stream addition.** x = distance between injection port and confluence site; A = peak height absorbance; a, b = 25 and 1250 μL sample injected volumes. Other experimental conditions as in Ref. [84]. *Adapted from Anal. Chim. Acta 198 (1987) 153, E.A.G. Zagatto, B.F. Reis, M. Martinelli, F.J. Krug, H. Bergamin-Filho, M.F. Gine, Confluent streams in flow injection analysis, with permission from Elsevier.*

SITE OF CONFLUENT STREAM ADDITION

The influence of the site where the confluent stream is added is shown in Fig. 5.17, which refers to two extreme situations. For large dispersion, the saturation index tends to unity and the influence of this parameter is irrelevant. On the other hand, when high sensitivity is required, the system is designed with a saturation index approaching zero, so that the site of the confluent stream addition becomes relevant, as described above.

ADDITIONAL COMPONENTS IN THE ANALYTICAL PATH

The presence of artefacts in the analytical path, such as mixing chambers, tubing connections, de-bubblers and other chamber-like components, can also affect sample dispersion in flow injection analysis. The effects of a mixing chamber and the detector inner volume are discussed in 3.1.2.2 and 6.3.2, respectively. The presence of devices for liquid—liquid extraction and gas diffusion (or dialysis) alters dispersion, and is dealt with in Chapter 8.

Finally, it should be stressed that there are some specific strategies that can strongly influence sample dispersion in flow injection analysis, *e.g.*, merging zones, zone sampling, stream splitting and closed-loop arrangements. These strategies are discussed in Chapter 7.

5.3. THE SEQUENTIAL INJECTION ANALYSER

The parameters inherent to the analytical procedure can be modified without alterations to the manifold architecture, and this aspect makes the sequential injection analyser very attractive.

5.3.1. Characteristics

Sequential injection systems were conceived [89] to simplify flow-based analytical procedures and enhance versatility and robustness for process monitoring. The confluence connections usually present in segmented flow and flow injection analysis are eliminated and different analytical applications can be implemented without the need for multiple manifold re-configuration.

The basic components of the system are a liquid driver with only one carrier stream, a multi-port selection valve and a detector (Fig. 2.9). The valve is the heart of the sequential injection system and normally comprises 6−10 peripheral ports and a central port in a multi-position valve configuration. The central port is linked to a holding coil and the peripheral ports are connected to different solution aspiration tubes and transmission lines that are linked to different manifold components, *e.g.*, detector and mixing chamber. Only one peripheral port is connected to the central port at any one time. Stream management inside the holding coil is accomplished by a bi-directional piston (or peristaltic) pump. The analyser is fully computer controlled and the injection volumes, residence times, delivery of solutions and analytical path lengths are selected based on a valve timing sequence and related flow rates.

In a typical sequential injection system (Fig. 2.9), the multi-port valve selects the sample, reagents and carrier/wash aliquots, which are sequentially aspirated towards the holding coil [89−91]. A stack of reproducible unsegmented plugs is then assembled inside this coil. The flow is thereafter reversed and the valve is switched in order to direct these plugs towards the detector. During transport of this stack, the sample and reagent plugs penetrate one another as a consequence of dispersion, and the interactions between their components give rise to one or more detectable species that can be monitored during passage of the processed sample through the detector. Analogously to the flow injection analyser, the different steps required by a specific application, such as dilution, reagent addition, chemical reactions and analyte separation/concentration, can be reproducibly accomplished while the sample flows through the analytical path.

Flow reversal plays a beneficial role in reducing sample dispersion [92]. The iso-concentration concentric cylinders typical of laminar flow are stretched during sample transport inside the holding coil and tend to be compressed again after flow reversal. As a result, a process analogous to peak de-convolution occurs (Fig. 5.18).

The sequential injection system is extremely versatile because modifications to the conditions required for development of the physico-chemical processes involved are achieved through software control of the system parameters rather than physical changes to the hardware. In fact,

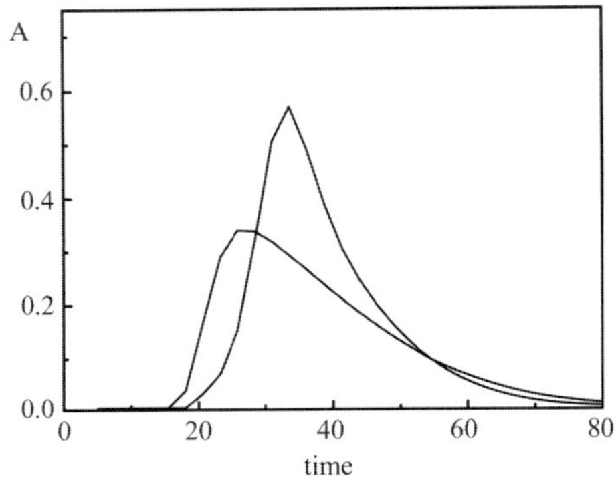

FIGURE 5.18 Influence of flow reversal on sample dispersion. The figure refers to a sample zone that passes through a 200 cm reactor (left) or through a 100 cm reactor under direct and then flow reversal conditions (right). For experimental details, see Ref. [92]. *Adapted from Talanta 79 (2009) 978, P.R. Fortes, M.A. Feres, M.K. Sasaki, E.R. Alves, E.A.G. Zagatto, J.A.V. Prior, J.L.M. Santos, J.L.F.C. Lima, Evidences of turbulent mixing in multi-pumping flow systems, with permission from Elsevier.*

the required solutions, volumes and timings (thus sample incubation time and analytical path length) are easily modified by a command from the computer. Each port of the selection valve has a specific purpose, and the combination of aspiration tubes for sample, standards, reagents, carrier/wash solutions, reactors, chambers and detectors around the valve is easily modified to suit a particular analysis.

Solid-phase analyte separation/concentration is efficiently implemented in the sequential injection analyser, without the need for a complex manifold, as stressed in a comprehensive review [93]. In most applications, a mini-column is inserted between the selecting valve and the detector and the solutions involved (sample, eluent, conditioning solutions) are aspirated towards the holding coil and, after flow reversal, pumped through the mini-column [94]. Gradient elution can also be directly exploited. More recently, the drawbacks associated with back-pressure effects arising from the column packing have been circumvented by taking advantage of monolithic columns [95]. Moreover, addition of a confluent stream immediately before detection to achieve derivatisation is also feasible, as discussed in Chapter 8.

Self-optimisation can be performed by taking advantage of the versatility inherent in the sequential injection system. The best analytical response can be attained by automatically varying the main parameters according to a previously defined optimisation algorithm [96].

A sequential injection system could, in principle, be modelled by considering the semi-volume ($S_{0.5}$) concept [97]. However, even in situations where the effects of reaction equilibria and kinetics are ignored, the profiles obtained using the dye approach are rather different from the experimental profiles [98]. Therefore, system dimensioning is better accomplished by varying the main parameters and applying optimisation algorithms such as simplex.

5.3.2. Sample Dispersion

The sequential injection concept relies on the single line configuration [99]; so knowledge of how the sample interacts with the reagents is of fundamental importance. As laminar flow dominates, Eq. 3.4 holds.

The original square input associated with sample insertion undergoes skewing and then tends towards a Gaussian profile; the recorded peaks are then indicative of the degree of mixing between the solutions. Peak skewing is due to the initial acceleration of the syringe piston, and is further amplified by Poiseuille flow as the mixed solution is forced by a pressure gradient in one direction and the fluid elements are retarded.

The random walk stochastic model [57,100] is useful for describing the mixing between the solutions involved as they are displaced through the analytical path (see also 3.1.3). This model assumes that molecules present in a fluid element do not all travel the same distance and therefore it is convenient to apply a fixed-step averaged displacement length (L). If a number of molecules are introduced into the system as a sample plug, they start together but travel in randomised steps and, as a consequence, a distribution profile tending towards Gaussian is observed. The spread of the sample zone, proportional to the peak width at half maximum, σ, and thus the degree of mixing with the neighbouring reagent or carrier solutions, is proportional to L and the square root of the number (n) of steps:

$$\sigma = \mathrm{L\,n}^{1/2} \tag{5.9}$$

The final displacement of any fluid element initially present in the sample plug is determined by the sum of displacements in all directions. Eq. 5.9 is then expanded to

$$\sigma^2 = \mathrm{L}^2\mathrm{n}_1 + \mathrm{L}^2\mathrm{n}_2 + \mathrm{L}^2\mathrm{n}_3\ldots = \sigma_1^2 + \sigma_2^2 + \sigma_3^2 \tag{5.10}$$

where 1, 2, 3, ... are the spatial coordinates.

As mixing depends on n, it is possible to get efficient sample/reagent homogenisation regardless of the distance involved. This can be attained by moving the sample zone forwards and backwards in several steps (n) of sufficient length (L). Mixing is then efficiently promoted due the combination of molecular diffusion and the flow imposed on the liquid that pushes/pulls the fluids. The potential of multiple flow reversals has seldom been exploited [101].

In fact, sample homogenisation in most sequential injection systems is based on zone penetration (overlap of adjacent zones), which depends on the sample and reagent volumes involved, manifold geometry and timing.

The degree of zone overlap, P, is defined [97] as

$$\mathrm{P} = 2\mathrm{W_p}/(\mathrm{W_s} + \mathrm{W_r}) \tag{5.11}$$

where w_P, w_S and w_r are the baseline widths of the formed product, sample and reagent zones, respectively.

The baseline widths related to the sample and reagent zones are normally estimated by sequentially replacing the sample and reagents by a coloured solution (dye approach [26]), so that chemical reactions are not involved. Eq. 5.11 provides only an estimate of P because the zone of the product formed depends on the chemical reactions involved, the related equilibria and kinetics and the molecular diffusion coefficients. So, its shape (thus the value of w_P) is dependent on the specific analysis involved [98]. A complete overlap between sample and reagent zones is attained for $P = 1$, and this can be achieved by increasing n in order to get a $(w_S + w_r)$ value equal to $2w_P$. This has seldom been exploited, however, because a higher value of n means a lower sampling rate.

Therefore, zone overlap in sequential injection systems is usually partial. In this situation, an iso-dispersion point, I_P, is observed (Fig. 5.19) where sample and reagent dispersions are identical. This point is not concentration dependent and, in the corresponding fluid element, the sample/reagent concentration ratio is the same as that observed before sample and reagent injections [97]. In the situation of complete reaction development, 1:1 analyte/reagent stoichiometry, reaction product monitoring and equivalent sample and reagent concentrations, the maximum signal is observed at the iso-dispersion point. As an excess of reagent is normally used, the position of the peak maximum is moved from I_D towards the centre of the sample zone until the reagent gradient leads to a condition where reagent concentration is no longer enough and the signal decreases.

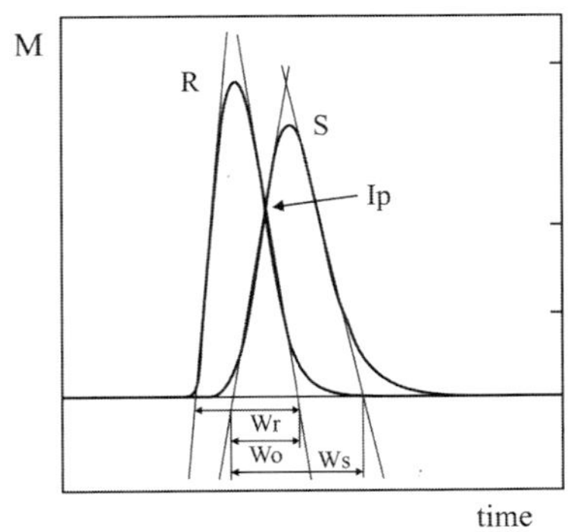

FIGURE 5.19 Recorded peaks for the reagent (R) and sample (S) solutions in a typical sequential injection system. M = monitored signal; I_P = iso-dispersion point; w_o, w_s, w_r, = baseline widths of the overlapped zone, sample zone and reagent zone respectively. Note that w_o is obtained by extrapolation. For details, see Ref. [97]. *Figure adapted with permission from "T. Guebeli, G.D. Christian, J. Ruzicka, Fundamentals of sinusoidal flow sequential injection spectrophotometry, Anal. Chem. 63 (1991) 2407". Copyright 1991, American Chemical Society.*

5.3.3. Controlling Sample Dispersion

To perform system optimisation, it is necessary to bear in mind how the dispersion and overlap processes depend on the sequentially injected volumes and total travelled distance. To this end, two distinct situations, involving large and limited sample dispersion, should be considered.

5.3.3.1. *Large Sample Dispersion*

Analogously to flow injection analysis, the inserted sample volume is a parameter of paramount importance in sequential injection analysis. A small zone volume permits high z values to be obtained and hence a high degree of sample dispersion. Very small sample and reagent aliquots are sampled, which promotes efficient interaction. However, a certain travelled distance is required to facilitate the dispersion process. To this end, a reaction coil with a suitable length and mixing efficiency is placed between the multi-position valve and the detector. Alternatively, the sample carrier stream can be inserted after sample/reagent mixing through a port in the multi-position valve, pushing the sample zone towards the central portion of the holding coil; after flow reversal, this path is travelled again, thus guaranteeing a suitable travelled distance.

5.3.3.2. *Limited Sample Dispersion*

Increasing the sample and reagent volumes, and thus the penetration zone volume, reduces the z value and lessens the dispersion of the sample portion associated with the analytical signal. The zone volume, however, cannot be increased indefinitely, as there is a compromise between this parameter and sample throughput. As a practical rule for system optimisation, it is adequate to keep the injected sample volume as large as possible (typically 3 $V_{0.5}$) and the reagent volume as at least 2 $V_{0.5}$. In this situation, the height of the recorded peak is less dependent on the inserted volumes and the system becomes more rugged. For still larger volumes, the influence of this parameter on sample dispersion is practically negligible, and this has recently been exploited in the development of the single interface flow system [102,103].

When two reagents (R_1 and R_2) are required to convert the analyte to a detectable species, and sensitivity is critical, a sandwich strategy [104] should be exploited, provided the order of reagent addition does not affect the rate of reaction. The sample plug is established between the R_i plugs and its volume should be as small as possible to permit a suitable R_1/sample/R_2 overlap to be attained but large enough to provide the required sensitivity. In analytical applications involving more than two reagent solutions, overlap is not easily attained and a better strategy is to include a mixing chamber in the manifold [105]. The selected volumes of sample and reagents are directed towards the chamber, where homogenisation occurs, and the resulting processed sample is directed towards the detector. This is the core feature of flow-batch analytical systems [106].

The addition of auxiliary components to perform specific steps, such as titration and microwave-assisted sample digestion, also influences sample dispersion, and is discussed in the subsequent chapters.

In summary, FIA and SIA systems have similar behaviour as a consequence of their common hydrodynamic foundation, but SIA systems include flow reversal, which modifies the intermixing of the injected plugs; the injected volumes in SIA systems should be dimensioned for each specific application as the $S_{0.5}$ values cannot be universally applied and should only be used as a general guide [98].

5.4. THE MULTI-COMMUTED FLOW ANALYSER

Addition, removal or exchange of manifold components, re-directing of flowing streams and establishment of tandem streams are easily and efficiently accomplished by multi-commutation.

5.4.1. Characteristics

Multi-commuted flow systems [107] have similar characteristics to the flow systems described above. They comprise several discrete commuting devices, *e.g.*, valves and pumps, strategically positioned in the manifold to provide enhanced versatility, and hence high performance [108], and even a novel type of flow system [109].

5.4.2. Sample Dispersion

Sample dispersion in multi-commuted flow systems is governed by the same parameters as in other flow systems. However, the manifold status can be modified by the operation of these discretely operated devices. Consequently, external timing of the various committed devices is an important aspect of controlled sample dispersion, as emphasised below.

5.4.3. Discretely Actuated Devices

Sliding bars and rotary valves were the first devices to be exploited for commutation. They operate in two positions (Fig. 6.10) and were initially used to insert a selected volume of sample into a carrier stream by means of a sampling loop. In the load position, the loop is in the sample-loading channel and the inlet and outlet ports of the carrier stream channel are directly connected. By switching (commuting) the sliding bar valve, the manifold status is modified, with the sampling loop now in the carrier stream channel and the inlet and outlet ports of the sample-loading channel directly connected (see also 6.2.2.3).

With two rotary valves or a sliding bar valve with two commutation sections [110], two solutions can be sampled and inserted, and this is the essence of flow systems exploiting, *e.g.*, merging zones or zone sampling (see 7.1.1 and 7.2.1).

Components other then the sampling loops or trespassing holes can also be attached to the movable portion of the valve (or commutater). The feasibility of reactor replacement was demonstrated in the spectrophotometric determination of nickel and cobalt in alloys relying on differential kinetic analysis [111]: the two different time intervals for reaction development were defined by inter-connecting two reactors with different lengths synchronously with two sample insertions. Resin mini-columns for analyte separation/concentration can be also attached to the valve in order to be displaced from the concentration to the eluent stream [112]. Analogously, the addition of components to the manifold and their further removal can be accomplished with a single valve [113]. The flow-through detector can also be attached to the central moving part of a commutater in order to be relocated between different manifold sites. This allows parallel or serial monitoring in assays requiring multi-site detection [114,115]. Electrochemical detectors can be re-conditioned when resting in the alternative site [116]. Displacement of a filtering device after passage of the flowing sample [117] is also feasible. The sample passes through a filtering device positioned in the analytical path, and the retained material is efficiently discarded by moving the filtering device to another site outside the analytical path, where a different solution passes through it under flow reversal conditions.

In spite of its favourable characteristics, relocation of manifold components has not been widely exploited, being restricted to some laboratories in Portugal, Spain and Brazil. This is perhaps due to a misunderstanding that the system is somewhat sophisticated and requires specific components. On the contrary, simple commercially available components (*e.g.*, the widely used six-port rotary valve) are used for accomplishing multi-commutation.

Valves (or commutaters) are also important for stream re-directing, which is essential in flow systems exploiting, *e.g.*, random access reagents [118], intermittent stream additions [119], sample stopping [120] and multiple flow reversals [121]. This latter strategy involves fast stream redirection towards two different sites along the analytical path. Moreover, a valve switched in a fast manner can act as a sampling device that selects the number of plugs of two different miscible solutions in order to establish a tandem stream (see also 3.1.1.2).

In most of the multi-commuted flow systems designed in the pumping mode, switching the valve directs the stream either towards the analytical path or towards its reservoir. Consequently, the liquid propelling device plus the re-directing valve can act as an intermittently operated pump. Discretely operated pumps can therefore be considered as commuting devices and are extensively exploited in multi-pumping and multi-syringe flow systems.

As the operation of multi-commuted flow systems depends on the operation of these discretely operated devices, timing is very important. Devices can be classified as either automated or automatic, depending on whether a concentration-oriented feedback mechanism is exploited.

5.4.3.1. *Automated Devices*

Automated or passive devices are those components that operate in a pre-defined manner. Inside a multi-commuted flow analyser incorporating automated devices, the conditions for sample handling are defined prior to sample introduction, according to the specific system design. Sample dispersion, timing, interactions with other solutions and monitoring are then maintained from one sample to another or, in other words, all samples are subjected to the same handling conditions. As reproducible timing is inherent to flow analysis [see Chapter 1], different time windows can be defined, allowing several time-dependent analytical steps to be efficiently accomplished.

Timing can also be set by means of an external device, usually the computer that operates the automated devices. Sample procedures requiring sample stopping [122] can be selected for illustrative purposes: the available time interval for reaction development is defined through actuation of a pump or valve during a pre-set time interval which defines the length of the STOP period.

Another example refers to flow systems with random reagent access [123], where the information needed for handling samples in a diverse manner, as well as the selection of the different reagent solutions needed for each specific determination, is provided via a keyboard.

5.4.3.2. *Devices with Feedback Mechanisms*

Operation of these devices is characterised by a concentration-oriented feedback mechanism, allowing "intelligent" flow analysers to be designed. Real-time alteration of the sample handling process dependent on a previous measurement expands the potential of the flow analyser, and the spectrophotometric determination of total nitrogen in Kjeldahl plant digests [124] is a good example. The digests presented high and variable acidity and the method was susceptible to pH variations; hence, every sample needed an individual pH adjustment. This individual sample conditioning was successfully performed in a flow-batch system.

Sample acidities were estimated and the results used as the basis for real-time sample conditioning through reagent additions, pH adjustment and/or sample dilution. Other applications involving real-time modifications of the system operation relying on feedback mechanisms refer to the need for, *e.g.*, multiple STOP periods [125], alterations in sample volumetric fraction [126] or cascade dilutions [127].

The number of discretely operated devices reflects the degree of system automation. As a rule, increasing the degree of automation improves system versatility but makes the manifold more complex. For optimum system design, a compromise between system versatility and complexity is therefore required.

5.4.2.3. *Trends*

The concept of multi-commutation has been recognised as relevant in flow analysis, as confirmed by the increasing number of referenced quotations in the literature. This recognition will certainly be increased by incorporating the concept in commercially available flow analysers.

5.5. OTHER FLOW ANALYSERS

Modern flow analysers with specific characteristics have also been proposed, as discussed below. They generally comprise discretely operated devices in the manifold, thus retaining the main characteristics of multi-commuted flow systems, and they often act as "smart" flow systems. This reinforces the belief that multi-commutation does not refer to a single generic mode of flow analyser but to different attributes of existing analysers.

These flow analysers are described concisely in a single section (5.5) because their development and applications are still evolving in individual laboratories.

5.5.1. The Mono-segmented Flow Analyser

This mode of flow analysis was proposed [128] as a means of easily and efficiently achieving extended sample handling times without excessive sample dispersion. The sample volume is inserted into an unsegmented carrier stream, and two air plugs are added at its ends in order to minimise sample broadening, and hence axial dispersion. The beneficial effects arising from the presence of air plugs at both ends of the sample bolus were already emphasised in 1972, in relation to a chemiluminometric determination of low concentrations of Cr(III) [129].

The approach is particularly attractive for implementing analytical procedures where sensitivity is critical and the chemical reactions involved are relatively slow [130–132], and can be also exploited for accomplishing differential kinetic analysis [133]. As the axial dispersion experienced by the sample is minimal, even for long analytical paths, sample integrity is preserved. Several samples can be simultaneously handled inside the analytical path. Analogously to segmented flow analysis, the air phase is generally removed immediately before sample detection by means of a gas-permeable membrane or a de-bubbler.

This innovation is also useful for attaining low sample dispersion with low injected sample volumes because the air phase hinders axial dispersion. This feature was demonstrated in the potentiometric determination of *d*-glucose in undiluted whole blood [134].

Only a few air plugs are present in the manifold at any one time; therefore, some processes inherent in unsegmented flow systems such as sample stopping, flow reversal and stream splitting can be exploited. It is also possible to precisely define the instant position of the air plugs, allowing, *e.g.*, the easy implementation of the bubble gating approach [5].

Gaseous samples can also be handled and the spectrophotometric determination of O_2 and NO_2 in the headspace of packages [135] is a good example of this innovation. The analyte was transported from the gaseous sample towards the liquid segment containing the colour-forming reagent via the thin film established on the tube wall. Sensitivity was enhanced by exploiting flow reversal and/or sample stopping.

The flow system is simplified with flame atomic absorption spectrometric detection [136] because an air plug is not placed at the front of the sample in order not to disturb the steady state of the flame. Consequently, the aqueous sample is inserted into the unsegmented carrier stream with only one air plug positioned after it. Tailing effects are therefore minimised and the sampling rate is significantly improved relative to ordinary flow injection systems with flame atomic absorption spectrometric detection. Moreover, removal of the gaseous phase is not needed.

The use of mono-segmented flow systems for improving liquid–liquid extractions and flow titrations is discussed in Chapter 8.

As stressed in Section 2.3, this innovation is also referred to as *segmental flow injection analysis* [137]. This broader term includes mono-segmented flow analysers and some specially designed flow systems involving, *e.g.*, injection of gaseous [138] or segmented [131] samples into a continuous flowing unsegmented carrier stream. Conversely, aqueous solutions can be inserted into a segmented stream. With this strategy, fluoride was potentiometrically determined in 0.1 mL of natural water samples at a sampling rate as high as 720 h^{-1}, with the r.s.d. $<1\%$ [22]. These examples emphasise that the borderline between segmented and unsegmented flow analysis is not distinct. The potential, limitations and applications of segmental flow injection systems have recently been reviewed [131].

5.5.2. The Discontinuous Flow Analyser

In flow analysis, the solutions are generally propelled by peristaltic or piston pumps operating in a discharge mode. An alternative arrangement involving one discharge and one suction pump is also possible and the difference between the suction and discharge flow rates allows the sample to be aspirated into the carrier stream. This feature has been referred to as *differential pumping*.

By maintaining the sample/reagent flow rates ratio, constant flow rates are attained, and this is inherent to, *e.g.*, stepper motors driving syringe pumps (see also 6.1.2). A conceptual advance on differential pumping is to vary the flow rates of one or both pumps in a reproducible manner and this is the essence of the discontinuous flow analyser [139].

The innovation is a powerful tool for controlling the flow rates and acceleration of the sample and reagent streams, especially in relation to the use of syringe pumps operated through a programmable, mechanically driven cam. The cyclic temporal variations in concentrations (improperly named *flow profiles*) for both suction and discharge pumping are attained in such a way that, at a given time, only the sample, the reagent or a pre-defined combination of both is conveyed via a mixing chamber towards the flow-through detector. Different cam shapes allow a wide variety of flow-based analytical procedures, including single or multiple reagent additions and automated flow titrations, to be efficiently implemented. In the original system [139], home-made syringe pumps driven by interchangeable cams linked to a common shaft were used, and the potential of the discontinuous flow analyser was demonstrated in relation to the argentometric titration of chloride in natural waters. Two years later, an efficient discontinuous flow analyser was designed for ultrafast titrations aimed at the determination of the titratable acidity of wines and total acidity of vinegars [140]. Instead of two pumps, a reciprocating pump was used. The feasibility of spectrophotometric determinations requiring reagent additions and spectrophotometric titrations was reported in relation to zinc and acidity determinations in electrolytic solutions [141]. Furthermore, a titrimetric procedure exploiting concentration gradients was proposed [142]: the sample travelled at a constant flow rate, being inserted by differential pumping into the carrier stream where a titrant concentration gradient had already been formed. In this way, ultrafast titrations (one titration every 25 s) were accomplished, yielding results comparable with classical titrations. The combination of discontinuous flow analysis with flame atomic absorption spectrometry was also proposed [143]: the abilities of the analyser to provide automatic calibration in approximately 30 s using a single solution, rapid automated standard additions and a sample bracketing procedure were emphasised. Another application of discontinuous flow analysis was reported in relation to the determination of sulphuric acid in metallurgical process streams by spectrophotometric flow titration [144]. This mode of flow analysis has been reviewed [145] with an emphasis on flow titrations, standard addition methods and atomic absorption spectrometric detection applied to industrial process control. Another application was the determination of silver in photographic emulsions, involving complexation of the silver bromide with cyanide followed by titration with sulphide [146].

During the development of discontinuous flow analysis in the 1990s, the ability of the analyser to insert different solutions as controlled and programmed pulses, thus establishing a tandem stream, was realised. Moreover, replacement of the syringe pump by a solenoid pump opened up the possibility of exploiting pulsed flows [147]. In spite of its favourable characteristics, discontinuous flow analysers have not evolved during the 2000s. This can perhaps be explained by recalling that tandem streams and pulsed flows have been extensively investigated in multi-commuted, multi-syringe and multi-pumping flow systems (see 2.5.1 and 2.5.2).

5.5.3. The Lab-on-Valve and Lab-on-Chip Flow Analysers

The Lab-on-Valve flow system resembles the sequential injection analyser but the main components are assembled on a multi-position valve. Exploitation of optical fibres permits spectrophotometric measurements to be performed very close to the main reaction coil. The compact geometry of the system makes it attractive for, *e.g.*, *in situ* assays. For more details of this system, see 2.4.2.

The Lab-on-Chip flow system comprises several miniaturised components, such as imprinted manifolds and electro-osmotic pumps, integrated within a single device. It is within the family of microfluidic techniques; further discussion is outside of the scope of this monograph.

5.6. DESCRIBING A FLOW ANALYSER

Terminology related to analytical methods based on flowing media, as well as terms describing the flow system and its components, is well documented [148,149], and a particular reference to process analytical chemistry is also available [150]. However, most of the proposed flow-based procedures and/or related instrumentation are not *fully* described in the literature or in commercial material. As a consequence, harmonisation of methods has not yet been achieved and adaptation of a given flow-based procedure to a different analyser is generally cumbersome. Some guidelines are therefore needed for the appropriate description of a flow analyser, related system design and methodological implementation. In this context, the following aspects should be considered [149].

5.6.1. Establishment of the Flowing Streams

For characterising the flowing streams, it is necessary to specify the flow patterns, the flow rates (and how to implement them) and the composition of the flowing solutions.

Information about flow patterns is of the utmost importance, particularly specifying if segmented or unsegmented flow is involved. Without this information, unsegmented flow is implied. Specifically for segmented flow systems, the nature of the segmenter should be highlighted or else air-segmentation is implied.

The tandem stream [118] is a unique stream initially comprising several neighbouring slugs of miscible solutions that undergo fast mixing while flowing through the analytical path (see 3.1.1.2). In describing flow systems with tandem streams, the way in which the involved solutions interact should be clearly indicated.

Flow rates should always be given. For specific applications involving intermittent streams [151], sinusoidal flows [152], flow reversal [153] or variable flow rates, *e.g.*, slow during sample transportation towards detection and fast after achievement of the analytical signal [80], the temporal variations in flow rates (see Fig. 3.4) should also be indicated. This is also true for sample stopping [122]. It is advisable to indicate less critical flow rates, *e.g.*, the sample aspiration rate in loop-based injections or the outlet stream flow rate in a de-bubbler unit.

It is important to describe how fluids are propelled, as well as the main liquid drivers such as peristaltic pump [1], piston pump [89], reciprocating pump [154], gas propeller [25] or other strategies for this task, *e.g.*, kinematic focussing [155] or gravity [156]. The peristaltic pump is by far the most common device for propelling fluids in flow analysis; therefore, the absence of any information assumes the use of such a pump. Characteristics of the pump tubing are mentioned in most publications, but this information is not needed unless there is an incompatibility with the propelled fluids.

The composition of the flowing streams is usually strictly dependent on the composition and flow rates of the individual pumped solutions and manifold design (see also Eqs. 3.10 and 3.11); hence, it is important to describe the preparation of the sample, carrier and reagent solutions.

5.6.2. Sample Introduction

How the sample is introduced and the related device should always be reported. Time-based introduction is essential in air-segmented and some unsegmented flow systems [89,107,157]: samples are aspirated into the manifold at a fixed flow rate during a pre-set specific period of time. In flow systems with loop-based injection [158], the samples are directed to fill a sampling loop that is further intercalated into the sample carrier stream. The sample selected volume should always be given, regardless of the use of time-based or loop-based injection. If pertinent, other related approaches such as simultaneous, sequential, tandem, hydrodynamic or nested injection (see 6.2.2) should be also described. The use of, *e.g.*, reagent injection or an auto-sampler should also be mentioned.

5.6.3. Manifold Characteristics

The type of tubing used for constructing the manifold should always be specified but less emphasis should be given to manifold components outside of the analytical path. There are different types of reactor (see 6.2.3) and they should be specified. The coiled reactor is by far the most common, and no further information means that this kind of reactor is used. Length and inner diameter of the tubing, winding diameter of the coiled reactor and the presence of different accessories (mixing chambers, solid-phase mini-columns, immobilised enzymes and connectors) in the analytical path should be reported.

5.6.4. Sample Processing

It is very important to emphasise the analytical steps to which the sample is subjected inside the flow manifold. The mean sample residence time should be given and in some special situations, *e.g.*, when a catalytic method is used, the sample residence time inside specific portions of the manifold should be stated.

The extent of overall sample dispersion should be provided, for which purpose the dispersion coefficient [159] has often been used. A more general index to describe the composition of any fluid element of the system in any part of the manifold at any time after sample introduction is the volumetric fraction X [160], and its main characteristics are discussed in 3.1.2.4.

Another relevant parameter is the degree of completion of the physico-chemical processes involved. This parameter is particularly important in relation to applications based on relatively slow chemical reactions but is often omitted. Its experimental determination usually involves two measurements, one carried out under actual conditions and the other after the sample has been processed in a similar way but without timing constraints, *i.e.*, the reaction has gone to completion. For this purpose, stopping of the pumping device is used, as shown in Fig. 5.20. This figure shows the recorded peaks after 3-fold injections of a typical sample; next,

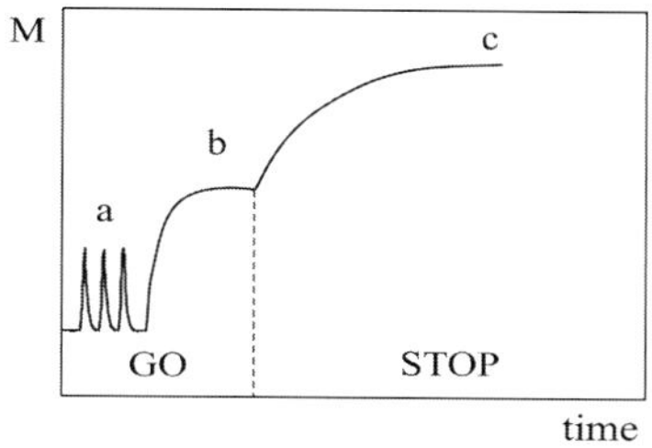

FIGURE 5.20 **Recorder tracing from an experiment carried out to quantify sample dispersion and the degree of completion of the physico-chemical process(es) involved.** a = recorded peaks after 3-fold sample injection; b = traces obtained by placing the sample in the situation of infinite volume; c = trace obtained after stopping the pump. For details, see text.

the sample carrier stream is replaced by the sample (infinite volume situation) and the recorder trace shows an asymptotic tendency towards a plateau that reflects the signal that could be in principle attained without sample dispersion; next, the pump is switched OFF and the reaction chemistry proceeds towards completion. The related plateau reflects the situation without sample dispersion and without any time constraints. It should be stressed that after switching the pump OFF, the recorded trace could be horizontal, indicating that the involved processes had already reached completion during the mean sample residence time in the manifold. A third possibility is to have an asymptotic tendency towards baseline restoration. This is an indication that an unstable chemical species was monitored. Better sensitivity could then be attained by reducing the sample residence time in the manifold by, *e.g.*, using a higher overall flow rate.

Worked example #4. A confluence flow injection system with a very low confluent stream flow rate is designed for the spectrophotometric determination of phosphate in plant digests. The linearity of the analytical calibration graph is good and the recorded absorbance corresponding to a 100.0 mg L^{-1} P (as phosphate) standard solution is 0.21. Replacing the sample carrier stream by this standard solution (sample infinite volume) yields a steady state situation and the related absorbance is 0.68. The pump is then turned off and an asymptotic increase in absorbance towards 0.95 is observed; determine the sensitivity improvement that in principle could be attained simply by increasing the sample volume and the mean sample residence time in the analytical path.

Answer: As linearity between measurement and concentration holds, the sample volumetric fraction X can be determined (Eq. 3.17) as 0.21/0.68 = 0.31. Analogously, the degree of reaction completion can be calculated as 0.68/0.95 = 0.72. The maximum increase in volumetric fraction, providing that the flow rate of the confluent stream addition could be neglected, is

$1/X = 1/0.31 = 3.24$. Sensitivity could then undergo a 324% enhancement simply by increasing the sample volume introduced. The maximum increase in reaction development is $1/0.72 = 1.40$; therefore, a 40% signal enhancement could be attained by letting the chemical reactions involved go to completion. This means a theoretical sensitivity enhancement of $3.24 \times 1.40 = 4.5$. From a practical point of view, and in order to avoid too large an injected volume and/or too long a sample resident time that could impair sampling rate, a 3-fold sensitivity enhancement would be realistic.

In flow systems comprising several active devices, *e.g.*, computer-controlled liquid driver and valves, it is mandatory to report the sequence of device operations related to one analytical cycle, including the exact timings and a full description of the functions of these devices.

When specific processes, *e.g.*, analyte separation/concentration are involved, the characteristics of the device needed for this task, as well its operating conditions and performance, should be specified.

5.6.5. Detection

A complete description of the flow-through detector, including operating conditions, should be provided. In *spectrophotometry,* the flow cell is usually distinct from the detector; therefore, its description, including light optical fibres (if relevant), is required. It is important to specify the inner volume and optical path length of the flow cell and, for specific applications, *e.g.*, multi-site detection [114], information about its illuminated portion is also relevant. In luminometric procedures, the shape of the flow cell as well as its arrangement in relation to the detector should also be reported.

In most applications, the flow-cell inner volume and/or detector response time are not critical. These parameters should therefore only be specified if they influence the recorded peak shape.

The shape and quality of the recorded signal reflect the performance of a flow analyser, and this aspect should be reported. Regarding interpretation of the analytical signal, it is essential to specify the measurement basis, *e.g.*, steady state, peak height, peak area or peak width at a defined height (see also 3.3). Relevant characteristics that should always be specified are the signal processing/recording device and the data treatment. Detailed descriptions of software for data acquisition (including signal filtering), feedback mechanisms and calculation algorithms are recommended if they are original, otherwise proper citation is preferred.

5.6.6. Performance of the Flow System

Besides the above-mentioned essential procedural aspects, it is also important to provide the main figures of merit for a proper description of a flow-based procedure. For didactic purposes, the figures of merit are herein classified into two groups depending on whether they are general or more specifically related to flow-based analytical procedures.

5.5.6.1. *General Figures of Merit*

Complete information on these figures is essential for any novel analytical procedure, regardless if it is implemented in a flow system or not.

ACCURACY

This expresses the agreement between estimated (or measured) value and the true or reference value [161,162]. Results obtained with flow-based analytical procedure should agree with the certified values (within 95% confidence intervals) associated with standard reference materials, and this is the ideal procedure for accuracy assessment. When certified values are not available, other values, *e.g.*, from intercomparison exercises involving several laboratories are acceptable. If such values are not available, results obtained with official or recommended analytical procedures can be used. If these procedures are not available or if official (or recommended) analytical procedures for the specific analyte do not exist, the results obtained with any properly validated procedure are a minimum requirement.

In the situation of special analytes for which there is no available analytical procedure, an indication of accuracy can be determined by spiking a known amount of the analyte to the sample and estimating the recovery.

Regardless of the protocol for accuracy assessment, several samples covering the entire dynamic range, particularly at the lower end of the range, should always be used [163].

PRECISION

This expresses the agreement between replicate results obtained for a single sample under fixed conditions [164]. Precision can be evaluated in terms of repeatability and/or reproducibility.

The repeatability indicates the spread of results, usually expressed in relative or absolute estimates of the standard deviations, after n-fold replicated processing of the sample. The higher the n value, the better the estimate of the standard deviation. Alternatively, the mean confidence interval can be used.

The reproducibility is an analogous figure, but refers to n-fold analyses of typical samples, ideally processed in different flow analysers operated by different people, at different times (hours, days). In addition to the information on analytical precision, reproducibility is also indicative of system ruggedness [165].

SELECTIVITY

This expresses the susceptibility of a given determination to the presence of potential interfering species in the sample [164,166]. The term *specificity* has been also used for this purpose, but this is strongly discouraged by IUPAC [166]. Selectivity data are usually gathered by comparing the results obtained by analysing synthetic samples prepared with and without the potential interfering species. Alternatively, selectivity can be evaluated as the ratio between sensitivities related to the analyte and to the potential interfering chemical species [164].

SENSITIVITY

This expresses the variation in analytical signal for a unit variation in analyte concentration. This figure of merit matches the slope of a linear analytical calibration graph. In situations of non-linearity, an algorithm for curve linearisation should be used [163]. Alternatively, the analytical sensitivity can be reported by expressing the lowest concentration difference that can be distinguished over the linear dynamic range. It is calculated as the relationship between sensitivity (or slope of the analytical curve) and instrumental noise, the latter being defined as the standard deviation of the blank [167].

DETECTION LIMIT

This expresses the lowest analyte concentration (or mass) that can be detected but not necessarily quantified [168]. The detection limit corresponds then to the lowest analyte concentration that provides a response different to that related to the blank value. One has to be careful about which formula is used; hence, different formulae with specific features [167,169] should be critically discussed. As a rule, the detection limit can be estimated as three times the blank signal. Consequently, a better detection limit is achieved with a lower signal-to-noise ratio.

DYNAMIC RANGE

This expresses the interval of analyte masses or concentrations over which the analytical calibration graph is linear [163]. The dynamic range then refers to the limits of the analyte masses or concentrations determinable with the proposed procedure. The lower limit, often called *the determination limit*, expresses the lowest analyte concentration (or mass) determinable with an uncertainty less than 10% [161]. This figure can be estimated by different approaches, and a good practice is to consider it as five times the blank signal. The upper limit corresponds to the highest allowed analyte concentration that can be properly estimated without limitations in detection and/or linearity of the analytical calibration graph. It is advisable to report the lower and higher concentration limits as well as the criteria for calculating them. In this regard, information on how the repeatability deteriorates at the vicinities of the limits of the analytical calibration graph is worthwhile.

It should be stressed that the working standard solutions should be prepared in order to cover the dynamic concentration range at regular intervals. So, for instance, if the dynamic range covers the 1–9 mg L^{-1} concentration range and 5 working standards are to be prepared, their concentrations should be 1.00, 3.00, 5.00, 7.00 and 9.00 mg L^{-1}.

5.6.6.2. *Figures of Merit Specific to Flow-Based Procedures*

The figures included in this group are those more specifically related to flow-based analytical procedures.

CARRYOVER

This expresses the contribution of one assay to the next in terms of analytical signal. It should be minimised and/or compensated in order to maintain analytical accuracy. In flow analysis, it is often the main limiting factor on the sampling rate, increasing reagent consumption, especially in procedures involving a continuously flowing reagent stream. Stating the degree of carryover is therefore recommended when specifying the sampling rate.

The overlap between successive sample zones, the presence of sample residues inside the analytical path and detector memory effects are important sources of carryover, the latter generally being of minor concern in flow-based *spectrophotometry*.

Sample overlap is a result of the continuous dispersion process occurring during sample transport towards the detector. The effect is minimised by improving the design of the flow manifold and/or by reducing the sampling rate.

Sample residues remaining in the analytical path are generally associated with the sampling step, especially when a sampler is available and system washing is not quantitative. Portions of the sample are then retained, usually inside the sampling probe, thereby influencing the next assay. Moreover, carryover can manifest itself in poorly rinsed systems where sample portions remain inside specific sites of the manifold [170]. Two kinds of carryover have been reported: within-method and between-method carryover.

The within-method carryover is inherent to single analyte determinations and can be efficiently compensated by applying Eq. 5.12 [7,10,171,172]:

$$R'_n = R_n - K_C R_{n-1} \tag{5.12}$$

where R_n and R_{n-1} are the detector responses related to n^{th} and $(n-1)^{th}$ samples; R'_n is the carryover-corrected response; and K_C is the carryover coefficient.

K_C is experimentally determined by processing two sample solutions with known analyte concentrations or a sample plus the blank [173]. Alternatively, only one sample solution can be used [174], and the difference between the first measurement and that obtained after several successive sample processing (Δh) is compared (Fig. 5.21).

The potential of Eq. 5.12 for compensating within-method carryover in flow injection analysis was recently demonstrated in a turbidimetric procedure for the determination of potassium in fertilisers [173]. The sample throughput underwent a 4-fold increase simply by setting a K_C value of 7%.

The between-method carryover is inherent to multi-parametric determinations and has been discussed mainly in relation to clinical analysis [175,176]. This kind of carryover is generally caused by partial removal of the reagents required for the previous determination. It can affect accuracy in, *e.g.*, sequential determinations, including speciation, particularly in connection with random reagent access, a universally accepted strategy for multi-parametric analysis [172]. In flow analysis, random reagent access has usually been accomplished by exchanging specific reagents after every analytical cycle. To this end, merging zones [177], sandwich techniques [178] or tandem streams [118] can also be exploited. This kind of carryover as a potential source of inaccuracy in flow analysis has not been studied in detail.

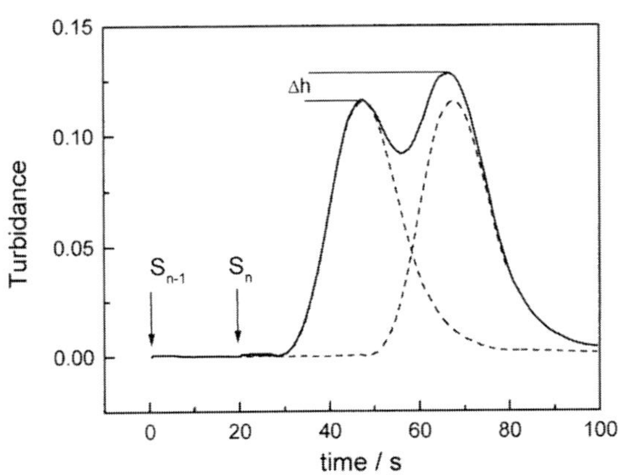

FIGURE 5.21 Recorded signal as influenced by the previous one. The overlapped peaks refer to two consecutive injections of the same sample. S_{n-1} and S_n = injection instants (S_n injected 20 s after S_{n-1} previous injection); full and traced lines = actual and ideal (no carryover) monitored signals; Δh = difference in recorded peaks due to carryover. *Adapted from Anal. Chim. Acta 438 (2001) 3, S. Vicente, E.P. Borges, B.F. Reis, E.A.G. Zagatto, Exploitation of tandem streams for carry-over compensation in flow analysis. I. Turbidimetric determination of potassium in fertilizers, with permission from Elsevier (Ref. [173]).*

SAMPLING RATE

Sampling or measurement rate (or frequency) expresses the number of samples introduced into the flow analyser per unit time. It should be reported in association with a specified carryover level. A lack of information implies that carryover is not measurable. It is recommended to specify the consumption of sample and/or reagents when discussing sampling rate, especially when less available or potentially environmentally hazardous reagents are used.

ROBUSTNESS

This expresses the susceptibility of the analytical procedure to variations in parameters such as flow rates, temperature, reagent concentrations and sample volume [165]. System robustness can be discussed from two perspectives. First, the dependence of the analytical signals on slight variations of relevant system parameters. Second, the dependence on instrumental stability, which can manifest itself as changes in the baseline and/or the analytical signal. The stability of the analytical curve over time and maintenance requirements should also be stated.

PORTABILITY

Another feature that can be useful in describing a flow system is its portability and its consequent suitability for, e.g., *in situ, in vivo* or on-site analyses, as a process analyser or for on-line monitoring [149].

References

[1] L.T. Skeggs Jr., Automatic method for colorimetric analysis, Am. J. Clin. Pathol 28 (1957) 311.

[2] L. Snyder, J. Levine, R. Stoy, A. Conetta, Automated chemical analysis: update on continuous-flow approach, Anal. Chem. 48 (1976) 943A.

[3] L.T. Skeggs, Persistence… and prayer: from the artificial kidney to the autoanalyzer, Clin. Chem. 46 (2000) 1425.

[4] E.A.G. Zagatto, P.J. Worsfold, Flow analysis: overview, in: P.J. Worsfold, A. Townshend, C.F. Poole (Eds.), Encyclopedia of Analytical Science, second ed., vol. 3, Elsevier, Oxford, 2005, p. 24.

[5] R.L. Habig, B.W. Schlein, L. Walters, R.E. Thiers, A bubble-gating flow cell for continuous-flow analysis, Clin. Chem. 15 (1969) 1045.

[6] J. Ruzicka, E.H. Hansen, Flow injection analyses. Part I. A new concept of fast continuous flow analyses, Anal. Chim. Acta 78 (1975) 145.

[7] G. Liu, B.J. van Wie, D. Leatzow, B. Weyrauch, T. Tiffany, Experimental design modeling of carryover to optimize air-segmented continuous flow analysis, Anal. Chim. Acta 408 (2000) 21.

[8] R.E. Thiers, K.M. Oglesby, Precision accuracy + inherent errors of automatic continuous flow methods, Clin. Chem. 10 (1964) 246.

[9] R.E. Thiers, J. Bryan, K. Oglesby, A multichannel continuous-flow analyzer, Clin. Chem. 12 (1966) 120.

[10] R.E. Thiers, R.R. Cole, W.J. Kirsch, Kinetic parameters of continuous flow analysis, Clin. Chem. 13 (1967) 451.
[11] M.A. Evenson, G.P. Hicks, R.E. Thiers, Peak characteristics and computers in continuous flow analysis, Clin. Chem. 16 (1970) 606.
[12] R.E. Thiers, A.H. Reed, K. Delander, Origin of lag phase of continuous-flow analysis curves, Clin. Chem. 17 (1971) 42.
[13] L.R. Snyder, H.J. Adler, Dispersion in segmented flow through glass tubing in continuous-flow analysis: the ideal model, Anal. Chem. 48 (1976) 1017.
[14] L.R. Snyder, H.J. Adler, Dispersion in segmented flow through glass tubing in continuous-flow analysis: The non ideal model, Anal. Chem. 48 (1976) 1022.
[15] W.J. Duncan, A.S. Thom, A.D. Young, Mechanics of Fluids, second ed., Edward Arnold, London, 1970.
[16] H. Janshekar, J.R. Mor, Automated determination of sugar in molasses, Anal. Chim. Acta 90 (1977) 201.
[17] R.D. Begg, Dynamics of continuous segmented flow analysis, Anal. Chem. 43 (1971) 854.
[18] C.E. Werkhoven-Goewie, U.A.T. Brinkman, R.W. Frei, The use of solvent segmentation in continuous-flow systems, and fluorescence labeling by derivatization, Anal. Chim. Acta 114 (1980) 147.
[19] M. de la Guardia, J. Ruzicka, Towards environmentally conscientious analytical-chemistry through miniaturization, containment and reagent replacement, Analyst 120 (1995) 17N.
[20] M. Cassaday, H. Diebler, R. Herron, M. Pelavin, D. Svenjak, D. Vlastelica, Capsule chemistry technology for high-speed clinical chemistry analyses, Clin. Chem. 31 (1985) 1453.
[21] P. Wilding, J. Pfahler, H.H. Bau, J.N. Zemel, L.J. Kricka, Manipulation and flow of biological fluids in straight channels micromachined in silicon, Clin. Chem. 40 (1994) 43.
[22] J.A. Borzitsky, A.V. Dvinin, O.M. Petrukhin, Y.I. Urosov, Segmental flow injection with ion-selective electrodes for the determination of fluoride in waters, Anal. Chim. Acta 274 (1993) 125.
[23] S. Vicente, E.A.G. Zagatto, P.C.A.G. Pinto, M.L.M.E.S. Saraiva, J.L.F.C. Lima, E.P. Borges, Exploiting gas diffusion for non-invasive sampling in flow analysis: determination of ethanol in alcoholic beverages, An. Acad. Bras. Cienc. 78 (2006) 23.
[24] J.P. Smith, V. Smith-Hinson, Flow injection analysis: quietly pushing ahead, Anal. Chem. 74 (2002) 385.
[25] K.K. Stewart, G.R. Beecher, P.E. Hare, Rapid analysis of discrete samples: the use of nonsegmented, continuous flow, Anal. Biochem. 70 (1976) 167.
[26] J. Ruzicka, E.H. Hansen, Flow Injection Analysis, John Wiley & Sons, New York, 1981.
[27] Z.-L. Fang, Nonequilibrated sample manipulation: the essence of flow-injection analysis, Microchem. J 45 (1992) 137.
[28] G. Taylor, Dispersion of soluble matter in solvent flowing slowly through a tube, Proc. Roy. Soc. London, Series A (Math. Phys. Sci.) A219 (1953) 186.
[29] G.I. Taylor, Diffusion and mass transport in tubes, Proc. Phys. Soc. London 67 (1954) 857.
[30] V. Ananthakrishnan, W.N. Gill, A.J. Barduhn, Laminar dispersion in capillaries. 1. Mathematical analysis, AIChE J 11 (1965) 1063.
[31] W.N. Gill, V. Ananthakrishnan, Laminar dispersion in capillaries. 2. Effect of inlet boundary conditions and Turner type of system capacitance, AIChE J 12 (1966) 906.
[32] N.S. Reejhsinghani, W.N. Gill, A.J. Barduhn, Laminar dispersion in capillaries. 3. Experiments in horizontal tubes including observations on natural convection effects, AIChE J 12 (1966) 916.

[33] W.N. Gill, V. Ananthakrishnan, Laminar dispersion in capillaries. 4. Slug stimulus, AIChE J 13 (1967) 801.
[34] M. Valcarcel, M.D. Luque de Castro, Flow Injection Analysis. Principles and Applications, Ellis Horwood, Chichester, 1987, p. 60.
[35] J.T. Vanderslice, A.G. Rosenfeld, G.H. Beecher, Laminar-flow bolus shapes in flow injection analysis, Anal. Chim. Acta 179 (1986) 119.
[36] H. Bate, S. Rowlands, J.A. Sirs, H.W. Thomas, The dispersion of diffusible ions in fluid flow through a cylindrical tube, J. Phys. D: Appl. Phys. 2 (1969) 1447.
[37] H. Bate, S. Rowlands, J.A. Sirs, Influence of diffusion on dispersion of indicators in blood flow, J. Appl. Physiol. 34 (1973) 866.
[38] R. Tijssen, Axial dispersion and flow phenomena in helically coiled tubular reactors for flow analysis and chromatography, Anal. Chim. Acta 114 (1980) 71.
[39] J.M. Reijn, W.E. van der Linden, H. Poppe, Dispersion in open tubes and tubes packed with large glass beads. The single bead string reactor, Anal. Chim. Acta 123 (1981) 229.
[40] J.T. Vanderslice, K.K. Stewart, A.G. Rosenfeld, D.J. Higgs, Laminar dispersion in flow-injection analysis, Talanta 28 (1981) 11.
[41] M.A. Gomes-Neto, M.D. Luque de Castro, A. Martin, M. Valcarcel, Prediction of the behaviour of a single flow-injection manifold, Talanta 32 (1985) 319.
[42] P.L. Kempster, H.R. van Vliet, J.F. van Staden, Prediction of FIA peak width for a flow-injection manifold with spectrophotometric or ICP detection, Talanta 36 (1989) 969.
[43] D. Betteridge, W.C. Cheng, E.L. Dagless, P. David, T.B. Goad, D.R. Deans, D.A. Newton, T.B. Pierce, An automated viscosimeter based on high-precision flow injection analysis. Part II. Measurement of viscosity and diffusion coefficients, Analyst 108 (1983) 17.
[44] J.F. Tyson, J.M.H. Appleton, A.B. Idris, Flow injection calibration methods for atomic absorption spectrometry, Anal. Chim. Acta 145 (1983) 159.
[45] K. Hofmann, I. Halasz, Mass transfer in ideal and geometrically deformed open tubes. 1. Ideal and coiled tubes with circular cross-section, J. Chromatogr. 173 (1979) 211.
[46] D.F. Leclerc, C.J. Smith, E.C. Toren, Axial dispersion in coiled tubular reactors: the effect of curvature at low Dean numbers, Anal. Chim. Acta 194 (1987) 109.
[47] U.D. Neue, H. Engelhardt, Reaction detector with three-dimensional coiled open tubes in HPLC, Chromatographia 15 (1982) 403.
[48] E.D. Katz, R.P.W. Scott, Low-dispersion connecting tubes for liquid chromatography systems, J. Chromatogr. 268 (1983) 169.
[49] M.A. Curtis, G.J. Shahwan, A high-efficiency open tubular post column HPLC reactor using PTFE tubing in a modified serpentine geometry, LC GC-Mag, Sep. Sci. 6 (1988) 158.
[50] R.D. Hull, R.E. Malick, J.G. Dorsey, Dispersion phenomena in flow-injection systems, Anal. Chim. Acta 267 (1992) 1.
[51] S.D. Kolev, E. Pungor, Description of an axially-dispersed plug flow model for the flow pattern in elements of fluid systems, Anal. Chim. Acta 185 (1986) 315.
[52] S.D. Kolev, E. Pungor, Influence of the main parameters of single-line flow-injection systems without chemical reaction on the output signal: Part 1. Dispersion characteristics of the main sections of the flow system, Anal. Chim. Acta 208 (1988) 117.
[53] S.D. Kolev, E. Pungor, Influence of the main parameters of single-line flow-injection systems without chemical reaction on the output signal: part 2. Simulation of the mathematical models, Anal. Chim. Acta 208 (1988) 133.
[54] S.D. Kolev, Mathematical modeling of flow-injection systems, Anal. Chim. Acta 308 (1995) 36.
[55] H. Wada, S. Hiraoka, A. Yuchi, G. Nakagawa, Sample dispersion with chemical reaction in a flow injection system, Anal. Chim. Acta 179 (1986) 181.

[56] C.L.M. Stults, A.P. Wade, S.R. Crouch, Experimental studies on the effect of temperature on dispersion in a flow injection system, Anal. Chim. Acta 192 (1987) 301.
[57] D. Betteridge, C.Z. Marczewski, A.P. Wade, A random walk simulation of flow injection analysis, Anal. Chim. Acta 165 (1984) 227.
[58] M. Sperling, Z.-L. Fang, B. Welz, Expansion of dynamic working range and correction for interferences in flame atomic-absorption spectrometry using flow-injection gradient ratio calibration with a single standard, Anal. Chem. 63 (1991) 151.
[59] G. Favaro, G.A. Sacchetto, P. Pastore, M. Fiorani, Liquid chromatographic determination of non-volatile nitrosamines by post-column redox reactions and voltammetric detection at solid electrodes. Study of a flow reactor system based on Ce(IV) reagent, Anal. Chim. Acta 273 (1993) 457.
[60] D. Betteridge, J. Ruzicka, The determination of glycerol in water by flow-injection analysis. A novel way of measuring viscosity, Talanta 23 (1976) 409.
[61] D. Betteridge, W.C. Cheng, E.L. Dagless, P. David, T.B. Goad, D.R. Deans, D.A. Newton, T.B. Pierce, An automated viscometer based on high-precision flow-injection analysis. Part I. Apparatus for high-precision flow-injection analysis, Analyst 108 (1983) 1.
[62] A. Rios, M.D.L. Castro, M. Valcarcel, Determination of viscosity with an open-closed flow-injection system, Talanta 34 (1987) 915.
[63] B.F. Rocks, R.A. Sherwood, L.M. Bayford, C. Riley, Zinc and copper determination in microsamples of serum by flow injection and atomic absorption spectroscopy, Ann. Clin. Biochem. 19 (1982) 338.
[64] J.J. Harrow, J. Janata, Heterogeneous samples in flow injection systems. Part 2. Standard addition, Anal. Chim. Acta 174 (1985) 123.
[65] L. Risinger, G. Johansson, Effect of sample viscosity on the behavior of thin-channel flow-through dialyzers in flow-injection analysis, Anal. Chim. Acta 261 (1992) 435.
[66] D.C. Stone, J.F. Tyson, Flow cell and diffusion coefficient effects in flow injection analysis, Anal. Chim. Acta 179 (1986) 432.
[67] M. Hangos-Mahr, E. Pungor, Electrochemical Detection in Flow Analysis, Akademia Kiado, Budapest, Hungary, 1983.
[68] E.R. Alves, M.A. Feres, E.A.G. Zagatto, J.L.F.C. Lima, Exploiting pulsed flows for heating improvement: application to determination of total reducing sugars in molasses and sugar-cane juices, Curr. Anal. Chem. 5 (2009) 65.
[69] E. Grushka, E.J. Kikta, Extension of the chromatographic broadening method of measuring diffusion coefficients to liquid systems. I. Diffusion coefficients of some alkylbenzenes in chloroform, J. Phys. Chem. 78 (1974) 2297.
[70] M. Ibrahim, Z. Gongwei, Z. Junjie, Determination of diffusion coefficients of proteins by flow injection analysis and its application to estimation of molecular masses of proteins, Instrum. Sci. Technol. 26 (1998) 333.
[71] A.D. McBrady, R. Chantiwas, A.K. Torgerson, K. Grudpan, R.E. Synovec, An absorbance-based micro-fluidic sensor for diffusion coefficient and molar mass determinations, Anal. Chim. Acta 575 (2006) 151.
[72] G. Gerhardt, R.N. Adams, Determination of diffusion coefficients by flow injection analysis, Anal. Chem. 54 (1982) 2618.
[73] G.W. Zou, Z. Liu, C.X. Wang, Flow injection analysis methods for determination of diffusion coefficients, Anal. Chim. Acta 350 (1997) 359.
[74] J.N. Yang, Z.H. Yao, Z.P. Gao, Flow-injection amperometric determination of diffusion coefficients, Anal. Chim. Acta 246 (1991) 341.
[75] S.-F. Li, H.A. Mottola, Continuous flow determination of relative diffusion coefficients of iron complexes with ligands of the 1,10-phenanthroline family and with 3-(2-pyridyl)-5,6-diphenyl-1,2,4-triazine in acetonitrile-water solutions, Anal. Chim. Acta 289 (1994) 79.

[76] L.D. Coo, M. Sadek, R.T.C. Brownlee, T.J. Cardwell, R.W. Cattrall, S.D. Kolev, Determination of the diffusion coefficient of 1-(2′-pyridylazo)-2-naphthol in ethanol-water solutions using flow injection and nuclear magnetic resonance techniques, Anal. Chim. Acta 386 (1999) 137.
[77] W.A. Boyle, R.F. Buchholz, J.A. Neal, J.L. McCarthy, Flow-injection analysis estimation of diffusion coefficients of paucidisperse and polydisperse polymers such as polystyrene sulfonates, J. Appl. Polymer Sci. 42 (1991) 1969.
[78] D. Robinson, J.E. Anderson, J.L. Lin, Measurement of diffusion coefficients of some indoles and ascorbic acid by flow injection analysis, J. Phys. Chem. 94 (1990) 1003.
[79] C.C. Paiton, H.A. Mottola, Kinetics in continuous flow sample processing. Chemical contributions to dispersion in flow injection techniques, Anal. Chim. Acta 158 (1984) 67.
[80] J. Ruzicka, E.H. Hansen, Flow injection analysis. Part X. Theory, techniques and trends, Anal. Chim. Acta 99 (1978) 37.
[81] F.S. Oliveira, M. Korn, Employment of a single standard solution for analytical curves in flow injection analysis system coupled to solid phase spectrophotometry, Quim. Nova 26 (2003) 470.
[82] R.S. Schifreen, D.A. Hanna, L.D. Bowers, P.W. Carr, Analytical aspects of immobilized enzyme columns, Anal. Chem. 49 (1977) 1929.
[83] A.U. Ramsing, J. Ruzicka, E.H. Hansen, The principles and theory of high speed titrations by flow injection analysis, Anal. Chim. Acta 129 (1981) 1.
[84] E.A.G. Zagatto, B.F. Reis, M. Martinelli, F.J. Krug, H. Bergamin-Filho, M.F. Gine, Confluent streams in flow injection analysis, Anal. Chim. Acta 198 (1987) 153.
[85] M.A. Segundo, H.M. Oliveira, J.L.F.C. Lima, M.I.G.S. Almeida, A.O.S.S. Rangel, Sample introduction in multi-syringe flow injection systems: comparison between time-based and volume-based strategies, Anal. Chim. Acta 537 (2005) 207.
[86] Y.-H. Li, H.-C. Ma, Two trends of sample dispersion variation with carrier flow rate in a single flow-injection manifold, Talanta 42 (1995) 2033.
[87] Y.-S. Li, Y. Narusawa, Zone circulating flow-injection analysis: theory, Anal. Chim. Acta 289 (1994) 355.
[88] A.C.B. Dias, J.L.M. Santos, J.L.F.C. Lima, C.M. Quintella, A.M.V. Lima, E.A.G. Zagatto, A critical comparison of analytical flow systems exploiting streamlined and pulsed flows, Anal. Bioanal. Chem. 388 (2007) 1303.
[89] J. Ruzicka, G.D. Marshall, Sequential injection: a new concept for chemical sensors, process analysis and laboratory assays, Anal. Chim. Acta 237 (1990) 329.
[90] C.E. Lenehan, N.W. Barnett, S.W. Lewis, Sequential injection analysis, Analyst 127 (2002) 997.
[91] R.B.R. Mesquita, A.O.S.S. Rangel, A review on sequential injection methods for water analysis, Anal. Chim. Acta 648 (2009) 7.
[92] P.R. Fortes, M.A. Feres, M.K. Sasaki, E.R. Alves, E.A.G. Zagatto, J.A.V. Prior, J.L.M. Santos, J.L.F.C. Lima, Evidences of turbulent mixing in multi-pumping flow systems, Talanta 79 (2009) 978.
[93] P. Chocholous, P. Solich, D. Satinsky, An overview of sequential injection chromatography, Anal. Chim. Acta 600 (2007) 129.
[94] A. Herbelin, J. Ruzicka, Pulse modulation. A novel approach to gradient-based flow injection techniques, Collect. Czech. Chem. Commun. 66 (2001) 1219.
[95] D. Satinsky, P. Solich, P. Chocholous, R. Karlicek, Monolithic columns. A new concept of separation in the sequential injection technique, Anal. Chim. Acta 499 (2003) 205.
[96] A. Rius, M.P. Callao, F.X. Rius, Self-configuration of sequential injection analytical systems, Anal. Chim. Acta 316 (1995) 27.

[97] T. Guebeli, G.D. Christian, J. Ruzicka, Fundamentals of sinusoidal flow sequential injection spectrophotometry, Anal. Chem. 63 (1991) 2407.
[98] A. Cladera, E. Gomez, J.M. Estela, V. Cerda, Effect of variables influencing S1/2 in sequential injection analysis. Extrapolability of S1/2 based results between SIA designs, Talanta 43 (1996) 1667.
[99] J.C. Masini, Demonstrating the fundamentals, potentialities and limitations of sequential injection analysis, Quim. Nova 31 (2008) 704.
[100] C.D. Crowe, H.W. Levin, D. Betteridge, A.P. Wade, A random-walk simulation of flow-injection systems with merging zones, Anal. Chim. Acta 194 (1987) 49.
[101] J.F. van Staden, A. Botha, Spectrophotometric determination of Cu(II) with sequential injection analysis, Talanta 49 (1999) 1099.
[102] M.F.T. Ribeiro, J.L.M. Santos, J.L.F.C. Lima, A.C.B. Dias, E.A.G. Zagatto, Single reaction interface in flow analysis, Talanta 68 (2005) 351.
[103] C.I.C. Silvestre, J.L.M. Santos, J.L.F.C. Lima, E.A.G. Zagatto, Single reaction interface flow system for chemiluminescent monitoring of mannitol based on its hydroxyl radical scavenger activity, Talanta 77 (2008) 518.
[104] A. Cerda, M.T. Oms, R. Forteza, V. Cerda, Sequential injection sandwich technique for the simultaneous determination of nitrate and nitrite, Anal. Chim. Acta 371 (1998) 63.
[105] A.N. Araújo, J.L.F.C. Lima, M.L.M.F.S. Saraiva, R.P. Sartini, E.A.G. Zagatto, A SIA system with mixing chamber for handling high concentrated solutions: Spectrophotometric catalytic determination of iodide in nutrition salts, J. Flow Injection Anal. 14 (1997) 151.
[106] R.S. Honorato, M.C.U. Araujo, R.A.C. Lima, E.A.G. Zagatto, R.A.S. Lapa, J.L.F.C. Lima, A flow-batch titrator exploiting a one-dimensional optimisation algorithm for end point search, Anal. Chim. Acta 396 (1999) 91.
[107] B.F. Reis, M.F. Giné, E.A.G. Zagatto, J.L.F.C. Lima, R.A.S. Lapa, Multicommutation in flow analysis. Part 1. Binary sampling: concepts, instrumentation and spectrophotometric determination of iron in plant digests, Anal. Chim. Acta 293 (1994) 129.
[108] M.A. Feres, E.A.G. Zagatto, J.L.M. Santos, J.L.F.C. Lima, The concept of multicommutation in flow analysis, in: M. Trojanowicz (Ed.), Advances in Flow Analysis, Wiley VCH, Weinheim, 2008, p. 167. Chapter 7.
[109] M.C. Icardo, J.V.G. Mateo, J.M. Calatayud, Multicommutation as a powerful new analytical tool, Trends Anal. Chem. 21 (2002) 366.
[110] B.F. Reis, H. Bergamin-Filho, E.A.G. Zagatto, F.J. Krug, Merging zones in flow injection analysis. Part 3. Spectrophotometric determination of aluminium in plant and soil materials with sequential addition of pulsed reagents, Anal. Chim. Acta 107 (1979) 309.
[111] M.A.Z. Arruda, E.A.G. Zagatto, N. Maniasso, Kinetic determination of cobalt and nickel by flow-injection spectrophotometry, Anal. Chim. Acta 283 (1993) 476.
[112] H. Bergamin-Filho, B.F. Reis, A.O. Jacintho, E.A.G. Zagatto, Ion exchange in flow injection analysis. Determination of ammonium ions at the μg/l level in natural waters with pulsed Nessler reagent, Anal. Chim. Acta 117 (1908) 81.
[113] M.F. Gine, H. Bergamin-Filho, E.A.G. Zagatto, B.F. Reis, Simultaneous determination of nitrate and nitrite by flow injection analysis, Anal. Chim. Acta 114 (1980) 191.
[114] E.A.G. Zagatto, H. Bergamin-Filho, S.M.B. Brienza, M.A.Z. Arruda, A.R.A. Nogueira, J.L.F.C. Lima, Multi-site detection in flow analysis. Part 1. Relocation of spectrophotometric detector, Anal. Chim. Acta 261 (1992) 59.
[115] V. Grassi, E.A.G. Zagatto, J.L.F.C. Lima, Flow-injection systems with multi-site detection, Trends Anal. Chem. 24 (2005) 880.
[116] R.I.L. Catarino, A.C.L. Conceição, M.B.Q. Garcia, M.L.S. Gonçalves, J.L.F.C. Lima, M.M.C. Santos, Flow amperometric determination of pharmaceuticals with on-line electrode surface renewal, J. Pharm. Biomed. Anal. 33 (2003) 571.

[117] O. Fatibello-Filho, J.A. Nobrega, Flow injection potentiometric determination of saccharin in dietary products with relocation of filtration unit, Talanta 41 (1994) 731.
[118] D.J. Malcome-Lawes, C. Pasquini, A novel approach to non-segmented flow analysis. 3. Nitrate, nitrite and ammonium in waters, J. Autom. Chem. 10 (1988) 192.
[119] E.A.G. Zagatto, A.O. Jacintho, J. Mortatti, H. Bergamin-Filho, An improved flow injection determination of nitrite in waters by using intermittent flows, Anal. Chim. Acta 120 (1980) 399.
[120] I.V. Toth, A.O.S.S. Rangel, J.L.M. Santos, J.L.F.C. Lima, Determination of aluminum(III) in crystallized fruit samples using a multicommutated flow system, J. Agric. Food Chem. 52 (2004) 2450.
[121] J.Y. Neira, E. Gonzalez, J.A. Nobrega, Improvement of sensitivity in flow analysis by exploiting a multi-reversed software-assisted system, Talanta 73 (2007) 400.
[122] J. Ruzicka, E.H. Hansen, Stopped-flow and merging zones: a new approach to enzymatic assay by flow injection analysis, Anal. Chim. Acta 106 (1979) 207.
[123] N. Bennaoui, C. Periou, C. Harault, G. Lemoel, Evaluation of the Technicon DAX 48, a multiparametric biochemical analyzer, Ann. Biol. Clin. 51 (1993) 713.
[124] J.M.T. Carneiro, R.S. Honorato, E.A.G. Zagatto, Individual sample conditioning in flow analysis. Determination of N-total in plant materials, Fresenius J. Anal. Chem. 368 (2000) 496.
[125] R.A.S. Lapa, J.L.F.C. Lima, J.L.M. Santos, Dual stopped-flow spectrophotometric determination of amiloride hydrochloride in a multicommutated flow system, Anal. Chim. Acta 407 (2000) 225.
[126] C.M.N.V. Almeida, R.A.S. Lapa, J.L.F.C. Lima, Oxidability determination in waste waters using an automatic titrator based on a multicommutated unsegmented flow system, Int. J. Environ. Anal. Chem. 78 (2000) 315.
[127] D.A. Whitman, G.D. Christian, Cascade system for rapid on-line dilutions in flow-injection analysis, Talanta 36 (1989) 205.
[128] C. Pasquini, W.A. Oliveira, Monosegmented system for continuous-flow analysis. Spectrophotometric determination of chromium(VI), ammonia, and phosphorus, Anal. Chem. 57 (1985) 2575.
[129] W.R. Seitz, W.W. Suydam, D.M. Hercules, Determination of trace amounts of chromium(III) using chemiluminescence analysis, Anal. Chem. 44 (1972) 957.
[130] C. Pasquini, Mechanical removal of the central sample zone to avoid air bubbles in monosegmented continuous flow analysis, Anal. Chem. 58 (1986) 2346.
[131] Z.-L. Zhi, Segmental flow injection analysis, a hybrid technique of segmented continuous flow analysis and flow injection analysis, Trends Anal. Chem. 17 (1998) 411.
[132] I.M. Raimundo-Junior, C. Pasquini, Automated monosegmented flow analyser. Determination of glucose, creatinine and urea, Analyst 122 (1997) 1039.
[133] Y.-S. Hsieh, S.R. Crouch, Flow reversal and flow recycle air-segmented flow injection for simultaneous determination of a binary mixture, Anal. Chim. Acta 304 (1995) 333.
[134] T. Buch-Rasmussen, Determination of d-glucose in undiluted whole blood using chemically modified electrodes and segmented sample injection in a flow system, Anal. Chim. Acta 237 (1990) 405.
[135] M.C.H. Silva, C. Pasquini, Bisegmented system for determination of low concentrations of gaseous constituents in gaseous samples, Anal. Chim. Acta 393 (1999) 121.
[136] B.F. Reis, M.A.Z. Arruda, E.A.G. Zagatto, J.R. Ferreira, An improved monosegmented continuous-flow system for sample introduction in flame atomic spectrometry, Anal. Chim. Acta 206 (1988) 253.
[137] L.C. Tian, X.P. Sun, Y.Y. Xu, Z.L. Zhi, Segmental flow-injection analysis. Device and applications, Anal. Chim. Acta 238 (1990) 183.

[138] F. Canete, A. Ríos, M.D. Luque de Castro, M. Valcarcel, Analysis of gaseous samples by flow injection, Anal. Chim. Acta 224 (1989) 127.
[139] D.P. Arnold, M.P. Russell, J.D. Petty, D.R. Sweatman, Automated titrations using a discontinuous programmed flow analyzer, Anal. Chem. 67 (1989) 2109.
[140] T.J. Cardwell, R.W. Cattrall, G.J. Cross, J.R. O'Cornell, J.D. Petty, G.R. Scollary, Determination of titratable acidity of wines and total acidity of vinegars by discontinuous flow analysis using photometric end-point detection, Analyst 116 (1991) 1051.
[141] R.I. Mrzljak, A.M. Bond, T.J. Cardwell, R.W. Cattrall, O.M.G. Newman, G.R. Scollary, Determination of zinc and acid in zinc plant electrolyte by discontinuous flow analysis, Analyst 117 (1992) 1845.
[142] J.D. Petty, D.A. Petty, R.M. Peachey, T.K. Smith, B.L. Krieger, High-accuracy preformed gradient discontinuous flow titrimetry, Talanta 41 (1994) 359.
[143] B.L. Krieger, G.M. Kimber, M. Selby, F.O. Smith, E.E. Turak, J.D. Petty, R.M. Peachey, Automatic sample handling and calibration methods for analysis by flame atomic absorption spectrometry using discontinuous flow analysis, J. Anal. At. Spectrom. 9 (1994) 267.
[144] T.J. Cardwell, R.W. Cattrall, G.J. Cross, G.R. O'Connel, J.D. Petty, G.R. Scollary, Determination of sulfuric acid in metallurgical process streams by discontinuous flow analysis, Anal. Chim. Acta 308 (1995) 197.
[145] R.W. Cattrall, G.R. Scollary, Discontinuous flow analysis: generation of fluid flows by differential pumping, Anal. Method. Instrum. 2 (1995) 61.
[146] A.J. Dole, T.J. Cardwell, R.W. Cattrall, R. di Giacomo, C.G. Rodrigues, G.R. Scollary, The determination of silver in photographic emulsions by discontinuous flow analysis, Electroanal. 10 (1998) 21.
[147] S.W. Lewis, P.S. Francis, K.F. Lim, G.E. Jenkins, X.-D. Wang, Pulsed flow chemistry: a new approach to solution handling for flow analysis coupled with chemiluminescence detection, Analyst 125 (2000) 1869.
[148] W.E. van der Linden, Classification and definition of analytical methods based on flowing media (IUPAC recommendations 1994), Pure Appl. Chem. 66 (1994) 2493.
[149] E.A.G. Zagatto, J.F. van Staden, N. Maniasso, R.I. Stefan, G.D. Marshall, Information essential for characterizing a flow-based analytical system. Recommendations for its use, Pure Appl. Chem. 74 (2002) 585.
[150] J.F. van Staden, Analytical aspects of chemical process control. Part 1. Fundamentals, Pure Appl. Chem. 71 (1999) 2303.
[151] J. Ruzicka, E.H. Hansen, Flow injection analysis, Chemtec. 9 (1979) 756.
[152] J. Ruzicka, G.D. Marshall, G.D. Christian, Variable flow-rates and a sinusoidal flow pump for flow-injection analysis, Anal. Chem. 62 (1990) 1861.
[153] J. Toei, Determination of glucose in clinical samples by flow reversal flow-injection analysis, Analyst 113 (1988) 475.
[154] Z.-L. Fang, M. Sperling, B. Welz, Comparison of three propulsion systems for application in flow-injection zone penetration dilution and sorbent extraction preconcentration for flame atomic absorption spectrometry, Anal. Chim. Acta 269 (1992) 9.
[155] V.P. Andreev, N.B. Ilyina, D.A. Holman, L.D. Scampavia, G.D. Christian, Experimental study of kinematic focusing. Comparison of electro injection and sequential injection determination of copper, Talanta 48 (1999) 485.
[156] G. Nagy, Z. Feher, E. Pungor, Application of silicone rubber-based graphite electrodes for continuous flow measurements. 2. Voltammetric study of active substances injected into electrolyte streams, Anal. Chim. Acta 52 (1970) 47.
[157] A.H. Kadish, D.A. Hall, A new method for continuous monitoring of blood glucose by measurement of dissolved oxygen, Clin. Chem. 11 (1965) 869.
[158] F.J. Krug, H. Bergamin-Filho, E.A.G. Zagatto, Commutation in flow injection analysis, Anal. Chim. Acta 179 (1986) 103.

[159] J. Ruzicka, E.H. Hansen, in: Flow Injection Analysis, second ed., Wiley-Interscience, NY, 1988.

[160] E.A.G. Zagatto, B.F. Reis, H. Bergamin-Filho, The concept of volumetric fraction in flow injection analysis, Anal. Chim. Acta 226 (1989) 129.

[161] ICH Expert Working Group, International Conference on Harmonisation (ICH), Validation of analytical procedures: Methodology, Q2(R1), step 4 version, 2005. http://www.ich.org (accessed 15.09.09).

[162] G.A. Shabir, Validation of high-performance liquid chromatography methods for pharmaceutical analysis: Understanding the differences and similarities between validation requirements of the US Food and Drug Administration, the US Pharmacopeia and the International Conference on Harmonization, J. Chromatogr. A 987 (2003) 57.

[163] M. Ribani, C.B.G. Bottoli, C.H. Collins, I.C.S.F. Jardim, L.F.C. Melo, Validation for chromatographic and electrophoretic methods, Quim. Nova 27 (2004) 771.

[164] A.C. Olivieri, N.K.M. Faber, J. Ferré, R. Boqué, J.H. Kalivas, H. Mark, Uncertainty estimation and figures of merit for multivariate calibration, Pure Appl. Chem. 78 (2006) 633.

[165] Y.V. Heyden, A. Nijhuis, J. Smeyers-Verbeke, B.G.M. Vandeginste, D.L. Massart, Guidance for robustness/ruggedness tests in method validation, J. Pharm. Biomed. Anal. 24 (2001) 723.

[166] J. Vessman, R.I. Stefan, J.F. van Staden, K. Danzer, W. Lindner, D.T. Burns, A. Fajgelj, H. Muller, Selectivity in analytical chemistry (IUPAC recommendations 2001), Pure Appl. Chem. 73 (2001) 1381.

[167] J.N. Miller, J.C. Miller, in: Statistics and Chemometrics for Analytical Chemistry, fourth ed., Prentice Hall, New York, 2000, Ch. 5.

[168] L.A. Currie, Nomenclature in evaluation of analytical methods including detection and quantification capabilities (IUPAC recommendations 1995), Pure Appl. Chem. 67 (1995) 1699.

[169] G.D. Christian, in: Analytical Chemistry, fifth ed., John Wiley & Sons, New York, 1994.

[170] J. Ruzicka, J.C. Tjell, Ion-selective electrodes in continuous-flow analysis. Determination of calcium in serum, Anal. Chim. Acta 47 (1969) 475.

[171] H.S. Strickler, P.J. Stanchak, J.J. Maydak, Application of correction equations for sample interactions in complex autoanalyzer methods, Anal. Chem. 42 (1970) 1576.

[172] K. Dixon, A theoretical study of carryover in selective access analyzers, Ann. Clin. Biochem. 27 (1990) 139.

[173] S. Vicente, E.P. Borges, B.F. Reis, E.A.G. Zagatto, Exploitation of tandem streams for carry-over compensation in flow analysis. I. Turbidimetric determination of potassium in fertilizers, Anal. Chim. Acta 438 (2001) 3.

[174] S. Angelova, H.W. Holy, Optimal speed as a function of system performance for continuous flow analyzers, Anal. Chim. Acta 145 (1983) 51.

[175] T.W. Stephens, Performance of two new algorithms for estimating within-method and between-method carryover evaluated statistically, Clin. Chem. 34 (1988) 1805.

[176] B. Schlain, H. Frush, C. Pennington, G. Osikowicz, K. Ford, Two stage procedure for evaluating interassay carryover on random access instruments, Clin. Chem. 42 (1996) 725.

[177] H. Bergamin-Filho, E.A.G. Zagatto, F.J. Krug, B.F. Reis, Merging zones in flow injection analysis. Part 1. Double proportional injector and reagent consumption, Anal. Chim. Acta 101 (1978) 17.

[178] J. Alonso, J. Bartrolí, M. Del Valle, M. Escalada, R. Barber, Sandwich techniques in flow injection analysis. Part 1. Continuous recalibration techniques for process control, Anal. Chim. Acta 199 (1987) 191.

CHAPTER

6

Instrumentation

Tubing and pumps are the essence of flow analysis. **W. van der Linden**

OUTLINE

Home-made components were dominant in the early development of flow analysis, as can be seen in Fig. 2.2 showing the prototype air-segmented flow analyser assembly. In the initial development of flow injection analysis, excellent results were obtained with very simple instruments, and devices were often assembled with pieces originally designed for children's toys [1,2] (see also Fig. 2.5). Nowadays, there are several instrument manufacturers [3–6], and the use of commercial equipment for high-throughput analysis is a reality. Novel, dedicated components for specific tasks are still being proposed and the use of computer-controlled instrumentation is increasing.

A flow system comprises three units: fluid propulsion, sample handling and detection and the components necessary to assemble and integrate these units, *e.g.*, transmission lines and connectors. This chapter describes all of these units and components.

6.1. FLUID PROPULSION

An ideal propulsion system should ensure reproducible flow rates on a short-term (hours) and long-term (days) basis, multi-channel capability (at least four parallel pumping channels to provide system versatility), resistance to aggressive reagents and solvents, readily adjustable flow rates and low initial investment and running costs [7]. The maintenance of a consistent flow is essential to obtain good analytical reproducibility. Flow rates are typically in the 0.2–5.0 mL min^{-1} range so that the system operates under low pressure, normally lower than 10 psi [0.689 bar]. The number of channels used depends on the manifold complexity.

There are various strategies for manipulating the flow during the course of an analytical cycle (see 3.1.1.3). Regarding temporal variations in flow rates, the most common schemes for handling the flowing samples exploit:

- Constant flow, also called linear flow (Fig. 3.4a). This is the most widely used in flow injection and segmented flow analysis, ideally presenting no temporal variations in flow rate. If there is a slight sinusoidal variation, especially with peristaltic pumps (see 6.1.1), a pulsating flow is established.
- Intermittent flow (Fig. 3.4f). As the name *intermittent flow* implies, this approach requires consistent and immediate stopping and restarting of the flow. Its exploitation permits the sample zone to remain stationary in the manifold for a pre-set time interval in order to provide additional time for development of a chemical reaction, resulting, for example, in enhancement of the sensitivity. If the sample zone is stopped inside the flow cell, matrix interferences can be circumvented by measuring the rate of reaction, *e.g.*, in the determination of ethanol in diluted whole blood [8]. The use of intermittent flow also permits the addition of optional confluent streams, *e.g.*, a washing stream to clean the analytical path after the analytical signal has been obtained [9]. This strategy is also useful in sequential injection analysis (see 5.3.1).
- Reversed flow (Fig. 3.4e). This is a powerful tool for kinetic methods (see also 8.6.4). It requires a stop/start feature in addition to gearing for reverse drive, and has been exploited mainly in sequential and bead injection determinations. An example is the kinetic turbidimetric determination of phytic acid in food samples relying on the influence of the analyte concentration on the rate of calcium oxalate crystallisation [10]. The sample and reagents were aspirated and then the flow was reversed, directing the sample zone towards the flow cell, where its central portion was stopped for 30 s, monitored and discarded.
- Pulsed flows (Fig. 3.4g). In the recent past, several researchers have highlighted the inability of some conventional pumps to generate a smooth and consistent flow but it is possible to take advantage of a reproducible pulsed flow for, *e.g.*, liquid-liquid extraction procedures and enzymatic reactions [11]. Pulsed flows are the essence of multi-pumping flow systems [12].

An elegant way to acquire all of the above-mentioned flow patterns under continuous pumping is to increase the degree of commutation of the system (see 5.4). As an example, the addition of stream re-directing valves strategically positioned in the manifold allows streams to be directed towards the main analytical stream or not, without modifying the operating conditions of the pumping device [13].

The specific temporal variation in flow rates adopted in a flow system imposes certain requirements on the propulsion system. Various devices for fluid propelling are therefore described below.

6.1.1. Peristaltic Pumps

The multi-roller peristaltic pump (Fig. 6.1) is the most widely used device for fluid propulsion in segmented flow and flow injection analysis. It generally delivers a unidirectional and constant flow that results in a highly reproducible sample residence time. Some pumps provide facilities for implementing variable flow rates and/or flow reversals, allowing the efficient implementation of flow-based analytical procedures relying on, *e.g.*, sample stopping, iterative flow reversals and time-based injections. More sophisticated models of peristaltic pumps are designed to allow very precise control of the flow rate, incorporating devices such as servo systems for adjusting the roller speed [14]. Peristaltic pumps, however, provide a pulsating flow, and a pump head with 8–10 closely positioned rollers is recommended to reduce fluctuations in flow rate. Half of the rollers should always be in contact with the pump tubing. Some multi-channel pumps have separate bridges for each channel, allowing independent compression control. Two colour-coded collars are usually glued near the ends of each pump tube, acting as points for fixing and tensioning the tube and as identifiers of the nominal flow rate.

Peristaltic devices with a variable drive speed can provide a wide range of flow rates, typically 0.2–5.0 mL min^{-1}, in several channels, each with a different flow rate if necessary. Partial damping is inherent in these fluid propulsion devices due to tube elasticity.

If a constant flow is required, dampers can be used to suppress the ripple component. The pulsating flow enters the device, either from the top or laterally, and leaves it at a constant flow rate, because the air

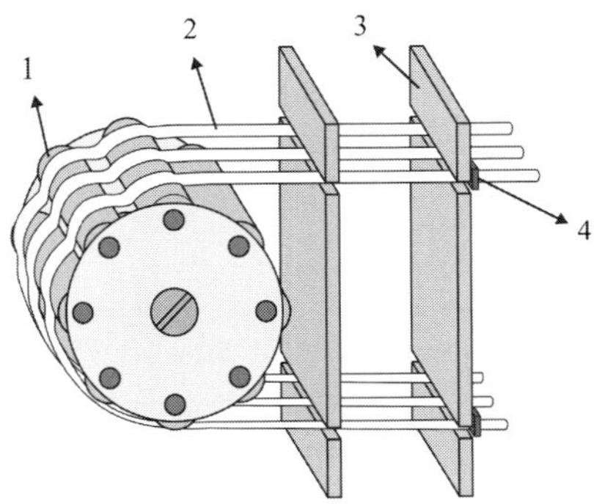

FIGURE 6.1 **Schematic representation of a peristaltic pump.** 1 = roller; 2 = pumping tubes; 3 = tube support; 4 = colour-coded collar.

trapped between the flowing liquid surface and the chamber top acts as a pulse suppresser. Another beneficial aspect is that gas bubbles are removed from the main stream during its passage through the damper. As a rule, dampers should be placed before the sample insertion device, otherwise they can act as chamber-like elements, impairing sensitivity and sampling rate.

The flow rate is governed by the internal diameter of the pump tubing and the pump rotation speed. For a given tubing material, the tube wall thickness is constant regardless of its inner diameter. Use of pump tubing has the advantage that no other parts of the pump come into direct contact with the liquids being pumped. The delivery rate, however, may gradually decrease as a result of tubing degradation due to reduced flexibility with use. Wear is dependent on the tubing material, the amount of compression and the nature of the pumped solutions. Changes usually occur very slowly and are compensated by regular calibration and periodic (normally monthly) replacement [15]. If the exact flow rate is required, it should be experimentally determined [16]. As tubing is relatively inexpensive, its replacement is straightforward and recommended as part of a routine maintenance programme.

A variety of tubing materials is commercially available (Table 6.1) and the correct choice is important for safe system operation and for sustaining a consistent flow [15].

The most widely used material is poly(vinyl chloride), PVC, which comes in various grades for different applications. Standard PVC (*e.g.*, Tygon®) is the general choice for aqueous solutions and dilute acids and bases, but is usually not suitable for organic solvents or concentrated acids. Aliphatic hydrocarbons and alcohols can be pumped using

TABLE 6.1 Pump Tubing Materials for Propulsion of Different Streams

Classes of Liquids or Solutions	Tubing Material (trade name)
Aliphatic hydrocarbons, carbon tetrachloride	Modified PVC (Solvaflex™)
Aromatic hydrocarbons, concentrated acids and bases, chloroform	Fluoroplasts (Acidiflex™)
Aqueous and dilute ethanol solutions, formaldehyde, acetaldehyde	PVC (Tygon®)
Dilute acids and bases	PVC (Tygon®), silicone rubber
Alcohols	Modified PVC (Acidiflex™)
Lower alcohols, acetone, acetic acid, acetic anhydride	Silicone rubber

Source: Adapted from Ref. [14].

modified PVC but this is not always suitable for other organic solvents or concentrated strong acids or bases. Thermally set fluorinated plastics (*e.g.*, Viton™) are recommended for concentrated mineral acids and most organic solvents, although they have a limited lifetime and are unsuitable for certain organic liquids such as ketones [17]. Other tubing materials are commercially available and should be used in accordance to the manufacturers' specifications [15].

A good peristaltic pump generally has no inherent inertia, allowing instant stop/restart in procedures requiring sample stopping, intermittent flow or flow reversal. The tubing internal volume should be as small as possible to allow rapid system start-up and flushing. Finally, and very importantly, peristaltic pumps are generally insensitive to minor variations in system back-pressure and stream viscosity, which are critical parameters for reproducible sample residence times and hence analytical precision [15].

It is advisable to include a special device for controlling the flow delivery for use in segmented flow analysis, and with some special strategies involving the confluent addition of air (or immiscible solvents). It consists of a clamp actuating at a regular frequency and imposing an artificial pulsation on the delivered flow, allowing reproducible addition of the confluent stream (see 3.2.2).

Some peristaltic pumps can be operated in synchronisation with other components of the flow system through a triggering element associated with the pump rollers. This permits, *e.g.*, highly precise injections of very small sample volumes [18,19]. As the position of the rollers determines the exact instant for switching the injector, the sample is inserted under constant flow geometry and more precise results are obtained. For pulsating flow systems without synchronisation, the small sample plug could be inserted at any time between the maximum and minimum flow rate. As dispersion is strongly dependent on flow rate, synchronisation avoids potential degradation of precision.

The peristaltic pump can be located before the injection port and in this situation the system is pumped; alternatively, it can be located after the detection unit in which case the system is aspirated (see 2.5). For pumped systems, the internal pressure is positive, which minimises the liberation of dissolved gases but pump pulsation is observed in the flow. For aspirated systems, the pulsation is attenuated but the system can more easily liberate gas from solution.

Another important parameter is the rotation of the rollers. In most pumps, the rollers are free to rotate and this can reduce the lifetime of the pump tubing. This effect is minimised by imposing a roller rotation speed that exactly matches its translational movement. In this way, frictional forces are minimised and tube lifetime is increased.

6.1.2. Syringe (Piston) Pumps

Syringe pumps have historically been associated with autoburettes and are currently experiencing a renaissance in flow analysis [15] due to the emergence of sequential injection [20], bead injection [21], lab-on-valve [22] and multi-syringe [23] analytical systems. They are durable devices, which can be produced from different materials, in order to be chemically inert to the different organic solvents involved. The main attractive features are the high reproducibility for time-based sample introduction and the absence of pump tubing, which results in low maintenance operation. Regarding piston displacement, different approaches have been exploited, and are discussed below.

6.1.2.1. Use of Large Volume Pistons (Syringes)

The piston is directly attached to a device that is pushed inside a cylinder by means of a constant speed rotating motor. The pump delivers a constant flow (without any pulsation) and check valves are not needed. The performance of a syringe pump is dependent on flow rate, syringe size and the architecture involved [24]. In the more sophisticated models, the motor speed and rotation can be varied, which offers real advantages for analytical applications involving multi-reagents, sample stopping or flow reversal.

Refilling of the syringe can restrict system performance because this step reduces the sampling frequency. The drawback can be minimised by increasing the syringe volume, which increases the number of samples that can be handled with a single refill. This approach is, however, limited because high precision piston drivers and uniform cylinder dimensions with a zero-tolerance plunger are required [25].

An attractive alternative is to exploit several syringes with different inner diameters for propelling different solutions at the same time, and this is the essence of the multi-syringe flow systems [23]. The syringe racks usually contains 2–4 syringes, which are simultaneously propelled through a single moving element (Fig. 6.2), providing facilities for multi-channel analysis or for increasing the capacity of the reservoir of the pumped solution. Each syringe is attached to a three-way stream-directing valve that either directs the pumped solution towards the analytical path or alternatively permits the syringe to be refilled.

6.1.2.2. Use of Small Volume Pistons

The piston is moved back and forth by a cam-driven plunger powered by a computer-controlled motor (Fig. 6.3); one-way directional check valves (usually ball-style valves) are needed. The cam driver can accelerate and change the flow rate smoothly throughout the entire length of each stroke. Reproducible pulseless flows are generated and compatibility

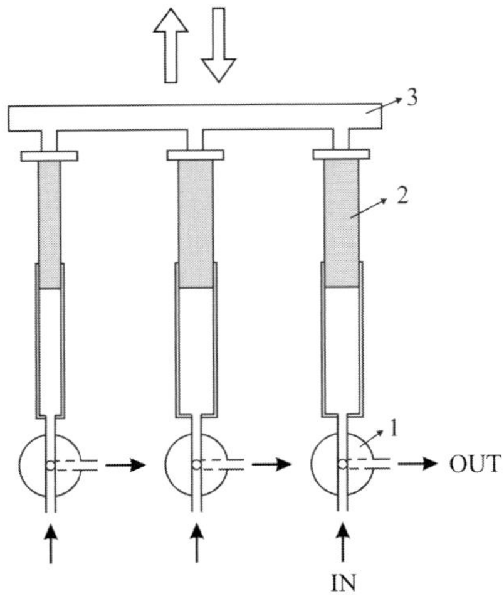

FIGURE 6.2 **Schematic representation of a multi-syringe pump.** 1 = three-way valve; 2 = piston; 3 = single moving element (displacement specified by empty arrows). *Adapted from Talanta 50 (1999) 695, V. Cerda, J.M. Estela, R. Forteza, A. Cladera, E. Becerra, P. Altimira, P. Sitjar, Flow techniques in water analysis, with permission from Elsevier (Ref. [26]).*

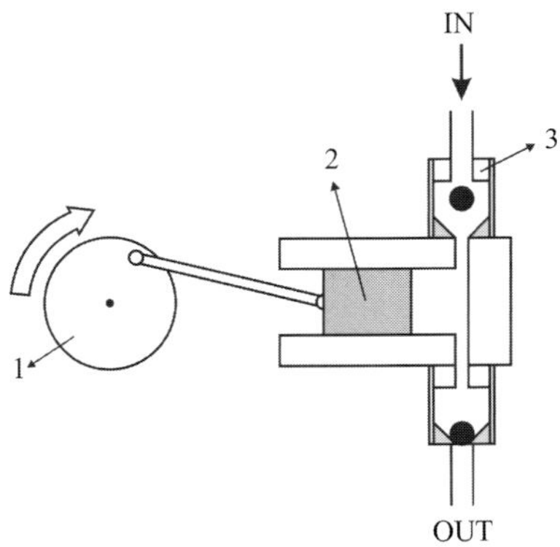

FIGURE 6.3 **Schematic representation of a piston reciprocating pump.** 1 = rotating element attached to the plunger (movement indicated by empty arrow); 2 = piston; 3 = inlet check valve.

with flow reversal is attained. The flow rate is determined as the product of the pump operating frequency times the pump stroke volume.

For pumps relying on back and forth piston movements, the regularity of the flow obtained depends on the number of pistons. Single piston pumps supply a markedly pulsed flow with a relatively long GO period, whereas dual-piston pumps give a significantly smoother flow [14]. Alternatively, a dual plunger can be exploited and this is inherent in reciprocating pumps (Fig. 6.4).

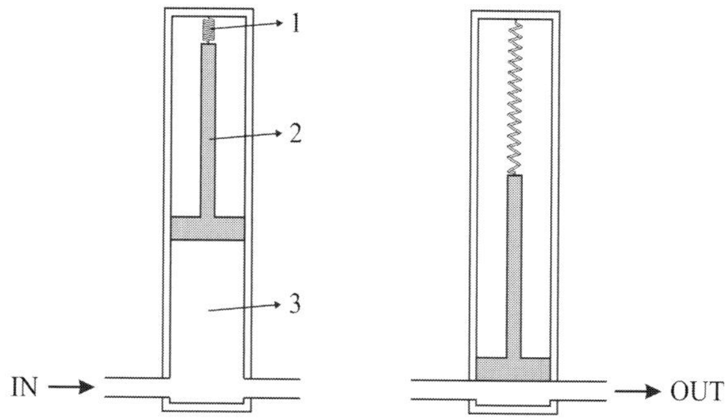

FIGURE 6.4 **Schematic representation of the operation of a solenoid pump.** The figure refers to a pump without a diaphragm. 1 = spring; 2 = piston around which the solenoid is wound; 3 = pump chamber. Left and right portions = full and empty chamber (stroke volume). The check valves are not shown.

Reciprocating piston pumps, often used in HPLC, are markedly more complex than peristaltic pumps and are another option for propulsion in flow systems. These pumps are made with inert materials and provide a suitable means of organic solvent delivery [15]. Their operation is analogous to that of the piston pumps, relying on moving a piston back and forth in a pump head. The piston movement is produced by rotation of a high precision motor-driven circular cam. A flexible seal around the piston prevents leakage of the pumped stream into the pump. Alternatively, a diaphragm can be used and the pump is then called a *diaphragm-type reciprocating pump*. Check valves open and close in order to maintain a one-way flow of solution. Pulsation can be reduced by a damping system.

The flow rate can be varied by modifying the cam rotation speed or the position of the piston/cam attachment and, hence, the pumped stroke. This kind of pump, especially the diaphragm-type pump, has often been used in microfluidic systems [27]. The main drawbacks are the high purchase cost, flow rates typically in the nL min^{-1} range, the need for one pump per channel and the susceptibility to air locks and suspended particulate matter [15]. Flow reversal is not possible.

6.1.3. Diaphragm Pumps

Another strategy for exploiting piston movements in flow systems involves the attachment of the piston to a diaphragm. The moving piston then acts as a pivot and its movements press and release the diaphragm. The wetted surface of the pump is normally made of PEEK or Viton™ [28]. One-way directional check valves (usually ball-style valves) are required.

6.1.3.1. *Solenoid Pumps*

These pumps can be designed with or without a diaphragm. Their operation involves the same mechanism as a cam-drive piston pump, but the movements of the piston are achieved by passing (or not) an electric current through a coiled resistor around the piston [29]. As the solenoid-driven displacement is very abrupt, the pump delivers a pulsed flow [30].

The use of miniaturised solenoid pumps reduces the pump weight and power requirements and opens up the possibility for computer control of each individual reagent and sample flow stream. The versatility of the flow analyser is then significantly increased. These characteristics have been exploited for, *e.g.*, the *in situ* spectrophotometric determination of nitrite in seawater [28]: the peristaltic pump of an ordinary flow injection system was replaced by four solenoid-driven diaphragm pumps with Viton check valves. The pumps required about 5% of the power used by a peristaltic pump, could operate for at least one year, and did not expose the pump tubing to reagents and carrier solutions, which is generally not as inert as the Teflon tubing used in the rest of the system. The main analytical characteristics were maintained, namely, 60 injections per hour, 0.5 mL min^{-1} sample and reagent flow rates, a 1.0% relative standard deviation for 10 μmol L^{-1} nitrite, and a detection limit of 0.1 μmol L^{-1} nitrite. This system was the precursor of the multi-pumping flow systems [12].

6.1.3.2. *Piezoelectric Pumps*

Piezoelectric materials are another alternative for inducing diaphragm movement (Fig. 6.5). The piezoelectric pump is characterised by low power requirements, operational simplicity, and portability. Operation of this pump without the need for check valves is feasible by exploiting diffusion nozzles [31], special pipes [32] or unsymmetrical ridges [33]. The pump delivers a unidirectional pulsed flow at flow rates typically ranging from 0.1–5.0 mL min^{-1}. Furthermore, the pump is made entirely from high precision micro-injection-moulded plastic components, which ensures high reliability and reproducibility and minimises the risk of failure or malfunction. The potential of this kind of pump in flow analysis was recently demonstrated by the spectrophotometric determination of gabapentin in pharmaceutical formulations [34].

6.1.4. Gas-pressurised Reservoirs

Another alternative for fluid propulsion is the gas-pressurised reservoir (Fig. 6.6) which was exploited in the very first flow injection analysers [35,36]. The strategy offers mechanical simplicity due to the absence of any moving part, delivers a pulseless flow and does not require a power supply. The pressure involved is usually <1 bar; therefore, use of

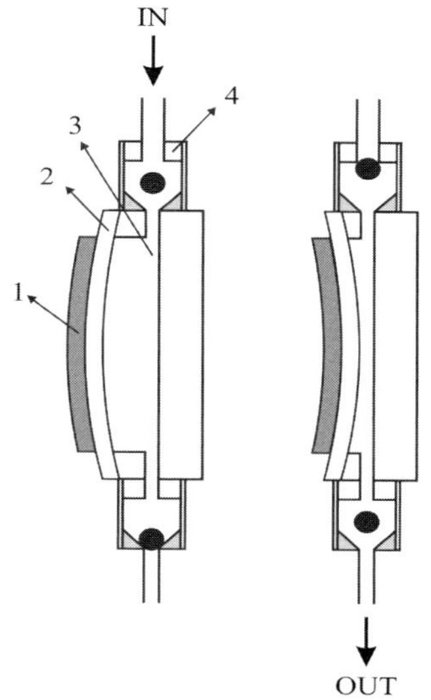

FIGURE 6.5 **Schematic representation of the operation of a piezoelectric pump.** 1 = piezo-actuator; 2 = diaphragm; 3 = pump chamber; 4 = inlet check valve. Left and right portions = full and empty chamber (stroke volume).

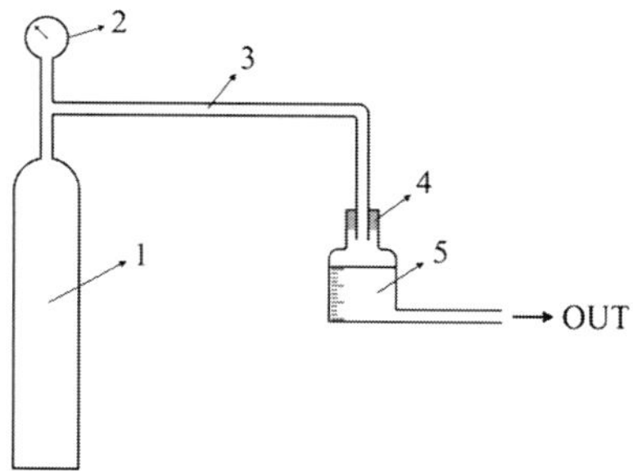

FIGURE 6.6 **Schematic representation of fluid propulsion by a gas-pressurised reservoir.** 1 = gas cylinder; 2 = manometer; 3 = gas transmission line; 4 = stopper; 5 = fluid to be delivered

a pressurised constant-head vessel becomes attractive as a portable system for teaching purposes [36]. The flow rate is, however, difficult to control, especially in multiple line systems. Consequently, these systems rarely have long-term flow rate stability, as variations in back-pressure due to thermal oscillations promote changes in fluid viscosity, thereby altering the volumetric flow rates [37]. Incorporation of pressure transducers helps to overcome this drawback via feedback and pressure control devices [15]. Flow reversal is not possible.

A similar strategy relying on fluid displacement by pressurised gas involves the exploitation of aquarium pumps [38,39]. These pumps are diaphragm piston pumps conceived for blowing air into a domestic aquarium. As the pump has blowing and sucking orifices, the flow system can be operated in the pumped or the aspirated mode. The resulting system is versatile and low cost but has seldom been used in flow analysis, probably because of irregular flow and back-pressure problems.

6.1.5. Osmotic Pumps

Another option for fluid propulsion in flow analysis relies on the osmotic and electro-osmotic processes [40,41]. These processes exploit the osmotic pressure created across a semi-permeable membrane separating a saturated salt solution from a lower salinity solution [42]. Water from the more dilute solution diffuses into the more saline solution, squeezing a flexible but impermeable internal bag containing the fluid to be pumped, thus generating a flow. Temperature dependence and low flow rates limit the use of these propelling devices in flow analysis.

With electro-osmotic flow, some solids carry a net charge when in contact with an aqueous medium due to chemical associations/dissociations and physical adsorption/desorption from their surfaces. Under slightly acidic or alkaline conditions, surfaces of quartz, ceramics, clay and sand are negatively charged, thus attracting positively charged counter ions present in the aqueous solution, establishing a diffuse double layer that is partly mobile. The application of an external electric field results in a net migration of these counter ions in the double layer towards the oppositely charged electrode. Due to its viscosity, the solution inside the micropores or the capillary structure is pulled and moves in the same direction. The flow rate is proportional to the current and the pressure is linearly proportional to the voltage. Another type of electro-osmosis involves the passage of a solvated ion from one side of a compartmented chamber to another. Electro-osmotic pumps generate very low flow rates (nL min^{-1}) in micro-channels. Implementation requires advanced electronics, as a stable voltage around 30 kV is usually needed. A noteworthy feature is that a different linear flow velocity is induced for each chemical species involved. The approach has been exploited in connection with miniaturised flow systems [43,44] and is commonly associated with capillary electrophoresis and related techniques.

6.1.6. Gravity

It is not surprising that fluids can be propelled by gravity [45], which provides a pulseless flow. The flow rate is highly dependent on back-pressure effects that are difficult to control due to variations in temperature and fluid viscosity. Such systems are simple, inexpensive and do not contain any moving parts, leading to minimal implementation and maintenance costs. The approach is attractive for *in situ* analysis and/or in connection with flow systems relying on a hanging drop [46].

6.2. SAMPLE HANDLING

The sample handling unit includes tubes, injection devices, reactors, mixing chambers and connectors, all of which are designed to permit efficient sample manipulation for accomplishing the different analytical steps involved. There are instruments where the entire unit is replaced in order to implement a different analytical procedure. The unit is then called a *manifold*. In this monograph, the manifold is defined as the portion of the flow system comprising the components related to stream manipulation as well as sample introduction and processing.

6.2.1. Tubes

Normally, the tubes used in flow analysis have internal diameters between 0.3 and 2.0 mm. Most analytical applications implemented in unsegmented flow systems are carried out inside 0.5 or 0.8 mm i.d. tubes.

Tubing for constructing the manifold (transmission lines, reactors) should be chemically inert and resistant to temperature changes. In segmented flow analysers, glass tubes are generally preferred in view of the characteristics of the air–liquid interfaces [47]. In unsegmented flow analysers, the materials of choice are polytetrafluoroethylene, polyethylene and polypropylene, mainly due to their chemical stability; tubing with noncollapsible walls is preferred in order to avoid blockages.

Some types of PTFE are not suitable for specific analytical applications involving gas generation, because the produced gaseous species may permeate through the tube wall. Moreover, materials melting at relatively low temperatures, such as polyethylene and polypropylene, should be avoided if the related physico-chemical processes are carried out at temperatures above 50°C. For these applications, PTFE should be used.

6.2.2. Sample Introduction

Precise introduction of a well-defined portion of the aqueous sample into the carrier stream is fundamental to flow analysis. This is generally accomplished by continuously pumping the sample into the manifold for a fixed period of time (time-based introduction) or by inserting the sample aliquot via a fixed-volume loop attached to a rotary or sliding bar valve (loop-based introduction). Hydrodynamic [48] and nested [49] sample introduction have also been used.

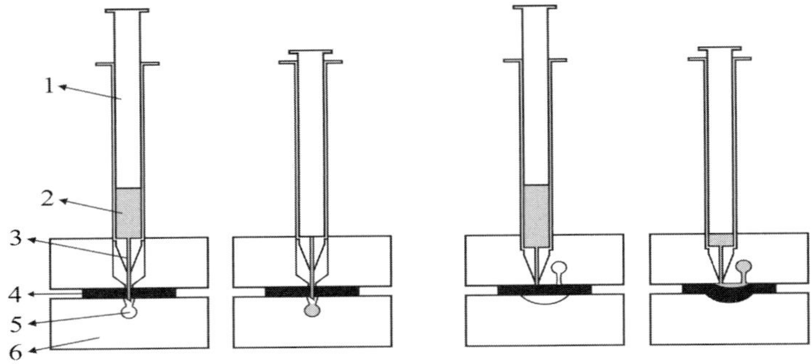

FIGURE 6.7 **Schematic representation of manual sample introduction using syringes with (left) or without (right) a needle.** 1 = syringe piston; 2 = sample inserted volume; 3 = needle; 4 = flexible plastic septum; 5 = main analytical channel (top view); 6 = Perspex™ support.

Sample insertion relying on time-based introduction is inherent to segmented flow analysis and was the preferred approach before the advent of flow injection analysis. Manual injection using syringes with [50] or without [1] needles was used in the earliest flow injection systems (Fig. 6.7) but is now rarely used, except when very small sample volumes are available.

In view of the excellent performance of a sampling loop as the volume-selecting element, loop-based introduction dominated during the fast development of flow injection analysis in the 1980s. Time-based introduction became popular again after the inception of sequential injection analysis and related techniques and the development of multi-commutation. Nowadays, both time-based and loop-based introduction are exploited and use of an autosampler is generally advisable.

6.2.2.1. Sampler

The sampler should be fast, as its maximum sampling frequency is often the limiting factor for the sampling rate. The availability of this component also permits carryover compensation according to Eq. 5.12, as the frequency of sample insertions, and consequently the carryover coefficient, is maintained. The sampling rate is then improved, as discussed in 5.6.6.2.

It is good practice to install a computer-controlled feedback mechanism in order to reduce the number of sample replicates required [51]. The minimum number of replicates (for instance, three) should be set *a priori* and more replicates are only needed if the estimation of the relative standard deviation of the analytical signal is higher than a pre-set value. Real-time data treatment is then required.

In more elaborate versions of the sampler, the sample and the reagent solutions are accommodated in sample and reagent trays, thus facilitating the implementation of analytical protocols involving random-access reagent selection [52].

6.2.2.2. *Time-based Introduction*

Time-based introduction is better implemented automatically, *i.e.*, without human intervention. In this mode, the sampler arm selects the sample, the carrier/wash and eventually the reagent solutions (Fig. 6.8) to be directed towards the manifold. The sampling time is the main parameter governing the volumes of these solutions introduced, which can also be modified by changing the aspiration rate in the fluid propulsion unit. This is usually an option in modern instruments.

Manual operation of the time-based flow analyser is also feasible, but the aspirated sample volume should always be in excess in order to guarantee the establishment of the flat peaked signal typical of segmented flow analysers (see *e.g.*, Fig. 2.5). In this way, variations in the sample volume introduced do not affect the recorded peak height. Carryover should be kept to a minimum, as its compensation is not feasible.

Time-based introduction also permits the implementation of tandem streams. This is done by placing a fast-switching valve at the end of the sampling tube and allowing the selection of successive plugs of sample, reagent and carrier/wash solutions. The potential of this approach is described in 3.1.1.2. It should be kept in mind, however, that in unsegmented flow-based procedures requiring the introduction of very small sample volumes, time-based injection may suffer from poor precision owing to the reliance on precise volumetric flow rates.

Another drawback is that the sample portions remaining inside the sampling tube after sample introduction can act as a carryover source [53]. In segmented flow analysis, this is circumvented by increasing the introduced sample volume, whereas in sequential injection analysis, these portions are directly discarded by aspirating the next sample towards the holding coil and, after flow reversal, directing them directly towards the waste container.

An attractive alternative is the use of the multi-position selecting valve (Fig. 2.10), usually with 6, 8, 10, or 12 outer ports and a common central port [15]. Selection valves are inherent to sequential injection analysis and

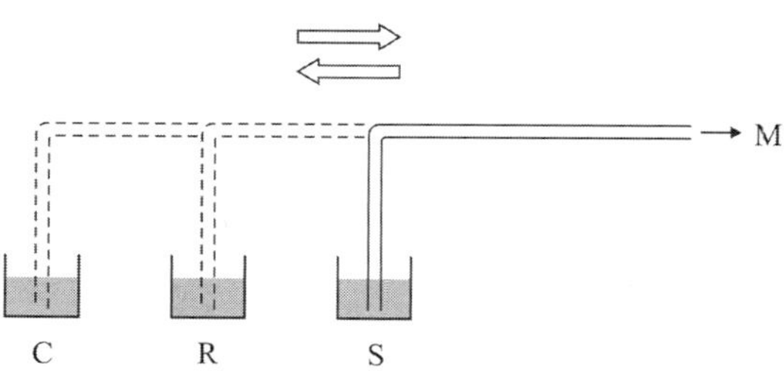

FIGURE 6.8 **Time-based introduction.** C, R and S = carrier/wash, reagent and sample solutions; M = towards manifold; arrows = movements of the sampler arm; dashed line = alternative positions of the sampler arm. For details, see text.

utilise the reproducible aspiration of different liquid solutions in a pre-set sequence towards the holding coil (see 5.3.1). They are electronically actuated and capable of long-term, maintenance-free operation. As time-based introduction is involved, it is advisable to use piston pumps to ensure long-term reproducibility of the volumes introduced. The reagent is aspirated immediately before and/or after the sample, and this strategy keeps its consumption to a minimum, thus reducing waste generation.

6.2.2.3. Loop-based Introduction

Loop-based sample introduction is mostly used in flow injection analysis and the design of different injection devices has contributed to the development of flow analysers. Reagent introduction into a flowing sample stream is sometimes adopted, especially when hazardous or expensive reagents are used. This strategy has confusingly been referred to as *reverse flow injection analysis* but the term *reagent injection flow analysis* [54] is more appropriate. The simultaneous introduction of sample and reagent aliquots is characteristic of flow injection systems with merging zones (see 7.1), and one of the main factors governing the performance of these systems is the precision of sample and reagent introduction. Several devices available for loop-based introduction are discussed below.

Low pressure rotary injection valves are commonly used, *e.g.*, those from Rheodyne™, typically fabricated from PTFE, with six ports and an external sampling loop (Fig. 6.9). Rotary valves with more ports, *e.g.*, 8–12, have also been used.

Good analytical precision is achieved with these valves, which can be manually, electrically or pneumatically operated. Moreover, they are capable of long-term, maintenance-free operation. The selected volume to be introduced is dependent on the length and inner diameter of the sampling loop, which can be readily changed.

Valve operation involves the transfer of the sampling loop between two channels inside which the sample and the carrier stream solutions are flowing. In the sampling (or load) position (Fig. 6.9, left), the sample is aspirated to fill the sampling loop, and excess sample is directed towards the waste container. At the same time, the carrier/wash stream is pumped through the valve towards the manifold. The selected sample volume is introduced into the carrier stream by switching the rotary valve to the alternative (inject) position (Fig. 6.9, right). Rapid switching of the valve ensures minimal flow perturbation as the carrier stream moves through the loop carrying the sample zone towards the manifold.

In early applications [2], a by-pass transmission line was incorporated in the valve in order to avoid disruption in the manifold during the time interval needed for the sampling loop to be displaced from the load to the inject position. With currently available valves and commutators, this by-pass line is no longer necessary, even if manual sample insertion is required.

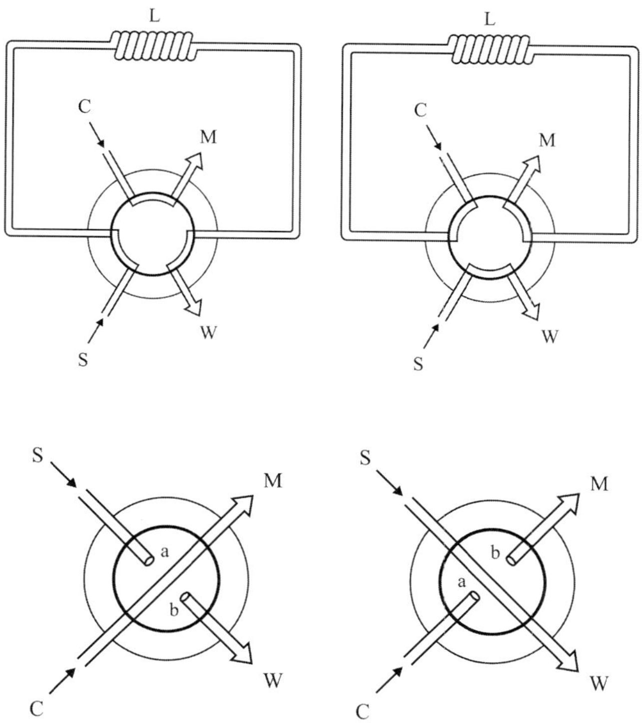

FIGURE 6.9 Schematic representation of loop-based introduction involving a six-port (upper) or a four-port (lower) rotary valve. Left and right portions = load and inject positions; S = sample; L = sampling loop; a, b (lower portion) = sites for loop insertion; C = carrier/wash stream; M = towards manifold; W = waste.

The repeatability of the introduced volume is very good provided the solutions are free of particulate matter, but sample consumption is higher than the volume actually injected (see below).

A cylindrical hole drilled inside the rotor can be used as the volume-selecting element instead of the sampling loop. By changing the rotor, different sample volumes can be selected, although the stator stays the same.

An alternative to a rotary valve is the sliding-bar valve, often called an *injector-commutator* (Fig. 6.10), that is manually or electronically operated by means of a lever [55]. The operation of sliding-bar valves is analogous to that of rotary valves, and also involves the sampling loop being positioned either in the sampling (load) or in the inject position. The loop is attached to a moving side or to the moving central bar, depending on whether a two-piece or a three-piece injector-commutator is used.

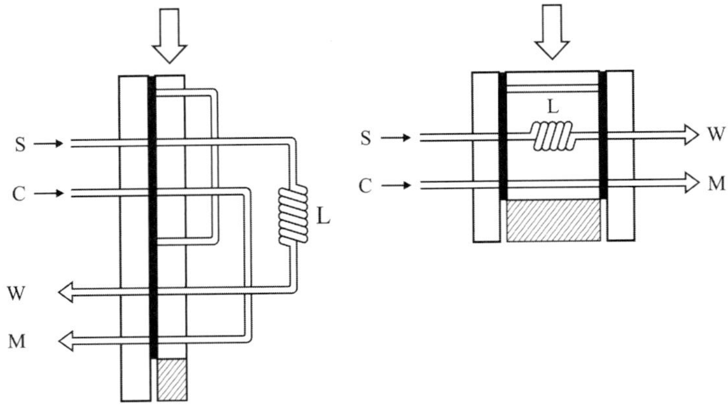

FIGURE 6.10 **Schematic representation of loop-based introduction involving two-piece (left) and three-piece (right) sliding bar injection valves.** Figure refers to load positions. S = sample; L = sampling loop; C = carrier/wash stream; M = towards manifold; W = waste; by the shaded areas. For details, see text.

A critical comparison between loop-based and time-based introduction shows that:

- Loop-based introduction is more suitable for inserting small sample volumes because the analytical repeatability tends to be preserved when the loop dimensions are reduced. For very small inserted sample volumes and the use of pulsed flows, pumping and injection should be synchronised in order to avoid sample insertion under different instantaneous flow rates (see also 5.2.3.2).
- Time-based introduction is more suitable for inserting large sample volumes for several reasons: (i) poorer repeatability is observed for low sample volumes [56]; (ii) tailing effects are less pronounced. Sample broadening starts at the injection port, in contrast to loop-based introduction, where sample broadening starts immediately after inserting the sampling loop into the carrier/wash stream; the trailing sample portion undergoes pronounced dispersion inside the loop or, in other words, before being effectively inserted into the main carrier stream; and (iii) a higher sampling rate is attained, as sample broadening is lower during the insertion process and the length of the analytical path is maintained. This length is increased for loop-based injection, especially if large sample volumes are used [57] (see also Fig. 5.12).
- Sample replacement is more cumbersome for time-based introduction, as the sample remaining inside the sampling tube can act as a carryover source. It is possible to circumvent this drawback using other valves and aspirating lines [58], but system complexity increases.

- Loop-based introduction approaches time-based introduction by sampling a very large volume and inserting only part of it [59]. Partial loop discharge is beneficial for reducing carryover effects, as the sample trailing edge formed inside the loop is directed directly towards waste during sampling of the next sample [55]. Loop filling and discharging should be performed in a reverse manner, with the sample entering one side of the loop and leaving it from the same side. This is achieved using opposing sampling and discharging flows.
- Loop-based introduction requires a larger volume of sample relative to time-based introduction, because the sampling loop should be quantitatively filled with the sample. To this end, an additional sample volume is required for quantitatively expelling the carrier/wash stream trapped inside the loop and any residue of the previous sample inside the sampling tube towards waste. This drawback is seldom emphasised in the literature. In contrast, for time-based introduction, the sample volume consumed is only the volume needed to fill the sampling tube plus the effective inserted sample volume (see above).

In order to improve system performance with loop-based introduction of large sample volumes, different approaches can be exploited:

1. Introduction of an air plug into the sampling loop just before the sample solution. Tailing effects related to the solution leaving the loop are minimised and a smaller sample volume is required.
2. Use of a syringe to assist filling the loop. The solution inside the loop immediately after switching the valve to the sampling position is washed out and, after the loop is emptied, the sample solution is carefully introduced into it by a manually operated syringe. This is very effective in reducing the required sample volume and can be done with most commercial valves.
3. Exploitation of partial and reproducible sampling. The sample is aspirated to fill the sampling loop but, as the total aspirated volume is limited, a steady state situation is not established. This approach is easily accomplished, but requires timed operation of the injection valve and is associated with an unavoidable loss in sensitivity.

6.2.2.4. Hydrodynamic Injection

Hydrodynamic injection relies on the forces exerted by stationary liquid columns and was implemented in a flow injection analyser in 1983 [48] as an efficient means of sample introduction. The volume to be injected is precisely defined in a region without moving parts, usually an H-shaped device. To this end, two pumps are normally used: one pump propels the sample towards the volume-selecting device while the second one is stopped (Fig. 6.11). The pump actions are then alternated and the flowing carrier stream pushes the selected sample portion towards the

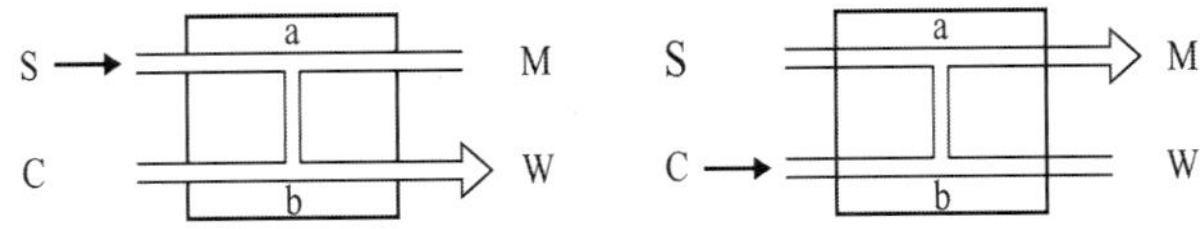

FIGURE 6.11 Schematic representation of hydrodynamic injection involving stream stopping. Left and right portions = load and inject positions; S = sample; C = carrier/wash stream; M = towards manifold; W = waste; a-b = sample inserted volume. For details, see text.

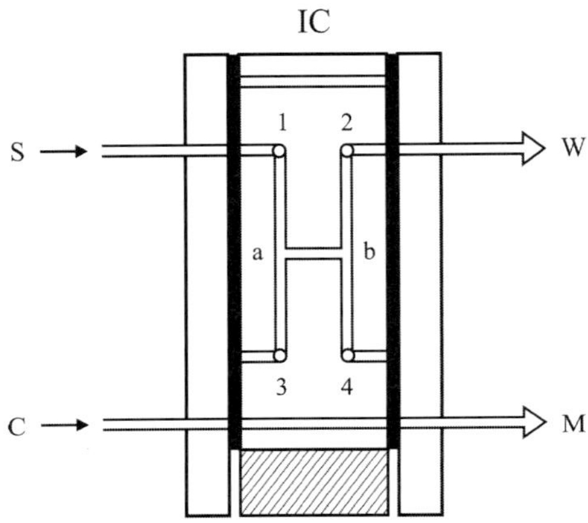

FIGURE 6.12 Schematic representation of hydrodynamic injection using an injector-commutater. The figure refers to the load position. S = sample; a-b = sample inserted volume; W = waste; C = carrier/wash stream; M = towards manifold; IC = injector-commutater; 1,2,3,4 = sites where the H-shaped device is attached; shaded area = alternative position of the injector-commutater central sliding bar. For details, see text.

main reactor. Compared with rotary valve injection, hydrodynamic injection is generally less precise and the main carrier stream is stopped during the sampling step, but the benefit of mechanical simplicity may outweigh this for certain applications. Moreover, two identical flow rates are required. Sample volumes ranging from nanolitres to millilitres can be introduced.

An alternative way to implement hydrodynamic injection is to use an injector–commutator [60]. With this strategy (Fig. 6.12), volumes as low as 600 nL can be precisely injected and the sample carrier stream is not stopped, resulting in a more stable flow. The other advantages of hydrodynamic injection are maintained.

6.2.2.5. *Nested Injection*

Nested injection involves the insertion of a sampling loop into a second loop which is simultaneously inserted into the main channel of the flow analyser [49,61]. The sample is then inserted between two plugs of a different solution into the sample carrier/wash stream, and interaction between them occurs during transport towards the detector. Alternatively, a reagent solution can be inserted into the sample, which is stopped inside a sampling loop, and the entire reaction zone is intercalated into the main channel. This latter possibility was demonstrated in the spectrophotometric flow injection determination of cellulase activity in precipitated bioreactor preparations and culture filtrates [62]: the substrate was inserted into the sample aliquot trapped in a sampling loop and the entire contents of this loop (sample plus substrate) were inserted into a buffered carrier stream; the *p*-aminobenzoylhydrazide (PABH) reagent was further added by confluence, allowing the main reaction to occur inside a heated (76 °C) knitted reactor; the coloured product was thereafter monitored, and the height of the recorded peak reflected the cellulose activity. This avoided the need for a separate blank injection.

This approach is also useful for implementing sandwich techniques [63], provided the second solution is the reagent. The sample zone is then sandwiched between two reagent plugs in a geometry analogous to the single line merging zones configuration (see also 7.1.1.2). Formation of pH gradients is also facilitated by taking advantage of nested injection, as demonstrated in the simultaneous fluorimetric determinations of ammonia and hydrazine involving reactions with *o*-phthalaldehyde and mercaptoethanol to form fluorescent derivatives at different pH values [64]. Moreover, this mode of sample introduction is inherent to monosegmented flow systems [65]: air is used instead of the second solution, and the sample is then inserted into the carrier/wash stream between two plugs of air.

In order to accomplish nested introduction, two loop-based rotary valves with coincident movements [64] or a two-section injector–commutator have been used (Fig. 6.13). In the load position, specified in the figure, the first and second loops are simultaneously filled with the sample and with the reagent (or air, if a mono-segmented flow analyser is used). Switching the injector inserts the selected sample volume between two reagent (or air) plugs into the carrier/wash stream, and the complex zone established is directed towards the detector.

6.2.3. Reactors

Analogously to process engineering, the reactor is one of the most important components of the flow system, inside which almost all the sample processing steps take place. When this component is a simple conduit to promote the transport of a sample to the detector without any chemical reaction, the device is also called a *transmission line* or *sample delivery line*. If only physical mixing is performed inside a coiled reactor,

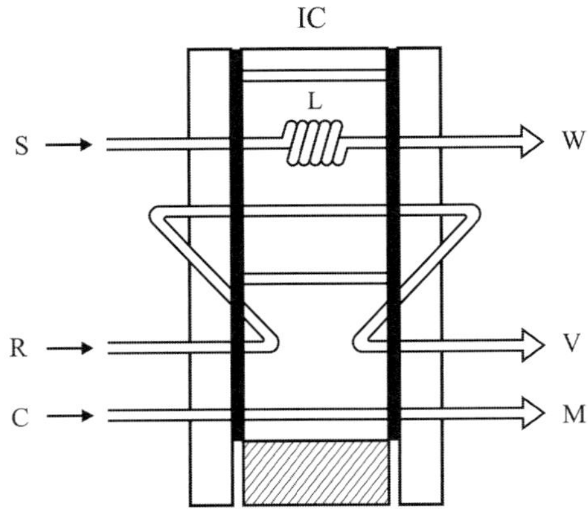

FIGURE 6.13 **Schematic representation of nested injection.** The Figure refers to the load position. S = sample; L = sampling loop; R = reagent (or air); C = carrier/wash stream; W = waste; V = reagent recovery vessel; M = towards manifold; IC = injector-commutater; shaded area = alternative position of the central sliding bar. For details, see text.

it is called a *mixing coil*. In flow analysis with *spectrophotometric* detection, analyte/reagent interactions usually take place, and optimisation of reactor design is crucial. There are different kinds of reactor (see below) and their performances have been experimentally evaluated [66].

6.2.3.1. *Coiled Reactors/Mixing Coils*

This is an open tube, which is coiled to get a more compact manifold and enhance radial homogenisation in unsegmented streams, thus improving the mixing conditions and reducing sample broadening. The parabolic profile of linear velocities typical for laminar flows through open tubes is then skewed, leading to circulation between the centre of the tube and the walls and *vice versa*. This effect is known as secondary flow and is caused by the centripetal forces acting perpendicular to the direction of the main flow. The extent of mixing depends on the aspect ratio (ratio of tube diameter to coil diameter). It should be stressed that the holding coil, the main reactor in sequential injection and bead injection systems, is usually a coiled large volume reactor. Use of this coil avoids the processed solutions coming into contact with the pump.

Glass coiled reactors are the reactors of choice in unsegmented flow analysis in view of the beneficial features arising from the air–liquid interfaces involved [47].

6.2.3.2. *Packed Bed Reactors*

This reactor is a tube (typically 2–10 cm long, i.d. 1.0 mm) filled with small particles, thus resembling a short mini-column. The particle size should be uniform so that a constant bed geometry is established and efficient mixing is attained. The streamlined flow deteriorates because the solution is forced through the small channels between the beads (the void volume), an effect known as *channelling*. Consequently, dispersion is much lower than in coiled open tubes and is similar to that of SPE mini-columns (see 8.5.2.3). The main drawback is the increased hydrodynamic pressure, which can be a limiting factor when the size of the beads is too small. In a different application of these reactors, it is sometimes necessary to insert a packed bed mini-column after the detector in order to increase the system back-pressure, thus minimising the formation of air bubbles in unsegmented flow systems (see also 4.3).

Column packing is very important during construction of the reactor. When the mini-column is not properly packed and/or the particles used are too heterogeneous, preferential pathways are established, through which most of the sample flows without exploiting the benefits of the packed bed.

6.2.3.3. *Single Bead String Reactors*

The single bead string reactor is a tube packed with particles (spheres) having a diameter that is a little larger than one half of the tube diameter [67,68]; in such a tube, the particles have a regular arrangement resembling a string. Dispersion in this type of reactor is less than in the coiled reactor but not as low as for a packed bed reactor. The advantage over packed bed reactors is that single bed string reactors generally have a lower pressure drop and are less susceptible to preferential pathway effects.

6.2.3.4. *Knitted (or Knotted) Reactors*

This reactor consists of an open tube, which is tightly knitted to ensure the presence of sharp bends [69]. Flexible tubing is better suited for this type of reactor, as it is easily knitted without creasing the tubing. Thus, high local aspect ratios and enhanced radial mixing are attained. Its performance is similar to the single bead string reactor.

6.2.3.5. *Reactor-like Artefacts*

Although not formally classified as reactors [70], a mixing chamber or a mini-column in the analytical path of a flow analyser can be regarded as a reactor-like artefact.

Inside a mixing chamber (Fig. 3.7), the inlet solutions are thoroughly mixed by the action of centripetal forces, usually assisted by a stirring device, *e.g.*, a small magnetic stirring bar [71]. The mathematical function describing the analyte concentration as a function of time is that of the tanks-in-series model with $N = 1$, i.e., the concentration at the chamber outlet exhibits an exponential response to a stepwise change in the incoming concentration (Eq. 3.9). The potential and limitations of a mixing chamber in relation to the improvement of the mixing conditions, achievement of high degree of sample dispersion, exponential dilution, time-based measurements and sample broadening, as well as the analytical applicability, are addressed in 3.1.2.2. Some phase separating devices and the flow cell can also be regarded as chamber-like components [72].

Mini-columns for analyte separation/concentration can also behave as reactors, resembling the packed bed reactor. In this context, organic polymer monoliths, largely used in the medical and biological fields [73], should be highlighted. Monolithic mini-columns consist of continuous beds with macropores and mesopores which are characterised by low back-pressure effects. These columns offer several other advantages [74], as emphasised in Chapter 8. In the context of flow analysis, monolithic mini-columns were implemented in a sequential injection analyser in 2003 [75] and the potential and limitations of the approach, called *Sequential Injection Chromatography,* were recently reviewed [76].

In summary, reactors in flow analysis can be either an open tube or a tube with particles inside it. The straight open tube reactor has been used in relation to theoretical studies of dispersion, in order to simplify the mathematical models involved. The coiled reactor is the most commonly used reactor due to its simplicity, low maintenance, compactness and ease of construction. One of its variants, the knitted reactor, has a superior ability to minimise sample broadening but is rarely used. The packed bed reactor has often been used, as it is compact and efficiently minimises sample dispersion. This kind of reactor is also used in some analytical procedures involving immobilised reagents (including solid phase extraction). In fact, the solid particles are tightly fixed together in a geometry similar to that of a typical packed bed reactor. Finally, the single bed string reactor offers interesting advantages, but has not been widely used.

6.2.4. Accessories

The manifold also comprises different accessories such as connectors, phase separating devices, and containers, some of which are described below.

6.2.4.1. Connectors

Fittings, barbed fittings and unions from different suppliers or home-made devices are used to connect the components of the manifold. They should provide suitable linking conditions to avoid loss of the handled solutions and/or the ingress of air bubbles. Glue, heat shrink tubing and push fit connections are often used to guarantee a complete seal. A small inner volume is a desirable manifold feature; the connectors should therefore be as small as possible, especially those placed along the analytical path.

Two types of connector are discussed below, namely, those used for linking manifold tubes together and those for attaching tubes to other manifold components, *e.g.*, injectors, valves, chambers and mini-columns.

CONNECTORS FOR LINKING TUBES TOGETHER

For linking manifold tubes to pumping tubes, or two similar pieces of tubing, external gloves or sleeves, usually made of Tygon® or silicone, have been used. If the tubes to be connected have different external diameters (Fig. 6.14, upper), the thinner one should be inserted into the thicker one (push fit). As a rule, the manifold tube is normally cut at a 45° angle to make the insertion easier. Alternatively, if the tubes to be connected have similar diameters, an external sleeve can be used (Fig. 6.14, lower).

CONNECTORS FOR LINKING MANIFOLD TUBES WITH OTHER COMPONENTS

Triple and quadruple unions, generally constructed in Perspex and allowing different merging angles [77] are commonly used for combining convergent streams or attaching tubes to manifold devices. These unions are usually built in the external sides of the manifold devices and often have special fittings, *e.g.*, screws and flanges (Fig. 6.15), to provide more reliable connections.

For the addition of confluent streams, badly designed connectors can lead to fragmented mixing that manifests itself as undulations in the recorded baseline. This drawback becomes more severe when there are

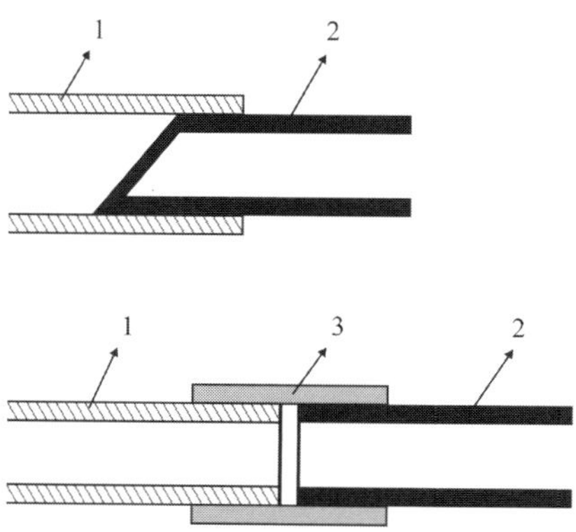

FIGURE 6.14 **Connecting a pump tube to a manifold tube.** Upper and lower portions = diameter of the manifold tube different from (upper) and similar to (lower) the pumping tube; 1 = pumping tube; 2 = manifold tube; 3 = sleeve.

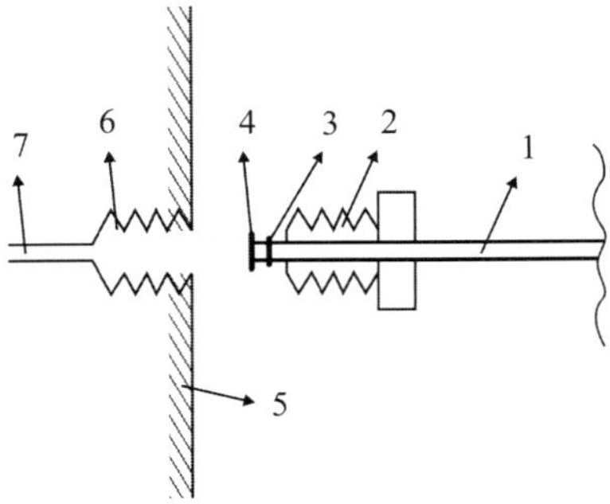

FIGURE 6.15 **Schematic representation of the attachment of a tube to a manifold component.** 1 = manifold tube; 2 = screw; 3 = "O" ring for sealing; 4 = flanged tube end; 5 = surface of the manifold component; 6 = screw fitting; 7 = outlet.

pronounced differences in the flow rates of the convergent streams [78]. This effect can be circumvented by:

- Using different inner diameters for each of the convergent channels so as to equalise the mean linear speed of the convergent streams.
- Adding the convergent streams at the opposite sides of a T-shaped triple union, *i.e.*, at 180°, in order to get a frontal interaction between the solutions involved.
- Using a connector that includes a needle which delivers the confluent stream at the central portion of the main stream, this approach often being exploited in segmented flow analysis.
- Using special strategies such as those involving, *e.g.*, a coaxial jet mixer [79], a packed bed connector [80] or a metal helix insertion [66].

Computer-operated valves and other active components can also behave as connectors, as at least three tubes are attached to them. The potential of this innovation is discussed in 5.4.3.

6.2.4.2. Solution Containers

The reagent, diluent, and carrier/wash solutions should be accommodated in specific flasks made from materials compatible with the solutions to be pumped, into which the sampling tubes are permanently immersed. Alternatively, the sampling tube can be attached near the flask bottom, in order to permit the placement of the reagent containers above the height of the manifold. This is recommended for minimising the eventual formation of air bubbles under reduced hydrodynamic pressure.

It is a good practice to place a filtering device (ceramic or synthetic material) between the solution to be pumped and the sampling tube to prevent the influx of particulate matter into the manifold.

Reagent containers should preferably have a cap in order to avoid contamination of the laboratory environment by volatile chemical compounds and/or contamination of the solutions to be pumped. An orifice for maintaining the container inner pressure during solution pumping is then required.

The containers for waste collection and eventual treatment and disposal should be similarly made and situated close to and above the height of the manifold. For waste needing immediate treatment, a specific solution (*e.g.*, with a suitable precipitating agent) should be added to the container for waste disposal prior to system operation and periodically replaced.

6.3. DETECTION AND DATA PROCESSING

The flow system is an effective means of precise sample presentation to the detector. The detection unit contains a flow cell with a fixed location and orientation; the handled sample is then always monitored with a fixed geometry.

Relevant parameters such as the characteristics of the flow cell, type of detector and data acquisition and treatment are briefly presented here. For a further discussion of developments in instrumentation related to the detection unit, overview articles [81–83] are recommended.

6.3.1. Flow Cells

For absorbance measurements, the flow cell is generally designed to permit the utilisation of a radiation beam parallel to its longitudinal axis (Fig. 6.16). Regardless of the flow cell type (classical, Z-shaped, spiral and waveguide-based), an averaged measurement (integration of measurements related to the fluid elements along the optical path) is always obtained. This is also true for luminometric measurements, which are not associated with an incident radiation beam.

A U-shaped glass tube was used in the first segmented flow analysers with spectrophotometric detection but is now rarely used due to the pronounced radiation losses at the curved portions and in the cylindrical walls of the tube. A single tube axially traversed by the radiation incident beam can be used instead of a typical flow cell in situations requiring a short optical path, as originally demonstrated in the spectrophotometric determination of the major constituents of fertilisers in a flow injection system [84]; this approach avoided the need for manual sample dilutions.

The flow cell inner volume should be selected as a compromise between sample broadening, collimation of the radiation beam and measurement repeatability. For too large a volume, the flow cell can act as

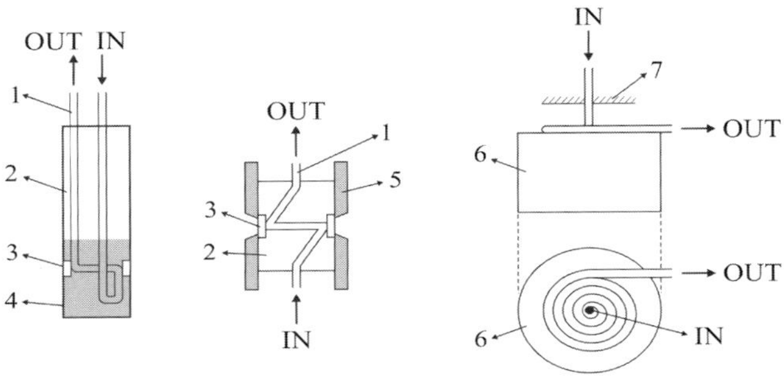

FIGURE 6.16 **Schematic representations of classical (left), Z-shaped (centre) and spiral (right) flow cells.** 1 = manifold tube; 2 = flow cell body; 3 = transparent window; 4 = body portion protected from light; 5 = support; 6 = photomultiplier tube; IN and OUT = stream inlet and outlet. The spiral flow cell is shown in side (upper) and plan (lower) view. For details, see text.

a chamber-like component, leading to excessive broadening of the flowing sample. The recorded concentration/time function would then not be truly monitored. On the other hand, too small an inner volume can lead to difficulties in collimation of the radiation beam, this aspect becoming more severe for shorter wavelengths. In addition, the number and volume of fluid elements involved might not be high enough to guarantee good measurement repeatability, especially for luminometric measurements where repeatability tends to be more dependent on the counting statistics. Different kinds of flow cells have been used, as discussed below.

6.3.1.1. Classical and Z-shaped Flow Cells

For flow-based spectrophotometric or turbidimetric analytical procedures, a general-purpose spectrophotometer can be used with either a classical or a Z-shaped flow cell (Fig. 6.16, left). Classical flow cells require a cell holder to be incorporated into the spectrophotometer, and this holder is specific for the instrument concerned. Z-shaped cells (Fig. 6.16, right) are more amenable to optical fibre technology. Connecting the flow cell to the manifold tubing is generally done as specified in 6.2.4.1, provided the flow cell necks can connect in the same way as manifold tubing or that special tube fittings can be placed over the flow cell inlet and outlet. In order to prevent the ingress of air bubbles, the stream should enter the flow cell compartment via a J-trajectory or feed the Z-shaped cell compartment from the bottom [85]. At present there is a tendency to use Z-shaped flow cells, probably due to their less complicated construction and, consequently, lower cost and maintenance.

Deposition of material or adherence of bubbles on the flow cell windows adversely affects the detector response (see also 4.3). The flow cells should therefore exhibit stay-clean properties. Flow perturbation inside the flow cell is often sufficient to prevent material accumulation provided there is adequate flushing but a secondary clean-up step involving, *e.g.*, surfactant or solvent addition to reagents, acid/base washing or back-flushing with water and/or air may be necessary.

The external parts of the flow cell to be traversed by the incident/emergent radiation (windows) are made from suitably transparent material (see Table 4.1), whereas the other external portions are generally protected from stray light. It is good practice to use black (or black painted) materials for building the flow cell and to close the cell compartment of the spectrophotometer during measurements. In modern instruments for absorbance/turbidance measurements exploiting optical fibres, this protection is not so important in view of the excellent collimation of the radiation beam and the possibility of chopping the incident radiation beam. For fluorimetry, chemiluminescence and bioluminescence, placement of the flow cell between a mirror and the detector improves the collection efficiency of the emitted radiation by reflection towards the detector. For these techniques, spiral flow cells are mostly commonly used.

6.3.1.2. *Spiral Flow Cells*

The main reactor is constructed in a planar spiral geometry and tightly attached to the detector (Fig. 6.16, right). The flowing sample generally enters the centre of the spiral, undergoes circular motion without leaving the detector window and exits from the outer coil of the spiral. In this way, a 2π geometry is attained. As emphasised above, placement of a mirror outside of the spiral increases the amount of radiation reaching the detector.

6.3.1.3. *Long Optical Path Length Flow Cells*

The analytical sensitivity is enhanced by increasing the b value (Eqs. 4.4, 4.9, 4.17); hence, flow cells can be designed to provide long optical path lengths. However, the b value in classical, Z-shaped or spiral flow cells cannot be increased indefinitely due to limitations in cell geometry and radiation beam collimation.

A very long optical path (one or two orders of magnitude longer than conventional flow cell geometries) can be established by channelling the radiation beam inside a long coiled tube which acts as a liquid core waveguide (LCW) [86]. For absorbance or turbidance measurements, the optical fibres associated with the incident and emergent light beams are connected axially to the optical path near the flow cell ends (Fig. 6.17). For luminescence measurements, only one optical fibre is needed for channelling the emitted radiation towards the detector.

The LCW tube is made from a material with a refractive index lower than that of the flowing solution. The solid/liquid interface (tube inner

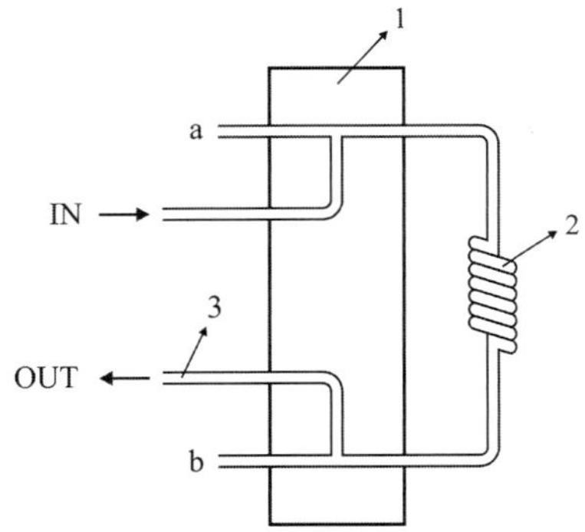

FIGURE 6.17 **Schematic representation of a long optical path flow cell.** 1 = support; 2 = coiled liquid core waveguide; 3 = manifold tubing; a and b = points for connecting optical fibres. For details, see text.

wall) acts as a mirror provided the radiation beam reaches the tube walls at angles >9.9–14.1° (depending on the tube material) measured from the normal of the LCW tubing [87]. As the photons involved undergo multiple reflections at the tube inner wall, the radiation is constrained inside the liquid core and a very long effective optical path length is established, thus increasing sensitivity.

Very saline samples and/or the addition of organic solvents to the samples were used in early applications in order to provide high refractive indices [88], but the applicability of the LCW flow cell was limited because, in general, only dilute aqueous samples with low refractive indices have been analysed. The development of different kinds of amorphous fluoropolymers (Teflon AFs) with refractive indices lower than that of pure water enabled the design and overall acceptance of flow cells relying on LCWs [89,90].

LCW tubing can be made either from Teflon AF-2400 or from a silica capillary covered with Teflon AF-1600 or Teflon AF-2400 [87]. With this latter arrangement, there is no contact between the flowing sample and the porous fluoropolymer, thus the adsorption of chemical species and retention of micro-air bubbles are minimised.

The analytical sensitivity tends to be linearly dependent on the optical path length [91] and absorbance values of about 80% of the expected ones have been reported [87,92]. This does not hold for the Schlieren effect, as the increase in the Schlieren noise with optical path length is less than proportional. This is due to the excellent collimation of the radiation beams involved [93] and is another favourable characteristic of LCW flow cells.

Excellent collimation is inherent in the use of optical fibre technology, which has had a major impact on flow analysis. As mentioned above, better collimation means less Schlieren noise and less stray light and, hence, a better signal-to-noise ratio. The potential exists for developing different flow cell configurations (or optrodes) connected to a remote spectrometer by optical fibres, with the option of multiplexing a number of flow cells.

LCW flow cells are compatible with compact solid-state photometers relying on light-emitting diode (LED) sources and photo-diode-resistive detectors [94]. Recently, a total internal reflection flow cell highly tolerant to Schlieren effects was designed [95]. The incident radiation beam from an optical fibre was introduced into a short-length quartz capillary through the sidewall at an incident angle of 53°.

Other strategies for presenting the handled sample to the detector, involving the use of beads for solid phase spectrophotometry or hanging drop approach for measurements without a flow cell, are discussed in the Sections 4.1.1.4 and 2.6.

6.3.2. Detectors

The detector should be compatible with the monitored flowing sample, and general analytical requirements such as sensitivity, geometry and response time have to be considered from the outset. For spectrophotometry, the modulation frequency, which is used to discriminate the analytical signal from the continuous background signal due to stray light, is also important. The signal integration time and the detector time constant can also be adjusted to suit the timescale of the peak width.

Classical transducers such as the photomultiplier tube and the photodiode have been used in most the flow systems [96,97]. Since the 1980s, the tendency to use diode array spectrophotometers has increased, mainly for simultaneous determinations [98] and/or for implementing dual wavelength spectrophotometry [99], the latter being more exploited in segmented flow analysis.

More recently, charge coupling devices (CCDs) have been used as detectors in flow analysis [100], especially in relation to flow cytometry [101].

6.3.3. System Control, Data Acquisition and Data Treatment

Reproducible timing is fundamental to flow analysis and is most effectively achieved by automation, which also allows unattended operation. An on-board computer can be used to control pumps, injection devices and ancillary devices such as switching valves and also for system operation, data acquisition and processing.

In addition to a powerful central processing unit, sufficient memory for program execution and raw data storage and manipulation is required. In order to acquire data from the detector, an analogue-to-digital converter with high resolution and a sampling frequency compatible with the detector response time, as well as appropriate data output facilities, are required. To enable rapid and consistent system start-up, especially in the event of a power failure, the incorporation of a battery-backed RAM is recommended.

The long-term accuracy of remotely deployed flow systems can be maintained by incorporating automated self-calibration procedures. Frequent recalibration allows the system to adapt to temporal changes in analytical sensitivity, detector response and system wear and tear, *e.g.*, pump tube degradation. If the detector response is demonstrably linear, a single point calibration is adequate for slope adjustment [102] and this is easily achieved via a two-way switching valve connected to the sample and standard flow channels.

Once stored, the data can be manipulated by a variety of algorithms to enhance the quality of the output information. General smoothing functions can be used to improve precision in addition to peak-find routines, procedures to determine peak height and width and area integration methods [103]. Algorithms based on differential rates of change in detector response can be used to eliminate detector spikes due to entrained air, and various statistical process control parameters can be monitored to identify unacceptable results. In the event of such a result, the flow system can easily be reset for repeat analysis and/or calibration; depending on the complexity of the manifold and the chemistry involved; this task can usually be accomplished in less than one minute.

6.4. MINIATURISATION OF THE FLOW SYSTEM

A full discussion of miniaturised flow system is outside of the scope of this monograph. Due to the importance of miniaturisation for the future of flow analysis, however, general comments on downsizing in flow analysis are included here.

The development of flow systems has contributed to a decrease in the volume of samples and reagents needed for a particular chemical analysis. This tendency was reinforced with the advent of micro-scale analytical methods relying on, *e.g.*, electrophoresis and micro-total analysis systems (μTAS), which influenced the introduction of miniaturised flow systems. Miniaturisation achieves a significant reduction in sample and reagent consumption and allows the design of compact and portable systems. As a result, the generation of chemical waste and the cost of analysis are reduced, and the possibility for *in situ* or *in vivo* assays is increased. Thus, in recent years, several miniaturised flow systems have been proposed.

One of the first prototypes involved the use of integrated micro-conduits [104]. The flow channels, the injection device and the outlet to the detector were mechanically engraved into a flat PVC block using channels with a cross sectional area of 0.8 mm^2 and covered with a flat plate glued on top of the machined block. The system was operated with a conventional peristaltic pump. A similar system, but using a piston pump, was proposed in the 1990's [105].

Other miniaturised devices have used machined modules made of polymeric materials such as poly(methyl methacrylate) with diameters of 0.8 mm coupled with peristaltic or syringe pumps [106] or exploiting electrokinetic pumping [107], as well as silicon devices with piezo electric pumping [108].

At the end of the 1990's, the fabrication of glass micro-scale manifolds using photolithographic etching was proposed. Electro-osmotic flow could be exploited due to the presence of SiOH groups (glass); the flow was established by applying a voltage along the channel [109–111].

Currently, μTAS [112] and sequential injection systems using the lab-on-valve approach [22] are more popular. The former exploits hydrodynamic injection and is now known as Lab-on-Chip due to the extreme reduction in dimensions. Reproducible analytical channels are made by micro-drilling, etching, photolithography or laser erasing, and different channel profiles can be obtained at a reasonable cost.

On the other hand, sequential injection with the lab-on-valve approach (5.5.3) does not exploit the strategy of downscaling the channel diameter but decreases the channel length and maintains the same approach as used in conventional sequential injection systems.

Regardless of the strategy utilised in the miniaturisation process, it should be noted that miniaturisation does not necessarily involve simple downscaling of the dimensions of the flow system and related parameters. For this reason, the concept of similarity for scaling and modelling the behaviour of fluids has been proposed based on three cornerstones: geometric similarity, kinematic similarity and dynamic similarity [113].

References

[1] J.W.B. Stewart, J. Ruzicka, H. Bergamin-Filho, E.A.G. Zagatto, Flow injection analysis. Part III. Comparison of continuous flow spectrophotometry and potentiomety for the rapid determination of total nitrogen content in plant digests, Anal. Chim. Acta 81 (1976) 371.

[2] D. Betteridge, Flow injection analysis, Anal. Chem. 50 (1978) 832A.

[3] M. Trojanowicz, Flow Injection Analysis. Instrumentation and Applications, World Scientific Publishing Co., Singapore, 2000.

[4] J.P. Smith, V. Smith-Hinson, Flow injection analysis: quietly pushing ahead, Anal. Chem. 74 (2002) 385A.

[5] J. Ruzicka, Flow Injection Analysis. From Beaker to Microfluidics (E-book), fourth ed., self published byavailable free of charge at, Jarda Ruzicka, Seattle, 2009, http://www.flowinjection.com/freecd.aspx, 2009.
[6] W. Lipps, Flow injection or segmented flow analysis – what are the similarities and differences?, 18 Jun. 2009. EzineArticles.com. 17 Sep 2009 <http://ezinearticles.com/?Flow-Injection-Or-Segmented-Flow-Analysis—-What-Are-the-Similarities-and-Differences?&id=2492372 >.
[7] Z.L. Fang, M. Sperling, B. Welz, Comparison of three propulsion systems for application in flow-injection zone penetration dilution and sorbent extraction preconcentration for flame atomic absorption spectrometry, Anal. Chim. Acta 269 (1992) 9.
[8] P.J. Worsfold, J. Ruzicka, E.H. Hansen, Rapid automated enzymatic method for the determination of alcohol in blood and beverages using flow injection analysis, Analyst 106 (1981) 1309.
[9] E.A.G. Zagatto, A.O. Jacintho, J. Mortatti, H. Bergamin-Filho, An improved flow injection determination of nitrite in waters by using intermittent flows, Anal. Chim. Acta 120 (1980) 399.
[10] J.G. March, B.M. Simonet, F. Grases, Kinetic-turbidimetric determination of phytic acid by sequential injection analysis, Anal. Chim. Acta 409 (2000) 9.
[11] P. Francis, S.W. Lewis, K.F. Lim, K. Carlsson, B. Karlberg, Flow analysis based on a pulsed flow of solution: theory, instrumentation and applications, Talanta 58 (2002) 1029.
[12] R.A.S. Lapa, J.L.F.C. Lima, B.F. Reis, J.L.M. Santos, E.A.G. Zagatto, Multi-pumping in flow analysis: concepts, instrumentation, potentialities, Anal. Chim. Acta 466 (2002) 125.
[13] K. Yasuda, T. Kuroishi, Magic of multi-port valve for flow injection analysis, J. Flow Injection Anal 6 (1989) 5.
[14] R.C. Prados-Rosales, J.L. Luque-Garcia, M.D. Luque de Castro, Propelling devices: the heart of flow injection approaches, Anal. Chim. Acta 461 (2002) 169.
[15] P. MacLaurin, K.N. Andrew, P.J. Worsfold, Flow injection analysis, in: F. McLennan, B.R. Kowalski (Eds.), Process Analytical Chemistry, Blackie, Glasgow, 1995, p. 159.
[16] B. Karlberg, G.E. Pacey, Flow Injection Analysis: A Practical Guide, Elsevier, 1989.
[17] M. Valcarcel, M.D. Luque de Castro, Flow Injection Analysis. Principles and Applications, Ellis Horwood, Chichester, (1987) 101.
[18] B.F. Reis, M.F. Gine, E.A.G. Zagatto, J.L.F.C. Lima, R.A.S. Lapa, Multicommutation in flow analysis. Part 1. Binary sampling: concepts, instrumentation and spectrophotometric determination of iron in plant digests, Anal. Chim. Acta 293 (1994) 129.
[19] I.L. Garcia, P. Vinas, N. Campillo, M.H. Cordoba, Use of submicroliter-volume samples for extending the dynamic range of flow-injection flame atomic absorption spectrometry, Anal. Chim. Acta 308 (1995) 85.
[20] J. Ruzicka, G.D. Marshall, Sequential injection: a new concept for chemical sensors, process analysis and laboratory assays, Anal. Chim. Acta 237 (1990) 329.
[21] J. Ruzicka, L. Scampavia, From flow injection to bead injection, Anal. Chem. 71 (1999) 257A.
[22] J. Ruzicka, Lab-on-valve: universal microflow analyzer based on sequential and bead injection, Analyst 125 (2000) 1053.
[23] F. Albertus, B. Horstkotte, A. Cladera, V. Cerda, A robust multisyringe system for process flow analysis. Part I. On-line dilution and single point titration of protolytes, Analyst 124 (1999) 1373.
[24] S.B. Neff, T.A. Neff, S. Gerber, M.M. Weiss, Flow rate, syringe size and architecture are critical to start-up performance of syringe pumps, Eur. J. Anaesthesiol. 24 (2007) 602.
[25] G.D. Christian, J. Ruzicka, Exploiting stopped-flow injection methods for quantitative chemical assays, Anal. Chim. Acta 261 (1992) 11.

[26] V. Cerda, J.M. Estela, R. Forteza, A. Cladera, E. Becerra, P. Altimira, P. Sitjar, Flow techniques in water analysis, Talanta 50 (1999) 695.
[27] P. Woias, Micropumps: past, progress and future prospects, Sensors Actuator B – Chem. 105 (2005) 28.
[28] D.A. Weeks, K.S. Johnson, Solenoid pumps for flow injection analysis, Anal. Chem. 68 (1996) 2717.
[29] G.D. Owens, R.W. Taylor, T.Y. Ridley, D.W. Margerum, Pulsed-flow instruments for measurements of fast reactions in solution, Anal. Chem. 52 (1980) 130.
[30] P.R. Fortes, M.A. Feres, M.K. Sasaki, E.R. Alves, E.A.G. Zagatto, J.A.V. Prior, J.L.M. Santos, J.L.F.C. Lima, Evidences of turbulent mixing in multi-pumping flow systems, Talanta 79 (2009) 978.
[31] E. Stemme, G. Stemme, A valveless diffuser/nozzle-based fluid pump, Sensor Actuator A – Phys. 39 (1993) 159.
[32] J.H. Zhang, J.H. Lu, Q.X. Xia, Research on the valveless piezoelectric pump with Y-shaped pipes, Front. Mech. Eng. China 2 (2007) 144.
[33] J.H. Zhang, J.H. Lu, Q.X. Xia, H. Li, Theoretical analysis and experimental investigation of valveless piezoelectric pumps with unsymmetrical ridges, Front. Mech. Eng. China 2 (2007) 13.
[34] M.F.T. Ribeiro, J.L.M. Santos, J.L.F.C. Lima, Piezoelectric pumping in flow analysis: application to the spectrophotometric determination of gabapentin, Anal. Chim. Acta 600 (2007) 14.
[35] K.K. Stewart, G.R. Beecher, P.E. Hare, Rapid analysis of discrete samples: the use of nonsegmented, continuous flow, Anal. Biochem 70 (1976) 167.
[36] J. Ruzicka, E.H. Hansen, A.U. Ramsing, Flow injection analyzer for students, teaching and research. Spectrophotometric methods, Anal. Chim. Acta 134 (1982) 55.
[37] R.Y. Wang, J.A. Jarrat, P.J. Keay, J.J. Hawkes, W.T. Coakley, Development of an automated on-line analysis system using flow injection ultrasound filtration and CCD detection, Talanta 52 (2000) 129.
[38] K. Grudpan, C. Taylor, H. Sitter, C. Keller, Flow injection analysis using an aquarium air pump, Fresenius J. Anal. Chem. 346 (1993) 882.
[39] R.C. Matos, I.G.R. Gutz, L. Angnes, R.S. Fontenele, J.J. Pedrotti, A versatile and pulsation free pneumatic impeller for flow analysis systems, Quim. Nova 24 (2001) 795.
[40] P.K. Dasgupta, S. Liu, Electroosmosis: a reliable fluid propulsion system for flow injection analysis, Anal. Chem. 66 (1994) 1792.
[41] B.J. Harmon, D.H. Patterson, F.E. Regnier, Mathematical treatment of electrophoretically mediated micro-analysis, Anal. Chem. 65 (1999) 2655.
[42] F. Theeuwes, S. Yum, Principles of design and operation of generic osmotic pumps for delivery of semisolid or liquid drug formulations, Ann. Biomed. Eng. 4 (1976) 343.
[43] F.-Q. Nie, M. Macka, B. Paull, Micro-flow injection analysis system: on-chip sample preconcentration, injection and delivery using coupled monolithic electroosmotic pumps, Lab Chip 7 (2007) 1597.
[44] H.W. Jannasch, K.S. Johnson, C.M. Sakamoto, Submersible, osmotically pumped analyzers for continuous determination of nitrate in situ, Anal. Chem. 66 (1994) 3352.
[45] J.C. Andrade, M. Ferreira, N. Baccan, A constant-flow gravity liquid-feed system for use in flow injection analysis, Quim. Nova 9 (1986) 123.
[46] H.H. Liu, P.K. Dasgupta, Analytical chemistry in a drop, Trends Anal. Chem 15 (1996) 468.
[47] L.R. Snyder, H.J. Adler, Dispersion in segmented flow through glass tubing in continuous-flow analysis: The ideal model, Anal. Chem 48 (1976) 1017.
[48] J. Ruzicka, E.H. Hansen, Recent developments in flow-injection analysis. Gradient techniques and hydrodynamic injection, Anal. Chim. Acta 145 (1983) 1.

[49] P.K. Dasgupta, H. Huang, Application of a nested loop system for flow injection analysis of trace aqueous peroxides, Anal. Chem. 57 (1985) 1009.
[50] J. Ruzicka, E.H. Hansen, Flow injection analyses. Part I. A new concept of fast continuous flow analysis, Anal. Chim. Acta 78 (1975) 145.
[51] B.F. Reis, M.F. Gine, E.A.M. Kronka, Low-cost microcomputer for system control and data treatment in flow injection analysis, Quim. Nova 15 (1992) 231.
[52] N. Bennaoui, C. Periou, C. Harault, G. Lemoel, Evaluation of the Technicon DAX 48, a multiparametric biochemical analyzer, Ann. Biol. Clin. 51 (1993) 713.
[53] S. Vicente, E.P. Borges, B.F. Reis, E.A.G. Zagatto, Exploitation of tandem streams for carry-over compensation in flow analysis. I. Turbidimetric determination of potassium in fertilizers, Anal. Chim. Acta 438 (2001) 3.
[54] A.G. Fogg, A.E. Davies, Reverse-flow injection or reagent injection? Normal flow-injection, reverse flow-injection and single-boundary measurements, Analyst 117 (1992) 1055.
[55] F.J. Krug, H. Bergamin-Filho, E.A.G. Zagatto, Commutation in flow injection analysis, Anal. Chim. Acta 179 (1986) 103.
[56] S.D. Rothwell, A.A. Woolf, A timed solenoid injector for flow analysis, Talanta 32 (1985) 431.
[57] A.U. Ramsing, J. Ruzicka, E.H. Hansen, The principles and theory of high speed titrations by flow injection analysis, Anal. Chim. Acta 129 (1981) 1.
[58] M.A. Feres, P.R. Fortes, E.A.G. Zagatto, J.L.M. Santos, J.L.F.C. Lima, Multi-commutation in flow analysis: recent developments and applications, Anal. Chim. Acta 618 (2008) 1.
[59] H.L. Wu, Y. Hayashibe, Y. Sayama, M. Shibukawa, K. Oguma, A computer-controlled variable-volume injector for flow injection analysis, Anal. Sci. 11 (1995) 401.
[60] E.A.G. Zagatto, O. Bahia-Filho, M.F. Gine, H. Bergamin-Filho, A simple procedure for hydrodynamic injection in flow injection analysis applied to the atomic absorption spectrometry of chromium in steels, Anal. Chim. Acta 181 (1986) 265.
[61] J. Ruz, A. Rios, M.D. Luque de Castro, M. Valcarcel, Determination of glucose in alcoholic beverages by flow injection with two internally coupled injection valves and an enzyme reactor, Anal. Chim. Acta 211 (1988) 281.
[62] P.J. Worsfold, I.R.C. Whiteside, H.F. Pfeiffer, H. Waldhoff, An automated spectrophotometric flow-injection procedure for the determination of cellulase activity in bioreactor preparations, J. Biotechnol. 14 (1990) 81.
[63] J. Alonso, J. Bartroli, M. Del Valle, M. Escalada, R. Barber, Sandwich techniques in flow injection analysis. Part 1. Continuous recalibration techniques for process control, Anal. Chim. Acta 199 (1987) 191.
[64] A. Rios, M.D. Luque de Castro, M. Valcarcel, Simultaneous flow-injection fluorimetric determination of ammonia and hydrazine with a novel mode of forming pH gradients, Anal. Chim. Acta 187 (1986) 139.
[65] C. Pasquini, W.A. Oliveira, Monosegmented system for continuous-flow analysis. Spectrophotometric determination of chromium(VI), ammonia, and phosphorus, Anal. Chem. 57 (1985) 2575.
[66] C. Silfwerbrand-Lindh, L. Nord, L.-G. Danielsson, F. Ingman, The analysis of aqueous solutions with ethanol soluble reagents in a flow injection system. Spectrophotometric determination of uranium, Anal. Chim. Acta 160 (1984) 11.
[67] J.M. Reijn, W.E. van der Linden, H. Poppe, Dispersion in open tubes and tubes packed with large glass beads. The single bead string reactor, Anal. Chim. Acta 123 (1981) 229.
[68] J.M. Reijn, W.E. van der Linden, H. Poppe, Transport phenomena in flow injection analysis without chemical reaction, Anal. Chim. Acta 126 (1981) 1.
[69] U.D. Neue, H. Engelhardt, Reaction detector with three-dimensional coiled open tubes in HPLC, Chromatographia 15 (1982) 403.

[70] W.E. van der Linden, Classification and definition of analytical methods based on flowing media – (IUPAC recommendations 1994), Pure Appl. Chem. 66 (1994) 2493.
[71] D.T. Burns, N. Chimpalee, M. Harriott, Flow-injection spectrophotometric determination of cobalt by extraction as tetrathiocyanatocobaltate(II) with the ethylenebis(triphenylphosphonium) cation, Anal. Chim. Acta 225 (1989) 123.
[72] E.A.G. Zagatto, J.M.T. Carneiro, S. Vicente, P.R. Fortes, J.L.M. Santos, J.L.F.C. Lima, Mixing chambers in flow analysis: a review, J. Anal. Chem. 64 (2009) 540.
[73] J.F. Wang, Y.J. Xu, S.G. Liu, S. Jiang, C.P. Pan, The application of monolithic columns in pesticides analysis, J. Sep. Sci. 30 (2006) 3.
[74] K.C. Saunders, A. Ghanem, W.B. Hon, E.F. Hilder, P.R. Haddad, Separation and sample pre-treatment in bioanalysis using monolithic phases: a review, Anal. Chim. Acta 652 (2009) 22.
[75] D. Satinsky, P. Solich, P. Chocholous, R. Karlicek, Monolithic columns. A new concept of separation in the sequential injection technique, Anal. Chim. Acta 499 (2003) 205.
[76] M. Fernandez, H.M. Gonzalez-San Miguel, J.M. Estela, V. Cerda, Contribution of multi-commuted flow analysis combined with monolithic columns to low-pressure, high-performance chromatography, Trends Anal. Chem. 28 (2009) 336.
[77] G.D. Clark, J.M. Hungerford, G.D. Christian, Optimization of confluent mixing in flow-injection analysis, Anal. Chem. 61 (1989) 973.
[78] E.A.G. Zagatto, B.F. Reis, M. Martinelli, F.J. Krug, H. Bergamin-Filho, M.F. Gine, Confluent streams in flow injection analysis, Anal. Chim. Acta 198 (1987) 153.
[79] L.D. Scampavia, G. Blankenstein, J. Ruzicka, G.D. Christian, A coaxial jet mixer for rapid kinetic analysis in flow injection and flow injection cytometry, Anal. Chem. 67 (1995) 2743.
[80] K.L. Marques, C.K. Pires, J.L.M. Santos, E.A.G. Zagatto, J.L.F.C. Lima, A multi-pumping flow system for chemiluminescent determination of ammonium in natural waters, Int. J. Environ. Anal. Chem. 87 (2007) 77.
[81] A.V. Medvedovici, V. David, Spectrophotometry: diode array, in: P.J. Worsfold, A. Townshend, C.F. Poole (Eds.), second ed., Encyclopedia of Analytical Science, vol. 8, Elsevier, Oxford, 2005, p. 321.
[82] J.N. Miller, Fluorescence: instrumentation, in: P.J. Worsfold, A. Townshend, C.F. Poole (Eds.), second ed., Encyclopedia of Analytical Science, vol. 3, Elsevier, Oxford, 2005, p. 106.
[83] C. Burgess, Optical spectroscopy: detection devices, in: P.J. Worsfold, A. Townshend, C.F. Poole (Eds.), second ed., Encyclopedia of Analytical Science, vol. 6, Elsevier, Oxford, 2005, p. 438.
[84] E.H. Hansen, F.J. Krug, A.K. Ghose, J. Ruzicka, Rapid determination of K, P and K in fertilizers by flow injection analysis, Analyst 102 (1977) 714.
[85] B. Karlberg, G.E. Pacey, Flow Injection Analysis. A Practical Guide, Elsevier, Amsterdam, 1989, pp. 70–71.
[86] M.R. Callahan, J.B. Rose, R.H. Byrne Rose, Long pathlength absorbance spectroscopy: trace copper analysis using a 4.4 m liquid core waveguide, Talanta 58 (2002) 891.
[87] C.M.C. Infante, F.R.P. Rocha, A critical evaluation of a long pathlength cell for flow-based spectrophotometric measurements, Microchem. J. 90 (2008) 19.
[88] K. Fuwa, W. Lei, K. Fujiwara, Colorimetry with a total-reflection long capillary cell, Anal. Chem. 56 (1984) 1640.
[89] R.D. Waterbury, W.-S. Yao, R.H. Byrne, Long pathlength absorbance spectroscopy: trace analysis of Fe(II) using a 4.5 m liquid core waveguide, Anal. Chim. Acta 357 (1997) 99.
[90] L.J. Gimbert, P.J. Worsfold, Environmental applications of liquid-waveguide-capillary cells coupled with spectroscopic detection, Trends Anal. Chem. 26 (2007) 914.
[91] W. Lei, K. Fujiwara, K. Fuwa, Determination of phosphorus in natural waters by long-capillary-cell absorption spectrometry, Anal. Chem. 55 (1983) 951.

[92] K.O. Lupetti, F.R.P. Rocha, O. Fatibello-Filho, Talanta 62 (2004) 463.

[93] A.C.B. Dias, E.P. Borges, E.A.G. Zagatto, P.J. Worsfold, A critical examination of the components of the Schlieren effect in flow analysis, Talanta 68 (2006) 1076.

[94] P.K. Dasgupta, H.S. Bellamy, H. Liu, J.L. Lopez, E.L. Loree, K. Morris, K. Petersen, K.A. Mir, Light emitting diode based flow-through optical absorption detectors, Talanta 40 (1993) 53.

[95] P.S. Ellis, B.S. Gentle, M.R. Grace, I.D. McKelvie, A versatile total internal reflection photometric detection cell for flow analysis, Talanta 79 (2009) 830.

[96] A. Cerda, V. Cerda, An Introduction to Flow Analysis, Universitat de les Illes Balears, Mallorca, Spain, 2009.

[97] H.-H. Liu, P.K. Dasgupta, Dual-wavelength photometry with light emitting diodes. Compensation of refractive index and turbidity effects in flow-injection analysis, Anal. Chim. Acta 289 (1994) 347.

[98] F. Lazaro, A. Ríos, M.D. Luque de Castro, M. Valcarcel, Simultaneous multiwavelength detection in flow injection analysis, Anal. Chim. Acta 179 (1986) 279.

[99] E.A.G. Zagatto, M.A.Z. Arruda, A.O. Jacintho, I.L. Mattos, Compensation of the Schlieren effect in flow-injection analysis by using dual-wavelength spectrophotometry, Anal. Chim. Acta 234 (1990) 153.

[100] A.C. Pappas, C.D. Stalikas, Y.C. Fiamegos, M.I. Karayannis, Determination of hydrogen peroxide by using a flow injection system with immobilized peroxidase and long pathlength capillary spectrophotometry, Anal. Chim. Acta 455 (2002) 305.

[101] F. Chen, J.R. Lu, B.J. Binder, Y.C. Liu, R.E. Hodson, Application of digital image analysis and flow cytometry to enumerate marine viruses stained with SYBR gold, Appl. Environ. Microbiol. 67 (2001) 539.

[102] E.A.G. Zagatto, F.R.P. Rocha, P.B. Martelli, B.F. Reis, Detecting and circumventing sources of inaccuracy in flow analysis, Pure Appl. Chem. 73 (2001) 45.

[103] R.G. Brereton, Chemometrics and statistics: signal processing, in: P.J. Worsfold, A. Townshend, C.F. Poole (Eds.), second ed., Encyclopedia of Analytical Science, vol. 2, Elsevier, Oxford, 2005, p. 46.

[104] J. Ruzicka, E.H. Hansen, Integrated microconduits for flow injection analysis, Anal. Chim. Acta 161 (1984) 1.

[105] K. Carlsson, H.S. Jacobsen, A.L. Jensen, T. Stenstrøm, B. Karlberg, Micro-continuous flow system for wet chemical analysis, Anal. Chim. Acta 354 (1997) 35.

[106] J.C. Fettinger, A. Manz, H. Ludi, H.M. Widmer, Stacked modules for micro flow systems in chemical analysis: concept and studies using an enlarged model, Sensor Actuator B 17 (1993) 19.

[107] A. Rainelli, R. Stratz, K. Schweizer, P.C. Hauser, Miniature flow-injection analysis manifold created by micromilling, Talanta 61 (2003) 659.

[108] B.H. van der Schoot, S. Jeanneret, A. van der Berg, N.F. de Rooij, Modular setup for a miniaturized chemical analysis system, Sensor Actuator B 15 (1993) 211.

[109] G.N. Doku, S.J. Haswell, Further studies into the development of a micro-FIA (μFIA) system based on electroosmotic flow for the determination of phosphate as orthophosphate, Anal. Chim. Acta 382 (1999) 1.

[110] G.M. Greenway, S.J. Haswell, P.H. Petsul, Characterisation of a micro-total analytical system for the determination of nitrite with spectrophotometric detection, Anal. Chim. Acta 387 (1999) 1.

[111] P.H. Petsul, G.M. Greenway, S.J. Haswell, The development of an on-chip micro-flow injection analysis of nitrate with a cadmium reductor, Anal. Chim. Acta 428 (2001) 155.

[112] A. Manz, N. Graber, H.M. Widmer, Miniaturized total chemical analysis systems: a novel concept for chemical sensing, Sensor Actuator B 1 (1990) 244.

[113] D. Vaireanu, P.R. Fielden, R.D. Snook, Similarity and scaling in flow injection analysis, Anal. Commun. 36 (1999) 403.

CHAPTER

7

Special Strategies for Flow Manipulation

However beautiful the strategy, you should occasionally look at the results.
Winston Churchill

OUTLINE

This chapter deals with special flow-based strategies such as merging zones, zone sampling, stream splitting, zone splitting, zone trapping, zone stopping, flow reversal and multi-detection. These strategies involve flow manipulations that are better implemented in the absence of stream segmentation. Without segmentation, the main flowing stream behaves as an uncompressible liquid column, allowing the sample to be handled more efficiently. The emphasis of this chapter is therefore on unsegmented flow analysis. There are many examples of these special strategies in the literature and the aim of this chapter is to use selected examples in the text and tables to illustrate particular features rather than providing a comprehensive literature review.

7.1. MERGING ZONES

The merging zones approach was conceived independently by Bergamin-Filho *et al.* [1] and by Mindegaard [2] as the first strategy exploiting multiple injections in the context of flow injection analysis. As the name implies, the strategy involves the establishment of two zones (usually the sample and the reagent) that interact with each other.

7.1.1. Implementation

The merging zones strategy has been implemented by exploiting two convergent carrier streams, a single carrier stream or an intermittent stream, and the main characteristics of the resulting systems are discussed below.

7.1.1.1. Merging Zones Relying on Convergent Carrier Streams

The sample and the reagent solutions are simultaneously inserted into convergent carrier streams, and merging of the established zones occurs at the point where the streams converge.

Most applications involve loop-based introduction using an injector-commutator (Fig. 7.1). The sample and reagent solutions are directed to fill their corresponding loops which define the aliquots to be inserted. Excess sample is directed to waste, whereas excess reagent can be directed either to waste or back into the reagent container. This reagent recycling leads to a slight dilution of the reagent by the carrier/wash solution remaining in the loop after reagent insertion. If the reagent concentration is not a critical parameter in the analytical procedure involved, recycling

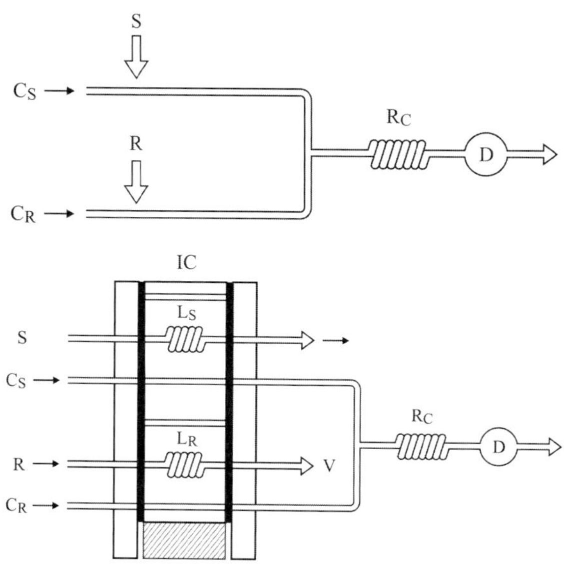

FIGURE 7.1 **Didactic representation (upper) and flow diagram (lower) of a typical flow system exploiting merging zones attained by stream convergence.** S = sample; R = reagent; C_S, C_R = convergent carrier streams; L_S, L_R = sampling loops; IC = injector-commutator; R_C = coiled reactor; D = detector; shaded area = alternative IC position; solid arrows = sites where pumping is applied.

can be exploited. Otherwise excess reagent should be collected in a recovery vessel for reuse or disposal.

Switching the injection device (two synchronised single injection valves or a double injection valve) inserts the sample and reagent aliquots into their convergent carrier streams and the established sample and reagent zones are pushed towards the stream convergence site. The chemical reactions involved proceed inside the reactor and the formed product is monitored by the flow-through detector.

The flow system in Fig. 7.1 is designed in the confluence configuration, with the sample and reagent injected into chemically inert carrier/wash streams. The main characteristics of confluence flow systems (see 3.2.2), *e.g.*, low Schlieren noise, no limitation on increasing the inserted volumes, good sample/reagent interaction and suitability for relatively slow chemical reactions, also hold for the implementation of merging zones by stream convergence. It should be recalled that the chemically inert carrier stream should have a matrix composition as similar as possible to that of the inserted solution (see 3.2.2). The reagent carrier stream should therefore be prepared to contain, *e.g.*, the chemical species yielding the Schlieren effect and/or coloured components, as demonstrated by the spectrophotometric determination of nitrite in natural waters relying on a modified Griess reaction [3]. The reagent solution comprised the colour-forming species (2.0% w/v sulphanilamide and 0.10% m/v N-(1-naphthyl)-ethylenediammonium dichloride) in the presence of 8.0% (v/v) phosphoric acid, whereas the chemically inert reagent carrier stream was an 8.0% (v/v) phosphoric acid solution.

Usually there is a pronounced difference in the flow rates of the convergent streams, especially if sensitivity is critical. In order to avoid excessive sample dilution at the convergence site (*cf.* Eq. 3.10), the flow rate of the reagent carrier stream is generally much lower than that of the sample carrier stream. The length of the reagent loop and the distance between the insertion port and the convergence site should be shorter than those associated with the sample in order to achieve a good overlap between the flowing zones.

7.1.1.2. Merging Zones Relying on a Single Carrier Stream

The sample and reagent aliquots are inserted into the same carrier/wash stream of a flow system designed in the single-line configuration (Fig. 7.2). Two zones initially separated by a transmission line that acts as a delay element are established and transported towards the detector. The interaction between them is a consequence of the axial dispersion involved and therefore strongly depends on the lengths of the initial sample and reagent plugs (inserted volumes), the initial distance between them and the carrier stream flow rate. Consequently, most applications involve large sample/reagent dispersion.

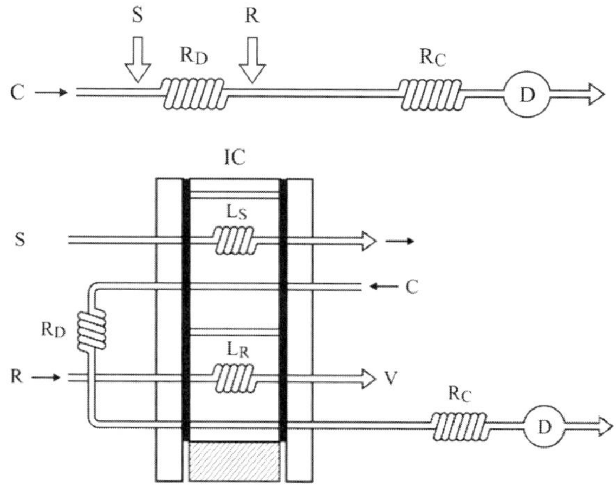

FIGURE 7.2 **Didactic representation (upper) and flow diagram (lower) of a typical single-line flow system with merging zones.** S = sample; R = reagent; C = carrier stream; L_S, L_R = sampling loops; IC = injector-commutator; R_D = delay coiled reactor; R_C = main coiled reactor; D = detector; shaded area = alternative IC position; solid arrows = sites where pumping is applied.

This type of flow system is simple and rugged, but its performance is subject to the same restrictions as ordinary single-line flow systems (see 3.2.1). In fact, the main drawback of merging zones implementation relying on a single carrier stream is that a variable baseline can adversely affect the measurement if the reagent zone is sensed by the flow-through detector. This mode of implementing merging zones is therefore most appropriate for *spectrophotometric* analytical procedures involving high dispersion of sample and reagents and/or when using a colourless reagent (or at least a reagent that does not absorb significantly at the wavelength set for monitoring the chemical species formed).

There are situations where an additional reagent solution is added by confluence. The site where the second reagent is added should be located as far from the injector as possible because stream addition impairs the axial dispersion of the sample [4].

Merging zones implemented in a single-line flow system are the essence of another strategy referred to as the *sandwich technique.*

THE SANDWICH TECHNIQUE

The sample aliquot is inserted between two solution aliquots or, in other words, "sandwiched" between two solutions. The strategy was originally implemented in a flow system dedicated to process analysis [5]. A large sample plug was inserted between two standard solutions by means of an 8-port rotary valve. Two sample/standard mixing zones were then established at the front and trailing edges of the flowing sample, allowing an ingenious recalibration procedure to be implemented. The theoretical basis was presented and demonstrated by the spectrophotometric determination of Fe(II) ions and the potentiometric determination of nitrate. A mathematical model of flow injection systems with sandwich techniques for simultaneous determinations was reported in 1990 [6].

This innovation was initially applied to iron speciation in ground waters [7] based on the interaction between Fe(II) and 1,10-phenanthroline. Sample was inserted between neighbouring plugs of water and reducing agent solution and the established sample zone merged with the main reagent stream. The coloured regions formed at the front, and rear sample interfaces reflected the Fe(II) and total Fe concentrations, respectively, in the injectate.

The feasibility of the innovation for saving expensive reagents was demonstrated by the spectrophotometric determination of glucose in serum [8] by an enzymatic method using *glucosidase* and *peroxidase*.

Spectrophotometric monitoring of Cr(VI) in the effluent from a pilot plant for the decontamination of waste waters from the leather industry was also reported [9]. As the monitored concentration of Cr(VI) was high and variable, a flow system was designed with the sandwich technique in order to widen the dynamic concentration range. To this end, the different volumetric fractions at the front and trailing edges of the flowing sample were exploited, resulting in two analytical curves covering two different concentration ranges, thus expanding the dynamic concentration range.

Simultaneous determinations relying on a sandwich technique were demonstrated by the spectrophotometric determination of Cr(VI) and iron in waste waters [10]. To this end, the sample plug was inserted between two aliquots of the reagents 1,5-diphenylcarbazide and 1,10-phenanthroline.

Sandwich techniques were further revised by Davies and Fogg [11], who proposed a flow system for the bromimetric and iodimetric determination of sulphite. The analytical signal related to the monitorand was gathered at the front portion of the sample zone whereas the analytical signal related to the determinand was obtained at the tailing portion. With this arrangement, the Schlieren effect was compensated. The authors provided a comprehensive review on the use of different injectors and valves as well as different injection strategies (*e.g.*, nested injection), for implementing the sandwich technique.

Sandwich techniques can also be implemented in a sequential injection analyser, as demonstrated by the spectrophotometric determination of nitrate and nitrite in natural waters [12]. The sample plug was set between two reagent plugs and the assembly was aspirated towards the holding coil of an ordinary sequential injection analyser. Two Griess reaction zones were established at both ends of the sample zone, and only one was allowed to interact with a copperised cadmium mini-column that reduced nitrate to nitrite.

It is important to stress that a typical sequential injection system cannot be considered as a flow system exploiting a sandwich technique, as sequential aspiration of different solutions, accommodation of the sampled aliquots inside a holding coil and flow reversal are inherent only to the sequential injection analyser.

7.1.1.3. Merging Zones Relying on an Intermittent Stream

The sample is inserted into a continuously flowing carrier stream and the reagent is added by confluence (Fig. 7.3). In order to avoid continuous pumping of the reagent solution, the reagent stream remains stopped during most of the analytical cycle and flows only during the passage of the sample zone through the convergence site. In this way, a strategy analogous to conventional merging zones is implemented.

It is interesting to note that in the pioneering application of this approach [13], loop-based sample injection was accomplished using a channel drilled into the rotor of a rotary injection valve, which acted as the volume-selecting element instead of an external sampling loop.

As an intermittent stream is involved, operation of the system requires either two pumps (one operating continuously and the other discretely) or increased commuting capability [14].

The reagent solution converges with the dispersing sample as a plug. Perfect zone overlap is therefore not attainable and lower

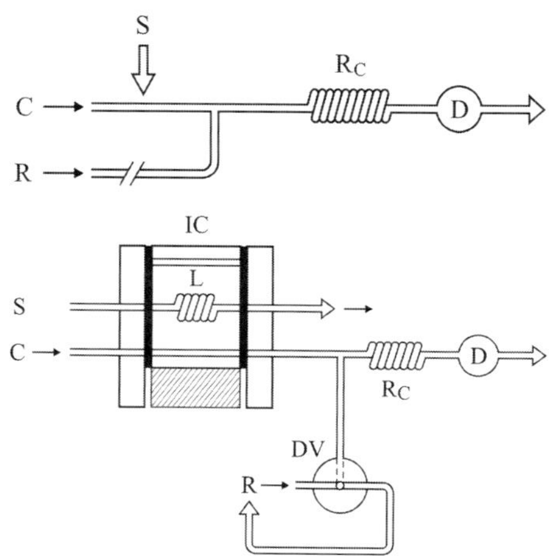

FIGURE 7.3 **Didactic representation (upper) and flow diagram (lower) of a typical flow system with merging zones relying on an intermittent stream.** S = sample; R = reagent; C = sample carrier stream; L_S = sampling loop; IC = injector-commutator; DV = directing valve; R_C = coiled reactor; D = detector; shaded area = alternative IC position; solid arrows = sites where pumping is applied.

sample-to-reagent volumetric ratios are observed at the less concentrated front and trailing portions of the sample zone. A critical comparison between reagent consumption associated with merging zones implemented by stream convergence and by intermittent pumping has not, however, been reported to date.

7.1.2. Applications

Exploitation of merging zones expands the potential of flow analysis, as highlighted in the following sections and in Table 7.1.

7.1.2.1. Reduction of Reagent Consumption

In contrast to ordinary flow systems where the reagent solution is continuously pumped, a reagent aliquot is inserted into a carrier stream. In this way, the reagent is consumed only in the presence of the sample, leading to a significant reduction in the volume of reagent required, hence minimising waste generation. This aspect was emphasised in the first flow injection systems with merging zones, designed for the spectrophotometric determination of phosphate in plant digests [1] and for flame atomic absorption/emission spectrometric determinations of calcium, magnesium and potassium in soil extracts and natural waters [15]: the consumptions of ascorbic acid and lanthanum were only 9% and 1%, respectively, of those inherent to the analogous flow systems without merging zones.

This aspect becomes more important in relation to analytical procedures requiring expensive reagents and the turbidimetric determination of yeast mannan [16] is a good example. The antigen reacted with the expensive antibody concanavalin A; therefore, the reagent addition via merging zones and a 1.0-min STOP period were required.

7.1.2.2. Matching the Analytical Signal with the Dynamic Concentration Range

The system is designed to provide a partial overlap of the sample and reagent zones. Only the trailing portion of the dispersing sample overlaps with the reagent zone, leading to a reduction in the recorded peak height. This approach is beneficial for the determination of high analyte concentrations without the need for a prior manual dilution of the assayed sample, as originally demonstrated by the spectrophotometric determination of human albumin using the bromocresol green method [2].

7.1.2.3. Single Point Determinations Incorporating Blank Determinations

With partial overlap of the sample and reagent zones, a portion of the sample zone may not interact with the reagent. The monitored signal associated with this region then reflects the blank value. As the analytical

TABLE 7.1 Selected Applications of Merging Zones

Analyte	Sample Matrix	Detection Technique	Analytical Range or Detection Limit	Remarks	Ref.
Acetylcysteine	Pharmaceutical formulations	Turbidimetry	5.0×10^{-5} mol L^{-1}	Water intermittent stream for system washing	[95]
Al	Plants, soils	UV–Vis	0–20 mg L^{-1}	Sequential merging of three zones	[96]
Al, Fe	Plant digests	UV–Vis, FAAS	0.00–15.00 mg L^{-1} (same range)	Simultaneous determinations/zone sampling	[29]
Albumin	Serum	Fluorimetry	3000–5000 mg dL^{-1}	Energy-transfer immunoassay/multi-commutation for sample stopping	[97]
Albumin	Serum	UV–Vis	10–60 g L^{-1}	Merging zones proposal/asynchronous merging zones for signal attenuation	[2]
Ammonium	Natural waters	Turbidimetry	10–200 μg L^{-1}	Original proposal for ion exchange in flow injection analysis/pulsed addition of the Nessler reagent	[19]
Azithromycin	Pharmaceutical formulations	UV–Vis	6.6 mg L^{-1}	Multivariate algorithm for system optimisation	[98]
Ca	Soils, waters, plants	UV–Vis	0–500 mg L^{-1} (three ranges)	Zone sampling for wide range determinations	[99]
Ca, Mg, K	Plant digests	FAAS, FP	0–350, 0–80, 0–350	ca 40-fold sample dilution	[15]
Cr(VI)	Waste waters	UV–Vis	0.01–1, 1–30 μg mL^{-1} Cr (two ranges)	Sandwich techniques for widening the dynamic concentration range	[9]
Cr(VI), Fe	Waste waters	UV–Vis	0.2–10 mg L^{-1} (both analytes)	Sandwich technique with different reagents	[10]

(*Continued*)

Table 7.1 Selected Applications of Merging Zones—cont'd

Analyte	Sample Matrix	Detection Technique	Analytical Range or Detection Limit	Remarks	Ref.
Cu, Zn	Serum	UV–Vis	0.0–2.5 μg mL^{-1} Cu + Zn (mixed standard solutions)	Sequential determinations/pulsed addition of a masking reagent for copper	[21]
DNA-binding with ethidium bromide, proflavine and cisplatin	Antitumour drugs	Fluorimetry	Not relevant	Zone stopping	[100]
Ethanol	Wines	UV–Vis	0.4% v/v	Gas diffusion/immobilised enzyme	[101]
Fe	Beers	UV–Vis	0.032 mg L^{-1}	Asymmetric merging zones	[102]
Fe(II), total Fe	Groundwaters	UV–Vis	0.1–9, 03–12 mg L^{-1}	Sandwich technique for iron speciation	[7]
Formaldehyde	Pale beers	UV–Vis	8–700 ng mL^{-1}	Catalytic method in an air-driven flow injection system	[103]
Glucose	Serum	UV–Vis	0.15–15 mmol L^{-1}	Proposal for zone stopping	[17]
Glucose, nitrite	Blood plasma, waters	UV–Vis	5 μg L^{-1} NO_2-N	Two zones and air plugs inserted into a single-line mono-segmented flow system	[104]
L-carnitine	Infant formulae	UV–Vis	10–80 mg L^{-1}	Bioreactor with immobilised enzyme/reaction with *carnitine acetyltransferase* coupled with acetyl coenzyme A and dithiobenzoate.	[105]
Mg	Serum	UV–Vis	5–20 μmol L^{-1}	Enzyme reactor	[106]
Mn	Silicate rocks	UV–Vis	0.5–2.0 mg L^{-1}	Pulsed addition of 2-aminoethanethiol for masking nickel	[107]

Table 7.1 Selected Applications of Merging Zones—cont'd

Analyte	Sample Matrix	Detection Technique	Analytical Range or Detection Limit	Remarks	Ref.
N, P	Plants	UV–Vis	0.05–0.5% , 0.2–5% (w/w)	Sequential determinations/indophenol blue and molybdenum blue methods	[20]
Nitrate	Plant extracts	UV–Vis	0.30–1.65 mg L^{-1}	Merging zones for standard addition/zone sampling for defining the added amounts/twelve addition levels	[24]
Nitrate, nitrite	Waters	UV–Vis	0.1–0.5, 1.0–5.0 mg L^{-1} N	Sequential determinations/solid reagent	[3]
Nitrate, nitrite	Natural waters	UV–Vis	2–100, 0.5–40 μmol L^{-1} N	Sandwich technique in a sequential injection system	[12]
Nitrite	Waters	UV–Vis	8–250 μg L^{-1} NO_2-N	Intermittent stream	[108]
Pb	Tap waters	UV–Vis	10 μg L^{-1}	Standard addition method	[109]
Phosphate	Plant digests	UV–Vis	10–60 mg L^{-1}	Proposal for merging zones/molybdenum blue method	[1]
Saccharin	Sweetener products	Turbidimetry	2.4 g L^{-1}	Previous saccharin extraction with ethyl acetate	[110]
Salicylic/acetylsalicylic acids	Pharmaceutical preparations	UV–Vis	100–600, 300–1800 mg L^{-1}	Sequential determinations required for a dissolution test	[111]
Sb(III), Sb(V)	Drugs	UV–Vis	30.0 μg L^{-1}	Triton X-100 addition for enhancing sensitivity	[112]
Sucrose	Sugarcane juices, molasses	UV–Vis	11–52% m/v	In-line sucrose hydrolysis	[113]
Yeast mannan	Blood serum	Turbidimetry	0–10 mg mL^{-1}	Immune precipitin interaction between concanavalin A (antibody) and analyte (antigen)/zone stopping	[16]

UV–Vis = spectrophotometry; FAAS = flame atomic absorption spectrometry; FP = flame photometry. Analytical range or detection limit given in original units.

and blank signals are obtained in a single sample run, the approach is useful for analytical procedures relying on reaction rate measurements. Its feasibility was demonstrated by the spectrophotometric determination of glucose in serum samples using *glucose oxidase* [17]: about 100 samples with variable blanks, which are typically encountered in clinical assays, were run per hour and the net absorbance value was available 30 s after sample insertion. Guidelines for attaining zone overlap were presented.

7.1.2.4. *Speciation*

With partial sample/reagent overlap, only a portion of the flowing sample interacts with a discriminating reagent, allowing speciation studies to be carried out. The feasibility of this innovation was demonstrated by the speciation of Fe(II) and Fe(III) in natural waters [18]. The first recorded signal reflected the Fe(II) concentration in the sample, whereas the second signal reflected the total iron content, as the Fe(III) originally present in the sample was converted in-line to Fe(II) by injecting a reducing reagent solution. To this end, synchronised injections of a large sample volume and a small reagent volume into separate carrier streams, stream convergence leading to asymmetric merging zones, and mixing downstream with the chromogenic reagent (1,10-phenanthroline) before detection were required.

7.1.2.5. *Reducing Undesirable Adsorption Processes*

Addition of the reagent as a plug minimises the adsorption of some reagents on the inner walls of the analytical path (main reactor and flow cell), thus avoiding baseline drift. The determination of ammonium in natural waters using the Nessler reagent [19] provides a good example. When the reagent was continuously pumped, the coiled tubular polyethylene reactor became brownish after about one hour of system operation and a pronounced baseline drift was observed. The drawback was eliminated by simultaneously injecting the sample solution and the Nessler reagent into the sample carrier stream and an alkaline reagent carrier stream, respectively, which merged together before reaching the main reactor.

7.1.2.6. *Sequential Determinations*

The sample is inserted simultaneously with the reagents required for a given determination; thereafter, it is inserted again but with the reagents required for the second determination. In this way, sequential determinations are performed in the same manifold. This approach was exploited initially in the spectrophotometric determinations of ammonium and phosphate in plant digests [20], which required rather different reaction conditions; the ammonium determination was carried out in a highly alkaline medium, whereas the phosphate determination was performed under acidic conditions. The analytical figures of merit were similar to those reported for the individual flow systems without merging zones, with the exception of the consumption of reagents, which was much lower.

Alternatively, the sample can be inserted twice and only one of the established zones merges with the specific reagent used for discriminating purposes, as demonstrated by the spectrophotometric determination of zinc and copper in serum [21].

7.1.2.7. *Standard Addition Method*

The flow system is designed to permit successive injections of the sample, each one accompanied by the injection of a different standard solution. Merging of the established zones permits the efficient implementation of the standard addition method involving a fixed sample dilution. The approach was initially exploited in the routine analysis of copper/nickel alloys by ICP-OES [22], the spectral interferences being minimised by applying the generalised standard addition method [23]. This innovation however required the preparation of a large set of standard solutions.

This drawback was circumvented by taking advantage of the zone sampling strategy (see 7.2), as demonstrated by the spectrophotometric determination of nitrate in plant extracts [24], which is susceptible to matrix effects. The standard solution was inserted into the first reagent carrier stream and twelve different aliquots with different but known concentrations were re-sampled and inserted into the second reagent stream in synchronisation with the sample injections. Merging of the sample/reagent zones led to the implementation of the standard addition method and accurate results were obtained.

7.2. ZONE SAMPLING

This strategy was conceived by Reis *et al.* [25] who demonstrated the feasibility of selecting a small portion of a flowing sample and introducing it into another carrier stream of a flow injection system.

7.2.1. Implementation

Zone sampling has usually been performed by using two loop-based injectors (Fig. 7.4) or a double injection valve/commutator furnished with external sampling loops [25,26].

The solution with the species of interest, usually the sample, is introduced via loop L_1 into a first carrier stream and the established zone flows through reactor R_1 towards a second injection port. After a time interval t_{ZS}, an aliquot of the dispersing zone is taken by the second sampling loop

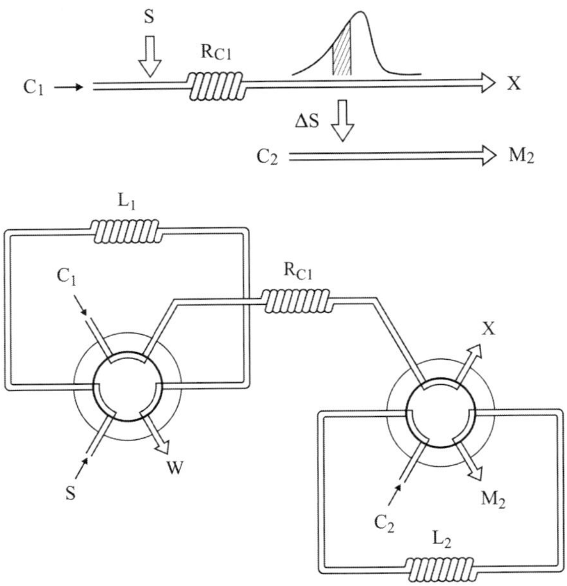

FIGURE 7.4 **Didactic representation (upper) and flow diagram (lower) of a flow system with zone sampling using two rotary valves for sample insertion.** S = sample; ΔS = re-sampled aliquot; R = reagent; C_1, C_2 = first and second carrier streams; L_1, L_2 = first and second sampling loops; R_{C1} = coiled reactor; X = towards first manifold or towards waste; M_2 = towards second manifold; W = towards waste; solid arrows = sites where pumping is applied.

(L_2) and inserted into a second carrier stream. A second sample zone is established, handled and monitored during passage through the flow cell. The remaining portion of the first sample zone is generally discarded. Precise timing control is essential for this application.

The feasibility of implementing zone sampling by taking advantage of multi-commutation has also been described [27].

7.2.2. Applications

A flow system with zone sampling comprises two almost independent subsystems, and the re-sampling loop (L_2) is the only component common to both. This architecture permits the time window associated with the t_{ZS} value to be more efficiently exploited. Implementation of zone sampling expands the applicability of flow analysis (see Table 7.2) and some applications of this strategy are highlighted in the following sections.

7.2.2.1. *High Sample Dispersion*

Sampling a small portion of a dispersing zone and introducing it into a second carrier stream in order to be re-dispersed is a very efficient

TABLE 7.2 Selected Applications of Zone Sampling and Zone Splitting

Analyte	Sample Matrix	Detection Technique	Degree of Dilution	Analytical Range or Detection Limit	Remarks	Ref.
Acyclovir	Pharmaceutical formulations	UV–Vis	4.8 to 17.3	50–1200 mg L^{-1}	Sequential injection chromatography with monolithic column/zone splitting	[114]
Al	Crystallised fruits	UV–Vis	N	0.1, 0.6, 0.8 mg L^{-1}	Multi-commuted flow system with three analytical ranges/zone splitting	[115]
Al, Fe	Plant digests	UV–Vis, FAAS	45.5, 1.51	0.00–15.00 mg L^{-1} (same range)	Simultaneous determinations/merging zones	[29]
Aminocaproic acid	Pharmaceutical formulations	Fluorimetry	100 to 1000	2.5×10^{-7} mol L^{-1}	Sequential injection analysis combined to tandem streams/zone splitting	[116]
Ammonium	Natural waters	UV–Vis	N	50.0–500.0 μg L^{-1}	Ion-exchange in-line concentration	[34]
C_1–C_5 n-mercaptans	Fuels	UV–Vis	Not relevant	0.0–2.0 mmol L^{-1}	Analyte extraction by an alkaline aqueous phase; intelligent zone sampling involving real-time definition of the t_s value/other systems investigated	[31]
Ca	Plants, milk, anti-acid tablets, fertilisers, rocks	UV–Vis	2.15 to 754	7.0 μg L^{-1}	Expert multi-commuted flow system/zone sampling combined with zone splitting	[27]
Ca	Waters, soil extracts, plant digests	UV–Vis	3.1, 110, 24.4	0–500 mg L^{-1} (three ranges)	Wide-range spectrophotometry/merging zones	[99]
Ca, Mg	Wines	UV–Vis	6 to 35	2.1, 1.8 mg L^{-1}	Magnesium masked with 8-hydroxyquinoline	[117]

(Continued)

Table 7.2 Selected Applications of Zone Sampling and Zone Splitting—cont'd

Analyte	Sample Matrix	Detection Technique	Degree of Dilution	Analytical Range or Detection Limit	Remarks	Ref.
Ca, Mg, Na, K	Wines	FAAS, FP	Up to 2000	10–140, 50–200, 50–250, 500–1500 mg L^{-1}	Expert system/cascade zone sampling	[118]
Captopril	Pharmaceutical preparations	UV–Vis	N	1.0 μmol L^{-1}	Iodimetric method/zone sampling for avoiding manual sample dilution	[119]
Chloride	Estuarine waters	UV–Vis	1250	0.0–2.5% w/v	Commutation for implementing zone splitting	[60]
Creatinine	Urine	UV–Vis	70	0.5–2.0 g L^{-1}	Kinetic determination based on Jaffe's reaction/zone splitting/zone stopping	[120]
Cu, Zn	Brass alloys	UV–Vis	100 to 3000	Not relevant	Simultaneous determinations/PAR reagent/PLS data treatment	[121]
Gemfibrozil	Drug dissolution samples	UV–Vis/ HPLC	6 to 10 (in the first subsystem)	2.28 mg L^{-1}	First and second subsystems: sequential injection analyser and HPLC with monolithic column	[122]
Glucose	Serum	UV–Vis	6.0 to 62.9	0–500 mg dL^{-1}	Mixing chamber as the main dispersing element	[123]
Isoniazid	Drug dissolution samples	Fluorimetry	10 to 240	34.3 ng mL^{-1}	Multi-commuted flow system/zone splitting	[62]
K	Plant digests	FAAS	131	50–500 mg L^{-1}	Original proposal for zone sampling	[25]
lactic acid	Kefir fermentation monitoring	Fluorescence in an oxygen optrode	N	0.02–60 mmol L^{-1}	Optical-fibre biosensor with immobilised *lactate oxidase*	[124]

Table 7.2 Selected Applications of Zone Sampling and Zone Splitting—cont'd

Analyte	Sample Matrix	Detection Technique	Degree of Dilution	Analytical Range or Detection Limit	Remarks	Ref.
L-alanine	Synthesis process	CL	100	0.5–25 mmol L^{-1}	Immobilised *L-amino acid oxidase*	[125]
L-malate	Wines, fruit juices	Fluorimetry	Up to 500	5.0–50000 μmol L^{-1}	*Malate dehydrogenase* and *aspartate aminotransferase* co-immobilised on controlled pore glass	[126]
Mexiletine hydrochloride	Pharmaceutical preparations	Fluorimetry	2 to 30	10–300 mg L^{-1}	Sequential injection chromatography with monolithic column/zone splitting	[127]
Nitrate	Plant extracts	UV–Vis	Enough for additions of 0.22–1.47 mg L^{-1}	0.30–1.65 mg L^{-1}	Standard addition method/zone sampling for defining the added amounts/merging zones for addition/twelve addition levels	[24]
Nitrosyl	Concentrated sulphuric acid	UV–Vis	100	0.16 mg L^{-1}	Comparison of different strategies for high sample dilution	[128]
Phosphate	Urine	UV–Vis	101	$0.05–3 \times 10^{-2}$ mol L^{-1}	Zone splitting/sequential injection analysis	[63]
Pseudoephedrine	Human blood plasma	UV–Vis	Not pertinent	12 μg L^{-1}	Zone sampling for selecting the central portion of the eluate zone and introducing it into the interface for separation by capillary zone electrophoresis	[129]
Rb	Drugs (Rb efflux assay)	FP (eventually FAAS)	170	25–750 mg L^{-1}	Expert flow system/real-time t_{sp} selection in accordance to the analyte concentration	[130]
V	Flue-gas scrubbers	UV–Vis with LED photometer	60 to 1242	0.1–1.2 g L^{-1}	Use of a mixing chamber/V(IV) masked with 1,2-cyclohexanediaminetetraacetic acid	[131]

UV–Vis = spectrophotometry; CL = chemiluminescence; FAAS = flame atomic absorption spectrometry; FP = flame photometry; N = not given by author(s). Analytical range or detection limit given in original units.

means of achieving a high degree of dispersion without the need to use a very small inserted volume and/or a very long reactor.

In ordinary flow systems, the inserted volume cannot be indefinitely reduced for practical reasons. Regarding the increase in reactor length, it should be recalled that beyond a certain length, any additional increase in length would promote a slight increase in dispersion (see 5.2.3.2), leading to pronounced adverse effects on the hydrodynamic pressure and the sampling rate.

With zone sampling, an aliquot of the less concentrated trailing edge of the flowing zone can be sampled. Also, the length of loop L_2 can be decreased in order to reduce the re-sampled aliquot and increase dispersion inside the second subsystem. A drastic reduction in the lengths of loop L_2 and the reactors is not needed to attain a high degree of dispersion, as the volumetric fraction is determined by the product of the volumetric fractions associated with both subsystems. A simple calculation clarifies this. If the volumetric fraction associated with the first subsystem ($X_{S,L1,ts} = C'/C_0$, C' = concentration inside second sampling loop during re-sampling; C_0 = initial concentration) is 0.05 and that associated with the second subsystem ($X_{S,D,tr} = C/C'$, C = concentration at the detector) is 0.1, a total volumetric fraction of 0.005 is attained, and this degree of dispersion would not be achieved in ordinary flow systems easily.

This aspect was originally demonstrated by the determination of potassium in plant digests by flame atomic absorption spectrometry, which required a high degree of sample dispersion [25]. The proposed system was simple and handled 120 samples per hour. In spite of the very low sample volumetric fraction involved ($X = 0.0076$), the results were precise and in agreement with those obtained by a conventional flame atomic absorption spectrometric procedure involving a previous hundred-fold manual sample dilution.

Further, Garn *et al.* [28] demonstrated the beneficial effect of using mixing chambers instead of coiled tubular reactors and exploiting two re-sampling steps. With the proposed innovation, named *cascade zone sampling*, volumetric fractions of the order of 10^{-6} were reported. The zone broadening typical of flow systems with mixing chambers (*cf.* Eq. 3.9) did not impair the sampling rate, as the sample aliquots were always taken as a short plug from the trailing edge of the flowing zone leaving the chamber.

It is important to emphasise that, if required, reagents should be added to the sample zone in the second subsystem in order to reduce their consumption.

7.2.2.2. *Simultaneous Determinations*

The sample is inserted into the first carrier stream and directed towards the detector where the first analyte is determined. During passage of the flowing zone through the re-sampling loop L_2, a small sample aliquot is taken and inserted into the second subsystem for the determination of the second analyte.

This innovation was originally proposed for the analysis of digests of plant leaves [29]. Iron was determined under conditions of limited sample dispersion ($X = 0.659$) by attaching the main coiled reactor to the nebuliser inlet of a flame atomic absorption spectrometer. Eight seconds after the first sample insertion, the commutator was switched, selecting a small and diluted sample aliquot and inserting it into the second subsystem where aluminium was spectrophotometrically determined under conditions of high dispersion ($X = 0.022$). Water was used as the first sample carrier stream, as the results obtained by flame atomic absorption spectrometry were not subject to Schlieren effects or variations in sample acidity; consequently, the need for concentrated buffer solutions in the determination of aluminium using Eriochrome cyanine R was avoided. About 120 samples were analysed per hour (240 determinations) and precise results in good agreement with ICP-OES were obtained.

7.2.2.3. *Variable Dispersions*

Each t_{ZS} value relates to a different portion of the zone flowing through the re-sampling loop L_2, and each re-sampled portion has a different yet known concentration. Consequently, several aliquots with different concentrations can be taken from the flowing zone simply by setting different t_{ZS} values.

This approach can be exploited to widen the dynamic concentration range, and the spectrophotometric determination of calcium in waters, plant materials, milk, antacid tablets, fertilisers and calcareous rocks [27] is a good illustration. A wide dynamic range (0.250–1000 mg L^{-1} Ca) was attained by using five analytical curves, each one covering a different concentration range. Reliable results were obtained using a multi-commuted flow system exploiting variations in the inserted sample volume, zone splitting and zone sampling involving three different t_{ZS} values.

Different aliquots with known mean concentrations can be sampled from the dispersed zone of a standard solution, allowing efficient implementation of the standard addition method [24]. Only one standard solution is needed, as emphasised in Section 7.1.

Analogously, an analytical curve relying on only one standard solution can be obtained by setting different t_{ZS} values. Thereafter, the t_{ZS} value is fixed and the samples are run. Other t_{ZS} values can be used for the analysis of those samples with concentrations outside of the original analytical range [30].

7.2.2.4. *Sampling the Central Portion of a Flowing Zone*

There are situations where the central portion of the sample zone should be sampled and inserted into a second carrier stream, and the determination of C_1–C_5 n-mercaptans in gasoline involving liquid–liquid extraction [31] is a good example. The aqueous phase (a sodium hydroxide solution) was inserted into a flowing organic sample (gasoline), and the analyte was extracted by the aqueous phase. When passing through the second injection port, the aqueous phase was selected, sampled and inserted into another aqueous phase (sodium hydroxide solution). For real-time defining the t_{ZS} value, the conductivity of the solution flowing through the re-sampling loop was monitored; a sudden increase triggered the injector, allowing only the central portion of the sodium hydroxide zone to be selected. With this "intelligent" zone sampling, cumbersome phase separation steps were avoided and the Schlieren noise was minimised.

Another situation refers to the monosegmented flow system into which the sample is inserted between two air plugs (see 5.5.1). In order to avoid the step of air removal, an analogous strategy for the mechanical removal of the central portion of the flowing sample can be implemented, as demonstrated by the spectrophotometric determination of ammonium ion involving relatively slow chemical reactions [32]. With this innovation, washing time was reduced and lower flow rates could be selected, thus increasing the mean sample residence time in the flow system and hence improving sensitivity.

7.2.2.5. *Detailed Study of Dispersion*

A t_{ZS} scan enables the entire dispersed zone to be transferred stepwise to the second carrier stream, allowing the detector to build an image of the sample zone at the re-sampling loop (see Fig. 3.13). This loop can then be regarded as a pseudo-sensor, allowing a detailed study of the sample zone to be accomplished. The general shape of the first sample zone is maintained regardless of the dimensions of the second subsystem. The inner volume of loop L_2 reflects the influence of the detector geometry in ordinary flow systems. If this loop is small enough, it can be regarded as a point sampler, permitting a detailed investigation of the boundary of the sample zone and the carrier stream. By comparing the concentration–time functions at the re-sampling loop and then replacing this loop by the inlet of an ICP-OES nebuliser, it was possible to conclude that the nebuliser has little influence on sample broadening, as reported in the landmark article dealing with flow injection analysis/ICP-OES hyphenation [33]. Moreover, zone sampling was exploited for assisting the design of a flow injection system for the spectrophotometric determination of ammonium ions in natural waters involving in-line SPE for analyte separation/concentration [34]. The shape of the eluted band was efficiently obtained and evaluated for optimising the elution conditions.

Analogously, a scan along the length of the R_{C1} reactor permits images of a flowing sample to be obtained. Fig. 3.13 shows images of the flowing sample and true concentration–time functions obtained by scanning the length of reactor R_{C1} or the t_{zs} value [35].

7.3. STREAM SPLITTING

This strategy involves the division of a flowing stream, thus producing two emergent streams. It has been exploited in both segmented and unsegmented flow analysis. As a rule, the original flowing stream contains the sample and reaches the splitting point without being segmented.

7.3.1. Segmented Flow Analysis

Stream splitting was originally exploited to improve the sampling step in segmented flow analysis for enzymatic assays or simultaneous determinations.

Regarding enzymatic assays, a very early procedure relying on a two-point analysis of serum samples [36] is a good example. The sample stream selected by the sampling arm was split into two streams with similar flow rates while flowing through a T-shaped connector. These streams were thereafter segmented and directed towards different analytical channels where the sample aliquots were incubated for 5-min and 15-min periods. The enzymatic reactions were then stopped by adding a suppressing reagent stream and both sample zones passed sequentially through a dialysis unit. The analyte was collected in two parallel reagent streams. The stream corresponding to the shorter incubation time passed through a 10-min delay coil, so that both sample zones passed simultaneously through a differential spectrophotometer where the longer incubation-time stream was read against the shorter incubation-time stream, yielding a direct measure of enzyme activity. The need to run serum blanks was therefore avoided. This approach was also applied to the spectrophotometric determination of *aspartate aminotransferase, lactate dehydrogenase* and *alkaline phosphatase* in serum samples [37].

The spectrophotometric determination of CO_2 and chloride in serum [38] was an early application of stream splitting for simultaneous determinations carried out in a segmented flow analyser. The sample stream was analogously split before segmentation, and the resulting streams were directed towards different analytical channels, where the analytes were dialysed in-line and collected in a buffered (pH 9.2) cresol red solution or in an acidic hexacyanomercurate(II) plus iron(III) nitrate solution, which acted as collecting streams. The transient absorbance changes in these streams were monitored by two spectrophotometers and recorded as peaks, the heights of which were proportional to the analyte concentrations in the sample.

As stream segmentation prior to stream splitting was not involved in either of these classical applications, the flow rates of the emergent streams were governed by the aspiration rates of a peristaltic pump. Although the splitting process is a simple concept, its initial development required much effort. After the experience with a two-channel system [39], it was several years before an eight-channel analyser was available [40], followed by a twelve-channel commercial version [41,42]. Nowadays, this approach is mainly exploited for simultaneous determinations involving multi-channel manifolds.

7.3.2. Unsegmented Flow Analysis

In unsegmented flow systems, stream splitting has mainly been exploited for wide-range *spectrophotometry*, differential kinetic analysis, simultaneous determinations or sample removal from the analytical path. After the splitting process, the resulting streams merge together immediately before detection or, alternatively, are individually handled and monitored by different flow-through detectors.

7.3.2.1. *Stream Splitting/Stream Merging*

The sample is inserted into the reagent carrier stream which undergoes splitting, thus separating the sample zone into two portions (Fig. 7.5). For better sensitivity, the sample can be inserted into a chemically inert carrier stream, the reagent being added by confluence immediately after the insertion port (see 3.2.2).

After splitting, the resulting streams are directed towards two parallel reactors at different flow rates, governed by the hydrodynamic pressures involved. Consequently, most of the original carrier stream (hence the

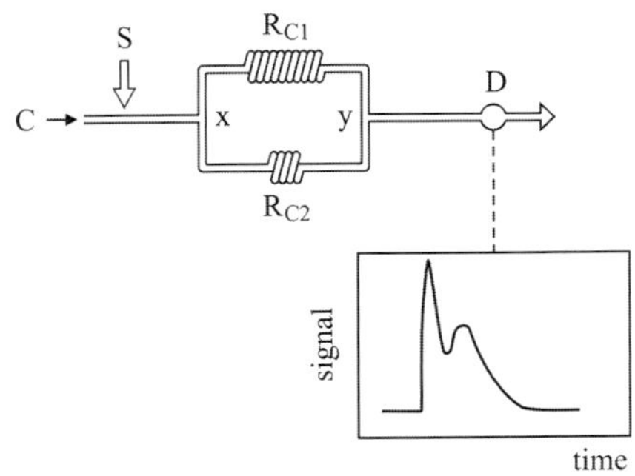

FIGURE 7.5 Didactic representation of a flow analyser with stream splitting/stream merging. C = reagent carrier stream; S = sample insertion; R_{C1} and R_{C2} = reaction coils; x and y = splitting and merging points; D = detector; solid arrow = site where pumping is applied. For details, see text.

larger sample portion) flows through the shorter (and/or thicker) reactor at a higher flow rate, whereas the smaller sample portion flows more slowly through the longer (and/or thinner) reactor. The streams leaving the reactors are then recombined. The sample portion handled in the shorter reactor reaches the mixing point before the portion handled in the longer reactor; therefore, an asynchronous merging of the sample portions is achieved. Due to the severe overlap involved, a complex sample zone is established, detected and recorded. The resulting concentration/time function is characterised by two local maximum values and one local minimum value.

The first and second maxima correspond to the sample portions handled inside the shorter and longer reactors which undergo low and high dispersion, respectively, and the minimum concentration value reflects the overlap between the two sample zones. It should be stressed that measurements performed at the maximum and minimum values are intrinsically more precise than those performed in regions of the flowing sample with pronounced concentration gradients [43,44].

EXPANSION OF THE DYNAMIC CONCENTRATION RANGE

The stream splitting/stream merging approach is useful for widening the dynamic concentration range of an analytical procedure, as originally demonstrated in relation to the spectrophotometric determination of chloride in river waters from the Amazon region [45]. The sample batches had highly variable analyte concentrations, as the samples collected near the Andes mountain chain were characterised by very low chloride concentrations, in contrast to those collected near Marajo Island (close to the Atlantic Ocean). Consequently, the flow system was designed to provide two analytical curves, associated with the two absorbance maxima of the recorded concentration/time function. With a split ratio of about two (the first peak height absorbance was double the second peak height), a sampling rate of 120 h^{-1} was attained with <1% carryover and the measurement precision was not influenced by the stream splitting and rerouting towards the flow cell. For some samples, the chloride concentrations could be determined using both analytical curves, and the results were not statistically different at the 99% confidence level. This feature is an additional accuracy assessment.

This innovation was critically investigated by Tyson and Bysouth, who demonstrated the feasibility of splitting the main stream into three channels [46].

DIFFERENTIAL KINETIC ANALYSIS

The portions of the flowing sample associated with the different peak maxima and minima are subject to different sample handling times and this can be exploited for differential kinetic analysis. This was demonstrated by Fernandez *et al.* [47] who investigated the main parameters involved and this was followed by a number of applications:

- The spectrophotometric determination of cobalt and nickel using 2-hydroxybenzaldehyde thiosemicarbazone [48].
- The catalytic fluorimetric determination of Mn^{2+} and Fe^{3+} using the oxidation of 2-hydroxybenzaldehyde thiosemicarbazone by hydrogen peroxide in the presence of amines as the indicator reaction [47].
- The fluorimetric determination of pyridoxal and pyridoxal-5-phosphate in blood serum based on their different oxidation rates in the presence of cyanide [49].
- The fluorimetric determination of Mn^{2+} and Fe^{3+} based on their catalytic effect on the oxidation of salicylaldehyde thiosemicarbazone by hydrogen peroxide in the presence of ammonia [50].
- The fluorimetric determination of silicate and phosphate in natural waters based on the different rates of formation of their molybdate heteropolyacids [51].
- The spectrophotometric determinations of furfural and vanillin in synthetic samples based on the different reaction rates of these compounds with *p*-aminophenol [52].

More recently, computer-assisted stream splitting was proposed [53] to minimise the variability in the separation of the sample zone [50], thus improving measurement precision and leading to a more versatile and rugged flow system. This innovation was accomplished by placing a three-way valve at the split point. Good results were obtained for the spectrophotometric determination of copper and zinc in plant digests involving metal complexation with cyanide, differential decomplexation kinetics and reaction with Zincon.

SIMULTANEOUS DETERMINATIONS

The spectrophotometric determination of nitrite and nitrate in natural waters [54] illustrates the exploitation of stream splitting/stream merging for simultaneous determinations. After the splitting process, one of the emergent streams travelled directly towards the convergence point, whereas the other stream passed through a copperised-cadmium mini-column to convert nitrate to nitrite, before reaching the convergence point. After stream convergence, the colour-forming reagent was added by confluence and the sample zone was directed towards the detector. The monitored absorbance/time function had two peak maxima proportional to nitrite and to nitrate plus nitrite concentrations.

In spite of the favourable characteristics, it should be stressed that stream splitting/stream merging has not often been used.

7.3.2.2. *Stream Splitting Without Stream Merging*

The sample is inserted into the reagent (or chemically inert) carrier stream which undergoes splitting and the resulting streams are directed towards different analytical channels and monitored by separate detectors. In the original approach [41], ammonium and phosphate in a single acidic plant digest were spectrophotometrically determined by the Berthelot and molybdenum blue methods, respectively. As measurements of a blue colour were involved in both methods, it was possible to use a single spectrophotometer with two similar flow-through cells placed in the same optical path. In view of the expected analyte concentrations in the digests and the analytical sensitivities involved, a 4:1 sample split ratio was used. Accurate results were obtained at a sampling rate of 200 h^{-1}.

7.3.3. Sample Removal from the Analytical Path

There are situations where a portion of the flowing sample is removed from the analytical path, and stream splitting can be exploited for accomplishing this task.

An important example is high sample dilution using confluent streams, often required in segmented [55] or unsegmented [56] flow analysis (see also 5.1.3 and 5.2.3). An excessive increase in the length of the sample zone resulting from confluence stream addition (Eq. 3.12), which might impair the sampling rate, is avoided by splitting the sample carrier stream and discarding one of the emergent streams, thus removing a portion of the processed sample and reducing its mean axial length. To this end, an additional T-shaped connector is placed in the analytical path in order to direct part of the flowing sample towards waste. Alternatively, the outlet flow rate of the de-bubbler can be increased.

The determination of sodium, potassium, calcium and magnesium in soil extracts [57] illustrates the implementation of this approach in a flow injection system. As these analytes are present in relatively high concentrations in the assayed samples, additions of diluent confluent streams were needed to increase dispersion. However, the total flow rate of the stream reaching the spectrometer was also increased, thus impairing nebulisation efficiency. This was circumvented by removing part of the sample carrier stream and the sampling rate was increased. In *spectrophotometry*, similar restrictions have been observed, especially when the increase in total flow rate impairs the development of relatively slow chemical reactions.

7.4. ZONE SPLITTING

This strategy is similar to zone sampling but involves cleavage of the flowing zone. Consequently, it is not possible to select any portion of the dispersing zone, only its trailing (or front) edge. In general, one of the resulting portions of the split zone is directly discarded, whereas the other is handled and monitored [58]. The feasibility of multiple zone splitting has also been demonstrated [59].

A flow system with zone splitting is generally designed in the single-line configuration and therefore tends to be simpler than a typical system exploiting zone sampling.

7.4.1. Implementation

Zone splitting is usually accomplished by resorting to flow reversal or stream directing valves. In general, the sample is inserted into a backwards flowing carrier stream and the established zone flows through reactor R_{C2} towards waste (Fig. 7.6, upper). The sample zone undergoes pronounced dispersion while flowing inside this reactor. After a pre-set delay time interval, t_{sp}, when the front portion of the sample zone has been discarded, the remaining sample portion (usually

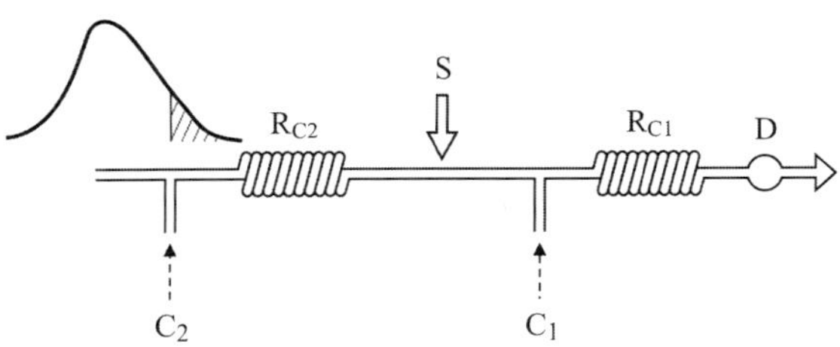

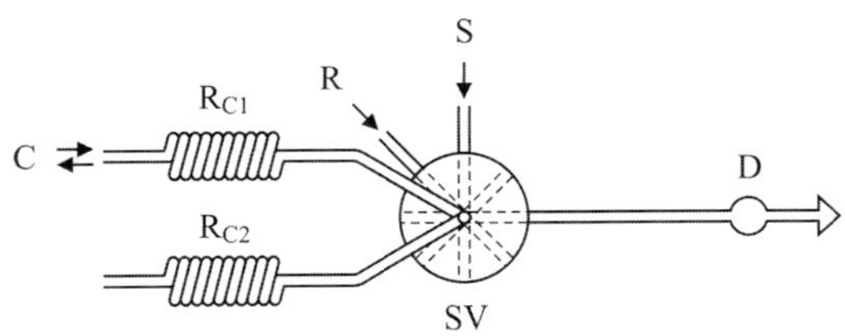

FIGURE 7.6 **Didactic representation and flow diagram of a flow system with zone splitting. Upper**: S = sample; C_1, C_2 = sample carrier streams (eventually the reagent), R_{C1}, R_{C2} = coiled reactors; shaded area = re-sampled aliquot; D = detector; traced arrows = intermittent pumping. **Lower: a typical sequential injection system**. S = sample; C = sample carrier stream; R = reagent; SV = multi-port selecting valve, R_{C1} = holding coil; other symbols as in the upper portion.

its trailing edge) is directed towards the detector. This sample portion undergoes further dispersion when flowing again through coil R_{C2} and through coil R_{C1}. Analogously to zone sampling, each t_{sp} value corresponds to a different sampled aliquot; hence, precise timing control is also needed.

In the original proposal [58], three pumps with intermittent stop/go actions were used, and the feasibility of replacing the reactor R_{C2} by a mixing chamber was discussed. Further, use of a single commutator to implement the innovation was demonstrated [60]. Another possibility is to exploit discretely operated pumps and valves [61]. With multi-commutation facilities, implementation of the zone splitting strategy is easily accomplished [62].

This strategy is more easily implemented in a sequential injection system [63]. To this end, only one bi-directional pump is required (Fig. 7.6, lower). The sample is aspirated towards the holding coil R_{C1} and, after flow reversal, directed towards the diluting coil R_{C2} and then to waste. After the t_{sp} pre-defined time interval, when most of the sample zone has gone to waste, the flow is reversed and the selected sample portion is again aspirated towards the holding coil R_{C1}. The total aspirated volume should be lower than the inner volume of R_{C1} in order to avoid aspiration of air. Sequential aspirations of the other solutions involved and further sample handling are accomplished as in an ordinary sequential injection system.

7.4.2. Applications

Zone splitting is a simple and efficient strategy for attaining high (and variable) degrees of sample dilution, as demonstrated by applications listed in Table 7.2.

7.5. SAMPLE INCUBATION

The sample residence time in the flow manifold, often associated with the expression *sample incubation time*, is an important parameter in flow-based analytical procedures involving relatively slow chemical reactions and/or physicochemical processes, *e.g.*, dialysis, gas diffusion or liquid–liquid extraction. This parameter may become a limiting factor in the system design, especially if sensitivity is critical. Moreover, the susceptibility to biased results is less pronounced when the chemical reactions and/or the involved physico-chemical processes tend towards completion. Fig. 1.4 refers to a hypothetical situation where biased results are obtained when chemical equilibrium is not reached.

The most efficient strategy for attaining a long sample residence time without excessive sample dispersion is to take advantage of stream segmentation (see 5.1.2). In fact, sample handling times of up to 10 min have often been reported in relation to segmented (and mono-segmented) flow analysis.

Regarding unsegmented flow analysis, zone stopping and zone trapping are efficient strategies for increasing the mean sample residence time.

7.5.1. Zone Stopping

The strategy was proposed by Ruzicka and Hansen in relation to blood analysis [17]. The sample is inserted into the carrier stream and, after additions of the required reagents, the fluid propulsion unit is switched off in order to stop the sample zone. The sample dispersion almost ceases during the STOP period (see 3.1.1.3), whereas the chemical reactions or the concentration/separation processes continue. In this way, a higher degree of completion is attained, thus improving sensitivity. Zone stopping under continuous pumping can also be achieved by resorting to commuting facilities, as demonstrated by the spectrophotometric determination of boron in plants involving the relatively slow interaction of the analyte with the Azomethine-H reagent [64]. This latter possibility has been critically evaluated elsewhere [65].

In general, the sample is inserted into a rapidly flowing carrier stream and the sample zone is stopped inside the main reactor. After flow restoration, the handled sample is monitored as in ordinary flow systems. Alternatively, the sample can be stopped inside the flow cell and the temporal increase in analytical signal during the STOP period is used as the measurement basis. As a rule, the pump is stopped when maximum analytical signal under flowing conditions is reached. The fluid element yielding the analytical signal is then monitored. Sample stopping at the detection unit has been exploited in several different ways:

- Single analyte kinetic determinations. The analytical signal is linearly time dependent and the slope of the recorded curve is proportional to the analyte concentration in the injectate. Using this approach, the injection of blanks is not necessary [66].
- Multi-parametric determinations. The increase in analytical signal is monitored and specific values related to different stopping times are considered. With this three-dimensional approach, multivariate calibration procedures can be efficiently implemented [67].
- Additional information about the chemistry involved. The STOP period can be started at different times, thus monitoring different sections of the sample zone with different sample/reagent volumetric fractions. The recorded curves then have different slopes and can

provide additional information regarding reaction kinetics, thus expanding the potential of this strategy [68,69].

Another possibility is to take advantage of dual zone stopping. The sample with the required reagents is inserted twice and a three-way valve selects the reactor towards which the sample zone is directed and stopped. Two samples are simultaneously handled and the sampling rate is almost doubled. This innovation was applied to the spectrophotometric determination of amiloride hydrochloride in pharmaceutical formulations [70] using a flow system with commuting facilities. The feasibility of multiple zone stopping was foreseen in 1981 [71] and applied to the spectrophotometric determination of *prostatic acid phosphatase* in human blood sera in 1986 [72]. A distribution valve directed the sample towards one of four coiled reactors, where it remained for a pre-set time interval. Each parallel reactor was filled with one sample at a time, while the others held other samples. The sampling rate then increased fourfold and the other favourable characteristics of the flow based procedure were maintained.

A shortcoming of the zone stopping strategy is that the sample is handled under static conditions; rates of chemical reactions can therefore be reduced, particularly if relatively dilute solutions and reactants with low transport numbers are involved. Moreover, undesirable particle settlement can occur when slightly soluble chemical species are formed, as sometimes observed in the authors' laboratories. These drawbacks can be minimised by letting the sample zone flow very slowly during the original STOP period. Although this zone *quasi*-stopping strategy is well known and offers interesting possibilities, no systematic investigation of its capabilities has yet been carried out.

7.5.2. Zone Trapping

This strategy is another option for increasing the sample handling time without increasing dispersion. It was proposed by Krug *et al.* [73] who demonstrated the feasibility of removing the sample zone from the analytical channel, keeping it under static conditions during a pre-set time interval and inserting it again into the analytical channel for further monitoring. The strategy was recently called *stopped-in-loop flow analysis* [74]. It was originally applied to the spectrophotometric determination of ammonium in natural waters with sample heating during the TRAP period. As zone trapping is accomplished after a pre-set time interval, t_{tp}, and involves a reproducible trapping period, precise timing control is essential. Consequently, a computer-operated injector-commutator or a timed six-way rotary valve is required.

The sample aliquot selected by the sampling loop L is inserted into a chemically inert carrier stream and the reagents are admitted by confluence (Fig. 7.7). After the t_{tp} time interval, when the most concentrated portion of the sample zone is flowing inside the reactor R_C, the injector-commutator is switched to the alternative position. The reactor with the sample zone inside it is then trapped between the two fixed bars of the injector-commutator. After the trapping period, the commutator is switched back to the position shown in Fig. 7.7 (lower), re-inserting R_C into the main analytical channel and thus allowing the handled sample to flow towards the detector. Simultaneously, the loop L is moved to the sampling position in order to permit the next sample to be inserted.

The less concentrated leading and trailing edges of the sample zone are usually not trapped and the main carrier/wash stream is always flowing through the detector at a relatively high flow rate. The sampling rate is therefore improved. On the other hand, the reagent stream is continuously pumped to waste and this limits the applicability of the strategy to analytical procedures involving less expensive reagents.

It is interesting to emphasise that the zone stopping strategy was conceived in relation to merging zones; to the best of the authors' knowledge, exploitation of zone trapping and merging zones in order to save expensive reagents has not yet been reported.

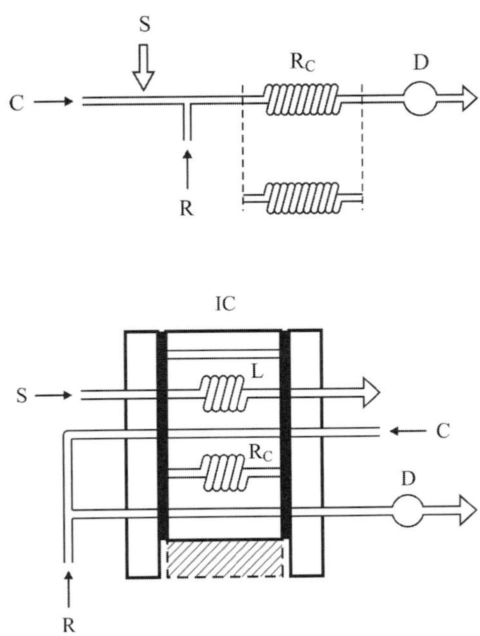

FIGURE 7.7 Didactic representation (upper) and flow diagram (lower) of a flow system with zone trapping relying on the use of an injector-commutator. S = sample; R = reagent; C = sample carrier stream; L = sampling loop; IC = injector-commutator; R_C = coiled reactor; D = detector; traced line = alternative R_C position; shaded area = alternative IC position; solid arrows = sites where pumping is applied.

With multi-commutation, the potentialities of zone trapping are expanded [14], opening up the possibility of trapping the original sample zone and releasing it stepwise as different aliquots, each one corresponding to a different mean concentration and a different trapping period. This innovation is a useful approach for widening the dynamic concentration range, for implementing the standard addition method, for using a single standard solution in order to attain the analytical curve, and to implement prior assays [30].

Dual zone trapping is feasible too, and the innovation is worthwhile for *e.g.*, real-time detecting and circumventing potential sources of inaccuracy. This potential was demonstrated by the spectrophotometric determination of phosphate in plant digests involving relatively slow chemical reactions [75]. As full completion of the chemical reactions was not attained and variations in acidity affected reaction rate, the analytical results were susceptible to matrix effects (Fig. 1.4). Two different trapping periods (10 and 20 s) were then set and both analytical signals were considered. If the ratio of these signals was similar (within an acceptable deviation) to that verified for the standard solutions, accuracy was assessed. On the other hand, if a different signal ratio was verified, calculations were needed for proper correction.

Selected applications of zone stopping and zone trapping are given in Table 7.3.

7.6. MULTI-DETECTION

In flow analysis, multi-detection is generally accomplished by resorting to multichannel flow analysers, and each channel incorporates a different dedicated detector. Another possibility is to take advantage of multi-parametric techniques such as ICP-OES, anodic stripping voltammetry and UV–visible spectrophotometry with diode array detection. A deeper presentation of this aspect is outside of the scope of this monograph.

Multi-detection relying on merely a single detector can also be accomplished. To this end, flow-based strategies exploiting concentration gradients and/or multi-peak recording such as stream splitting/stream merging, merging zones, zone sampling, zone recycling or multi-site detection have been proposed. The potential, limitations and applications of flow systems exploiting these features have already been discussed in this chapter, with the exception of zone recycling and multi-site detection, which are presented below.

7.6.1. Zone Recycling (Multi-peak Recording)

Sample insertion, handling and monitoring are accomplished as in ordinary flow analysers. The main difference is that the sample zone is

TABLE 7.3 Selected Applications of Zone Stopping and Zone Trapping

Analyte	Sample Matrix	Detection Technique	Analytical Range or Detection Limit	Incubation Time/s	Remarks	Ref.
Ag, Fe, Mn	Alloys	UV–Vis	20–160 ng mL^{-1}	60	Simultaneous catalytic determinations/Kalman filter/sample stopping at the detector	[132]
Amiloride hydrochloride	Pharmaceutical tablets	UV–Vis	0–120 μg mL^{-1}	120	Dual zone stopping/binary sampling	[70]
Ammonium	Natural waters	UV–Vis	5 μg L^{-1}	30	Sample zone heating during trapping period	[73]
Ammonium, phosphate	Natural waters	UV–Vis	7.0, 17.0 μg L^{-1}	130	Four parallel trapping reactors/two detectors/merging zones for reagent addition	[133]
Ammonium, phosphate	River and marine waters	Fluorimetry	1, 0.3 μmol L^{-1}	60	*In situ* analysis/sequential injection system/four parallel incubation coils	[134]
Amylase, xylanase, polygalacturonase, protease	Bioreactor preparations	UV–Vis	0.1 U mL^{-1}	40–900	Flow injection system with parallel incubation coils	[135]
Aniline, cyclohexylamine	Commercial sweeteners	UV–Vis	4×10^{-6}, 8×10^{-6} mol L^{-1}	150	Simultaneous determinations/sample "infinite volume"/zone stopping at the detector	[136]
Ascorbic acid	Wines, beers, urine, pharmaceutical preparations	Fluorimetry	0.025–1.0 μg L^{-1}	60	Kinetic flow injection enzymatic procedure/zone stopping/merging zones	[137]
B	Plants	UV–Vis	0.25–6.00 mg L^{-1}	200	Three zones simultaneously trapped in a multi-commuted flow system	[138]

Table 7.3 Selected Applications of Zone Stopping and Zone Trapping—cont'd

Analyte	Sample Matrix	Detection Technique	Analytical Range or Detection Limit	Incubation Time/s	Remarks	Ref.
Cd	Alloys, waters	UV–Vis	0.8–56.2 mg L^{-1}	40	Optode membrane incorporating PAN/reflectance measurements/zone stopping at the detector/comparison of different flow-based procedures	[139]
Captopril	Pharmaceutical preparations	CL	0.011 μg L^{-1}	12	Zone stopping at the detector/situation of sample "infinite volume"	[140]
Creatinine	Urine	UV–Vis	0.5–2.0 g L^{-1}	48	Kinetic determination based on Jaffe's reaction/zone splitting/zone stopping	[120]
Cu, Ni	Electroplating solutions	UV–Vis	0.5–6.0 mg L^{-1} Cu 1.0–15.0 mg L^{-1} Ni	120	Artificial neural networks calibration/catalytic method/zone stopping	[141]
Cu, Zn, Ni, Mn	Natural waters	UV–Vis	0.1–2.0 mg L^{-1}	ca 100	Sequential injection system/artificial neural networks calibration	[142]
DNA-binding with ethidium bromide, proflavine and cisplatin	Antitumour drugs	Fluorimetry	Not relevant	24	Merging zones/zone stopping at the detector	[100]
Ethanol	Whole blood	UV–Vis	0–32.0 mg L^{-1}	14	Zone stopping at the detector/slope estimation/no blank run	[66]
Ga, Al	Synthetic solutions	UV–Vis	1–5, 20–100 mg L^{-1}	60	Zone stopping at the detector/differential kinetic analysis/multivariate calibration	[143]

(Continued)

Table 7.3 Selected Applications of Zone Stopping and Zone Trapping—cont'd

Analyte	Sample Matrix	Detection Technique	Analytical Range or Detection Limit	Incubation Time/s	Remarks	Ref.
Gaseous formaldehyde	Indoor and ambient air	UV–Vis	0.08 ppbv	300	Alternated gas collection (STOP)/sample handling (GO) involving two parallel diffusion scrubbers	[144]
Lactate dehydrogenase activity	Synthetic suspensions	UV–Vis	0–1430 U L^{-1}	80	Zone stopping at the detector/different delay times for wide range determination	[69]
Levodopa, benserazide	Pharmaceutical formulations	UV–Vis	4.1–20.3 × 10^{-4} mol L^{-1} 0.85-4.25 × 10^{-4} mol L^{-1}	168	Multidimensional partial least-square regression/zone stopping	[67]
Morphine, naloxone	Narcotics	CL	0.01–1.00 mg L^{-1} 0.10–1.50 mg L^{-1}	30	Simultaneous time-resolved CL/zone stopping at the detector	[145]
Nitrate, nitrite	Natural waters, foods	Fluorimetry	62–620, 0.9–13.8 ng mL^{-1}	45	Catalytic procedure/zone stopping inside the flow-cell/possibility for monitoring different regions of the flowing zone	[146]
Pd	Airborne particulate matter, automobile catalysts	UV–Vis	0.007 mg L^{-1}	45	Zone stopping/liquid-liquid extraction	[147]
Paracetamol	Pharmaceutical preparations	Fluorimetry	0.5–15 μg mL^{-1}	5.8	Zone stopping inside the flow-cell	[148]
Paraoxon	Waters	UV–Vis	9 × 10^{-9}–5 × 10^{-7} mol L^{-1}	60	Use of immobilized *acetylcholinesterase*/zone stopping/in-line reactivation of the enzyme reactor	[149]

Table 7.3 Selected Applications of Zone Stopping and Zone Trapping—cont'd

Analyte	Sample Matrix	Detection Technique	Analytical Range or Detection Limit	Incubation Time/s	Remarks	Ref.
Phenol, resorcinol	Smoke	UV–Vis	64, 7.5 ng mL^{-1}	2700 (for phenol)	Zone trapping inside 4 parallel coils	[150]
Prostatic acid phosphatase	Blood serum	UV–Vis	2-28 IU L^{-1}	300	Multiple zone stopping/time-based injection/ proposal of parallel incubation coils	[72]
Proteins	Fermentation medium	Turbidimetry	Not relevant	Not available	Zone stopping/merging zones/antigen-antibody interaction	[151]
Phenolic compounds	Waters	UV–Vis	2–9 μg mL^{-1}	600	Zone stopping/situation of sample "infinite volume"/simultaneous kinetic determinations using partial least squares (PLS-2)	[152]
2-Thiobarbituric acid reactive substances	Olive oil	UV–Vis	0.4–10 × 10^{-4} mol L^{-1}	1800	Zone trapping inside 10 parallel heated reactors	[153]
2-Thiobarbituric acid reactive substances	Foods	UV–Vis	1–10 μmol L^{-1}	1200	Long incubation time obtained with a sampling – delay unit with 8 hanging reactors	[154]
V	Drinking water	UV–Vis	0.1–2.0 μg L^{-1}	180	Catalytic method/zone trapping inside two heated parallel reactors	[155]
Vitamins B_1, B_2, B_6	Tablet dissolution tests	Fluorimetry	Not relevant	135	Chemometrics involving an artificial neural networks calibration/zone stopping at the detector	[156]

UV Vis – spectrophotometry; CL = chemiluminescence; SPS = solid phase spectrophotometry. Analytical range or detection limit given in original units.

not directly discarded after being monitored but, instead, is allowed to pass again through the flow cell, yielding another recorded signal. The process can be repeated many times. Consequently, several peaks are recorded after a single sample insertion, each one associated with a different sample handling time and a different sample dispersion, hence with different sample and reagent volumetric fractions, X_S and X_R.

7.6.1.1. *Implementation*

Multi-peak recording was originally implemented by exploiting flow reversal, and this innovation was applied to the determination of glucose in blood serum using a two-point rate assay [76]. A two-peak recording was sufficient. After being monitored, the sample zone was passed again through the detector, under reverse flow conditions, and the difference in heights of both recorded peaks was used as the measurement basis. In this way, blank injections were not necessary. The need for a bi-directional pump was avoided by placing a rotary valve immediately after the flow cell. Switching this valve allowed the already-monitored sample zone to pass again through the detector.

Another possibility is to recycle the sample zone. To this end, the handled sample zone is passed through a 4-way rotary valve before reaching the detector (Fig. 7.8, left). After it has left the detector, the valve is switched to the alternative position (Fig. 7.8, right) and the sample zone starts to be recycled. Once several peaks have been recorded, each one associated with a different sample handling time and different X_S and X_R values, the valve is switched back to the position shown in Fig. 7.8 (left), directing the sample zone towards waste. The successive passage of the sample zone through the pump can be avoided by exploiting multi-commutation [77].

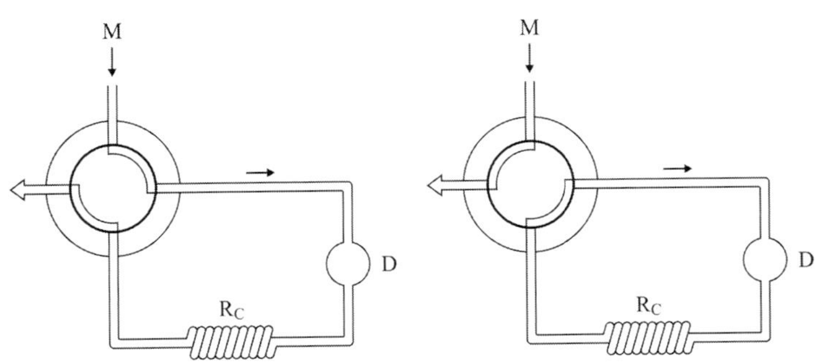

FIGURE 7.8 **Flow diagram of a flow system with multi-peak recording.** M = from manifold; D = detector; RC = reaction coil; arrows = sites where pumping is applied.

7.6.1.2. *Applications*

This strategy is especially attractive in relation to analytical procedures relying on relatively slow chemical reactions (catalytic procedures and differential kinetic analysis) and/or exploiting different degrees of sample dispersion. The influence of zone recycling on sample dispersion and the potential, limitations and application range of the strategy are presented elsewhere [78,79].

7.6.2. Multi-site Detection

With this strategy [80], a single detector can access different manifold sites. These sites can be located in the same or in different analytical channels; thus, the strategy can be implemented in relation to serial or parallel monitoring. The related flow systems are simple and exhibit improved performance.

Loop-based injection is used for sample insertion and a similar procedure is used for displacing the tubular detector D between monitoring sites a and b (Figs 7.9 and 7.10, lower). Consequently, this strategy has usually been accomplished using two loop-based injectors or a double injection valve/commutator [81], where a flow cell with two small transmission lines replaces one of the sampling loops. Although the flow system can be manually operated, computer-controlled valves or commutators have been preferred.

The solution inside the detector during valve switching is inserted into another manifold site through which a different solution is flowing.

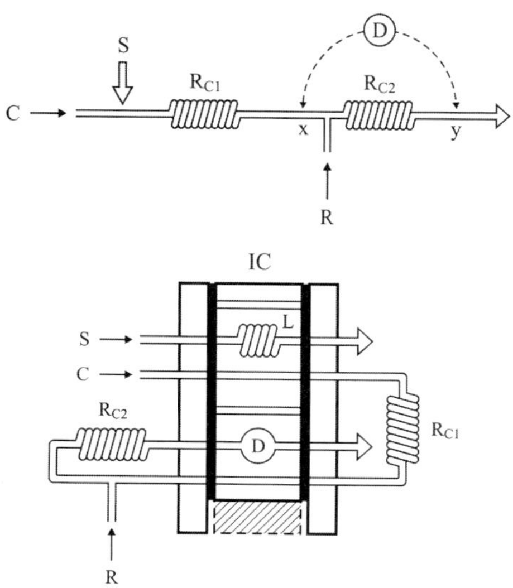

FIGURE 7.9 **Didactic representation (upper) and flow diagram (lower) of a flow system with multi-site detection designed for serial monitoring.** S = sample; C = carrier/wash stream; R = reagent; L = sampling loop; R_{C1}, R_{C2} = coiled reactors; x, y = monitoring sites; D = detector; IC = injector-commutator; shaded area = alternative IC position; solid arrows = sites where pumping is applied.

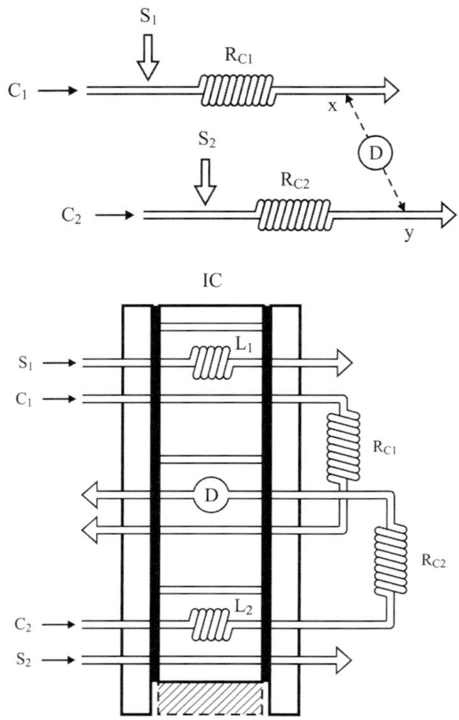

FIGURE 7.10 **Didactic representation (upper) and flow diagram (lower) of a flow system with multi-site detection designed for parallel monitoring.** S_1, S_2 = samples; C_1, C_2 = reagent carrier/wash streams; L_1, L_2 = sampling loops; R_{C1}, R_{C2} = coiled reactors; x, y = monitoring sites; D = detector; IC = injector-commutator; shaded area = alternative IC position; solid arrows = sites where pumping is applied.

Consequently, the strategy is prone to Schlieren effects. Moreover, the available time for monitoring at a specific site is lower relative to ordinary flow systems. Consequently, a small inner volume tubular flow cell and a detector with a fast response time are preferred.

Improvement in sampling rate is one of the main advantages of this strategy. In ordinary flow systems, a relatively long time interval is needed after obtaining the analytical signal in order to avoid carryover, thus limiting the sampling rate. With multi-site detection, this time interval undergoes a dramatic reduction because the detector is moved to another site in the manifold after obtaining the analytical signal. Consequently, the wash time then refers only to the manifold portions related to the detection unit and not to the entire analytical path. As the inner volume of the detector plus accessories is smaller than that of the analytical path, the sampling rate is improved.

The main characteristics, advantages, limitations and applications of flow systems with multi-site detection have been reported elsewhere [81].

7.6.2.1. *Implementation*

Regarding multi-site detection with serial monitoring (Fig. 7.9), the sample aliquot is introduced into the sample carrier/wash stream C and, after being handled inside reactor R_{C1}, reaches the first monitoring site x where it is quantified. After leaving the detector, the sample zone undergoes further handling inside reactor R_{C2} and is monitored again, downstream, by the same detector that was relocated.

Regarding parallel monitoring (Fig. 7.10), different sample aliquots are injected into different carrier/wash streams (C_1, C_2) which share the same detector D. This innovation is possible because the detector is moved between the different analytical channels involved.

7.6.2.2. *Applications*

As flow systems with multi-site detection involve either serial or parallel monitoring, the applications of the strategy highlighted in the following sections are classified by these different monitoring modes.

SERIAL MONITORING

The strategy was originally applied to iron speciation in natural waters [80]. The sample was inserted into a chemically inert carrier/wash stream, and the colour-forming reagent (1,10-phenanthroline) was added by confluence. As this reagent reacted with Fe(II), the analytical signal gathered at the first monitoring site was proportional to the Fe(II) content in the injectate. After leaving the detector, a confluent stream (ascorbic acid solution) was added to the sample zone in order to convert Fe(III) to Fe(II) ions. The second passage of the sample through the detector led to a recorded peak proportional to the total iron content in the injectate. The inner volume of the flow cell plus accessories was however relatively large (420 μL) because a conventional spectrophotometer was used and long transmission lines between the flow cell and the switching valve were needed. Therefore, the sampling rate was not very high. This innovation was further improved [82] by reducing the inner volume of the detector plus accessories to 64 μL. To this end, an optical system exploiting LED-based technology was designed: both the light source and the detector were integrated within the flow cell that was attached to the moveable portion of an injector-commutator. Transmission lines between the detector and the moving components were therefore not needed. The procedure was applied to drug analysis: Schlieren noise was not observed and the sampling rate was estimated to be 133 h^{-1}. It is important to stress that a similar arrangement can be achieved using optical fibres.

This innovation is useful in relation to spectrophotometric procedures limited by the colour and/or turbidity of the samples (*e.g.*, red wines, waste waters, beverages and biological fluids) which normally require blank compensation. As a rule, the first measurement is performed immediately after sample injection, whereas the second one is generally done after reaction development. The net analytical signal is then obtained by subtracting (after proper dispersion matching) the first measurement – proportional to the blank value – from the second one. In this context, the determination of uric acid in urine using tripyridiltriazine as the colour-forming reagent [83] is a good example. The method involved Fe(III) reduction by the analyte and evaluation of the remaining Fe(II) through reaction with tripyridyltriazine. As the sample zone was quantified at two distinct sites in the manifold, interference due to the presence of other reducing agents or oxidising agents, as well as the inherent sample colour, was compensated. The system was favourably compared with others exploiting zone stopping in terms of cost, simplicity and ruggedness.

The difference in sample handling times associated with both monitoring sites can also be efficiently exploited, as the sample zone can be quantified under different reaction conditions and timing without requiring pronounced modifications to the manifold architecture. The potentiometric determination of sucrose in sugar cane juices and syrups [84] demonstrated this approach. The first analytical signal reflected the concentrations of the reducing sugars; an acidic stream was then added in order to promote sucrose hydrolysis. As the contribution of sucrose to the analytical signal was dominant in the second recorded signal, the difference in recorded peak heights was used as the measurement basis and the blank signals were easily compensated. Another example is the amperometric determination of salicylic and acetyl salicylic acids (SA and ASA) in pharmaceutical products [85]. The first measurement referred to the SA content of the sample. Sodium hydroxide was added immediately after the first monitoring site, allowing partial but reproducible SA hydrolysis. In this way, the contribution of ASA was highlighted in the second readout.

Differential kinetic analysis can be analogously implemented, as demonstrated by the spectrophotometric determination of nickel and copper in plant digests using 2-(5-bromo-2-pyridylazo)-5-(diethylamino)-phenol (Br-PADAP) as the colour-forming reagent [86]. The sample was inserted and the reagent was immediately added by confluence. Passage of the sample zone through the detector led to a recorded peak reflecting mainly the copper contribution to the analytical signal, as complexation of Cu^{2+} by Br-PADAP is relatively fast. After leaving the flow cell, the sample zone was directed to flow inside the heated reactor R_{C2} (Fig. 7.9) where the relatively slow complexation of Ni^{2+} occurred. The second peak then reflected the contribution of this analyte to the analytical signal. Enhanced figures of merit were reported. This innovation can be regarded as pseudo-differential kinetic analysis, because the sample was handled under different conditions. This is a favourable characteristic for enhanced discriminating power.

The standard addition method can be efficiently implemented in a flow system with multi-site detection, as demonstrated by the amperometric determination of ascorbic acid in fruit juices [87]. Only one sample aliquot was inserted, and the sample zone was monitored under two different conditions, before and after the standard addition by confluence. Measurements related to the first and second detection sites were normalised and the matrix effect was efficiently circumvented. About 55 samples were run per hour, yielding precise results in agreement with a reference procedure.

PARALLEL MONITORING

As the flow system comprises two analytical channels sharing a single relocating detector, different sample zones can be simultaneously handled and successively monitored. The feasibility of the approach was demonstrated by the spectrophotometric determination of aluminium and iron in plant digests [80]. This innovation was further applied to the spectrophotometric determination of calcium and magnesium in soil extracts and natural waters [88]. *o*-Cresolphthalein complexone was used as the chromogenic reagent, ammonium-ammonia as the buffer system, and EGTA or 8-hydroxyquinoline as the masking agents for Ca and Mg, respectively. The authors emphasised the improvement in sampling rate after implementation of multi-site detection.

Immobilised enzymes were used in connection with multi-site detection for the spectrophotometric determination of ethanol and glycerol in wines [89]. The analytical channels comprised a mini-column filled with a different enzyme (*ethanol dehydrogenase* or *glycerol dehydrogenase* immobilised onto beads) and shared a single relocating detector. The cost of analysis was lower than that of using commercial kits.

With parallel monitoring, a noteworthy improvement in sampling rate can be attained by designing the flow system with two almost identical analytical channels sharing the detector, as demonstrated by the determination of glycerol in soaps, detergents and lixivia using a small internal volume periodate selective electrode [84]. The method involved carbohydrate oxidation by periodate and further evaluation of the remaining periodate. The sampling rate was improved by *ca* 85% relatively to an analogous flow injection system without multi-site detection.

Multi-site detection is also attractive in relation to monosegmented flow systems, into which the sample zone is established between two air bubbles (see 5.5.1). Removal of the air phase is not necessary, as the detector is displaced from the main analytical channel after the peak maximum has been registered but before passage of the air bubble, which is then easily discarded. The feasibility of this approach was demonstrated by the determination of boron in plants using the Azomethine-H condensation reaction [90]. Four sample zones were always present in the analytical path, allowing a high sampling frequency (120 h^{-1}) in spite of the 105-s sample handling time. A minimal distance between successive zones was maintained in order to avoid carryover effects. Analogously, the spectrophotometric determinations of triglycerides and cholesterol in blood serum exploiting relocating flow cells and monosegmented streams were also proposed [91,92].

Detector relocation between two different flowing streams also permits in-line detector reconditioning. This reconditioning step is often a limiting factor in flow-based electroanalytical procedures because adsorption effects can be significant. The sample/electrode contact time is reduced in flow systems, and the carrier stream usually facilitates reconditioning, but progressive poisoning has often been reported [93] and in-line conditioning generally requires a complex manifold. With multi-site detection, detector relocation between two different streams permits efficient detector re-conditioning. The chemical composition and flow rate of the conditioning stream can be selected independently of the composition of the carrier/wash stream. This was demonstrated by the amperometric determination of diltiazem and nimesulide in pharmaceutical preparations using a tubular glassy carbon electrode [94]. Adsorption of organic compounds on the electrode active surface, which impairs the applicability of the batch procedure to large-scale analysis, was avoided.

7.7. FINAL REMARKS

The majority of applications related to the different strategies highlighted in this chapter were proposed during the 1980's and 1990's and less emphasis is now given to them, partly because most commercially available instruments do not provide facilities for their implementation. Moreover, they are very suitable for unsegmented flow analysis but not for segmented-flow or microfluidic analysis. In the near future, however, interest in these flow-based strategies should increase because the fundamental principles underpinning the methods for flow manipulation are well established and can provide enhanced information compared with conventional approaches.

References

[1] H. Bergamin-Filho, E.A.G. Zagatto, F.J. Krug, B.F. Reis, Merging zones in flow injection analysis. Part 1. Double proportional injector and reagent consumption, Anal. Chim. Acta 101 (1978) 17.

[2] J. Mindegaard, Flow multi-injection analysis - a system for the analysis of highly concentrated samples without prior dilution, Anal. Chim. Acta 104 (1979) 185.

[3] M.F. Gine, H. Bergamin-Filho, E.A.G. Zagatto, B.F. Reis, Simultaneous determination of nitrate and nitrite by flow injection analysis, Anal. Chim. Acta 114 (1980) 191.
[4] E.A.G. Zagatto, B.F. Reis, M. Martinelli, F.J. Krug, H. Bergamin-Filho, M.F. Gine, Confluent streams in flow injection analysis, Anal. Chim. Acta 198 (1987) 153.
[5] J. Alonso, J. Bartroli, M. Valle, M. Escalada, R. Barber, Sandwich techniques in flow injection analysis. Part 1. Continuous recalibration techniques for process control, Anal. Chim. Acta 199 (1987) 191.
[6] J.L. Montesinos, J. Bartroli, M. Poch, M. Valle, J.L.F.C. Lima, A.N. Araujo, Mathematical modelling of sequential determinations by flow-injection sandwich techniques, Anal. Chim. Acta 234 (1990) 67.
[7] J. Alonso, J. Bartroli, M. Valle, R. Barber, Sandwich techniques in flow-injection analysis. Part 2. Simultaneous determination of iron(II) and total iron, Anal. Chim. Acta 219 (1989) 345.
[8] A.N. Araujo, J.L.F.C. Lima, J. Alonso, J. Bartroli, M. Poch, Flow injection system based on the sandwich technique for saving expensive reagents, Clin. Chim. Acta 203 (1991) 67.
[9] J. Alonso, J. Bartroli, R. Barber, Sandwich techniques in flow-injection analysis. Part 3. Simultaneous determination of Cr(VI) in two concentration ranges, Anal. Chim. Acta 261 (1992) 219.
[10] A.N. Araujo, J.L.F.C. Lima, A.O.S.S. Rangel, J. Alonso, J. Bartroli, R. Barber, Simultaneous determination of total iron and chromium(VI) in wastewater using a flow injection system based on the sandwich technique, Analyst 114 (1989) 1465.
[11] A.E. Davies, A.G. Fogg, Flow injection sandwich technique applied to indirect visible spectrophotometric determinations. Bromimetric and iodimetric determination of sulphite, Anal. Chim. Acta 318 (1995) 55.
[12] A. Cerda, M.T. Oms, R. Forteza, V. Cerda, Sequential injection sandwich technique for the simultaneous determination of nitrate and nitrite, Anal. Chim. Acta 371 (1998) 63.
[13] J. Ruzicka, E.H. Hansen, Flow injection analysis. Principles, applications and trends, Anal. Chim. Acta 114 (1980) 19.
[14] M.A. Feres, P.R. Fortes, E.A.G. Zagatto, J.L.M. Santos, J.L.F.C. Lima, Multicommutation in flow analysis: recent developments and applications, Anal. Chim. Acta 618 (2008) 1.
[15] E.A.G. Zagatto, F.J. Krug, H. Bergamin-Filho, S.S. Jørgensen, B.F. Reis, Merging zones in flow injection analysis. Part 2. Determination of calcium, magnesium and potassium in plant material by continuous flow injection atomic absorption and flame emission spectrometry, Anal. Chim. Acta 104 (1979) 279.
[16] P.J. Worsfold, The bio-analytical potential of flow injection analysis, Anal. Chim. Acta 145 (1983) 117.
[17] J. Ruzicka, E.H. Hansen, Stopped flow and merging zones. A new approach to enzymatic assays by flow injection analysis, Anal. Chim. Acta 106 (1979) 207.
[18] T. Yamane, E. Goto, A flow injection system for the simultaneous determination of two components in a single injection and with a single detector, Anal. Sci. 5 (1989) 221.
[19] H. Bergamin-Filho, B.F. Reis, A.O. Jacintho, E.A.G. Zagatto, Ion exchange in flow injection analysis. Determination of ammonium ions at the μg/l level in natural waters with pulsed Nessler reagent, Anal. Chim. Acta 117 (1980) 81.
[20] B.F. Reis, E.A.G. Zagatto, A.O. Jacintho, F.J. Krug, H. Bergamin-Filho, Merging zones in flow injection analysis. Part 4. Simultaneous spectrophotometric determination of total nitrogen and phosphorus in plant material, Anal. Chim. Acta 119 (1980) 305.
[21] R.M. Liu, D.J. Liu, A.L. Sun, Simultaneous determination of multicomponents by flow injection analysis. Determination of copper and zinc in serum by using zincon as colouring reagent, Talanta 40 (1993) 511.
[22] E.A.G. Zagatto, A.O. Jacintho, F.J. Krug, B.F. Reis, R.E. Bruns, M.C.U. Araujo, Flow-injection systems with inductively-coupled argon plasma atomic emission

spectrometry. Part 2. The generalized standard addition method, Anal. Chim. Acta 145 (1983) 169.
[23] B.E.H. Saxberg, B.R. Kowalski, Generalized standard addition method, Anal. Chem. 51 (1979) 1031.
[24] M.F. Gine, B.F. Reis, E.A.G. Zagatto, F.J. Krug, A.O. Jacintho, A simple procedure for standard additions in flow injection analysis. Spectrophotometric determination of nitrate in plant extracts, Anal. Chim. Acta 155 (1983) 131.
[25] B.F. Reis, A.O. Jacintho, J. Mortatti, F.J. Krug, E.A.G. Zagatto, H. Bergamin-Filho, L.C.R. Pessenda, Zone sampling processes in flow injection analysis, Anal. Chim. Acta 123 (1981) 221.
[26] J. Moeller, New injection valves and their application in flow injection analysis, Fresenius Z. Anal. Chem. 329 (1988) 685.
[27] F.R.P. Rocha, P.B. Martelli, R.M. Frizzarin, B.F. Reis, Automatic multicommutation flow system for wide range spectrophotometric calcium determination, Anal. Chim. Acta 366 (1998) 45.
[28] M.B. Garn, M. Gisin, H. Gross, P. King, W. Schmidt, C. Thommen, Extensive flow-injection dilution for in-line sample pretreatment. Comparison between single-stage and dual-stage modules, Anal. Chim. Acta 207 (1988) 225.
[29] E.A.G. Zagatto, A.O. Jacintho, L.C.R. Pessenda, F.J. Krug, B.F. Reis, H. Bergamin-Filho, Merging zones in flow injection analysis. Part 5. Simultaneous determination of aluminium and iron in plant digests by a zone-sampling approach, Anal. Chim. Acta 125 (1981) 37.
[30] F.R.P. Rocha, C.M.C. Infante, W.R. Melchert, A multi-purpose flow system based on multi-commutation, Spectrosc. Lett. 39 (2006) 651.
[31] P.K. Dasgupta, W. Lei, Solvent extraction in continuous flow systems with intelligent zone sampling, Anal. Chim. Acta 226 (1989) 255.
[32] C. Pasquini, Mechanical removal of the central sample zone to avoid air bubbles in monosegmented continuous flow analysis, Anal. Chem. 58 (1986) 2346.
[33] A.O. Jacintho, E.A.G. Zagatto, H. Bergamin-Filho, F.J. Krug, B.F. Reis, R.E. Bruns, B.R. Kowalski, Flow injection systems with inductively-coupled argon plasma atomic emission spectrometry. Part 1. Fundamental considerations, Anal. Chim. Acta 130 (1981) 243.
[34] M.M. Santos-Filha, B.F. Reis, H. Bergamin-Filho, N. Baccan, Flow-injection determination of low levels of ammonium ions in natural waters employing preconcentration with a cation-exchange resin, Anal. Chim. Acta 261 (1992) 339.
[35] E.A.G. Zagatto, O. Bahia-Filho, H. Bergamin-Filho, Recording the real sample distribution and concentration time functions in flow injection analysis, Anal. Chim. Acta 193 (1987) 309.
[36] H.C. Pitot, N. Pries, M. Poirier, A. Cutler, Rate analysis of enzyme reactions by continuous and interrupted flow procedure, in: Automation in Analytical Chemistry, Media Press, New York, 1966, pp. 555–558. in A.W. Skillen, Automation of enzyme assays, J. Clin. Pathol. S1-4 (1970) 31.
[37] A.W. Skillen, Automation of enzyme assays, J. Clin. Pathol. S1-4 (1970) 31.
[38] M.A. Kenny, M.H. Chengt, Rapid, automated, simultaneous determination of serum CO_2 and chloride with the "AutoAnalyzer I," Clin. Chem. 18 (1972) 352.
[39] L.T. Skeggs Jr., Automatic method for colorimetric analysis, Am. J. Clin. Pathol. 28 (1957) 311.
[40] L.T. Skeggs Jr., H. Hochstrasser, Multiple automatic sequential analysis, Clin. Chem. 10 (1964) 918.
[41] J.W.B. Stewart, J. Ruzicka, Flow Injection Analysis. Part V. Simultaneous determination of nitrogen and phosphorus in acid digests of plant material with a single spectrophotometer, Anal. Chim. Acta 82 (1976) 137.

[42] L.T. Skeggs, Persistence and prayer: from the artificial kidney to the autoanalyzer, Clin. Chem. 46 (2000) 1425.
[43] M. Gisin, C. Thommen, K.F. Mansfield, Hydrodynamically limited precision of gradient techniques in flow injection analysis, Anal. Chim. Acta 179 (1986) 149.
[44] E.A.G. Zagatto, M.F. Gine, E.A.N. Fernandes, B.F. Reis, F.J. Krug, Sequential injections in flow systems as an alternative to gradient exploitation, Anal. Chim. Acta 173 (1985) 289.
[45] J. Ruzicka, J.W.B. Stewart, E.A.G. Zagatto, Flow Injection Analysis. Part IV. Stream sample splitting and its application to the continuous spectrophotometric determination of chloride in brackish waters, Anal. Chim. Acta 81 (1976) 387.
[46] J.F. Tyson, S.R. Bysouth, Network flow injection manifolds for sample dilution and calibration in flame atomic absorption spectrometry, J. Anal. At. Spectrom. 3 (1988) 211.
[47] A. Fernandez, M.A. Gomes-Nieto, M.D. Luque de Castro, M. Valcarcel, A flow-injection manifold based on splitting the sample zone and a confluence point before a single detector unit, Anal. Chim. Acta 165 (1984) 217.
[48] A. Fernandez, M.D. Luque de Castro, M. Valcarcel, Comparison of flow injection analysis configurations for differential kinetic determination of cobalt and nickel, Anal. Chem. 56 (1984) 1146.
[49] P. Linares, M.D. Luque de Castro, M. Valcarcel, Simultaneous determination of pyridoxal and pyridoxal-5-phosphate in human serum by flow injection analysis, Anal. Chem. 57 (1985) 2101.
[50] F. Lazaro, M.D. Luque de Castro, M. Valcarcel, Sequential and differential catalytic-fluorimetric determination of manganese and iron by flow injection analysis, Anal. Chim. Acta 169 (1985) 141.
[51] P. Linares, M.D. Luque de Castro, M. Valcarcel, Fluorimetric differential kinetic determination of silicate and phosphate in waters by flow-injection analysis, Talanta 33 (1986) 899.
[52] P. Linares, M.D. Luque de Castro, M. Valcarcel, Differential kinetic determination of furfural and vanillin by flow injection analysis, Microchem. J. 35 (1987) 120.
[53] C.C. Oliveira, R.P. Sartini, B.F. Reis, E.A.G. Zagatto, Multicommutation in flow analysis. Part 4. Computer assisted splitting for spectrophotometric determination of copper and zinc in plants, Anal. Chim. Acta 332 (1996) 173.
[54] R. Burakham, M. Oshima, K. Grudpan, S. Motomizu, Simple flow-injection system for the simultaneous determination of nitrite and nitrate in water samples, Talanta 64 (2004) 1259.
[55] R. Khouel, A. Aminot, Fluorometric determination of ammonia in sea and estuarine waters by direct segmented flow analysis, Mar. Chem. 57 (1997) 265.
[56] D.A. Whitman, G.D. Christian, Cascade system for rapid on-line dilutions in flow-injection analysis, Talanta 36 (1989) 205.
[57] A.M.R. Ferreira, A.O.S.S. Rangel, J.L.F.C. Lima, Flow injection systems with stream splitting and a dialysis unit for the soil analysis of sodium and potassium by flame emission spectrometry, and calcium and magnesium by atomic absorption spectrophotometry, Commun. Soil Sci. Plant Anal. 26 (1995) 1532.
[58] G.D. Clark, J. Ruzicka, G.D. Christian, Split zone flow injection analysis: an approach to automated dilutions, Anal. Chem. 61 (1989) 1773.
[59] B.F. Reis, E.A.G. Zagatto, P.B. Martelli, S.M.B. Brienza, Trial measurements in flow analysis, Analyst 118 (1993) 719.
[60] S.M.B. Brienza, E.A.G. Zagatto, O.M. Matsumoto, A simplified split zone flow injection system, Lab. Robot. Autom. 3 (1991) 115.
[61] L. Tymecki, K. Strzelak, R. Koncki, A single standard calibration module for flow analysis systems based on solenoid microdevices, Talanta 79 (2009) 205.

[62] R.A.S. Lapa, J.L.F.C. Lima, J.L.M. Santos, Fluorimetric determination of isoniazid by oxidation with cerium(IV) in a multicommutated flow system, Anal. Chim. Acta 419 (2000) 17.
[63] D.G. Themelis, A. Economou, A. Tsiomlektsis, P.D. Tzanavaras, Direct determination of phosphate in urine by sequential-injection analysis with single on-line dilution-calibration method and photometric detection, Anal. Biochem. 330 (2004) 193.
[64] M.A.Z. Arruda, E.A.G. Zagatto, A simple stopped-flow method with continuous pumping for the spectrophotometric flow-injection determination of boron in plants, Anal. Chim. Acta 199 (1987) 137.
[65] M. Blanco, J. Coello, H. Iturriaga, S. Maspoch, J. Riba, Modification of a commercial dual channel valve to be used in a stopped-flow injection analysis system, J. Flow Injection Anal. 7 (1990) 3.
[66] P.J. Worsfold, J. Ruzicka, E.H. Hansen, Rapid automated enzymatic method for the determination of alcohol in blood and beverages using flow injection analysis, Analyst 106 (1981) 1309.
[67] M. Pistonesi, M.E. Centurion, B.S.F. Band, P.C. Damiani, A.C. Olivieri, Simultaneous determination of levodopa and benserazide by stopped-flow injection analysis and three-way multivariate calibration of kinetic-spectrophotometric data, J. Pharm. Biomed. Anal 36 (2004) 541.
[68] J. Ruzicka, E.H. Hansen, Flow injection analysis. Part X. Theory, techniques and trends, Anal. Chim. Acta 99 (1978) 37.
[69] S. Olsen, J. Ruzicka, E.H. Hansen, Gradient techniques in flow injection analysis. Stopped-flow measurement of the activity of lactate dehydrogenase with electronic dilution, Anal. Chim. Acta 136 (1982) 101.
[70] R.A.S. Lapa, J.L.F.C. Lima, J.L.M. Santos, Dual-stopped-flow spectrophotometric determination of amiloride hydrochloride in a multicommutated flow system, Anal. Chim. Acta 407 (2000) 225.
[71] J. Ruzicka, E.H. Hansen, Flow Injection Analysis, J. Wiley & Sons, NewYork, 1981.
[72] B.F. Rocks, R.A. Sherwood, M.M. Hosseinmardi, C. Riley, The use of holding coils to facilitate long incubation times in unsegmented flow analysis. Determination of serum prostatic acid phosphatase, Anal. Chim. Acta 179 (1986) 225.
[73] F.J. Krug, B.F. Reis, M.F. Gine, J.R. Ferreira, A.O. Jacintho, E.A.G. Zagatto, Zone trapping in flow injection analysis. Spectrophotometric determination of low levels of ammonium ion in natural waters, Anal. Chim. Acta 151 (1983) 39.
[74] N. Teshima, S. Ohno, T. Sakai, Stopped-in-loop flow analysis of trace vanadium in water, Anal. Sci. 23 (2007) 1.
[75] E.A.G. Zagatto, F.R.P. Rocha, P.B. Martelli, B.F. Reis, Detecting and circumventing sources of inaccuracy in flow analysis, Pure Appl. Chem. 73 (2001) 45.
[76] J. Toei, Determination of glucose in clinical samples by flow reversal flow injection analysis, Analyst 113 (1988) 475.
[77] R.A.S. Lapa, J.L.F.C. Lima, B.F. Reis, J.L.M. Santos, Continuous sample recirculation in an opened-loop multicommutated flow system, Anal. Chim. Acta 377 (1998) 103.
[78] M. Valcarcel, M.D. Luque de Castro, F. Lazaro, A. Rios, Multiple peak recordings in flow injection analysis, Anal. Chim. Acta 216 (1989) 275.
[79] Y. Narusawa, Y. Miyamae, Radial dispersion by computer-aided simulation with data from zone circulating flow-injection analysis, Anal. Chim. Acta 296 (1994) 129.
[80] E.A.G. Zagatto, H. Bergamin-Filho, S.M.B. Brienza, M.A.Z. Arruda, A.R.A. Nogueira, J.L.F.C. Lima, Multi-site detection in flow analysis. Part 1. Relocation of spectrophotometric detector, Anal. Chim. Acta 261 (1992) 59.
[81] V. Grassi, E.A.G. Zagatto, J.L.F.C. Lima, Flow-injection systems with multi-site detection, Trends Anal. Chem. 24 (2005) 880.

[82] S.R.B. Santos, M.C.U. Araujo, R.S. Honorato, E.A.G. Zagatto, J.L.F.C. Lima, R.A.S. Lapa, An improved leaping detector for flow analysis applied to iron speciation in drugs, J. Autom. Meth. Manag. Chem. 22 (2000) 83.

[83] A.N. Araujo, J.A.M. Catita, J.L.F.C. Lima, Kinetic determination of uric acid in urine based on single-line flow-system with multi-site detection, Anal. Sci. 14 (1998) 809.

[84] J.A. Gomes-Neto, A.R.A. Nogueira, H. Bergamin-Filho, E.A.G. Zagatto, J.L.F.C. Lima, M.C.B.S.M. Montenegro, Multi-site detection in flow analysis. Part 3. Periodate tubular electrode with low inner volume as a relocatable detector, Anal. Chim. Acta 285 (1994) 293.

[85] R.I.L. Catarino, M.B.Q. Garcia, R.A.S. Lapa, J.L.F.C. Lima, E. Barrado, Sequential determination of salicylic and acetylsalicylic acids by amperometric multisite detection flow injection analysis, J. AOAC Int. 85 (2002) 1253.

[86] D. Vendramini, V. Grassi, E.A.G. Zagatto, Spectrophotometric flow injection determination of copper and nickel in plant digests exploiting differential kinetic analysis and multi site detection, Anal. Chim. Acta 570 (2006) 124.

[87] R.I.L. Catarino, M.B.Q. Garcia, J.L.F.C. Lima, E. Barrado, M. Vega, Relocation of a tubular voltammetric detector for standard addition in FIA, Electroanal. 14 (2002) 741.

[88] A.R.A. Nogueira, S.M.B. Brienza, E.A.G. Zagatto, J.L.F.C. Lima, A.N. Araújo, Flow injection system with multisite detection for spectrophotometric determination of calcium and magnesium in soil extracts and natural waters, J. Agric. Food Chem. 44 (1996) 165.

[89] A.O.S.S. Rangel, I.V. Toth, Enzymatic determination of ethanol and glycerol by flow injection parallel multi-site detection, Anal. Chim. Acta 416 (2000) 205.

[90] A.R.A. Nogueira, S.M.B. Brienza, E.A.G. Zagatto, J.L.F.C. Lima, A.N. Araujo, Multisite detection in flow analysis. Part 2. Monosegmented systems with relocating detectors for the spectrophotometric determination of boron in plants, Anal. Chim. Acta 276 (1993) 121.

[91] A.N. Araújo, J.A.M. Catita, J.L.F.C. Lima, E.A.G. Zagatto, , Monosegmented flow-analysis of slow enzymatic reactions: determination of triglycerides in serum, Fresenius J. Anal. Chem. 360 (1998) 100.

[92] A.N. Araújo, J.A.M. Catita, J.L.F.C. Lima, Monosegmented flow-analysis of serum cholesterol, Farmaco 54 (1999) 51.

[93] A. Izquierdo, M.D. Luque de Castro, M. Valcarcel, Simultaneous determination of copper, lead, cadmium and zinc in previously lyophilized biological tissues using flow-injection analysis anodic-stripping voltammetry, Electroanal. 6 (1994) 894.

[94] R.I.L. Catarino, A.C.L. Conceição, M.B.Q. Garcia, M.L.S. Gonçalves, J.L.F.C. Lima, M.M.C. Santos, Flow amperometric determination of pharmaceuticals with on-line electrode surface renewal, J. Pharm. Biomed. Anal. 33 (2003) 571.

[95] W.T. Suarez, H.J. Vieira, O. Fatibello-Filho, Flow injection turbidimetric determination of acetylcysteine in pharmaceutical formulations using silver nitrate as precipitant reagent, J. Braz. Chem. Soc. 18 (2007) 1028.

[96] B.F. Reis, H. Bergamin-Filho, E.A.G. Zagatto, F.J. Krug, Merging zones in flow injection analysis. Part 3. Spectrophotometric determination of aluminum in plant and soil materials with sequential addition of pulsed reagents, Anal. Chim. Acta 107 (1979) 309.

[97] C.S. Lim, J.N. Miller, J.W. Bridges, Automation of an energy-transfer immunoassay by using stopped-flow injection analysis with merging zones, Anal. Chim. Acta 114 (1980) 183.

[98] J.L. Rufino, H.R. Pezza, L. Pezza, Flow-injection spectrophotometric determination of azithromycin in pharmaceutical formulations using p-chloranil in the presence of hydrogen peroxide, Anal. Sci. 24 (2008) 871.

[99] A.O. Jacintho, E.A.G. Zagatto, B.F. Reis, L.C.R. Pessenda, F.J. Krug, Merging zones in flow injection analysis. Part 6. Determination of calcium in natural waters, soil and plant materials with glyoxal bis(2-hydroxyanil), Anal. Chim. Acta 130 (1981) 361.

[100] A. Alonso, M.J. Almendral, Y. Curto, J.J. Criado, E. Rodríguez, J.L. Manzano, Determination of the DNA-binding characteristics of ethidium bromide, proflavine and cisplatin by flow injection analysis. Usefulness in studies on antitumor drugs, Anal. Biochem. 355 (2006) 157.
[101] A.O.S.S. Rangel, I.V. Tóth, Determination of ethanol in wines by flow injection spectrophotometry using gas-diffusion and an immobilized enzyme reactor, Am. J. Enol. Vitic. 50 (1999) 259.
[102] S.M.V. Fernandes, A.O.S.S. Rangel, J.L.F.C. Lima, Colorimetric determination of iron in beer by flow-injection analysis using the merging zones technique, J. Inst. Brew. 101 (1995) 281.
[103] X.F. Yue, Y.N. Zhang, Z.Q. Zhang, An air-driving FI device with merging zones technique for the determination of formaldehyde in beers, Food Chem. 102 (2007) 90.
[104] V.O. Brito, I.M. Raimundo Jr., Simultaneous multiple injection in monosegmented flow analysis, Anal. Chim. Acta 371 (1998) 317.
[105] I.M.P.L.V.O. Ferreira, M.N. Macedo, M.A. Ferreira, Enzymic flow injection determination of free L-carnitine in infant formulae, Analyst 122 (1997) 1539.
[106] J.M. Fernandez-Romero, M.D. Luque de Castro, M. Valcarcel, R. Quiles-Zafra, Spectrophotometric determination of magnesium in serum by using a flow-injection system with an immobilized enzyme reactor, Anal. Chim. Acta 283 (1993) 447.
[107] K. Oguma, K. Nishiyama, R. Kuroda, Spectrophotometric flow injection analysis of silicates for manganese, Anal. Sci. 3 (1987) 251.
[108] E.A.G. Zagatto, A.O. Jacintho, J. Mortatti, H. Bergamin-Filho, An improved flow injection determination of nitrite in waters by using intermittent flows, Anal. Chim. Acta 120 (1980) 399.
[109] J.A. Schneider, J.F. Hornig, Spectrophotometric determination of lead in tap water with 5,10,15,20-tetra(4-N-sulfoethylpyridinium)porphyrin using merging zones flow-injection, Analyst 118 (1993) 933.
[110] C.B. Mendes, E.P. Laignier, M.R. Pereira, L. Brigagao, P.O. Luccas, C.R.T. Tarley, A simple turbidimetric flow injection system for saccharin determination in sweetener products, Chem. Pap. 64 (2010) 285.
[111] J.M.L. Fernandez, M.D. Luque de Castro, M. Valcarcel, Automatic continuous on-line monitoring of salicylic acid and acetylsalicylic acid in pharmaceuticals, J. Automat. Chem. 12 (1990) 263.
[112] M.F. Lima, V.G.K. Almeida, R.J. Cassella, Merging zones FIA system for the spectrophotometric determination of Sb(III) and total Sb in drugs used in the treatment of Leishmaniasis, Spectrosc. Lett. 39 (2006) 769.
[113] E.A.G. Zagatto, I.L. Mattos, A.O. Jacintho, Determination of sucrose in sugar-cane juice and molasses by flow-injection spectrophotometry, Anal. Chim. Acta 204 (1988) 259.
[114] C.K. Zacharis, A. Verdoukas, P.D. Tzanavaras, D.G. Themelis, Automated sample preparation coupled to sequential injection chromatography: On-line filtration and dilution protocols prior to separation, J. Pharm. Biomed. Anal. 49 (2009) 726.
[115] I.V. Tóth, A.O.S.S. Rangel, J.L.M. Santos, J.L.F.C. Lima, Determination of aluminum (III) in crystallized fruit samples using a multicommutated flow system, J. Agric. Food Chem. 52 (2004) 2450.
[116] P.C.A.G. Pinto, M.L.M.F.S. Saraiva, J.L.M. Santos, J.L.F.C. Lima, Fluorimetric determination of aminocaproic acid in pharmaceutical formulations using a sequential injection analysis system, Talanta 68 (2006) 857.
[117] D.G. Themelis, P.D. Tzanavaras, A.V. Trellopoulos, M.C. Sofoniou, Direct and selective flow-injection method for the simultaneous spectrophotometric determination of calcium and magnesium in red and white wines using online dilution based on "zone sampling," J. Agric. Food Chem. 49 (2001) 5152.

[118] R.A.S. Lapa, J.L.F.C. Lima, J.L.M. Santos, Determination of calcium, magnesium, sodium and potassium in wines by FIA using an automatic zone sampling system, Food Chem. 55 (1996) 397.

[119] E. Schmidt Jr., W.R. Melchert, F.R.P. Rocha, Flow-injection iodimetric determination of captopril in pharmaceutical preparations, J. Braz. Chem. Soc. 20 (2009) 236.

[120] A.N. Araujo, J.L.F.C. Lima, B.F. Reis, E.A.G. Zagatto, Multicommutation in flow analysis. Part 3. Spectrophotometric kinetic determination of creatinine in urine exploiting a novel zone sampling approach, Anal. Chim. Acta 310 (1995) 447.

[121] M. Blanco, J. Coelho, H. Iturriaga, S. Maspoch, M. Redon, J. Riba, Multi-component analysis of concentrated solutions by flow-injection analysis with zone sampling and partial least-squares resolution, Anal. Chim. Acta 259 (1992) 219.

[122] P.D. Tzanavaras, D.G. Themelis, P. Rigas, Automated zone-sampling dilution by coupling sequential injection analysis to high-throughput HPLC for the direct determination of gemfibrozil, J. Sep. Sci. 32 (2009) 2819.

[123] J. Toei, An improved zone sampling method for flow injection analysis, Anal. Lett. 21 (1988) 1633.

[124] B.A.A. Dremel, W.-S. Yang, R.D. Schmid, On-line determination of lactic acid during kefir fermentation based on a fibre-optic lactic acid biosensor and flow-injection analysis, Anal. Chim. Acta 234 (1990) 107.

[125] E.P. Borges, B.F. Reis, An enzymatic flow-injection procedure with chemiluminescence detection for on-site determination of L-alanine in synthesis process, J. Braz. Chem. Soc. 16 (2005) 1226.

[126] G.C. Chemnitius, R.D. Schtnid, L-Malate determination in wines and fruit juices by flow injection analysis adaptation of coupled dehydrogenase/transferase system, Anal. Lett. 22 (1989) 2897.

[127] P.D. Tzanavaras, C.K. Zacharis, P. Rigas, Novel automated assay for the quality control of mexiletine hydrochloride formulations using sequential injection and on-line dilution, J. Pharm. Biomed. Anal. 48 (2008) 1254.

[128] U.V. Jørgensen, S. Nielsen, E.H. Hansen, Dilution methods in flow injection analysis, evaluation of different approaches as exemplified for the determination of nitrosyl in concentrated sulphuric acid, Anal. Lett. 31 (1998) 2181.

[129] H.-W. Chen, Z.-L. Fang, Combination of flow injection with capillary electrophoresis. Part 5. Automated preconcentration and determination of pseudoephedrine in human plasma, Anal. Chim. Acta 394 (1999) 13.

[130] C.M.P.V. Lopes, A.A. Almeida, J.L.M. Santos, J.L.F.C. Lima, Automated multi-commutated flow system for flame atomic spectroscopy determination of rubidium at high concentrations, At. Spectrosc. 27 (2006) 13.

[131] M.J.C. Taylor, G.D. Marshall, S.J.S. Williams, J.F. van Staden, C. Saling, The determination of vanadium (V) in the presence of vanadium (IV) using 4-(2-pyridylazo) resorcinol in a flow-injection manifold, Anal. Chim. Acta 329 (1996) 275.

[132] Y.Z. Ye, H.Y. Mao, Y.H. Chen, Catalytic kinetic simultaneous determination of iron, silver and manganese with the Kalman filter by using flow injection analysis stopped-flow spectrophotometry, Talanta 45 (1998) 1123.

[133] R.N. Fernandes, B.F. Reis, Flow system exploiting multicommutation to increase sample residence time for improved sensitivity. Simultaneous determination of ammonium and ortho-phosphate in natural water, Talanta 58 (2002) 729.

[134] C. Frank, F. Schroeder, R. Ebinghaus, W. Ruck, A fast sequential injection analysis system for the simultaneous determination of ammonia and phosphate, Microchim. Acta 154 (2006) 31.

[135] H.F. Pfeiffer, H. Waldhoff, P.J. Worsfold, I.R.C. Whiteside, Automated flow-injection procedures for the determination of hydrolytic enzymes in bioreactor preparations, Chromatographia 33 (1992) 49.

[136] J. Saurina, S. Hernandez-Cassou, Flow-injection and stopped-flow completely continuous flow spectrophotometric determinations of aniline and cyclohexylamine, Anal. Chim. Acta 396 (1999) 151.
[137] H.P. Huang, R.X. Cai, Y.M. Du, Y.N. Zeng, Flow-injection stopped-flow spectrofluorimetric kinetic determination of total ascorbic acid based on an enzyme-linked coupled reaction, Anal. Chim. Acta 309 (1995) 271.
[138] C.A. Tumang, G.C. Luca, R.N. Fernandes, B.F. Reis, F.J. Krug, Multicommutation in flow analysis exploiting a multizone trapping approach: spectrophotometric determination of boron in plants, Anal. Chim. Acta 374 (1998) 53.
[139] C. Sanchez-Pedreno, M.S. Garcia, J.A. Ortuno, M.I. Albero, R. Exposito, Kinetic methods for the determination of cadmium(II) based on a flow-through bulk optode, Talanta 56 (2002) 481.
[140] J.A.M. Pulgarın, L.F.G. Bermejo, P.F. Lopez, Sensitive determination of captopril by time-resolved chemiluminescence using the stopped-flow analysis based on potassium permanganate oxidation, Anal. Chim. Acta 546 (2005) 60.
[141] D.M. Magni, A.C. Olivieri, A.L. Bonivardi, Artificial neural networks study of the catalytic reduction of resazurin: stopped-flow injection kinetic-spectrophotometric determination of Cu(II) and Ni(II), Anal. Chim. Acta 528 (2005) 275.
[142] N. Chu, C. Ding, S.-H. Fan, Stopped-flow sequential injection spectrophotometry for simultaneous determination of Cu, Zn, Ni, and Mn in environmental water with artificial neural networks calibration, Anal. Lett. 43 (2010) 335.
[143] M. Blanco, J. Coello, H. Iturriaga, S. Maspoch, J. Riba, E. Rovira, Kinetic spectrophotometric determination of Ga(III)-Al(III) mixtures by stopped-flow injection analysis using principal component regression, Talanta 40 (1993) 261.
[144] K. Toda, K.I. Yoshioka, K. Mori, S. Hirata, Portable system for near-real time measurement of gaseous formaldehyde by means of parallel scrubber stopped-flow absorptiometry, Anal. Chim. Acta 531 (2005) 41.
[145] J.A.M. Pulgarin, L.F.G. Bermejo, J.M.L. Gallego, M.N.S. Garcia, Simultaneous stopped-flow determination of morphine and naloxone by time-resolved chemiluminescence, Talanta 74 (2008) 1539.
[146] T. Perez-Ruiz, C. Martinez-Lozano, V. Tomas, Kinetic spectrofluorometric determination of nitrite and nitrate at the nanograms per milliliter level by manual and flow-injection methods, Anal. Chim. Acta 265 (1992) 103.
[147] A.N. Anthemidis, D.G. Themelis, J.A. Stratis, Stopped-flow injection liquid-liquid extraction spectrophotometric determination of palladium in airborne particulate matter and automobile catalysts, Talanta 54 (2001) 37.
[148] J.A.M. Pulgarin, L.F.G. Bermejo, Flow-injection stopped-flow spectrofluorimetric kinetic determination of paracetamol based on its oxidation reaction by hexacyanoferrate(III), Anal. Chim. Acta 333 (1996) 59.
[149] C.G. Maria, T.M. Munoz, A. Townshend, Reactivation of an immobilized enzyme reactor for the determination of acetylcholinesterase inhibitors. Flow-injection determination of paraoxon, Anal. Chim. Acta 295 (1994) 287.
[150] K.D. Khalaf, B.A. Hasan, A. Morales-Rubio, M. de la Guardia, Spectrophotometric determination of phenol and resorcinol by reaction with p-aminophenol, Talanta 41 (1994) 547.
[151] C. Fenge, E. Fraune, R. Freitag, T. Scheper, K. Schuegerl, On-line monitoring of monoclonal-antibody formation in high density perfusion culture using FIA, Cytotechnology 6 (1991) 55.
[152] M. de la Guardia, K.D. Khalaf, B.A. Hasan, A. Morales-Rubio, J.J. Arias, J.M. Garcia-Fraga, A.I. Jimenez, F. Jemenez, Simultaneous kinetic spectrophotometric determination of five phenolic compounds by reaction with p-aminophenol, using partial least squares data treatment, Analyst 121 (1996) 1321.

[153] P.G. Nouros, C.A. Georgiou, M.G. Polissiou, Determination of olive oil 2-thiobarbituric acid reactive substances by parallel flow injection, Anal. Chim. Acta 417 (2000) 119.

[154] S.S. Jørgensen, G. Sørensen, A combined sampling and delay unit for flow injection analysis. The automated determination of 2-thiobarbituric acid reactive substances in foods, Anal. Chim. Acta 322 (1996) 69.

[155] N. Teshima, M. Kuno, M. Ueda, H. Ueda, S. Ohno, T. Sakai, Automated stopped-in-dual-loop flow analysis system for catalytic determination of vanadium in drinking water, Talanta 79 (2009) 517.

[156] W. Li, J. Chen, B.-R. Xiang, D.-K. An, Simultaneous on-line dissolution monitoring of multicomponent solid preparations containing vitamins B_1, B_2 and B_6 by a fiber-optic sensor system, Anal. Chim. Acta 408 (2000) 39.

CHAPTER

8

Sample Handling

The flow analyser is an excellent tool for sample management. **F.J.Krug**

OUTLINE

A typical quantitative analysis involves selection of the method, collection of a representative sample, sample treatment, circumventing interferences, acquisition of the analytical signal, calculation of the results and estimation of their reliability [1]. Time consuming and labour intensive sample treatment steps such as dilution, extraction, digestion, separation / concentration and reagent addition are often required, particularly for detection techniques needing considerable sample handling and rigid timing control, such as spectrophotometry. This may explain why *spectrophotometric* analytical procedures were considered somewhat old fashioned before the advent of flow analysis [2].

The flow analyser is an excellent tool for the precise handling of solutions, allowing the different analytical steps to be carried out efficiently and automatically, under reproducible conditions, inside a closed environment. These advantages are particularly evident in relation to *spectrophotometric* detection, as emphasised in this Chapter, which focuses on the ability of the flow analyser to carry out the different sample handling steps required in an analytical method.

8.1. SAMPLING

Flow analysis is associated with wet chemical methods and samples are generally collected and transported to the laboratory for analysis. After optional preparative step(s), e.g., dilution, dissolution, extraction, deproteinisation, or analyte separation/concentration, the resulting aqueous test sample is accommodated in a cup in the sampler tray of the flow analyser for further handling. This practice has been adopted since the appearance of the first commercially available flow analysers, as shown in Fig. 2.3.

There are however situations where the sample cannot be transported to the laboratory due to the high possibility of sample contamination, analyte losses and/or conversion to other chemical species by processes such as precipitation, adsorption and volatilisation. *In situ* analysis overcomes many of these issues and is therefore becoming increasingly popular. Moreover, some samples must be assayed in real time, *e.g.,* for industrial process monitoring. Transport of unstable (including gaseous) samples is also difficult. As a result, various approaches have been proposed for directly integrating the sampling step with flow analysis. For didactic purposes these approaches are classified by the physical nature of the sample matrix.

8.1.1. *In Situ* Analysis of Aqueous Samples

The flow analyser is well suited for handling aqueous samples, which are directly inserted into the manifold by taking advantage of loop-based or time-based sample introduction. Due to the portability, ruggedness, low power requirement and low maintenance requirement inherent to the unsegmented flow systems, specific approaches for *in situ* sample collection and handling have been proposed.

To this end, loop-based injection is preferred and the sample aspiration tube (Fig. 6.9) behaves as a sampling probe, as demonstrated in the landmark work of Thomsen *et al.* who determined reactive silicate in coastal waters during a cruise from Monterey Bay to San Francisco Bay in the USA [3]. The sample was continuously aspirated from a water layer about 2 m below the sea surface and a situation of "infinite sample volume" was attained in the main channel of a shipboard reagent injection (reversed flow) flow injection system. A reagent aliquot was injected into the flowing sample every 45 s and the resulting coloured zone was quantified by spectrophotometry. The height of the recorded peak was directly related to the reactive silicate concentration at the specified location and depth.

The feasibility of using a portable flow injection system for the *in situ* monitoring of river water was demonstrated by the design and deployment of a field instrument with spectrophotometric detection for the determination of phosphate [4]. Solid-state technology (LED source and photodiode detector) was used to construct a compact, battery-operated analyser that could operate without maintenance for several weeks, yielding reliable, pseudo-continuous results. The monitor was further applied to the determination of nitrate [5] and ammonium [6] in river waters. Details of the instrumentation were also reported [7,8].

For seawater analysis, it is possible to pump the sample directly from the sea to a shipboard analyser. The sample can then be de-gassed in-line and handled in a shipboard flow system. This possibility, and the main practical, physico-chemical and interpretative aspects, were discussed in relation to the *in situ* spectrophotometric determination of seawater pH [9]. The attractive features and applications of shipboard flow injection spectrophotometric determinations are discussed elsewhere [10,11]. A compact, shipboard flow analysis system incorporating a UV photoreactor has also been reported for the determination of total phosphorus in estuarine and marine waters [12]. Another way to sample seawater is to use a towed torpedo-shaped "fish," deployed off the crane arm of a hydrographic winch at a distance of ~5 m from the ship, connected to a deck mounted pump by appropriate tubing. Filtration can be incorporated in-line and the filtered sample collected or pumped directly to the flow analyser [13]. This approach (Fig. 8.1) minimises contamination and preconditions the sample, as demonstrated in the chemiluminometric determination of iron(II) in surface seawater.

An alternative sampling approach is to deploy a submersible analyser [14,15]. Water is sucked from the exterior environment by an on-board pump in order to fill the sampling loop, with excess sample directed back to the exterior. The flow system is remotely controlled and the results can be either stored in the analyser or transmitted back to the ship via a cable. In this context, a small submersible flow injection analyser, with solid-state spectrophotometric detection, was conceived for the *in situ* determination of nitrate (Fig. 8.2). Its performance and versatility were assessed by results from laboratory, shipboard (North Sea IMPACT Cruise) and *in situ* (Tamar Estuary, UK) analysis. Excellent temporal and spatial resolutions were reported and this aspect is crucial for investigating dynamic processes in estuarine, coastal and open ocean waters.

Another application of *in situ* analysis is the quality control of industrial processes. In general, the sample is taken from the process (batch or continuous) and directed towards the analytical path of the flow system; a reagent aliquot is inserted into it (reagent injection flow injection configuration – see also 3.2.1). The sample volume extracted is usually minimal compared with the scale of the process. Alternatively, the process

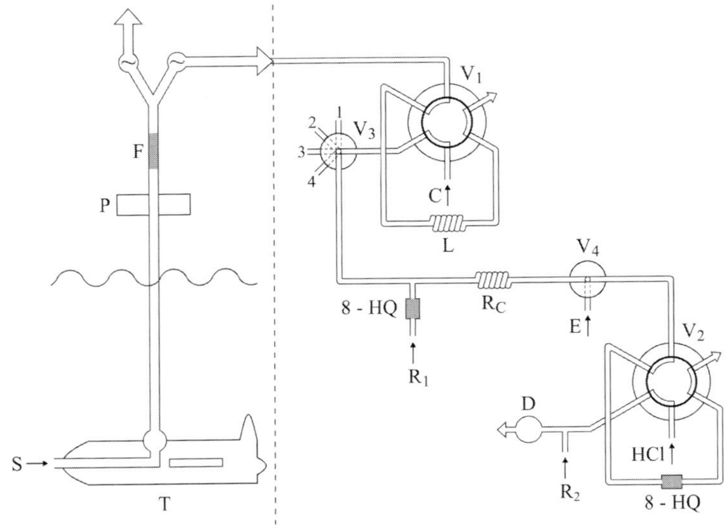

FIGURE 8.1 **Schematic representation of the flow system (right) with a towed torpedo-shaped "fish" (left).** S = seawater inlet; T = towed torpedo (1−2 m below sea level); P = high volume pump; F = filtration cartridge; ~ = flow regulators; L = sampling loop; C = carrier stream (low iron seawater) = 1,2,3,4 = standard solutions; R_1 = buffer solution; 8-HQ = 8-hydroxyquinoline mini-column; R_C = coiled reactor; E = eluent/wash solution; HCl = eluent towards detection; R_2 = luminol reagent; D = detector; V_1, V_2 = six-way injection valves, V_3 = ten-way selecting valve; V_4 = three-way directional valve; sinusoidal line: sea level. For details, see Ref. [13]. *Figure adapted with permission from "A.R. Bowie, E.P. Achterberg, P.N. Sedwick, S. Ussher, P.J. Worsfold, Real-time monitoring of picomolar concentrations of iron(II) in marine waters using automated flow injection-chemiluminescence instrumentation, Environ. Sci. Technol. 36 (2002) 4600". Copyright 2002, American Chemical Society.*

sample can flow directly through a sampling loop, as in ordinary flow-injection systems. Details of unsegmented flow analysis of industrial samples are provided elsewhere [16].

It is interesting to note that in-line sample filtration can be implemented in all of the above configurations. The characteristics and potential of in-line sample filtration are emphasised in Section 8.4.

Another potential *in situ* application of flow analysis, as yet unexploited, is the nondestructive analysis of liquid samples such as old wines. The sampling tube of the flow analyser can be designed to be very thin and hard in order to act as a sampling probe that can penetrate, *e.g.*, bottle stoppers. After sampling, the stopper remains practically intact, and the contents of the bottle remain preserved.

8.1.2. Gaseous Samples

Analysis of gaseous samples is often required in studies related to, *e.g.*, atmospheric environmental monitoring and control, respirometry,

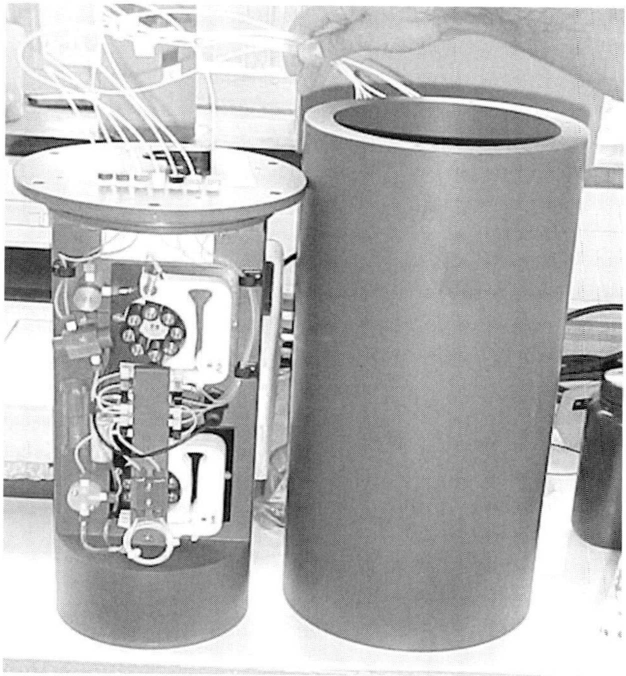

FIGURE 8.2 **A submersible flow injection analyser.** The housing (left) has been removed to show the main analyser components (right). The two peristaltic pumps (with white platens) are on the right hand side, the reaction coil (light blue) is on the lower left hand side and the sampling loop (light blue) is at the bottom of the instrument. Sample inlet and outlet are at the top of the instrument. For details see text. *Reprinted from Talanta 58 (2002) 1015, P.C.F.C. Gardolinski, A.R.J. David, P.J. Worsfold, Miniature flow injection analyser for laboratory, shipboard and* in situ *monitoring of nitrate in estuarine and coastal waters, with permission from Elsevier (Ref. [14]).*

production of gaseous mixtures for industrial or hospital use, food-packaging atmospheres and occupational exposure. Wide concentration ranges (from a few volumetric parts per billion to parts per cent) and low sample volumes are involved, and flow *spectrophotometry* can provide the required detectability and selectivity.

In general, the gaseous sample is collected by a suitable liquid solution in a specially designed gas-filtering device and the resulting solution is processed in a flow analyser. As the gas collection step is generally accomplished outside of the flow manifold, the gaseous sample is in fact converted to a liquid sample. Discussions focussing on this batchwise operation, as well as on special devices for gas filtering/collection are outside the scope of this monograph. Analytical procedures involving analyte conversion to a gaseous form that is collected in another medium, a type of in-line analyte separation/concentration, are discussed in Section 8.5.

For the direct analysis of gaseous samples, the flow system should provide a renewable contact area between the liquid acceptor stream and the gaseous sample, and renewal of this area is achieved by flushing the liquid phase at some stage of the process [17]. The possibility of reagent renewal is the main advantage of flow systems applied to the direct analysis of gaseous samples.

Regarding *in situ* analysis of gaseous samples, two strategies have been used.

8.1.2.1. The Sample Feeding the Flow System

A well-defined volume of the gaseous sample is directed towards the flow manifold, where it interacts with a liquid medium, and the chemiluminometric determination of nitrous acid in an indoor environment [18] is a good example. A calibrated vacuum pump aspirated the volume of air through a diffusion scrubber and the membrane-separated analyte was collected into a buffered stream that directed it towards an 8-port injection valve; after passing a de-bubbler, the analyte zone was directed towards the detector. With this approach, a 120 pp tv detection limit was attained. The characteristics, potential and limitations of analytical procedures involving diffusion scrubbers and other devices for gas sampling aimed at in-line flow-based determinations have been discussed elsewhere [19].

It is somewhat surprising that although the segmented flow analyser involves both liquid and gaseous phases (see 5.1.1), it has not been used for the analysis of gaseous samples [20]. On the other hand, in unsegmented and mono-segmented flow systems, this gas/liquid interaction has been used in different ways in order to add the gaseous analyte into a liquid phase for further handling and monitoring. One way to implement this approach relies on the exploitation of an advanced mono-segmented flow analyser involving two liquid and three gaseous segments per determination, as demonstrated in the spectrophotometric determinations of oxygen in food packages and nitrous oxide in synthetic air [21]. An aqueous plug of the analyte acceptor solution (eventually a reagent) was directed to flow inside a manifold section, followed by an inert gas plug, in order to establish a thin liquid film on the inner wall of the tubing manifold; the gaseous sample was then passed through this solution-coated manifold section and the analyte was collected; next, a liquid plug of the colour-forming reagent was then allowed to flow, removing the analyte and directing it towards the detector; the last air segment was used for system washing.

8.1.2.2. The Flow System Acting as a Sampling Probe

Another possibility for the *in situ* analysis of gaseous samples is to permeate the analyte directly towards the flow system using a membrane, soap bubble, hanging drop or similar system [17,19,20]. An interesting approach is to take advantage of tubular semi-permeable membranes that

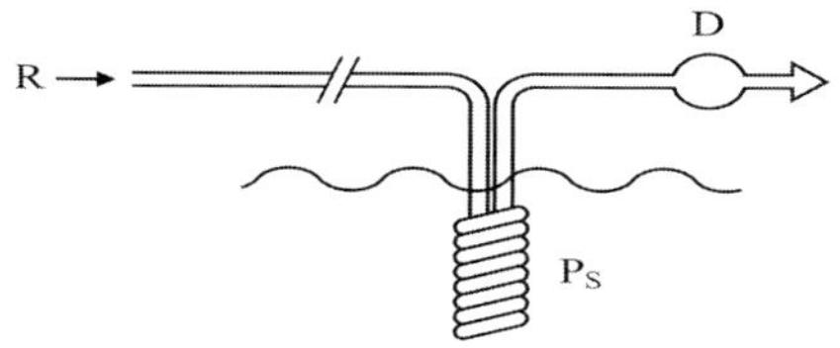

FIGURE 8.3 **A flow system with a sampling probe.** R = reagent acceptor stream; P_S = sampling probe; D = detector; || = stream stopping. Portion above the sinusoidal line = flow system; portion below the sinusoidal line = sample. For details, see text.

can act as the analyte sampling device (Fig. 8.3) associated with the flow system.

The tubular membrane is coiled on a suitable support, and the assembly can be regarded as a sampling probe. When the reagent is flowing, only a small amount of analyte crosses the membrane because the mean residence time of each element of fluid containing the acceptor reagent stream inside the probe is too low. A baseline signal is then attained. Stopping the reagent stream for a pre-set time interval permits the analyte to diffuse through the tubular membrane where it is collected. Restarting the flow directs the zone of collected analyte towards the detector: the transient signal is recorded as a peak, the height of which is proportional to the analyte concentration in the sample. This innovation has often been exploited for gaseous sample analysis and is also useful for liquid samples, as demonstrated in the spectrophotometric determination of ethanol in alcoholic beverages [22].

8.1.3. Solid Samples

Flow analysis is associated with wet chemistry; thus, solid samples of diverse origin, *e.g.*, alloys, soils, sediments, sludges and foodstuffs, are normally subjected to in-line treatment in order to form a liquid analyte zone. The two most common examples are in-line sample electrolytic dissolution, where the analyte zone is formed as a consequence of applying a direct electric current to the solid sample, and sequential extractions of soils and sediments.

8.1.3.1. Electrolytic Dissolution

Since the pioneering work of Chirnside *et al.* on the spectrophotometric determination of boron in metallic nickel [23], electrolytic dissolution has been exploited for the direct analysis of several solid conductive materials, including metals, ores and alloys.

IMPLEMENTATION

The sample (anode) is placed on top of an electrolytic chamber, a controlled direct current is applied and the oxidised material is

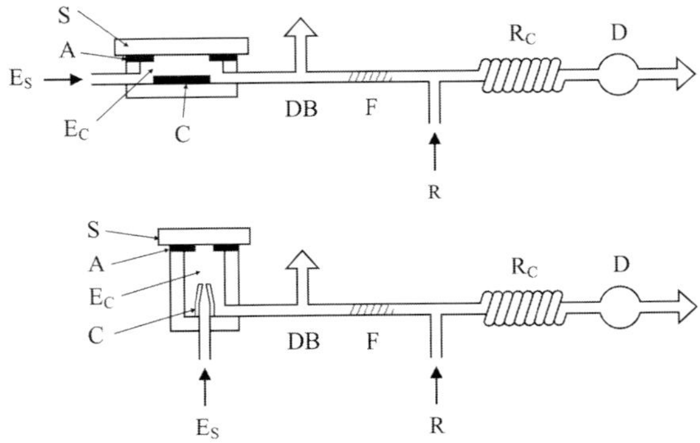

FIGURE 8.4 **Flow diagrams of flow systems exploiting sample electrolytic dissolution.** Upper figure = classical geometry, lower figure = jet geometry; S = solid sample (anode); C = cathode; E_S = electrolytic solution; E_C = electrolytic chamber; A = adapter; DB = de-bubbler; F = filter; R = optional reagent; R_C = coiled reactor; D = detector; arrows = sites where pumping is applied. Electrical contacts are not shown. For details, see text.

dissolved into the electrolytic solution (Fig. 8.4, upper). According to Faraday's law, the amount of dissolved sample is proportional to the applied current and electrolysis time. The sample dissolution step is very fast and compatible with flow analysis, as demonstrated in 1965 by Barabas and Lea, who adapted an electrolytic dissolution chamber in an air-segmented flow analyser for the spectrophotometric determination of phosphorus in metallic copper [24]. This innovation meant that thirty samples could be assayed per hour.

Alternatively, the electrolytic solution/sample carrier stream can enter through a thin tube (cathode) at the bottom of the chamber (Fig. 8.4, lower), forming a jet which is directed towards the solid sample at the top of the chamber [25,26]. With this jet geometry, sample surface area is reduced and the electric current density can be increased, which promotes nonselective dissolution.

APPLICATIONS

In 1983, electrolytic dissolution was exploited in the metallurgical industry for the batchwise determination of soluble aluminium in steels by flame atomic absorption spectrometry [27]. A specially designed spoon transferred a few millilitres of the red-hot molten material to a disk-shaped mould where the material solidified when cooled. The resulting pellet was placed on the electrolytic chamber and, after the dissolution step, an aliquot of the electrolytic solution was taken and analysed for aluminium. With an advanced electrolytic chamber design, electrolysis time was reduced to about 40 s. The feasibility of batchwise electrolytic dissolution for multi-elemental analyses of alloys was demonstrated in the simultaneous determination of Al, Si, Mn, Cr, Ni and Co by ICP-OES [28].

For flow analysis incorporating electrolytic dissolution, very small characteristic masses, often below the ng level, are reported for metal determinations. This is a consequence of the analytical sensitivity and the small sample volume required, and is an attractive feature of in-line electrolytic dissolution. As a very small dissolved mass is required, rapid electrolysis (a few seconds) under a moderate current (mA) is sufficient. This was demonstrated in the flow-based spectrophotometric determination of aluminium in steels [29]. The analyte was oxidised and dissolved in a flowing acidic electrolytic solution that also acted as the sample carrier stream of the flow analyser. This innovation was further applied to the spectrophotometric determination of molybdenum in alloys [30]. In both applications, the anode was the polished metallic sample, and the cathode was a gold or silver coated electrode placed at the bottom of the electrolytic chamber (Fig. 8.4). A silicone rubber sheet (adapter) was placed between the solid sample and the chamber walls in order to avoid leakage and to define the sample surface area to be dissolved. This classical geometry is the most commonly used.

Sampling by means of electrolytic dissolution permits concentration gradients along the metallic sample to be determined, thereby providing information on sample homogeneity and/or changes in analyte concentration near the alloy surface. Surprisingly, the potential of this approach has not been exploited.

Spectrophotometry was used in the original proposal for electrolytic dissolution [23] and in the first flow-based applications [24,29] but detection techniques such as ICP-OES [31–33], ICP-MS [34], FAAS [26,35] and ET-AAS [36] are now more commonly used. In order to avoid the need for certified reference materials to determine accuracy, alternative procedures for calibration have been proposed [25]. The fundamental aspects of the technique and applications have been reviewed elsewhere [37].

8.1.3.2. Sequential Extractions

Studies concerning soil fertility and plant nutrition usually involve the determination of the nutrient concentrations in a soil extract, which can be related to nutrient bioavailability. There is increasing interest in the chemical and physical characterisation of the different forms of nutrients in soils and a common approach is to fractionate the nutrient species according to their extractability by leaching reagents of increasing aggressiveness [38]. This provides relevant information on nutrient availability to plants, the nutrient capacity of soils and nutrient mobility.

With regard to phosphorus, knowledge of the distribution of the different physico-chemical forms of the element in soils is useful for understanding soil biogeochemistry, soil genesis and soil fertility. The characterisation and quantification of these forms by sequential extractions were initiated by Dean in 1938, as emphasised elsewhere [39]. The feasibility of in-line dynamic fractionation of macronutrients in environmental solids using flow analysis was recently demonstrated by the partitioning of inorganic phosphorus in agricultural soils [40].

The soil sample was placed inside a mini-column which was inserted into the analytical path of a sequential injection system. Three different phosphorus fractions in the soil, namely exchangeable, Al- and Fe-bound, and Ca-bound, were determined using 1.0 mol L^{-1} NH_4Cl, 0.1 mol L^{-1} NaOH and 0.5 mol L^{-1} HCl solutions respectively as the leaching reagents. These extracting solutions passed successively through the mini-column and the corresponding phosphate fractions that were displaced were spectrophotometrically determined after molybdenum blue formation. Despite the variable chemical composition of the different extracts, a single flow manifold was assembled that did not require re-configuration or modification of the composition of the reagents.

8.1.4. Slurry Samples

Slurry samples are complex matrices comprising at least two phases (generally solid/liquid), and their introduction into the flow analyser is difficult [41]. The flowing sample suspension needs to be modified to an optically transparent solution before further handling, for which in-line sample bleaching or digestion (see 8.4) can be used. Direct analysis of slurry samples has been often reported for flow systems with atomic absorption/emission spectrometry, inductively coupled plasma optical emission spectrometry, inductively coupled plasma mass spectrometry and related techniques [42], where the sample zone flows through a highly energetic environment (flame, plasma and furnace). The absence of such an environment is probably the main reason why flow systems with *spectrophotometric* detection are rarely used for slurry samples.

8.1.5. Highly Viscous Samples

The sampling step can be a limiting factor in the analysis of liquors, honeys, oils and other viscous samples, because a precise sample aliquot cannot easily be taken with volumetric pipettes. Therefore sampling under warm conditions and/or mass-based evaluation of the sampled amount is often required [43,44].

This limitation also holds in the analysis of sugar cane molasses undergoing fermentation for ethanol production. During the fermentation process, molasses is warm and hence has a low viscosity, and the sampling step is easily accomplished *in situ*. After collection and during transport to the analytical laboratory, the temperature of the molasses tends to equilibrate with the ambient temperature and the sample becomes more viscous. The aliquoting process for laboratory analysis therefore requires re-heating of the sample in order to permit efficient sample handling [43]. The analysis time and the susceptibility to sample contamination are increased and sample throughput decreases. This problem also occurs in the analysis of invert sugar syrups, an important raw material in the food industry, but can be circumvented by sampling the warm original molasses or syrup into dissolvable pre-weighed capsules that are brought to the laboratory and placed directly in the analytical path of a flow system.

The feasibility of analysing capsules by flow analysis was demonstrated in the spectrophotometric determination of glucose and fructose in invert sugar syrups previously collected inside rapidly dissolvable gelatine capsules [45]. Each capsule was placed in the chamber of a flow-batch system and dissolution was efficiently accomplished under acidic conditions, followed by sample dilution and determination of the analytes. This latter step relied on the oxidation of the reducing sugars by periodate and quantification of the remaining periodate as the tri-iodide complex formed after the addition of potassium iodide [46]. The sample was handled at two different alkalinities and glucose and fructose were discriminated from each other by taking account of their different oxidation rates. The flow system was very stable, handled 50 samples per hour, required small quantities of reagents (only 28.8 μg $NaIO_4$ and 2.5 mg KI per determination) and yielded precise results (r.s.d. $<2\%$) in agreement with HPLC for samples with 20–60% (m/m) glucose plus fructose.

8.2. SAMPLE DILUTION

The analyte concentration in the portion of the sample zone yielding the analytical signal should match the dynamic concentration range of the flow-based analytical procedure. To this end, in-line sample dilution is often required, and this step is efficiently accomplished in a flow analyser. As emphasised in Chapter 1, calibrated devices such as burettes, volumetric flasks and pipettes are not required, as the sample and reagent volumetric fractions are maintained for all of the assayed sample and standard solutions.

Expressions such as *sample dilution* and *dilution-like process* have often been used for quantifying the dispersion associated with the sample element of fluid yielding the analytical signal (see 3.1.2.3). Although not strictly correct, the term *dilution* is sometimes used here for didactic purposes. Depending on the specific analytical procedure, the flow system should provide high, low or even variable sample dilution.

8.2.1. High Dilution

A high degree of dilution is required when both the expected analyte concentration in the sample and the sensitivity of the analytical method involved are high. In-line dilution has been preferred in order to minimise the introduction of errors in the analytical results due to sample manipulations inside beakers, pipettes and/or volumetric flasks. The process is efficiently and rapidly accomplished, regardless of the flow analyser mode.

In segmented flow systems, high sample dilution can be attained by adding confluent streams at high flow rates in order to promote a sudden lowering of the sample concentration at the confluence points (*cf.* Eq. 3.10). Axial dispersion plays a minor role in this context. The pronounced enlargement of the sample zone (Eqs. 3.12 and 3.13), which might limit the sampling rate, is avoided by taking advantage of the stream splitting strategy for the removal of part of the sample zone (see 7.3.3). In procedures involving several confluence points along the analytical path, such as cascade or multi-stage dilutions, the reagent(s) should be added as the last confluent stream(s) in order to minimise consumption, reduce waste generation and avoid undesirable precipitation effects.

In mono-segmented flow systems, high sample dilution is seldom an issue because they were conceived to improve flow-based methods where sensitivity is critical [47]. High sample dilution is however required for the analysis of some highly coloured samples and can be accomplished by exploiting hydrodynamic injection [48]. A very small volume of sample is added to the diluent/reagent mono-segment, allowing high sample dilution to be achieved without pronounced axial dispersion. An almost homogeneous mono-segment is established; thus, the procedure is less susceptible to Schlieren noise. The feasibility of this innovation was demonstrated by the spectrophotometric determination of glucose in blood plasma [49]. The sample and reagent/diluting solution volumes were 5.0 and 220 μL, allowing a high degree of dilution. The flow system was characterised by a sample residence time of 6.5 min, and a 90 h^{-1} sampling rate and reliable results were obtained.

In unsegmented flow systems, the continuous sample dispersion process occurring during transport of the sample zone (*cf.* Eq. 3.4), the establishment of concentration gradients along the sample zone and the presence of artefacts in the analytical path (5.2.3.2) are important factors that determine the degree of in-line sample dilution. High dilutions are usually attained by inserting very small sample volumes, and hydrodynamic injection is useful in this context, as illustrated by the determination of chromium in steels [50]; with a 0.5 μL inserted volume, a sample volumetric fraction of 0.0072 (sample dilution = 138) was attained, and a prior manual dilution of the steel dissolutions was not required. Precise results (r.s.d. $<1\%$), in agreement with FAAS and ICP-OES, were attained at a sampling rate of 120 h^{-1}.

Concentration gradients in the sample zone can also be exploited to reduce the analytical sensitivity. Different approaches have been used but in general, the more diluted fluid elements at the front and trailing edges of the dispersing sample zone are associated with the analytical signal:

- Electronic dilution [51,52]. Each fluid element of the flowing sample is associated with a specific mean residence time and with a different degree of dilution. Selecting different monitoring times permits several analytical signals with different analyte concentrations to be obtained. If the analytical signal relates to a very dilute portion of the sample zone located at the trailing edge, a high degree of sample dilution is attained. As computer-controlled selection of the fluid element involved is required for better reproducibility, this strategy has incorrectly been called *electronic dilution.*
- Zone sampling [53]. A diluted portion of the sample zone, usually at the trailing edge, is selected, re-sampled and inserted into a second carrier stream for further analysis (see 7.2.1). As this strategy involves two *quasi*-independent subsystems, the sample volumetric fraction is determined as the product of the volumetric fractions associated with both subsystems; hence, high sample dilution is efficiently attained (see Table 7.2).
- Zone splitting [54]. Most of the flowing sample zone is discarded and only its trailing edge is handled (see 7.4.1). As the mean analyte concentration in this portion of the sample zone is very low, a high degree of sample dilution is efficiently attained (see Table 7.2).
- Partial zone overlap. Only the front and trailing portions of the sample zone interact with the reagent (see 7.1.1). The analytical signal then relates to the analyte diluted in-line. Most applications are related to asymmetrical merging zones [55], sandwich techniques [56] or sequential injection analysis [57,58] (see Table 7.1).
- Dispersion chamber. Placing a dispersion chamber in the analytical path is an alternative strategy for attaining high sample dilution in unsegmented flow analysis [59], and some applications are presented in Table 7.2.

In flow-based analytical procedures where the chemical reactions involved and/or the steps of analyte separation/concentration are relatively slow, the analytical signal can be attenuated by reducing the mean available time for the development of these physico-chemical processes. From a practical point of view, the result is equivalent to in-line analyte dilution. This possibility has been exploited in the spectrophotometric determination of hydrogen peroxide in contact lens care solutions using a sequential injection lab-on-valve system [60]. In view of the high sensitivity of the reaction between the analyte and 2,2′-azino-bis (3-ethylbenzothiazoline-6-sulfonic acid) in the presence of the enzyme *horseradish peroxidase*, a dialysis unit was placed in the analytical path for both sample clean-up and sensitivity adjustment, thus avoiding the need to implement in-line dilution. Dialysis efficiency was only 1.25%, resulting in an 80-fold increase in the upper limit of the dynamic concentration range.

8.2.2. Low Dilution

When sensitivity is critical, sample dilution should be minimised; moreover, sufficient time should be available for sample handling if the physico-chemical processes involved are relatively slow. These requirements are fulfilled in a segmented flow analyser with the confluent stream flow rates set as low as possible in order to minimise sample dilution at the confluent sites (*cf.* Eq. 3.10). Another option is the use of a monosegmented flow analyser which permits long sample residence times to be achieved without excessive sample dispersion.

In unsegmented flow analysis, low sample dilution is achieved by increasing the introduced sample volume (see 5.2.3.2), but this volume cannot be increased at will. Any increase beyond a given value (usually 4 $V_{0.5}$) would lead to a slight increase in the sample volumetric fraction and a pronounced increase in wash time (Fig. 5.11), and hence a decrease in sampling rate. This increase in wash time is due to tailing effects and can be minimised by:

- Improving system design. The manifold should be made with high-quality/high-performance components (especially the main reactors) and with as low an internal volume as possible.
- Allowing the trailing portion of the sample zone to bypass the detector. After the analytical signal is obtained, a stream directing valve, placed immediately before the detector, is switched in order to direct the trailing portion of the sample zone directly towards waste.
- Increasing the total flow rate. This practice is however not recommended when sensitivity is critical and the method involved relies on relatively slow physico-chemical processes or, in other words, when lower flow rates are required (in this situation, the strategies outlined below are useful).
- Adding an intermittent wash stream. The sample zone is transported towards the detector at a low flow rate to achieve an increased degree of reaction completion. After the analytical signal has been obtained, an intermittent wash stream is added to remove the trailing portion of the sample zone.
- Exploiting multi-site detection. After the analytical signal has been obtained, the detector is moved to another manifold site or to another analytical channel, so that the trailing portion of the sample does not flow through it.

- Exploiting zone trapping. The central portion of the sample zone is trapped inside the main reactor or inside the detector and allowed to stand under static conditions to allow reaction development. Its front and trailing portion are directly discarded at a high flow rate.

8.2.3. Variable Dilutions

Several measurements can be performed on the same sample for *e.g.*, wide-range *spectrophotometry*, simultaneous determinations, selectivity evaluation, accuracy assessment or compensation of matrix effects by the standard addition method. Each measurement normally relates to a different sample (or standard) volumetric fraction, hence a different degree of dilution. To this end, different approaches involving single or multiple sample (or standard) insertions and/or exploitation of concentration gradients have been used, as emphasised in Chapter 7.

The degree of in-line sample dilution can be defined in real-time and this is a key benefit of expert flow systems (see 8.7).

8.3. REAGENT ADDITIONS

In *spectrophotometry*, reagents are generally added to the sample in order to convert the analyte into a highly efficient radiation absorbing or emitting chemical species; in catalytic analytical procedures, they are required for the development of the indicator reaction (see 8.6). Moreover, some reagents are needed for, *e.g.*, adjusting the pH of the reaction medium and/ or minimising interferences caused by the presence of chemical species. The latter is achieved mainly through precipitation or complexation reactions and the added reagents are referred to as *masking agents*.

8.3.1. General Approach

Some reagent less flow-based analytical procedures have been proposed for the spectrophotometric determination of, *e.g.*, coloured organic analytes [61], UV-absorbing chemical species [62] or metal ions as aquo-complexes [63], but addition of at least one reagent is generally required. When sensitivity is critical, a large sample volume is needed; thus, the flow system should be designed in the confluence configuration (see 3.2.2) in order to permit efficient sample/reagent interaction.

In flow analysis, attainment of chemical equilibrium is not necessarily required, and this feature is of paramount importance, as emphasised in Chapter 1. Masking however should be quantitative to improve selectivity. In general, the masking agent is added prior to the main reagent, especially if the rate of the masking reaction is a limiting factor. This means that the flow analyser should be designed to provide a suitable time interval for masking before the addition of the main reagent, and this aspect is more critical if the main reactions involved are not reversible. As the masking reagent is normally a concentrated buffered solution, its addition is more efficiently accomplished by stream confluence, and the Schlieren effect tends to be less pronounced in this configuration (see 4.2.3.4).

Most *spectrophotometric* analytical procedures are pH-dependent, and the buffer system can be efficiently established in-line. In segmented flow systems and unsegmented flow systems with a chemically inert carrier stream, the main stream is characterised by a relatively steady acidity or alkalinity. The exact amount of the alkaline or acid component of the buffer system can then be added by confluence in order to attain a suitable pH value in the reaction medium. Moreover, if the main stream is initially too acidic or too alkaline, a strong base or acid can be added to the pH adjusting confluent stream.

The flow system should then provide facilities for reagent addition, and two approaches have generally been exploited.

8.3.1.1. The Single-reagent Approach

Only one solution, acting as the diluent, buffer forming, masking and main reagent, is combined with the sample. System design is therefore simplified, leading to a rugged flow system, often exploiting the single-line configuration. Moreover, this approach permits the direct implementation of automated analytical procedures originally developed for other types of analyser *e.g.*, the centrifugal analyser [64].

This approach is less useful however when one of the different chemical species involved yields a less stable solution: small volumes of the single reagent should then be prepared shortly before use. On the other hand, the use of several reagents permits the preparation and storage of large volumes of more stable solutions, and only reagents yielding less stable solutions should be prepared in low volumes immediately before use.

Moreover, the single-reagent approach cannot always be exploited, as several flow-based applications require successive additions of different reagent solutions to the sample, *e.g.*, those involving an indicator reaction and/or the addition of a masking agent prior to the main reagent (see above). Also, there are several analytical methods that rely on successive chemical reactions taking place under different conditions. In these situations, a multi-reagent approach is recommended.

8.3.1.2. The Multi-reagent Approach

The flow system is designed to permit different reagent streams to be successively added by confluence. This is the essence of segmented flow analysers and is inherent in unsegmented flow analysers designed in the confluent configuration.

It should be stressed that both the single-reagent and the multi-reagent approaches can generally be exploited for implementing a given *spectrophotometric* analytical procedure in the flow analyser. The approach is generally chosen as a compromise between system ruggedness, ease of operation and the management of solutions in a large-scale laboratory. The decision to use the single- or multi-reagent approach is therefore dictated by individual laboratory practice.

Other possibilities for reagent additions are discussed below.

8.3.2. Solution Addition During Sampling

A convergent stream (A – Fig. 8.5) merges with the sample prior to the sampling loop, permitting different solutions to be added without the need to prepare large numbers of solutions. This innovation is useful for:

- Sample conditioning. The added solution provides suitable conditions for analyte concentration/separation by, *e.g.*, solid-phase extraction, as originally demonstrated in the determination of ammonium in natural waters [65].
- Evaluation of selectivity. The blank and several solutions, each with a potential interfering chemical species at a given concentration, are successively placed as solution A. The different sample zones established are processed and the related recorded signals are used to obtain the selectivity chart.
- Implementation of the standard addition method. Different standard solutions are placed as solution A and this approach is one possibility for implementing the standard addition method (see 8.6.3).
- Simultaneous determinations. The flow system is designed with two sampling loops (Fig. 8.6) and a masking reagent is added between them, thus altering only one of the established sample zones, as demonstrated in the catalytic determination of molybdenum and tungsten in tool steel alloys [66]. The method exploited the effects of Mo(VI) and W(VI) on the rate of iodide oxidation by hydrogen peroxide in an acidic medium, and the contribution of W(VI) was

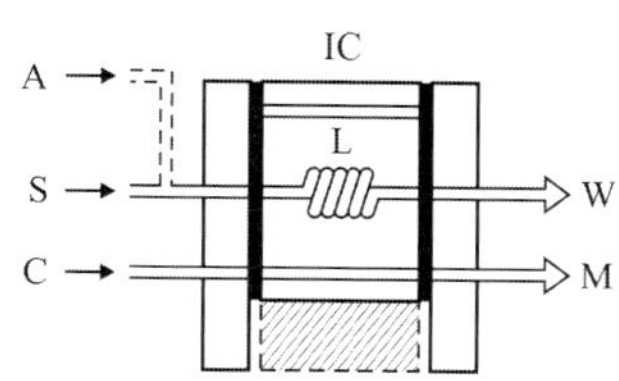

FIGURE 8.5 **Flow diagram of a flow system with optional solution addition prior to the sampling loop.** S = sample; A = solution to be optionally added; L = sampling loop; C = sample carrier stream; W = towards waste; M = towards manifold; IC = injector-commutator; shaded area = alternative IC position; solid arrows = sites where pumping is applied.

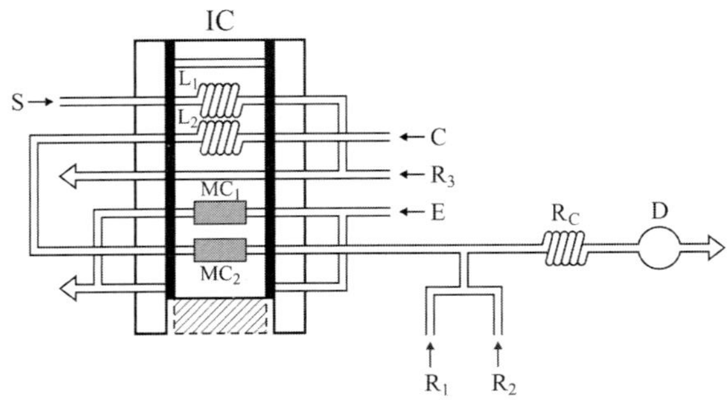

FIGURE 8.6 **Flow diagram of a flow system for sequential determinations involving optional masking addition.** S = sample; C = sample carrier stream; L_i = sampling loops; E = eluent, MC_i = mini-columns; R_1 and R_2 = reagents related to the indicator reaction; R_3 = masking reagent; R_C = coiled reactor; IC = injector-commutator; shaded area = alternative IC position; solid arrows = sites where pumping is applied. *Source: Figure adapted from Ref. [66].*

suppressed in the presence of citrate. Two analytical signals, proportional to the Mo + W and to the Mo content in the injectate, were obtained. SPE was exploited for the removal of potential interfering cations, and reliable results in agreement with ICP-OES were obtained.

- Improvement in the detection limit. There are analytical procedures where no analytical signal is observed below a given analyte concentration due to limitations associated with the chemical equilibria involved. The turbidimetric flow injection determination of sulphate in plant digests and natural waters [67] is a good example, as no analytical signal was observed for <20 mg L^{-1} SO_4–S (curve a, Fig. 8.7). The addition of sulphate to the sample zone improved the conditions for super saturation, thus allowing low analyte concentrations to be determined. This innovation is also useful for some spectrophotometric applications involving relatively high formation constants for the monitored chemical species.

8.3.3. Reagent Additions for Indicator Reactions

Catalytic flow-based analytical procedures normally involve the addition of two reagent streams for the development of the indicator reaction (see 8.6.4.3). In the absence of a catalyst, a low degree of reaction completion is attained during the residence time of the combined reagents in the flow manifold, and this gives the baseline. The addition of the catalyst (usually the analyte) accelerates the development of the indicator reaction, so that passage of the sample zone through the detector results in a transient variation in the monitored signal. As in ordinary flow systems, this signal variation is recorded as a peak, which is ideally proportional to the analyte concentration in the sample.

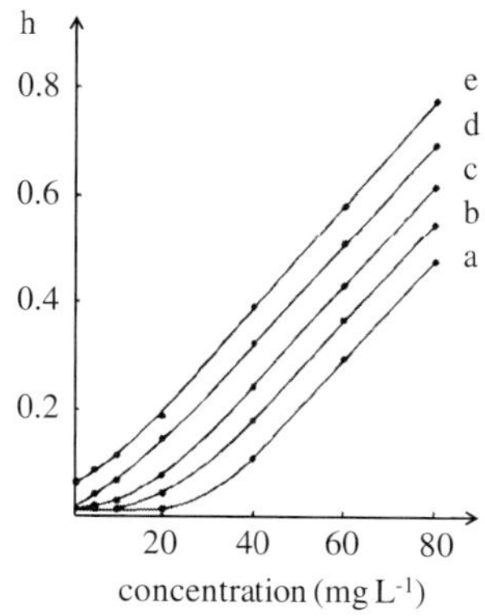

FIGURE 8.7 **Analytical curves associated with the turbidimetric flow injection determination of sulfate.** h = recorded peak height in turbidance, concentration = mg L^{-1} SO_4-S in the injectate, a—e = concentrations of the added solutions (0, 100, 200, 300, 400 mg L^{-1} SO_4-S). For details see text. *Adapted from Anal. Chim. Acta 145 (1983) 179, F.J. Krug, E.A.G. Zagatto, B.F. Reis, O. Bahia-Filho, A.O. Jacintho, S.S. Jørgensen, Turbidimetric determination of sulphate in plant digests and natural waters by flow injection analysis with alternating streams, with permission from Elsevier (Ref. [67]).*

Different strategies have been used for reagent additions in flow-based catalytic methods, depending on the rates of the indicator reaction under catalysed and uncatalysed conditions, as well as on the mode of the flow system involved.

If the development of the uncatalysed indicator reaction is very slow (or even negligible), restrictions on system design are less important, and manifold architectures analogous to that in Fig. 8.8, upper, are normally used. Both reagent streams are added to each other and the mixing conditions of the combined stream are improved inside reactor R_{C1} before reagent entry into the main analytical channel. The flow rate of the combined reagent stream is higher than those of the individual reagent streams (*cf.* Eq. 3.10) and this is a favourable aspect of reagent addition [68]. With a slow development of the uncatalysed indicator reaction, the baseline tends to be small, facilitating the implementation of catalytic methods in flow systems designed in the single-line configuration. Inverted peaks arising from direct sample introduction into a radiation absorbing (or emitting) stream are less important. This is confirmed by the number of catalytic methods implemented in sequential injection and similar analysers [69].

Limitations in system design become more severe when the indicator reaction proceeds in a pronounced manner in the absence of the catalyst. In order to improve the detection limit, the measurement repeatability and the ratio of development of the indicator reaction under catalysed and uncatalysed conditions should be as high as possible. An improvement in repeatability is generally attained in flow systems designed in the confluence configuration (see 3.2.2) and the highest ratio of indicator

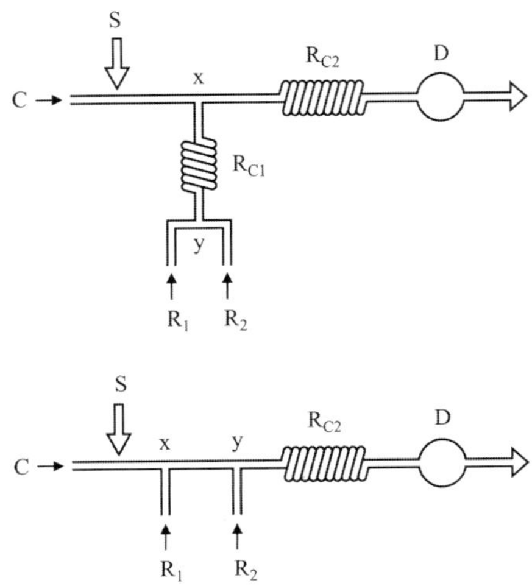

FIGURE 8.8 Didactic representation of typical flow systems for catalytic determinations involving a slow (upper) or a fast (lower) uncatalysed indicator reaction. S = sample; C = sample carrier stream; R_1 and R_2 = reagents; R_{C1} and R_{C2} = coiled reactors; D = detector; x and y = confluence sites; empty arrow = sample introduction. For details, see text.

reaction development is observed at the most concentrated region of the sample zone when the time available for the development of the indicator reaction inside this region under catalysed or uncatalysed conditions is the same. To this end, the manifold architecture in Fig. 8.8, lower, is the most widely used. The reagent streams are successively added to the sample zone and this provides the same time for the development of the indicator reaction under catalysed and uncatalysed conditions. Confluence sites x and y should be as close to each other as possible, which can be achieved using a four port connector. Prior mixing of both reagents is not recommended, as this would facilitate the development of the indicator reaction under uncatalysed conditions. Moreover, use of flow systems in the single-line configuration would be limited. In this regard, implementation of catalytic methods in equential injection analyser is better accomplished by aspirating the solutions involved in the sequence: reagent #1 – sample – reagent # 2. The indicator reaction then starts after the R_1 and R_2 reagent zones overlap or, in other words, when the three different fluid elements (sample, R_1 and R_2) are simultaneously put in contact with each other.

8.3.4. In-line Formation of Unstable Reagents

An important potential of a flow analyser resulting from its transient character (see Chapter 1) is that attainment of chemical equilibrium is not necessarily required. As a corollary, thermodynamically unstable reagents can be used, provided the rate of decomposition is reproducible. In-line formation of unstable reagents is then worthwhile, as demonstrated in 1974 by the spectrophotometric determination of boron in a segmented flow analyser using Azomethine-H as the colour-forming reagent [70]. Azomethine-H, the condensation product of the H-acid (8-amino 1-naphthol-3,6-disulphonic acid) and salicylaldehyde [71], is very sensitive to borates, forming a yellow ion association compound by a reversible reaction [72]; the presence of borates displaces the equilibrium, thus increasing the absorbance of the reaction medium. The reagent however tends to hydrolyse in aqueous media and should not be stored for long periods in order to avoid a decrease in sensitivity. In flow-based analytical procedures, this hydrolysis results in a pronounced baseline drift [70]. This drawback has been circumvented by separately adding the H-acid and the salicylaldehyde, thus forming Azomethine-H in-line [70,73].

During the 1980s, in-line formation of strongly reducing [74–79] or oxidising [80–83] agents in unsegmented flow analysis was critically discussed by Schothorst and co-workers.

Strongly reducing agents such as Cr(II), V(II) and U(III) are generally formed from metallic powders in a solid-phase reactor and, to this end, the Jones reductor is often used [84]. Their use in batchwise analytical procedures is however limited by the need to work under oxygen-free conditions, which makes storage difficult [74].

Strongly oxidising agents, *e.g.*, Ag(II) and Co(III), are generally formed inside electrochemical reactors [85] and their use in batchwise analysis is also limited. As these reagents tend to be reduced in aqueous media, their half-lives are short and, consequently, storage is also difficult.

These drawbacks are effectively circumvented in flow analysis due to the closed nature of the manifold and the relatively short sample residence time in the analytical path. Flow-based analytical procedures involving the short-lived Ag(II) species, characterised by a half-life of about 120 s [80], are a good example.

In-line formation of strongly reducing or oxidising chemical species in flow injection systems was initially exploited for applications involving spectrophotometric or amperometric detection [85]. As most of these species are characterised by a high molar absorptivity coefficient and/or by an uncommon redox state, highly sensitive analytical procedures relying on the direct interaction of the analyte with the chemical species formed in-line have been proposed. The determinations of nitrate and nitrite by reduction with Cr(II) [75], of nitrate, nitrite, hydroxylamine and hydrazine by reduction with V(II) [76] and of monosaccharides by oxidation with Cu(III) [83] are good examples.

Strongly oxidising agents such as Ag(II), BrO^-, bromine, Co(III), $[Cu(HIO_6)_2]^{5-}$, ClO^- and Mn(III) are especially attractive in relation to chemiluminometric analytical procedures, as demonstrated in the determinations of captopril [86,87], catecholamines [88], chlortetracycline [89], gentamycin [90], isoniazid [91,92], quinine [93,94] and tetracyclines [95] in pharmaceutical formulations and sulfidein air samples [96,97].

Other reagents have also been formed in-line and exploited for specific applications.

- Reduced organic compounds. The determination of copper in natural waters and biological materials [98] is a good example. The method relies on the influence of Cu^{2+} on the oxidation rate of the reduced form of 2,6-dichlorophenolindophenol (DCPI) by hydrogen peroxide, with spectrophotometric monitoring of the coloured product formed. In order to establish the unstable reduced form of the DCPI reagent, a very dilute (5.0×10^{-4} mol L^{-1}) ascorbic acid stream merged with the original DCPI stream before its confluence with the main stream. A detection limit of 0.09 μg L^{-1} Cu was obtained.
- Nitrite. This chemical species, formed in-line by passing a potassium nitrate stream through a copperised cadmium solid-phase reactor, was used in the spectrophotometric determination of sulfadiazine in pharmaceutical formulations [99]. A nitrite solution that would normally be stored in a refrigerator, standardised and frequently replaced was therefore not required.
- Bromine. Convergence of bromide, bromate and hydrochloric acid streams leads to formation of the unstable bromine reagent that can be added to the main analytical channel of a flow injection system. The approach was exploited for the spectrophotometric determination of acetylcysteine in sachets [100]. A noteworthy feature is that in-line waste treatment involving elimination of excess bromine was performed by adding an ascorbic acid stream to the main channel after detection.

8.3.5. In-line Reagent Dissolution

Another strategy for avoiding the storage of relatively unstable wet reagents is to take advantage of in-line reagent dissolution, which also reduces the possibility of human error and contamination. As the solid reagent is stored under less stringent conditions, its viability is maintained. This approach is especially attractive for managing a large-scale laboratory when the number of assayed samples and analytes is high, as the preparation of reagents is often the limiting step.

Placement of commercial cartridges with powdered reagents (often lyophilised or crystallised) into the flow manifold enables in-line reagent dissolution, as demonstrated by Marshall *et al.* [101] who designed a sequential injection analyser with a cartridge tray attached to the main selecting valve and applied it to the chemiluminometric determination of adenosine-5′-triphosphate (ATP). They described the preparation and

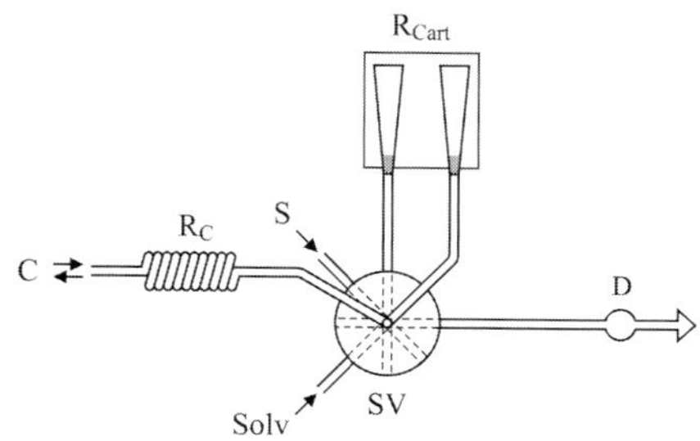

FIGURE 8.9 Flow diagram of a sequential injection system with in-line reagent dissolution. Solv = reagent solvent; S = sample; C = sample carrier stream; SV = selection valve; R_C = holding coil; R_{Cart} = reagent cartridge; D = detector. Dashed areas = connections between R_{Cart} and the flow system. *Source: Figure adapted from Ref. [101].*

homogenisation of the powdered reagents, as well as the different strategies for dissolution.

A sequential injection analyser with in-line reagent dissolution operates as follows. Initially, a large volume of the reagent (solvent) is aspirated towards the holding coil (Fig. 8.9) and further flow reversal permits the addition of a precisely metered diluent volume to one of the reagent cartridges. Reagent dissolution is improved by back-and-forth pumping of the diluent inside the cartridge. After complete dissolution, a liquid reagent aliquot and a sample aliquot are sequentially aspirated towards the holding coil and processed as in ordinary flow systems. With this innovation, one single reagent dissolution is enough for several samples and cartridge replacement is easy. Whenever required, the reagent concentrations can be altered in real time to better match the requirements of the assay [101].

Titrand generation associated with coulometric titrations is another type of in-line reagent dissolution, and this approach has mostly been exploited in flow titrations with electrochemical detection [102–105]. The characteristics, potential and limitations of this approach are presented in 8.6.2.

8.3.6. Solid Reagents

These substances comprise metal powders or filings, alloys, slightly soluble salts or oxides, resins, adsorbents, molecularly imprinted polymers, functionalised beads and enzymes. Their use in flow analysis offers interesting possibilities, because the chemical reactions involved occur at the solid/liquid interface [106]. The reagent is generally maintained during processing of an entire sample batch, and this is important for enhanced system ruggedness and improved laboratory management.

The reagent is generally packed in a mini-column strategically positioned in the analytical path. Alternatively, it can be fixed on the inner walls of the manifold or on fluidised beads, *e.g.*, in bead injection systems.

8.3.6.1. *Characteristics of Reagent Confinement*

The main characteristics, requirements, advantages and limitations of reagent confinement in flow analysis are as follows:

- Fast sample/reagent chemical interaction. This feature is essential, as the sample zone passes rapidly through the solid material regardless of its geometry.
- Reagent inertia. The reagent should be chemically inert in relation to the solutions flowing through it and resistant to microbial degradation in order to improve the mini-column (or reactor) lifetime.
- Minimal reagent consumption. Consumption is ideally restricted to the amount effectively reacted with the sample. For slightly soluble materials, additional reagent consumption is required as a consequence of the solubility product involved.
- Improved analytical sensitivity. Sample dilution caused by the addition of a reagent solution is avoided.
- Manifold simplicity. A specific channel for adding the reagent stream (or the reagent zone – merging zones configuration) is not required.
- The feasibility of using slightly soluble reagents. Reagents that cannot be prepared in solution can be used at relatively high "concentrations".
- The feasibility of using fluidised beads. Beads with reagents on their surface are efficiently managed.
- Improved mixing conditions. The use of packed mini-columns or open tubular reactors with a curled inner wall designed in different geometries (see 6.2.3) improves radial mixing.
- Lower sample/reagent interaction. The reactions involved occur only at the liquid–solid interface.
- Increased hydrodynamic pressure. This shortcoming is more evident in packed bed mini-columns (see 8.5.2.3).
- Compatibility with solid-phase spectrometry. Moving beads are inherent to solid-phase spectrometry (see 4.1.1.4) and have also been exploited in advanced flow-based chemiluminometric analytical procedures.
- Compatibility with solid-phase extraction. Immobilised reagents are the essence of flow-based strategies for solid-phase extraction (see 8.5.2).

8.3.6.2. *Strategies for Reagent Confinement*

Reagent immobilisation is generally accomplished by using slightly soluble substances or by anchoring specific chemical species on a solid support.

Metal powders or fillings, alloys, slightly soluble salts or oxides, resins, adsorbents, molecularly imprinted polymers (MIP's) and enzymes in their natural environment (*e.g.*, bacteria, cells and tissues) can be packed in mini-columns without any treatment, and this strategy is a natural immobilisation of the reagent. The main characteristics and performance of mini-columns packed with solid particles are discussed in 8.5.2.3.

To achieve higher flow rates without a pronounced increase in hydrodynamic pressure, the solid materials can also be incorporated on the tubing walls of the manifold through impregnation, coating and similar processes [107]. Another alternative is to melt the reagent inside a mini-column in such a way that a hollow space is left in the centre after cooling [108]. Fairly soluble organic compounds can then be immobilised without affecting the mini-column lifetime.

With the reagent immobilised on the tubing inner wall, sample/reagent interaction is limited due to the laminar flow regime involved. This drawback can be circumvented by tightly coiling or knitting the manifold tube (see 6.2.3), thus improving radial mass transport and consequently sample/reagent interaction. In this context, the use of pulsed flows is promising but this has not been exploited fully.

Anchoring chemical species on a solid support is another possibility for reagent immobilisation in flow analysis. This approach enables enzyme cartridges, ion-exchange resins and other functionalised beads to be effectively exploited. The beads are generally packed in mini-columns, and fluidisation can also be used.

8.3.6.3. *Applications of Confined Reagents*

Metal powders or filings, alloys and similar materials have been used for analyte conversion through redox reactions, and an early flow-injection system designed for the simultaneous spectrophotometric determination of nitrate and nitrite in river waters and soil extracts [109] is a good example. The sample was injected twice, generating two sample zones: one passed through a mini-column packed with copperised cadmium filings where nitrate was quantitatively reduced to nitrite, and the other bypassed the mini-column. Two analytical signals, corresponding to nitrate + nitrite and to nitrite concentrations, were then obtained. Other applications include mini-columns packed with the Jones reductor and with Devarda's alloy for the spectrophotometric flow-injection speciation of iron [110] and for the determination of dissolved organic nitrogen in natural waters using a segmented flow analyser [111].

Alternatively, the oxidising or reducing solid reagent can be immobilised on a suitable support, as demonstrated in the spectrophotometric and chemiluminometric flow injection determinations of manganese in natural waters using mini-columns of lead(IV) dioxide or sodium bismuthate immobilised on silica gel beads [112] and paper pulp [113], respectively.

Slightly soluble salts or oxides have mostly been used in analytical procedures relying on displacement reactions, as illustrated by the determination of sulphate in natural waters using a mini-column packed with barium chloranilate [114]. The analyte reacted with the confined salt, yielding the less soluble barium sulphate and releasing the highly coloured chloranilate ion which was quantified in-line. Another application is the spectrophotometric determination of total nitrogen in plant digests using a mini-column packed with $AgCl_{(s)}$ [115]. Ag^+ was sequestered by ammonia as diamminoargentate(I) and further reacted with bromopyrogallol red and 1,10-phenanthroline to form a coloured ternary complex. A flow injection procedure for the spectrophotometric determination of Cr(VI) involving in-line dissolution of the 1,5-diphenylcarbazide reagent [116] can also be highlighted. A mini-column packed with the solid reagent (25% m/m in silica) was incorporated in a single-line flow system, and the dissolution of the required amount of reagent was accomplished as the sample passed through it. Dispersion was reduced and the detection limit was improved due to the absence of the Schlieren effect. The use of an open tubular reactor with a reagent impregnated inner wall can be illustrated by the determination of reserpine in tablets [117]. Crystalline $MnO_{2(s)}$ was thermally embedded on the walls of Tygon™ tubing incorporated in the flow system. The immobilised Mn(IV) catalysed oxidation of the analyte by periodate and the product formed, 3,4-didehydroreserpine, was spectrophotometrically monitored.

Commercially available resins (*e.g.*, ion-exchange and chelating), adsorbents (*e.g.*, silica gel and activated carbon) and MIPs have been most commonly used for SPE (see 8.5.2).

Enzymes in their natural environment have been used as "solid reagents" and *Saccharomyces cerevisae* is a good example [118]. Enzyme cartridges have also been used, as demonstrated by the determination of glucose in plasma using flow injection manifolds with spectrophotometric and chemiluminometric detection and an in-line immobilised enzyme coil [119]. The coil consisted of glucose oxidase covalently bound to the inner surface of a nylon tube (25 cm long, 1.0 mm i.d.) which was coiled into a cylindrical acrylic exoskeleton.

8.3.7. Suspension Addition

The dynamicnature of flow analysers permits the addition of a flowing stream containing a suspension to the main analytical channel, as demonstrated by the turbidimetric determination of sulphate in plant digests involving in-line formation of slightly soluble lead sulphate [120]. In order to improve the super saturation conditions, two confluent streams merged together, allowing the controlled and reproducible formation of a lead phosphate suspension that was then added by confluence to the main analytical channel. The seed crystals acted as nuclei, leading to the formation of a more homogeneous suspension, hence increasing the turbidity and, consequently, the analytical sensitivity and repeatability.

The spectrophotometric determination of cadmium in natural waters involving in-line formation of an aqueous dithizone suspension [121] illustrates another application of in-line suspension addition. Solid dithizone reagent was packed in a mini-column through which a surfactant (Triton X-100) stream was allowed to flow. The emerging suspension formed was added by confluence to the main analytical channel of a flow-injection system. With this innovation, good sensitivity was achieved without the need for an analyte separation/concentration LLE step and the entire procedure was carried out in the aqueous phase.

8.4. SAMPLE TREATMENT

It is a difficult task to define broad terms related to sample preparation, sample (or analyte) extraction and analyte separation/concentration, which becomes even more difficult when physico-chemical processes, *e.g.*, microwave, UV or ultrasound irradiation, are involved. In this monograph, in-line sample digestion, bleaching, hydrolysis, oxidation and related approaches are therefore discussed under the general term *sample treatment*.

With regard to sample digestion, concentrated acids are usually added to the sample for dissolution and decomposition, thus simplifying the sample matrix. The chemical reactions involved are generally developed at a high temperature, and often a high pressure, leading to modification of the analyte to simpler chemical species and to the evolution of gaseous species, mainly CO_2 and H_2O. The digest is ideally an optically transparent solution. The digestion step is often the slowest and most critical part of an analytical method and accounts for a large proportion of the total analysis time. Errors can be introduced due to contamination during sample manipulation and analyte losses can occur due to volatilisation when the digestion is performed in a heated open flask. Moreover, precautions must be taken to prevent exposure to acid fumes. These drawbacks are minimised when the digestion step is incorporated in a flow analysis system.

8.4.1. Early Developments

Initial attempts to perform in-line digestion of plant tissues were made in1960 [122]. The Kjeldahl digestion procedure was implemented in a segmented flow analyser by adding the solid sample and the digestion mixture (500 mL L^{-1} sulphuric acid + 3.0 mL L^{-1} perchloric acid + 4.8 mL L^{-1} selenium oxychloride) to one end of a tubular-shaped digestion vessel with a spiral groove on its inner wall. Continuous rotation pushed the mixture towards the opposite end of the vessel and improved mixing conditions. The entire assembly was placed inside a furnace. The outlet flow was handled as in ordinary segmented flow analysis (see 5.1.1). Protein in plant materials was determined in the 0.1–10.0 mg L^{-1} range at a sampling rate of 10 h^{-1} and good repeatability of results (r.s.d. <1.5 %) was reported.

The spectrophotometric flow injection determination of chemical oxygen demand (COD) in wastewaters with in-line oxidation [123] was an early application of in-line treatment in unsegmented flow analysis. The sample was inserted into a water carrier stream which merged with an acidic dichromate confluent stream and the combined stream passed through a PTFE coiled reactor at 120 °C. Accurate results were obtained at a rate of 15 h^{-1}; the detection limit and precision were 5 mg L^{-1} COD and 0.4%, respectively. Another early application was the determination of orthophosphate and polyphosphates, such as diphosphate and triphosphate, with in-line hydrolysis [124]. A strongly acidic solution containing molybdenum(V) and molybdenum(VI) was used as the carrier so that hydrolysis of polyphosphates and colour development of the resultant orthophosphate were simultaneously accomplished. For speeding up the chemical reactions involved, the temperature of the reaction coil was maintained at 140 °C, and a back-pressure coil was required (see 4.3). A sampling frequency of 45 h^{-1} and r.s.d.[3] of <1% were reported. The efficiency of the system was comparable with that of an air-segmented flow system.

8.4.2. Auxiliary Energy Addition

Due to the relatively short time available for sample treatment in flow analysis, the chemical reactions involved often need to be accelerated. To this end, increasing the oxidising potential (nature/concentration of the oxidising agents, temperature and pressure), use of catalysts and/or exploitation of auxiliary energy are useful. The auxiliary energy behaves as a "clean reagent" which is not present in the waste that is generated and its addition does not cause any dilution of the sample. In this context, strategies exploiting microwave, UV or ultrasound irradiation are presented below.

8.4.2.1. Microwave Irradiation

CHARACTERISTICS

Microwaves are a form of electromagnetic radiation produced in magnetrons, which can be absorbed by a process called dielectric heating. Dipolar molecules, *e.g.*, water, undergo rotation in order to try to align themselves with the alternating electric field of the microwave. This movement generates heat which dissipates towards neighbouring molecules through collisions. Exploitation of microwaves as an energy source permits high temperatures and pressures to be attained in a short time, which reduces the total time required for sample preparation [125].

Domestic or focussed microwave ovens have mostly been used and deliver electromagnetic radiation characterised by a frequency around 2.45 GHz ($\lambda = 12.2$ cm). Consequently, the amount of energy absorbed is proportional to the power of the microwave source.

The sample is placed in a vessel strategically positioned inside the oven. A noteworthy feature, in contrast to classical heating strategies such as conduction, radiation or convection, is that microwave radiation is not absorbed by the container walls, only by the sample/reagent mixture itself [126]. Consequently, a more homogeneous heating is attained and less energy is consumed.

Related analytical procedures are characterised by simplicity, safe operation, low blank values, reduced contamination risk and compatibility with a wide-range of samples. A comprehensive review of microwave-assisted digestion is given in Ref. [127].

IMPLEMENTATION

The main reactor exposed to microwave irradiation is generally a coiled tube or a low inner volume chamber located inside the microwave oven. Winding the coiled tube around the oven output antenna can improve the irradiation geometry [128].

Digestion can be performed prior to or after sample introduction into the main analytical channel [126,129]. In the former configuration, the sample and the digestion solution are introduced into the microwave oven in a continuous manner and, following digestion, a discrete aliquot is selected and delivered to the detector. As an infinite sample volume situation is approached, variations in sample dispersion due to the release of gas bubbles are not as important, but the system becomes less versatile; moreover, a large sample volume is required. In the latter arrangement (Fig. 8.10), the sample is introduced together with the reagents and flows through the main reactor placed inside the oven, thereafter being cooled and de-gassed before reaching the detector. In both modes, sample stopping is often a useful strategy.

Flow-based analytical procedures with in-line microwave-assisted digestion generally exploit the slurry technique for introducing the sample into the flow analyser. Some slurry samples, *e.g.*, whole blood and milk, can be directly introduced into the main analytical channel whereas other materials need to be ground and the suspended particles stabilised by adding a suitable surfactant [130]. Details of the implementation of

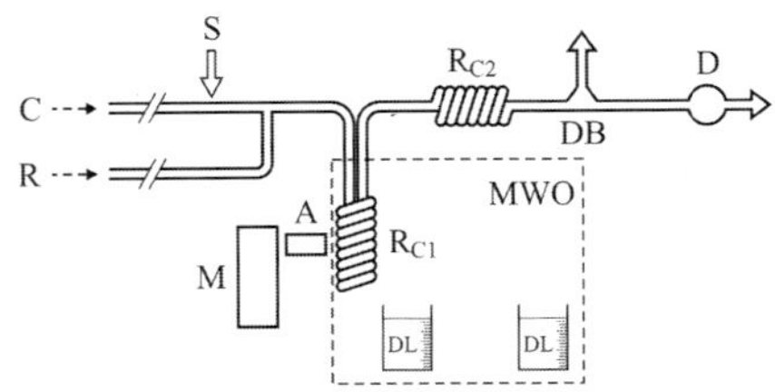

FIGURE 8.10 **Experimental setup for microwave irradiation.** The figure refers to a flow system with in-line irradiation after sample insertion into the analytical path. S = sample; C = sample carrier stream; R = reagent; R_{C1} = irradiation reactor; R_{C2} = mixing/cooling reactor; DB = de-bubbler; D = detector; MWO = microwave oven; M = magnetron; A = antenna; DL = dummy load (flasks with tap water for absorbing microwaves); || = optional flow stopping. For details, see text.

in-line microwave-assisted digestion in flow analysis are presented in *e.g.*, Refs. [126,127,129,131,132].

APPLICATIONS

Due to the pronounced heating effect of microwave irradiation, most applications refer to in-line sample preparation for the determination of metals in biological and clinical matrices. This approach has also been exploited to accelerate the development of, *e.g.*, hydrolysis, extraction and leaching, as well as for the photo-degradation of organic matter (Table 8.1).

The favourable characteristics of this approach were demonstrated in the pioneering work incorporating in-line digestion in flow injection analysis [133]. Iron, copper and zinc were determined in whole blood by inserting both the sample and the digestion solution into convergent carrier streams; the established zones merged and the combined flowing zone passed through a Pyrex™ coiled reactor located inside a domestic microwave oven; thereafter, the processed sample reached the nebuliser of a flame atomic absorption spectrometer. A 25-s sample residence time inside the oven was sufficient for quantitative digestion without the pronounced release of acidic fumes or water vapour.

There are many applications involving atomic absorption/emission spectrometry, ICP-OES or ICP-MS [129], probably because the high energy associated with the furnace, flame or torch can destroy undigested organic material. These techniques are not affected by the Schlieren effect; also, measurements are only slightly altered by the presence of gaseous bubbles.

In contrast, there are few applications using flow systems with *spectrophotometric* detection because the presence of undigested organic material may compete with the analyte for the reagent. This combination has been more exploited for analysis requiring mild reaction conditions.

TABLE 8.1 Selected Applications of Sample Treatment

Analyte	Sample	Step	Detection Technique	Analytical Range or Detection Limit	Remarks	Ref.
Acetylsalicylic acid	Tablets	Microwave-assisted analyte hydrolysis	UV–Vis	4.0 μg mL^{-1}	Flow injection system; merging zones; in-line microwave-assisted alkaline hydrolysis of acetylsalicylic acid to salicylic acid that reacts with Fe(III) to form a coloured complex	[414]
Ammonia	Cigarettes	Ultrasound-assisted bleaching	UV–Vis	0.05 μg mL^{-1}	Flow injection system with a pervaporation unit; iterative change of flow direction for improving extraction; Berthelot reaction	[415]
Asulam (carbamate pesticide)	Irrigation, tap waters	UV-assisted analyte photodegradation (pH 8.3)	CL	40 μg L^{-1}	Multi-commuted flow system; oxidation of the photoproducts by permanganate in acidic sulphuric medium or by hexacyanoferrate(II) in alkaline medium	[416]
β-Estradiol	Natural waters	Microwave-assisted template removal from MIP	Fluorimetry	1.12 μg L^{-1}	Flow injection system; SPE using a MIP mini-column	[417]
Bromoxynil	Waters, pesticide formulations	UV irradiation for improving analyte degradation	CL	5.0×10^{-3} mg L^{-1}	Flow injection system; manifold with a photoreactor; monitoring of the chemiluminescent photofragments	[418]
Boron	Soils	Ultrasound-assisted bleaching	UV–Vis	up to 35 μg g^{-1}	Flow injection system; bleaching unit in the sampling loop; bleaching solution under recycle; Azomethine-H as the colour-forming reagent	[419]

(Continued)

Table 8.1 Selected Applications of Sample Treatment—cont'd

Analyte	Sample	Step	Detection Technique	Analytical Range or Detection Limit	Remarks	Ref.
Chemical oxygen demand	Well, river, waste waters	Microwave-assisted oxidation by $K_2Cr_2O_7$/ H_2SO_4	UV–Vis	1.5 mg L^{-1}	Flow injection system; membrane de-bubbler to remove the gas phase formed during irradiation	[134]
Chemical oxygen demand	Freshwaters	UV irradiation for improving analyte degradation	UV–Vis	0.5 mg L^{-1}	Flow injection system; manifold incorporating UV-photocatalytic oxidation instead of conventional heating	[420]
Chloramphenicol	Pharmaceutical formulations	UV irradiation for analyte degradation; reaction of photofragments with permanganate yields CL	CL	30 ng mL^{-1}	Flow injection system; irradiation prior to sample introduction	[421]
Cr(VI)	Sediments, soils	Microwave-assisted analyte leaching	UV–Vis	0.012 μg mL^{-1} (extract) or 0.12 μg g^{-1} (dry basis)	Flow injection system with optional SPE; iterative change of flow direction for improving extraction; derivatisation with 1,5-diphenylcarbazide	[422]
Cr(VI)	Soils	Ultrasound-assisted analyte leaching	UV–Vis	1.23 ng g^{-1}	Flow injection system with optional SPE; iterative change of flow direction for improving extraction; derivatisation with 1,5-diphenylcarbazide	[423]
DIC, DOC	Freshwaters	UV irradiation for improving oxidation with peroxydisulfate in acidic medium	UV–Vis	0.05 mg L^{-1} (both analytes)	Sequential injection system; UV irradiation for DOC determination; in-line CO_2 diffusion; tandem stream with sample/digestion mixture segments	[424]

Table 8.1 Selected Applications of Sample Treatment—cont'd

Analyte	Sample	Step	Detection Technique	Analytical Range or Detection Limit	Remarks	Ref.
Diverse	Water, whole blood, culture cells	Ultrasound-assisted filtration	UV–Vis	Diverse	Multi-commuted flow injection system with three parallel analytical paths; pressurised gas as the fluid propellent	[150]
DOC	Estuarine waters, sediment pore waters	UV irradiation for improving oxidation with peroxydisulfate in acidic medium	UV–Vis	0.26 mg L^{-1}	Air-segmented flow analyser; in-line CO_2 dialysis	[425]
Fe	Multi-vitamin tablets	UV irradiation for sample dissolution and thermal digestion using a 100 W tungsten lamp, HCl and H_2O_2	UV–Vis	100–500 μg per tablet (1 g)	Multi-commuted flow system; in-line SPE for analyte separation/ concentration as chlorocomplex; analogous flow system for the determination of Cu and Zn by FAAS	[426]
Flavonoids	Medicinal plants	Ultrasound-assisted extraction	UV–Vis	1.2 μg mL^{-1}	Flow injection system; extraction vessel in a ultrasonic water bath	[427]
Fluoro-quinolone antibiotics	Soils	Microwave-assisted extraction	Fluorimetry	0.15 μg g^{-1}	Flow injection system; extracted species in-line inserted into a derivatisation / carrier stream	[428]
Formaldehyde	Foods	Microwave for accelerating the Hantzsch reaction with cyclohexane 1,3-dione	Fluorimetry	0.02 ng mL^{-1}	Flow injection system; sample stopping inside the microwave oven	[429]
Formetanate	Mineral, tap waters	Microwave-assisted analyte hydrolysis	UV–Vis	0.20×10^{-5} mol L^{-1}	Flow injection system; in-line waste treatment exploiting TiO_2 addition and UV irradiation	[430]

(Continued)

Table 8.1 Selected Applications of Sample Treatment—cont'd

Analyte	Sample	Step	Detection Technique	Analytical Range or Detection Limit	Remarks	Ref.
Glycerol (free, total)	Biodiesel	Ultrasound-assisted extraction into a $NaIO_4$ plus acetylacetone aqueous-ethanol solution	UV–Vis	0.0028, 0.0310% (m/m)	Flow injection system; LLE without phase separation; ethyl acetate for free/bound glycerol discrimination	[431]
Hydrogen peroxide	Tap, lake, rain, purified waters	Ultrasound-assisted luminol-H_2O_2-Co(II) reaction	CL	1.0×10^{-9} mol L^{-1}	Flow injection system; coiled reactor immersed in a water bath inside which the sonication probe is inserted	[147]
Magnesium, calcium, iron	Foodstuffs	Microwave-assisted sample digestion with concentrated nitric acid	UV–Vis	5.0–50.0 mg L^{-1}, 15.0–150 mg L^{-1}, 2.0–20.0 mg L^{-1}	Sequential injection system; direct sample introduction as a natural suspension or as a slurry; open digestion vessel inside a microwave oven	[432]
N-methyl-carbamates	Contaminated soils and fruits (apples, pears, cucumbers)	Ultrasound-assisted extraction	Fluorimetry	12 ng g^{-1}	Unsegmented flow system; in-line filtration; in-line SPE; HPLC; post-column derivatisation	[433]
Nitrogen(total)	Wastewaters from urban waste water treatment plant	Microwave-assisted oxidation of nitrogen-containing compounds to nitrate with an alkaline peroxydisulfate solution	UV–Vis	0.21 mg L^{-1}	Flow injection system; chamber-like de-bubbler for removal of the gaseous species formed during in-line oxidation; nitrate reduction to nitrite; Griess-Ilosvay reaction	[434]

Table 8.1 Selected Applications of Sample Treatment—cont'd

Analyte	Sample	Step	Detection Technique	Analytical Range or Detection Limit	Remarks	Ref.
Nitrogen(total)	Estuarine, marine, fresh, waste waters	microwave-assisted oxidation of nitrogen-containing compounds to nitrate with an alkaline peroxydisulfate solution	UV–Vis	0.03 mg L^{-1}	Flow injection system; optional in-line heating before UV-irradiation; Accurel microporous tubing for removal of the gaseous species formed during in-line oxidation; nitrate reduction to nitrite; Griess-Ilosvay reaction	[435]
Paracetamol	Pharmaceutical formulations	Microwave-assisted alkaline hydrolysis yielding *p*-aminophenol	UV–Vis	0.2 μg mL^{-1}	Flow injection system; reaction of *p*-aminophenol with 8-hydroxyquinoline in the presence of KIO_4 yielding a blue indophenol dye; de-bubbler prior to the flow cell	[436]
Paracetamol	Suppositories	Ultrasound-assisted LLE and alkaline hydrolysis	UV–Vis	0.38 μg mL^{-1}	Flow injection system; LLE without phase separation; iterative change of the flow direction to improve extraction/hydrolysis; organic phase inserted as a plug before detection	[149]
Paracetamol	Pharmaceutical formulations	Microwave-assisted alkaline hydrolysis yielding *p*-aminophenol	UV–Vis	0.2 μg mL^{-1}	Flow injection system; sample stopping inside the microwave oven; reaction of *p*-aminophenol with *o*-cresol in alkaline medium yielding a blue compound; the solution leaving the oven flows through a coil inside an ice-bath; gas-phase removed prior to the colour-forming reaction	[437]

(Continued)

Table 8.1 Selected Applications of Sample Treatment—cont'd

Analyte	Sample	Step	Detection Technique	Analytical Range or Detection Limit	Remarks	Ref.
Polyphenols	Extra virgin olive oils	Ultrasound-assisted LLE	UV–Vis	24 μg mL^{-1}	Flow injection system; LLE without phase separation; iterative change of the flow direction for improving extraction; alkaline aqueous phase as extractant; derivatisation with Folin-Ciocalteu reagent	[438]
Phosphorus (total)	Foodstuffs	Microwave-assisted digestion with concentrated nitric acid	UV–Vis	20.0–400.0 mg L^{-1}	Sequential injection system; digestion bomb inside the microwave oven as a peripheral component of the system	[135]
Phosphorus (total)	Waste waters, effluents	Microwave-assisted digestion with an acidic peroxydisulfate solution	UV–Vis	0.09 mg L^{-1}	Flow injection system; on-line filtration; two microporous polypropylene tubes as de-bubblers; orthophosphate detected as phosphomolybdenum blue	[439]
Phosphorus (total)	Waste waters	Microwave-assisted digestion with nitric acid	UV–Vis	0.10 mg L^{-1}	Flow injection system; addition of the *pyrophosphatase* enzyme prior to sample introduction	[440]
Phosphorus (total)	Waste waters	Microwave-assisted digestion with an acidic peroxydisulfate solution	UV–Vis	0.9 mg L^{-1}	Multi-syringe flow system exploiting multi-commutation; orthophosphate detected as phosphomolybdenum blue	[441]
Phosphorus (dissolved orthophosphate plus dissolved organic phosphorus)	Waste waters	UV-irradiation for organic phosphorus oxidation to orthophosphate	UV–Vis	0.08 mg L^{-1}	Multi-pumping flow system; vanadomolybdate method for detecting the orthophosphate	[442]

Table 8.1 Selected Applications of Sample Treatment—cont'd

Analyte	Sample	Step	Detection Technique	Analytical Range or Detection Limit	Remarks	Ref.
Reducing sugars	Sugarcane juices	Microwave-assisted sugar oxidation by hexacyanoferrate(III) in alkaline medium	UV–Vis	15.0 μmol L^{-1}	Flow injection system; derivatisation using the oxidising reagent; de-bubbler before measurement	[443]
Reducing sugars	Wines	Microwave-assisted sugar oxidation by hexacyanoferrate(III) in alkaline medium	UV–Vis	9 μmol L^{-1}	Flow injection system; derivatisation using the oxidising reagent; de-bubbler before measurement; coiled reactor around the oven antenna	[128]
Safflower yellow	Herbal medicine	Microwave-assisted extraction	UV–Vis	8.0 μg mL^{-1}	Flow injection system; in-line extract filtration and dilution	[444]
Starch	Flour, bread	Microwave-assisted free sugar washing and hydrolysis	UV–Vis	0.01 g L^{-1}	Flow injection system; analyte reaction with alkaline neocuproine reagent	[445]
Urea	Freshwaters	Microwave-assisted N-urea reduction to N-ammonia	UV–Vis	2.40 μmol L^{-1}	Flow injection system; ammonia permeation through a PTFE membrane towards bromothymol blue acceptor stream	[250]

UV–Vis = spectrophotometry; CL = chemiluminescence; FAAS = flame atomic absorption spectrometry; MIP = molecularly imprinted polymer; SPE = solid phase extraction; DIC = dissolved inorganic carbon; DOC = dissolved organic carbon.

A pioneering application of microwave-assisted oxidation in a flow-injection analyser with spectrophotometric detection was described by Balconi *et al.* [134] who determined the chemical oxygen demand in well, river and wastewaters. The sample was inserted into a water carrier stream and the potassium dichromate/sulphuric acid mixture was added by confluence; the sample zone passed through the coiled reactor which was wound around a strong microwave absorbing support rod to improve the oxidation rate; thereafter, the sample zone passed through a membrane degassing unit and a spectrophotometric detector. Precise and accurate results were obtained. In-line digestion has also been implemented in a sequential injection system for the spectrophotometric determination of phosphorus in foodstuffs [135]. The sample and the digestion solution were sequentially aspirated towards a chamber placed inside a domestic microwave oven. The flow was stopped and the sample was kept inside the chamber for mineralisation and gas-phase removal; the flow was then restarted in the reverse direction, transporting the sample towards the detector.

8.4.2.2. UV Irradiation

CHARACTERISTICS

The UV spectral region ranges from 40 to 400 nm and this range is divided into UV A (400–315 nm), UV B (315–280 nm) and UV C (<280 nm) [136]. Region A is generally referred to as near UV and region C as vacuum UV due to the need for air removal to avoid radiation absorption by O_2 at <200 nm.

Commercially available deuterium, xenon and mercury vapour lamps have been used as UV radiation sources, with a mercury lamp being by far the most commonly used source for sample preparation. This exploits the intense light that accompanies mercury evaporation in an electric arc, first observed by Wheatstone in 1835 [136]. Mercury lamps are characterised by low cost, ease of operation, high electric-to-electromagnetic energy conversion efficiency and preferential emission in the UV spectral region. The lamp also contains a noble gas, usually argon, which does not participate directly in photon generation but increases the number of electrons and ions involved, and this is beneficial for the processes taking place in the lamp. The spectrum of the emitted radiation is rich in UV lines and strongly dependent on the vapour pressure inside the lamp.

Mercury lamps are generally divided into two groups, depending on the pressure involved; low (10^{-6}–10^{-5} bar) and high (1–100 bar). Lamps with a pressure of *ca* 1 bar are sometimes referred to as medium pressure lamps. The characteristics of mercury lamps are summarised below [137]:

- Low-pressure lamps: good monochromaticity of the emitted radiation (two narrow bands at 254 nm (80%) and 185 nm [138]);

high electric-to-electromagnetic energy conversion efficiency; relatively long lifetime; a long voltaic arc required for higher intensity of the emitted radiation; typical working temperature 40 °C.

- Medium- and high-pressure lamps; several broad bands, with highest emission intensity at 365 nm (medium pressure), and 436 nm and 546 nm (high pressure); pronounced emission in the visible (15%) and infrared (35%) spectral regions for medium pressure lamps; relatively short lifetime; high emission intensity, thus potentially compact; typical working temperature 600−800 °C.

Interaction of the electromagnetic UV radiation with the sample depends on the formation of free radicals and the mechanisms involved are specific to a given situation, as emphasised elsewhere [136,137,139].

IMPLEMENTATION

Exploitation of UV irradiation permits clean sample preparation procedures to be designed because the addition of large amounts of reagents is usually avoided. Furthermore, UV irradiation is effective and easily incorporated in flow analysis [138].

The flow system comprises a photoreactor, usually a tubular coiled reactor or a low inner-volume chamber, which is positioned near the lamp. Alternatively, the tubular photoreactor can be wound around the lamp, a strategy that has been more exploited with low-pressure lamps characterised by lower radiation intensity.

For efficient exploitation of the incident UV radiation, the photoreactor wall should be transparent to the radiation; reflection and focussing of the radiation are also important [137]. Heat dissipation during lamp ON is intense, especially for high-pressure lamps, thus increasing the sample temperature. This effect is beneficial for accelerating some reactions but the boiling point of the sample can be rapidly reached. The temperature of the reaction medium should therefore be controlled. To this end, cooling strategies, *e.g.*, lamp ON/OFF switching and/or improved ventilation, are useful.

APPLICATIONS

Relatively mild conditions are inherent to sample treatment with UV radiation. This approach has mostly been applied to metal determinations by atomic absorption/emission spectrometry and other techniques relying on torches, flames or furnaces. In *spectrophotometry*, it has been useful in several situations (see Table 8.1):

- In-line digestion of waters, as initially demonstrated in the amperometric determination of phosphate involving oxidation by peroxydisulfate [140].

- Breakdown of metal complexes, decreasing the organic matter content and release of the metal from dissolved metal–organic complexes, as exemplified in the chemiluminometric determination of copper in waters with a high dissolved organic carbon content [138]. The treatment did not achieve complete dissolution of organic carbon during the sample residence time in the irradiated region but this was not a limiting factor because the analyte was quantitatively released from the complexant medium.
- Sample dissolution and analyte conversion to a more easily detected species, *e.g.*, the sequential injection determination of total nitrogen in wastewaters involving Kjeldahl oxidation and nitrate formation [141]. The spectrophotometric flow injection determination of organic phosphorus in natural waters [142], relying on the release of the different forms of phosphorus as orthophosphate, is another good example.
- Accelerating the rate of chemical reactions with oxidants, *e.g.*, hydrogen peroxide, oxygen, peroxydisulfate, dichromate and nitric acid [136]. The free radicals formed are used to break down organic molecules and the addition of a TiO_2 suspension as a photocatalyst [139] further enhances degradation.

8.4.2.3. *Ultrasound Irradiation*

CHARACTERISTICS

Ultrasound waves are mechanical vibrations (frequency 20 kHz–10 GHz) produced by a piezoelectric device. These waves can be established in a liquid sample and produce cavitation. Very high temperatures are associated with the locations where cavitation occurs, so the effect can be exploited to assist sample preparation [143].

Ultrasound irradiation has a beneficial influence on sample digestion, physical dispersion, the development of chemical reactions, analyte separation/concentration and precipitation reactions carried out in flow systems [144,145]. In general, sensitivity is improved, especially for heterogeneous systems, reactions with slow kinetics and low analyte concentrations.

IMPLEMENTATION

The implementation of in-line ultrasound for sample preparation in flow analysis is analogous to microwave or UV irradiation. The coiled reactor (or a mini-chamber) is immersed in a water bath to which ultrasound is applied. As mechanical waves are involved, the inner walls of the tubing should be thin and flexible to avoid wave damping, which would decrease the efficiency of the ultrasound. Alternatively, an ultrasound probe can be placed near the ultrasound reactor [143]. The use of an ultrasound probe is particularly attractive for the in-line preparation of solid samples, as demonstrated in the determination of herbicides in soil and sediment samples [146]. The sample was inserted in a mini-chamber, the ultrasound probe was placed 20 mm from the top surface of the chamber and a leaching stream was passed through it; continuous in-line extraction occurred during the sonication process. The leaching stream carried the extracted analytes downstream to a chromatograph with spectrophotometric detection. Glass beads were added to the chamber in order to avoid excessive particle compaction and to reduce temperature gradients.

APPLICATIONS

Ultrasound-assisted sample preparation in flow analysis is a relatively recent approach and the number of applications is much lower than those using microwave or UV radiation [145]. In general, a relatively long sonication time is needed. As mild conditions are involved, this approach has been more exploited in combination with atomic absorption/emission spectrometry and related techniques, as the partially prepared samples are more compatible with these techniques (see above). Apart from sample solubilisation/digestion, there are other potential applications:

- Breakdown of chemical bonds and/or dissolution of solid materials, especially viscous and/or dirty samples (see Table 8.1).
- Acceleration of chemical reactions due to the increased production of free radicals. An example is the chemiluminometric determination of hydrogen peroxide in lake, rain, tap and purified waters [147]. The reactions involved in the sample preparation step and in the chemiluminometric detection step were accelerated by ultrasound.
- Hydride generation, as demonstrated by the determination of toxic arsenic species in human urine by cold-vapour atomic absorption spectrometry [148]. Two advanced oxidation processes relying on high-intensity focussed ultrasound were compared and ultrasound played an important role in both. Accurate results were reported.
- Liquid–liquid extraction, as exemplified by the spectrophotometric determination of paracetamol in suppositories [149] without phase separation (8.5.1.3). A pronounced speeding up of analyte extraction and hydrolysis was attained with several iterative flow reversals.
- Filtration, as exemplified in a multi-purpose flow system for the spectrophotometric analysis of waters, blood and other biological tissues [150].

Other selected applications involving *spectrophotometric* detection are presented in Table 8.1.

8.5. ANALYTE SEPARATION/CONCENTRATION[1]

Flow-based analytical procedures, including in-line separation/concentration steps for improving sensitivity and/or selectivity, are generally characterised by excellent analytical figures of merit. A unique feature is that the equilibria of the physico-chemical processes involved do not necessarily need to be reached, as partial and reproducible development of these processes is inherent to flow analysis (see Chapter 1). Relatively slow and cumbersome processes, *e.g.*, dialysis and distillation, can therefore be efficiently implemented. Moreover, the degree of completion of the processes involved can be controlled in order to match the analytical signal with the dynamic concentration range of the detector [151], and this is a beneficial factor for expanding the applicability of flow analysis. Another favourable characteristic relates to the interface between the different phases (solid, aqueous, organic and gaseous) involved. In contrast to batchwise analytical procedures, where the phases are generally in contact with each other in a static manner, a continuous surface renewal is achieved in flow-based analytical procedures, and this advantage is highlighted in this section. Other characteristics of flow systems with in-line analyte separation/concentration are the same as those often reported for ordinary flow systems (see Chapter 1), such as a high sampling rate, low reagent consumption, low waste generation, the possibility of handling unstable chemical species, low susceptibility to sample losses and/or contamination, and excellent system ruggedness.

8.5.1. Liquid–liquid Extraction, LLE

Liquid–liquid extraction, formerly referred to as solvent extraction, involves the mass transference between two immiscible liquid phases, generally an aqueous phase and an organic solvent. This transference is driven by the differences in solubility of the chemical species involved in the two phases.

Rapid extraction, adaptability to a number of different solutes and solvents, the possibility of solvent recycling, enhancement in analytical selectivity and/or selectivity, ease of mechanisation and the possibility of exploiting different extraction strategies are favourable characteristics inherent to LLE [152]. Drawbacks relate to operator protection and environmental safety requirements, the need for toxic, harmful, flammable and/or volatile organic solvents, the formation of undesirable emulsions, solvent costs for some highly selective procedures and the requirement for large volumes of sample, extractant or solvent [153].

[1]The term *concentration* is used here in preference to the more commonly used term *preconcentration*.

Furthermore, the transference of a polar or charged solute involves dehydration; although trivial in some LLE procedures, this can be of paramount importance, particularly for the determination of metal ions [154,155].

Implementation of LLE in a flow analyser is advantageous, as originally and independently demonstrated during the 1960's by Stanton and McDonald [156] and by Pentz [157] in the spectrophotometric determination of zinc in soils and sediments and taurine in animal urine, respectively, using a segmented flow analyser. In 1978, LLE was implemented in flow injection analysis independently by Karlberg and Thelander [158], Bergamin-Filho *et al.* [159] and Kina *et al.* [160].

8.5.1.1. *The Classical Manifold Architecture*

This architecture (Fig. 8.11) was dominant during the nineteen seventies and eighties, and was also exploited when LLE was implemented in flow injection analysis. Nowadays, it is still widely used, as emphasised in recent comprehensive reviews [153,161]. Further innovations involving the use of a supported liquid membrane between two aqueous phases [162] and/or taking advantage of the high versatility of the unsegmented flow systems have also been proposed.

FLOW DIAGRAM

The sample and carrier/wash solutions are aspirated as in ordinary air-segmented flow systems (see 5.1.1). The main stream is segmented by air and merges with the solvent stream, yielding a segmented aqueous–organic mixture that is directed towards the extraction reactor. In applications involving analyte conversion to a chemical species that is highly soluble in the organic phase, a reagent is either added by stream confluence prior to LLE or otherwise be present in the organic phase. Analyte transference to the organic phase takes place during passage of the aqueous–organic mixture through the extraction reactor. Thereafter, the segmented sample reaches a phase-separating unit, and the organic phase is directed towards the flow-through detector.

The classical manifold architecture is fully compatible with segmented flow analysis. It has also been exploited in flow injection analysis, but the inherent segmentation alters the characteristics of the analyser [163].

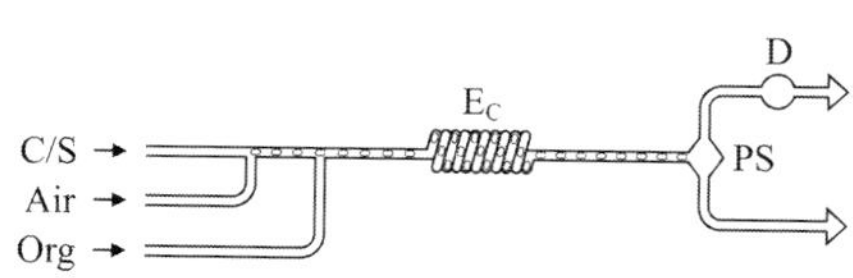

FIGURE 8.11 **A segmented flow system exploiting LLE designed with the classical manifold architecture.** C/S = carrier/sample stream; Air = air; Org = lower density organic solvent; E_C = extraction reactor; PS = phase separating device; D = detector.

FLUID PROPULSION

The streams are pumped as in ordinary flow systems, and peristaltic and syringe pumps (6.1.1 and 6.1.2) are most commonly used. Difficulties in establishing a constant flow rate for the organic solvent were highlighted in the pioneering implementations of LLE in flow analysis [156,157].

For the determination of zinc in soils and sediments [156], benzene was used as the organic solvent into which dithizone was dissolved and colour formation occurred inside the organic phase. A displacement bottle was used to propel the benzenic dithizone solution. This device is a large closed chamber initially filled with the organic solvent. A water stream delivered by the pump enters the displacement bottle via a small inlet orifice and displaces the organic solvent via an outlet orifice connected to the manifold transmission line. This outlet orifice is located in the bottom or the top of the chamber, depending on whether the solvent density is higher or lower than water. With this approach, the solvent does not interact with the pump tubing and a steady flow rate is attained. It is good practice to start the operation of the displacement bottle by filling it with a two-phase (aqueous and organic) fluid. Drawbacks related to phase miscibility effects are minimised because emulsion formation is less pronounced when water is added to an already water-equilibrated organic solvent.

Another possibility is to use solvent-tolerant tubing (Table 6.1), as demonstrated in the landmark article dealing with the determination of taurine in urine [157]. A Solvaflex™ pumping tube was used for directly pumping the chloroform solvent and only needed to be replaced only after at least one week of operation.

THE EXTRACTION PROCESS

In flow systems with the classical manifold architecture, the flow pattern inside the coiled extraction reactor is analogous to that shown in Fig. 5.2. Hence, mass transference occurs mostly through the thin liquid film on the inner wall of the tube and is favoured by the establishment of vortices (5.1.1). This film is predominantly composed of the aqueous or the organic phase, depending on the nature of the tubing and the hydrophobicity of the phases [163]. The continuous renewal of the surface available for mass transference is another favourable characteristic of this manifold architecture. Moreover, the mass transference can be accelerated by immersing the coiled extraction reactor inside a warm thermostated bath [164] or by exploiting ultrasound irradiation [144]. With careful selection of the organic phase and optimisation of the system dimensions, the mass fraction of the analyte transferred is typically 80–99% [165].

It should be stressed that mass transference through the liquid film on the tube wall is inherent to the classical manifold architecture and also holds for other architectures, as discussed below.

PHASE SEPARATION

The aqueous–organic mixture is continuously added to the phase-separating unit positioned after the extraction reactor. Quantitative separation of the immiscible phases is not attained, but generally >80% of the analyte mass is extracted [166] and directed towards the detector. Three different driving forces have been exploited for phase separation in relation to the classical manifold architecture:

1. **Gravity.** The segmented flowing stream enters a chamber (Fig. 8.12, upper) where the phases are physically separated [159] due to their immiscibility and different densities. The phase of interest, usually the organic one, is aspirated from this chamber towards the detector through an outlet orifice located at the chamber bottom or top, depending on the solvent density. The aqueous phase, together with residues of the organic phase, is discarded at a flow rate determined by the difference between the flow rates of the segmented inlet stream and organic outlet stream. Discarding a small fraction of the aqueous phase together with the organic phase leads to a rugged and efficient separation.
2. **Gravity/affinity**. Turnings or discs made from hydrophobic (or hydrophilic) materials can be placed on one side of the tube reaching the phase-separating unit (Fig. 8.12, centre) in order to assist the phase separation [158,166]. Consequently, a low inner-volume T-shaped separation unit can be designed, thus improving system performance.

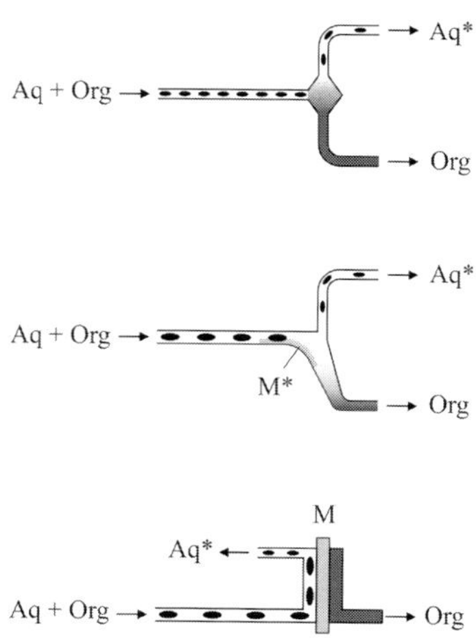

FIGURE 8.12 **Phase-separating devices relying on gravity (upper), gravity/affinity (centre) and porosity (bottom).** The figure refers to a flow system where the thin film on the tubing inner wall is aqueous (colourless) and the organic phase (black) is monitored; Aq + Org = aqueous/organic mixture; Org = organic phase; Aq* = aqueous phase with residual organic phase; M* = controlled-porosity turnings, sheets or disks; M = membrane; arrows = stream directions. For details, see text.

3. **Porosity**. The segmented flowing stream is forced to cross a porosity-controlled membrane (Fig. 8.12, bottom) through which only the organic phase is ideally allowed to flow [167]. Phase separation can be improved by the use of a double membrane phase separator [168] and/or by taking advantage of chromate membranes [161,169]. Quantitative separation is however not attained due to the slow nature of this process, and the need to discard a significant amount of the organic phase with the aqueous phase. This approach has often been exploited, as it is inherently simple, involves low inner volume and is independent of the solvent density. Long-term variation in membrane porosity caused by the inlet of the aqueous phase into the membrane is however the main limitation of this approach, but this can be minimised by cleaning the separating unit from time to time.

QUANTITATIVE ASPECTS

Flow systems with a classical manifold architecture for LLE generally have the same dimensions as ordinary flow systems. In general, a high sample volume is introduced in order to minimise sample dispersion inside the chamber-like manifold components such as the phase-separating chamber and the flow-through detector. This volume is however limited by the concomitant drop in sampling rate (see 5.2.3.2 and Fig. 5.8).

The total flow rate should be compatible with the rates of the chemical reactions and extraction involved. However, this parameter cannot be increased at will as it could impair phase separation.

The length of the coiled extraction unit has little influence on sample dispersion, due to the flow segmentation involved (5.1.1). Hence, this length should be increased in analytical procedures where the extraction step is relatively slow.

The aqueous-to-organic flow rate ratio is generally the most important parameter for determining the analyte distribution between phases and should be as high as possible if high sensitivity is aimed. Beyond a threshold value, any increase in this parameter does not enhance sensitivity because the mass transference also depends on the distribution coefficient [170]. This effect can be minimised by selecting an appropriate organic solvent. An increase in the aqueous-to-organic flow rate ratio is also restricted by other limiting factors:

- With too high an aqueous flow rate, the mean available time for extraction is reduced and mass transference might be negatively affected; moreover, phase separation might be impaired.
- With too low an organic flow rate, the mean length of the organic segments inside the coiled extraction unit is reduced; although this is beneficial in terms of an increased area for mass transference, reducing the segments might lead to a plug coalescence and the consequent

establishment of a continuous solvent film on the tubing inner wall, thus impairing extraction [166]. Moreover, the sampling rate might be unacceptably low, especially if the manifold components after the phase-separating device have a significant inner volume.

- With too high an aqueous-to-organic flow rate ratio, the stability of the organic segments is critical and can lead to a drop in measurement repeatability. Moreover, phase miscibility might be a limiting factor in system design; in the extreme situation, the organic phase might form an emulsion in the aqueous phase.

8.5.1.2. *Manifold Architecture Including Membrane Extraction*

In this architecture (Fig. 8.13), the solvent is immobilised in a liquid membrane and sandwiched between two aqueous (donor and acceptor) streams [162]. Drawbacks related to a flowing organic stream are therefore circumvented and the resulting flow system is generally very simple and rugged. Only a few solvents however can be immobilised in a liquid membrane [171]. An important characteristic of this architecture is that the phase-separation step is not required.

FLOW DIAGRAM

The sample is introduced into an unsegmented flowing carrier stream and any reagent needed for analyte conversion to a chemical species that is highly soluble in the organic phase is added by stream confluence prior to LLE. When passing close to the membrane, the analyte diffuses through the organic phase immobilised inside the liquid membrane towards the acceptor stream. The collected species is thereafter directed towards the flow-through detector. Analytical procedures relying on back extraction are then easily implemented.

This manifold architecture is more compatible with unsegmented flow analysis. Only aqueous flowing solutions are involved and hence the streams are pumped as in ordinary flow systems.

THE EXTRACTION PROCESS

In flow systems with a liquid membrane for analyte extraction into the immobilised organic solvent, extraction is still governed by the relative

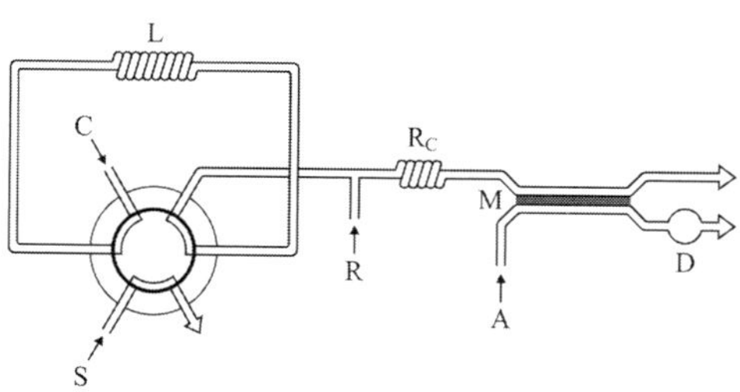

FIGURE 8.13 **A flow system with membrane extraction.** S = sample; L = sampling loop; C = carrier stream; R = optional make-up reagent; R_C = coiled reactor; A = acceptor stream; M = liquid membrane; D = detector. For details, see text.

solubilities of the chemical species involved (the analyte or its conversion product) in the donor stream and in the organic phase. This process is accelerated by removing the analyte from the organic medium towards the acceptor stream; this is best accomplished by appropriate selection of the nature and composition of this stream [162,172]. Hence, mass transference occurs between two aqueous phases via a liquid membrane. A thin layer of organic solvent can be established by immobilising the organic solvent within the pores of disposable hollow fibres, usually made of polypropylene [171], and this approach can potentially be exploited in flow analysis.

QUANTITATIVE ASPECTS

A typical manifold for analytical procedures relying on membrane extraction (Fig. 8.13) is analogous to that used for in-line dialysis (see 8.5.4). Consequently, the main parameters involved, *e.g.*, the geometry of the analytical path, the sample volume introduced, flow rates, timing and membrane characteristics (especially porosity and thickness), have an analogous influence on the analytical figures of merit.

If sensitivity is critical, the sample volume should be increased but the compromise with sampling rate (5.2.3.2) should be taken into account. Flow rates should be compatible with the manifold geometry, the rates of the chemical reactions involved and analyte diffusion. The flow system can be designed with counter-current acceptor and donor streams. The donor-to-acceptor streams flow rate ratio should be as high as possible but, again, any deterioration in sampling rate and/or in completeness of the processes involved should always be taken into account.

8.5.1.3. Manifold Architectures Without Stream Segmentation and/or Phase Separation

A noteworthy feature of flow systems exploiting in-line LLE is that stream segmentation and/or phase separation is not necessarily required for analytical procedures involving enhanced detection capabilities, wetting film[2] extraction and/or the establishment of a cloud point, as discussed below.

ENHANCED DETECTION CAPABILITIES

Initial attempts to perform LLE in flow analysis without phase separation were carried out by Kina *et al.* [160] who designed a flow injection system for the fluorimetric determination of potassium involving extraction into 1,2-dichloroethane containing a macrocyclic compound. The aqueous sample was injected into an unsegmented organic flowing

[2]The expressions *wetting film* and *wet film* have been indistinctly used. In this monograph, *wetting film* is used throughout.

stream and potassium underwent wetting film extraction. As the ion associate formed was almost nonfluorescent in water and strongly fluorescent in 1,2-dichloroethane, phase separation was not required. This was the first solvent extraction/on-tube detection system without any phase separation [161].

Another possibility is to use a detector able to discriminate measurements related to the different phases involved, *i.e.*, to monitor both the organic and the aqueous plugs. This was elegantly demonstrated in a model system with spectrophotometric detection involving ion pairing of anionic surfactants with methylene blue [173]. A flow-through cuvette was not required and measurements were made directly on the segmented flowing zone, in a geometry now called *on-tube measurements* or *capillary flow cell*. To this end, the incident radiation beam was perpendicular to the monitor and segments. A similar approach was used for pesticide determinations [174] and its performance was compared with that of an ordinary flow system with the classical LLE manifold geometry. These flow systems are characterised by a very low detector inner volume. The aqueous and organic phases are simultaneously monitored, so these flow systems can also be used to investigate extraction kinetics.

With this manifold architecture, however, a pronounced Schlieren effect was observed at multiple aqueous/organic interfaces. Use of dual-wavelength spectrophotometry allowed both the compensation of the effect and the confirmation of solvent arrival at the detection site [175]. Other flow systems exploiting measurements at three wavelengths, auxiliary detection for plug recognition or detector ON/OFF switching for monitoring only one phase have also been proposed [166].

This manifold architecture has been more widely used with flame atomic absorption spectrometry and other detection techniques using a nebuliser [176] which are less affected by the Schlieren effect. The reported sensitivity and/or selectivity enhancements are due to the beneficial influence of solvent addition on the nebuliser efficiency. The nebuliser also acts as a phase-separating device, predominantly directing one phase towards detection and the other towards waste; surprisingly, this latter aspect has not been fully investigated.

WETTING FILM EXTRACTION

The main characteristic of the classical manifold architecture for in-line LLE (Fig. 8.11) is that analyte extraction occurs mostly via the thin liquid film established on the inner wall of the tube, which is continuously renewed. Phase separation however is often a limiting factor in determining system performance. Nonetheless, the flow system can be designed to permit segmentation of the sample zone by the confluent organic stream, analyte extraction in the downstream reactor and formation of a single phase by the confluent addition of a specific solvent (Fig. 8.14) which is thereafter homogenised and directed towards the detector. Analyte transference between the phases occurs mostly via the thin film established on the tubing inner wall; therefore, the process is often referred to as wetting film extraction.

As the analyte enters the thin extracting film, its mean linear speed is reduced so that, when the single phase is formed, the analyte zone is already separated from the main sample zone. The specific solvent yielding the single phase consists of an organic liquid soluble in both phases [177]. With this architecture, it is possible to record two peaks, the first one relating to a chemical species present in the sample zone and the second one reflecting the analyte content. The feasibility of this approach was originally demonstrated in the spectrophotometric determination of active components in motion sickness tablets [178] involving analyte extraction in hexanol and single-phase formation after methanol addition.

An interesting possibility for improving the extraction efficiency is to exploit flow reversal [179–181] during analyte collection. With this approach, the aqueous phase in contact with the organic solvent is continuously renewed, leading to an increase in the aqueous-to-organic phase volumetric ratio. If the solvent plug undergoing forward/backward movement is monitored in a specially designed flow-through tubular cuvette, several peaks are recorded, each one related to a given sample residence time and a different extraction efficiency. This approach has been called *multi-peak recording*.

This concept led to the implementation of wetting film extraction in a model sequential injection system [182]. The solvent, followed by an air plug, was initially aspirated towards the holding coil, and a thin stationary organic wetting film was established on the wall of the extraction reactor. The sample plug was then introduced and the analyte was extracted into the coated solvent. After further addition of air, the single-phase forming stream, acting as an eluting solution, was introduced in order to displace the analyte towards the detector. An important feature of this architecture is that both the extractable and unextractable components of the sample can be efficiently separated and

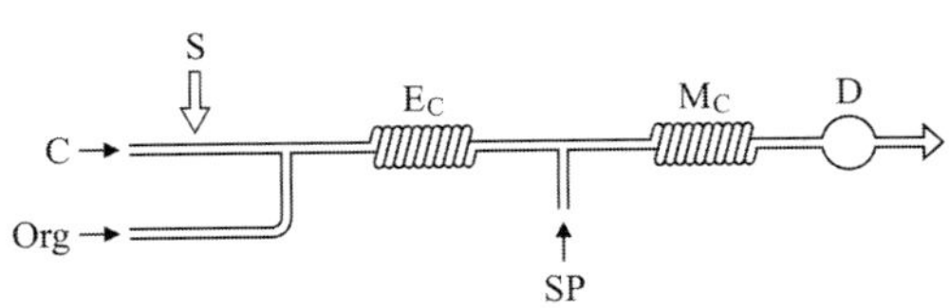

FIGURE 8.14 **Flow diagram of a flow injection system with wetting film extraction.** S = sample; C = aqueous carrier stream; Org = organic solvent; E_C = extraction coiled reactor; SP = solvent added for single phase establishment; M_C = mixing coiled reactor; D = detector. For details, see text.

simultaneously determined. Moreover, the consumption of organic solvent is minimal and stream segmentation and phase separation are not required.

Sample throughput can be increased by establishing two wetting films inside two parallel reactors and this innovation was demonstrated in the spectrophotometric determination of nitrophenols in a multi-syringe flow system [183].

CLOUD POINT EXTRACTION

In view of the benefits of green chemistry, replacement of organic solvents by less toxic substances, *e.g.*, surfactants, is desirable and cloud point extraction is one way to achieve this [184]. A surfactant is added to the sample at a concentration higher than the critical micelle concentration to yield an organised medium consisting of an anisotropic phase which is often considered as a two-phase system. The analyte is concentrated in the surfactant-rich phase through hydrophobic or hydrophilic interactions with the organised micellar assemblies. The efficiency of the concentration step can be improved by raising the temperature or by exploiting the salting-out effect, and the stability of the assembly is improved by cooling the reaction medium after the concentration step; centrifuging can be beneficial for separating the surfactant-rich phase. Thereafter, an aliquot of this phase is directed towards the detector. The pioneering work implementing cloud point extraction in flow injection analysis focussed on the determinations of coproporphyrin and gadolinium in urine samples by chemiluminescence [185] and by ICP-OES [186], respectively. The sample, complexant reagent and nonionic surfactant solutions were mixed together and the micellar medium was heated by immersing the main reactor in a thermostated water bath at 25 °C; thereafter, the surfactant-rich phase was retained in cotton wool, and the analyte was directly towards an ICP-OES detector. This strategy is an excellent alternative to LLE in flow analysis [185,187,188] and some applications are presented in Table 8.2.

8.5.1.4. *Other Manifold Architectures*

There are situations where specific strategies are needed to improve system performance, some of which are outlined below.

USE OF CHAMBERS

A chamber is a useful manifold component in some applications, *e.g.*, the spectrophotometric determination of copper with APDC (ammonium pyrrolidinedithiocarbamate) involving the production of the Cu-APDC complex and extraction into chloroform [189]. A dual-channel dropping segmentor was used for the dropwise addition of segments of the sample and reagent solutions into an unsegmented flow of organic solvent.

TABLE 8.2 Selected Applications of Liquid–Liquid Extraction

Analyte	Sample	Solvent	Detection Technique	Analytical Range or Detection Limit	Remarks	Ref.
2-, 3-, 4-Nitrophenol	Seawater, waste water	2:1 (v/v) butylchloride : 1-octanol	UV–Vis	36.1, 45.9, 4.9 $\mu g\ L^{-1}$	Sequential injection system; wet film extraction; acetonitrile as washing agent	[446]
2-, 3-, 4-Nitrophenol	Tap, waste, ground, sea water	2:1 (v/v) 1-chlorobutane : 1-octanol	UV–Vis	0.11, 0.46, 0.07 $\mu mol\ L^{-1}$	Multi-syringe flow system; dual wetting film; multivariate calibration	[183]
8-Chlorotheophylline, diphenhydramine	Dramamine tablets	Cyclohexane	UV–Vis	0.25–2.50 $mmol\ L^{-1}$ for both analytes	C-FIA; membrane for phase separation; both phases monitored for simultaneous determinations	[447]
^{90}Sr	Environmental samples	BCHC in octanol	Liquid-scintillation	1.9 Bq	Sequential injection system; wet film extraction; elution with acetonitrile	[448]
Aluminium	Potable, river, waste waters	Chloroform	Fluorimetry	0.2 $\mu g\ L^{-1}$	Comparison between different manifold architectures	[449]
Anionic surfactants	River water	Toluene + MIBK	UV–Vis	18 $\mu g\ L^{-1}$	C-FIA; membrane for phase separation; ion association with MG	[450]

Table 8.2 Selected Applications of Liquid–Liquid Extraction—cont'd

Analyte	Sample	Solvent	Detection Technique	Analytical Range or Detection Limit	Remarks	Ref.
Anionic surfactants	Water	Chloroform	UV–Vis	0.08–10 μg mL^{-1}	Flow injection system; sample "infinite volume"; ion pairing with methylene blue; flow reversals for widen the concentration dynamical range; no phase separation	[179]
Antidepressant drugs	Pharmaceutical formulations	Dichloromethane	Fluorimetry	0.25–3.0 mg L^{-1}	C-FIA; membrane for phase separation; ion association with DMAS	[451]
Barbiturates, serotonin re-uptake inhibitors	Urine	1-octanol : butylchloride /alkaline phosphate	HPLC	0–25 μg mL^{-1} for phenobarbital, amobarbital, pentobarbital, secobarbital	SIA system; back extraction; phase separation by differential flow velocities of the aqueous and organic phases	[452]
Benzalkonium, cetylpyridinium (quaternary ammonium salts)	Pharmaceutical formulations	1,2-dichloroethane	UV–Vis	5.0×10^{-7}–2.0×10^{-6} mol L^{-1}	C-FIA; membrane for phase separation; ion association extraction and thermochromism of the ion associates	[453]
Berberine	Pharmaceutical preparations	1,2-dichloroethane	Fluorimetry	4×10^{-9}–1×10^{-6} mol dm^{-3}	C-FIA; double membrane extraction; ion pairing with perchlorate	[454]
Bilirubin	Human serum samples	Triton X-114 (surfactant) + Na_2SO_4 (salt)	CL	1.8 μg L^{-1}	Flow injection system; cloud point extraction	[455]

(*Continued*)

Table 8.2 Selected Applications of Liquid–Liquid Extraction—cont'd

Analyte	Sample	Solvent	Detection Technique	Analytical Range or Detection Limit	Remarks	Ref.
Bromothymol blue, methylene blue (model system)	Synthetic solutions	Benzene	UV–Vis	12.5 μg L^{-1}	Sequential injection system; wetting film extraction	[182]
Cadmium, lead	Pottery glaze, petrol	Carbon tetrachloride	UV–Vis	50 ng mL^{-1}	C-FIA; membrane for phase separation; dithizone (the complexant agent) in the organic phase	[456]
Cadmium	Synthetic solutions	Chloroform	UV–Vis	85 ng L^{-1}	Mono-segmented flow system; wetting film extraction; no phase separation; 1-(2′-pyridylazo) naphthol in the organic phase	[195]
Calcium	River water	Benzene : chlorobenzene	UV–Vis	2.0×10^{-7} mol L^{-1}	C-FIA; membrane for phase separation; ion association between Ca-crown complex and propyl orange	[457]
Chlorobenzenes	Water	Toluene	GC	1.0 μg L^{-1}	Liquid–liquid micro-extraction in a single organic drop (see 2.6); injection into the gas chromatograph	[458]
Copper	Plant digests	Carbon tetrachloride	UV–Vis	5.0 μg L^{-1}	C-FIA with an open-phase separation chamber; dithizone (the complexant agent) in the organic phase	[459]

Table 8.2 Selected Applications of Liquid–Liquid Extraction—cont'd

Analyte	Sample	Solvent	Detection Technique	Analytical Range or Detection Limit	Remarks	Ref.
Copper	Bovine liver, river water, lake water	MIBK	UV–Vis	5.0 μg L^{-1}	Flow system with gaseous carrier streams; micro-batch chamber for chemical reactions, extraction and phase separation; l-(2′-pyridylazo) naphthol in the organic phase	[460]
Coproporphyrin	Urine	Triton X-114 or X-100 (surfactant) + $(NH_4)_2SO_4$ (salt)	CL	2.0 μg L^{-1}	Flow injection system; cloud point extraction	[185]
Cr(VI), Cr(III)	Natural waters	Octanol :/4-methyl-2-pentanone	UV–Vis	2.0 μg L^{-1}	Sequential injection system; wet film extraction; elution with acetonitrile; Cr(III) previously oxidised to Cr(VI) for sequential determinations	[461]
Enalapril	Pharmaceutical formulations	Dichloromethane	UV–Vis	1.5–60 μg mL^{-1}	C-FIA; membrane for phase separation; ion pair formation with BTB	[462]
Erythromycin	Pharmaceutical formulations	Chloroform	Fluorimetry	0.68 μg mL^{-1}	C-FIA; membrane for phase separation; ion pair formation with ETB	[463]
Fe(II), Fe(III), Al(III)	Natural waters	Chloroform	UV–Vis	0.06 mg L^{-1} (for all analytes)	C-FIA; membrane for phase separation; kinetic discrimination and multivariate calibration for speciation	[464]

(*Continued*)

Table 8.2 Selected Applications of Liquid–Liquid Extraction—cont'd

Analyte	Sample	Solvent	Detection Technique	Analytical Range or Detection Limit	Remarks	Ref.
Fe(III)	Tap, waste water	Chloroform	UV–Vis	29.0 μg L^{-1}	C-FIA with merging zones; membrane for phase separation; different ion pairs tested	[465]
Free cyanide, thiocyanate	Human saliva and pralidoxime solutions	Chloroform	UV–Vis	0.004, 0.007 mg L^{-1}	C-FIA; membrane for phase separation; formaldehyde addition for cyanide masking	[466]
Glycerol (free, total)	Biodiesel	$NaIO_4$ plus acatylacetoneaqueous-ethanol solution	UV–Vis	0.0028, 0.0310% (m/m)	Flow injection system without phase separation; ethyl acetate for free – bound glycerol discrimination; ultrasound-assisted extraction	[431]
Gold, thallium	Coal fly ash, ores	Benzene	UV–Vis	0–5 mg L^{-1} (for both analytes)	C-FIA; in-line SPE for separating the analytes as chloro-complexes; membrane for phase separation	[467]
Histamine	Whole blood	n-heptane	Fluorimetry	1–30 ng mL^{-1}	Air-segmented flow system; gravity for phase separation	[468]

Table 8.2 Selected Applications of Liquid–Liquid Extraction—cont'd

Analyte	Sample	Solvent	Detection Technique	Analytical Range or Detection Limit	Remarks	Ref.
Lead	Plant digests	Carbon tetrachloride	UV–Vis	12 μg L^{-1}	Multi-commuted flow system; air as carrier stream; dithizone (the complexant agent) in the organic phase; gravity-based phase separation	[469]
Lidocaine	Pharmaceutical formulations	Chloroform	UV–Vis	1.4–16.2 μg mL^{-1}	C-FIA; membrane for phase separation; ion association with BCG	[470]
Molybdenum	Aqueous samples	Toluene/THAB, methanol/DPC	UV–Vis	2.5×10^{-8} mol L^{-1}	sequential injection system; wetting film extraction; back extraction; acetone as washing agent	[471]
Neostigmine	Pharmaceutical formulations	TBPE-H/1,2-dichloroethane	UV–Vis	1.8×10^{-8} mol L^{-1}	C-FIA; ion pairing with TPBEH; double membrane phase separation	[472]
Nonionic surfactants	Synthetic solutions	1,2-dichloroethane	UV–Vis	2–60 mg L^{-1}	C-FIA; ion pairing with TBPE-K; improved phase separator based on gravity	[473]
Oil products, phenols	Natural waters	Hexane and tributyl phosphate	UV–Vis, fluorimetry	0.5–100 μg L^{-1}	Flow injection system; possibility of "infinite volume" sample; exploitation of chromatomembrane	[474]

(*Continued*)

Table 8.2 Selected Applications of Liquid–Liquid Extraction—cont'd

Analyte	Sample	Solvent	Detection Technique	Analytical Range or Detection Limit	Remarks	Ref.
Palladium	Airborne particulate matter, automobile catalysts	Chloroform	UV–Vis	0.007 mg L^{-1}	C-FIA with zone stopping; membrane for phase separation	[475]
Partition coefficients (octanol/water)	Vanillin, acetaminophen, metoprolol	Octanol	UV–Vis	–	Wet film extraction in a capillary flow system; measurements in both phases	[192]
Perchlorate	Salt samples	MIBK	UV–Vis	0.003 μg mL^{-1}	C-FIA; ion pairing with BCB	[476]
Phosphorus (as orthophosphate)	River waters	Benzene : MIBK	UV–Vis	0.1 ng mL^{-1}	C-FIA; ion pairing with MG; membrane for phase separation; detailed study of segmenter geometry	[477]
Picric acid	Waters, pharmaceutical formulations	Ethyl acetate	UV–Vis	1.06 mg L^{-1}	Sequential injection system; independent extraction and detection units; phase separation by gravity; flow reversal for improving extraction; ion association with MPVIL	[478]
Potassium	Synthetic samples	1,2-dichloroethane with dibenzo-18-crown-6 macrocyclic compound	Fluorimetry	2×10^{-4}–1×10^{-3} mol L^{-1}	Original flow system without phase separation	[160]
Potassium	River, tap water	1:1v/v benzene : chlorobenzene	UV–Vis	1×10^{-6} mol L^{-1}	C-FIA; membrane for phase separation; K-crown ether associated with anionic azo dye	[479]

Table 8.2 Selected Applications of Liquid–Liquid Extraction—cont'd

Analyte	Sample	Solvent	Detection Technique	Analytical Range or Detection Limit	Remarks	Ref.
Rhenium	Ores	Benzene	UV–Vis	4 mg kg^{-1} (dry basis)	C-FIA; membrane for phase separation; prior crown ether formation for Mo(VI) separation	[480]
Sb(V)	Anti-leishmaniotic drugs	Toluene	UV–Vis	7.0×10^{-5}–7.2×10^{-4} mol L^{-1}	Multi-commuted flow system; ion pair formation between hexachloroantimoniate anion and rhodamine B; gravity for phase separation	[481]
Solasodine	Fruits	Chloroform	UV–Vis	1.41 μg mL^{-1}	Sequential injection system; gravity for phase separation inside a conical chamber	[482]
Steroid and bile acid sulfate	Biological samples	1,2-dichloroethane	CL, fluorimetry	0.5 pmol L^{-1}(CL), 25 pmol L^{-1} (fluorimetry)	C-FIA; membrane for phase separation; ion pairing with lucigenin	[483]
Thiamine(vitamin B_1)	Pharmaceutical formulations	Chloroform	Fluorimetry	1 μg L^{-1}	C-FIA; organic solvent injected into the flowing sample (single solvent plug); no phase separation	[484]
Uranium	Ore leachate	Tributyl phosphate in heptane	UV–Vis	0.1 mg L^{-1}	C-FIA; gravity for phase separation	[485]

UV–Vis = spectrophotometry; CL = chemiluminescence; C-FIA = flow injection system with the classical manifold architecture; GC = gas chromatography; LC = liquid chromatography; BCB = brilliant cresylblue; BCG = bromocresol green; BCHC = 4,4′-(5′)-bis(*tert*-butylcyclohexano)-18-crown-6; BTB = bromothymol blue; DMAS = 9,10-dimethoxyanthracene-2-sulphonate; ETB = erythrosine B; MG = malachite green; MIBK = methylisobutyl ketone; MPVIL = 2-[2-(4-methoxy-phenylamino)-vinyl]-1,3,3-trimethyl-3H indollium; TBPE-H = tetrabromophenolphthalein ethyl ester; TBPE-K = tetrabromophenolphthalein ethyl ester, potassium salt.

A conical chamber acted as a confluence connector for the convergent solutions, also allowing the delivery of the organic phase towards on-tube detection. A similar approach was implemented in a lab-on-valve system for the determination of anionic surfactants with methylene blue [190]. The sample, reagent and organic solvent were sequentially aspirated into the holding coil, which also acted as the extraction reactor; better extraction efficiency was achieved by several flow reversals. Thereafter, the segments were directed towards a conical phase separator attached to the selection valve, where the aqueous and organic phases were separated and the organic phase in the lower section of the conical-type phase separator was monitored.

The exploitation of a mini-chamber is the essence of the flow-batch system (5.5.3), and hence the implementation of LLE in this system is best accomplished with this device. This was demonstrated in the spectrophotometric determination of cocaine relying on the formation of a cocaine–cobalt thiocyanate complex and extraction into chloroform [191]. The solutions involved were inserted into a reaction chamber and an optical fibre allowed absorbance measurements without the need for phase separation. Good analytical figures of merit were reported.

A chamber can also be used with micro-batch wise extraction to gather physico-chemical data, *e.g.*, the determination of octanol/water partition coefficients [192]. The chamber was added to the analytical path to mimic one inherent to the analogous LLE batchwise procedure.

Improved sensitivity can be achieved by keeping the aqueous phase inside the chamber and adding the organic phase into it from the top; the extracted analyte remains on the bottom and is then pushed forwards towards the detector. The feasibility of this approach was demonstrated by the determination of anionic surfactants in natural waters by reaction with methylene blue and extraction into chloroform [193]. The reported analytical figures of merit were very good.

Alternatively, the organic phase can be maintained at the bottom of the chamber, a water stream with a large sample zone passes over it and the organic phase is monitored in chamber [194]. The resulting system is simple and rugged and the aqueous-to-organic phase volumetric ratio is efficiently optimised.

MONO-SEGMENTATION

A mono-segmented flow system (see 5.5.1) provides good conditions for organic/aqueous phase interaction; the extraction process is carried out between two air bubbles and vortices are established inside the plug. An efficient extraction, with reduced sample dispersion, low carryover and thus a high sampling rate, is achieved, as demonstrated in the spectrophotometric determination of cadmium [195].

Another possibility is to exploit single-phase extraction. The slightly miscible aqueous and organic phases are completely emulsified after addition of a ternary solvent, which is very soluble in both phases, yielding a homogeneous medium [177]. As a consequence, the extraction process does not depend on the geometry of the aqueous/organic phase interface. Thereafter, a suitable volume of one of the original phases (usually the aqueous phase) is added to separate both phases again. The phase with the concentrated analyte is then quantified. Improved in-line extraction of Fe(III) in a mono-segmented flow system was achieved with this approach [196].

AQUEOUS BIPHASIC EXTRACTION

In analytical procedures with two immiscible aqueous phases, the traditional organic solvents used in LLE can be replaced by nontoxic, nonflammable and nonvolatile solvents [154,155,197]. The feasibility of implementing aqueous biphasic extraction in flow analysis was recently demonstrated by the fluorimetric determination of lead by reaction with 8-hydroxyquinoline-5-sulfonic acid [198]. The flow system was rugged and the interface formed between the different aqueous plugs acted as an extraction (enrichment) and reaction interface. This innovation is environmentally friendly as it relies on the utilisation of two immiscible phases that are intrinsically aqueous.

ZONE SAMPLING

Zone sampling (see 7.2) can be exploited in some expert flow systems with LLE in order to improve their performance, as initially demonstrated by the determination of C_1-C_5 n−mercaptans in gasoline [199] (see also 8.7). Another example is the determination of ammonium by exploiting ion pairing [200].

8.5.1.5. *Applications*

The number of LLE applications in the field of unsegmented flow analysis grew rapidly between the mid-80s and the early years of the 21st century, with about 250 articles being published [153]. About 75% of the applications refer to *spectrophotometry*, as it is more compatible with an organic phase than other instrumental analytical techniques. LLE applications listed in Table 8.2 are predominantly associated with unsegmented flow analysis, as most of the recent applications involving air-segmented flow analysis have been published in less accessible sources.

8.5.2. Solid-Phase Extraction, SPE

Separation/concentration steps relying on SPE are efficiently implemented in flow analysis because successive additions of the sample, eluent and conditioning solutions are easily made and different manifold configurations can be exploited. Comprehensive reviews on the theme have recently been published [161,201]. Mini-columns packed with solid materials are most commonly used [202], and the solid phase is conditioned to permit selective separations of either the species to be determined or the potential interferents present in the sample, thus enhancing sensitivity and/or selectivity. As a rule, flow analysers with SPE are simple and robust, yielding reliable results quickly.

8.5.2.1. Similarities with Liquid Chromatography

The boundary between flow analysis exploiting SPE and liquid chromatography can be perceived as somewhat blurred [203], but there is a clear distinction between the two techniques due to the different operating conditions involved.

In a flow system including a mini-column for SPE, larger particles (usually >50 mesh <400) are used to keep the hydrodynamic pressure low (typically a few bar) and achieve good column packing. The flow analyser generally operates at ambient temperature and hence isothermal conditions are maintained. Moreover, counter-flow elution is preferred [204]. Each fluid element of the sample passes rapidly through the mini-column; consequently, the mean times available for sorption and elution are short (usually 1–120 s) and high sampling rates, often >100 h^{-1}, are achieved.

In view of the short timescale of the analytical cycle, step changes in the concentrations of species in the stream flowing through the mini-column, and the low pressure and temperature involved, several chemical species can be simultaneously retained and eluted, i.e. the process involves a *non-chromatographic separation* [205]. These are the main differences between flow analysis exploiting SPE and liquid chromatography.

As the sorption and elution steps are accomplished by changing the nature or the concentrations of the solutions involved, the formation of concentration gradients is unavoidable. These gradients can give rise to occurrences of the Schlieren effect [206], and this is the main reason why detection techniques less affected by the Schlieren effect, *e.g.*, atomic absorption/emission spectrometry, are most commonly used in flow systems with in-line SPE.

During the development of unsegmented flow analysis with SPE, several strategies involving sequential elution for inter-analyte separation leading to simultaneous determinations have been reported, *e.g.*, [207–210]. With the recent use of monolithic columns in flow analysis (sequential injection chromatography [211], see 5.3.1), the use of sequential elution is becoming more popular [212].

8.5.2.2. *The Solid Phase*

Synthetic ion-exchange resins [213], silica-based exchangers [214] and activated alumina [209] have been most commonly used; other materials used include cross-linked dextran gel (Sephadex) [215], fullerenes (especially C_{18} and C_{60})[216], polyurethane foam [217], PTFE turnings [218] or fibres [219], cigarette filters [220], functionalised cellulose [221], activated carbon [222], grape bagasse [223], rice husks [224] and bio-organisms (*Saccharomyces cerevisiae*) [225].

Prior conditioning of these materials is generally required, which depends on the specific analytical application involved [161]. In this regard, the immobilisation of 2-(6′-methyl-2′-benzothiazolylazo)chromotropic acid on polyurethane foam for the determination of copper in food samples [226], copper diethyldithiocarbamateon cigarette filters for mercury speciation in fish tissue [220] and 2-mercaptothiazole on silica gel for gold and palladium determinations in geological samples and alloys [214] are good examples. In some applications, the analyte is converted into a more suitable chemical species and the sorption of Zn and Cd as chlorocomplexes on an anion exchange resin [207,227] is a classical example. Exploitation of molecularly imprinted polymers is advantageous, as these materials permit highly selective analyte sorption [228,229]; moreover, chiral separations are possible. There are critical reviews that focus on the types of material suitable for SPE in flow analysis [213,230].

As a rule, the solid material is packed inside a mini-column; the simplicity of this setup is due to the ease of manufacture, manipulation and packing, as well as the practicality of inserting the mini-column at any site in the manifold.

Implementing SPE in flow analysis without mini-columns is also feasible. To this end, other materials, *e.g.*, commercially available ion-exchange or chelating extraction disks [231], disposable extracting cartridges [232], polymer-composite membranes [233] and monolithic columns [212], have been used. Analytical procedures exploiting addition, transport and disposal of the solid particles, analyte sorption/elution in flow reversal, and the use of fluidised beads have also been proposed, as highlighted at the end of this section.

8.5.2.3. *The Mini-column*

The mini-column should be as small as possible to minimise dispersion and increase sample throughout, yet large enough to ensure efficient mass transfer. The aspect ratio (length:internal diameter) is therefore an important parameter to optimise. In general, a mini-column is assembled in a glass or plastic cylindrical container (length typically 2–10 cm, i.d. typically 1–3 mm). Special materials such as foams, porous membranes, sintered glass or glass wool are placed at the ends of the mini-column to avoid losses of the solid phase during operation of the flow system. Sample filtration is sometimes required, and the mini-column can provide this simultaneously with in-line analyte separation/concentration or sample clarification steps, although pre-filtration is recommended to maximise column lifetime.

The volume of the liquid phase should be kept as small as possible relative to the volume occupied by the solid particles; otherwise, the mini-column might behave as a chamber-like element, leading to a pronounced dispersion of the analyte zone (see 3.1.2.2). To minimise the liquid volume, the size of the solid particles and/or the inner volume of the mini-column should be reduced. The use of smaller solid particles is also beneficial for increasing the total particle surface area available for SPE.

The particle size cannot be reduced indefinitely however due to increasing hydrodynamic pressure and difficulties in column packing. The use of small particles can also lead to the formation of preferential pathways, through which the sample flows rapidly, thus reducing contact between the aqueous and solid phases. Moreover, limitations associated with particle swelling [204] and the increased hydrodynamic pressure due to progressively tighter particle packing with column use [234] are also important considerations. Elution of the analyte in the opposite flow direction to loading is therefore a recommended practice.

8.5.2.4. *Implementation*

Different manifold geometries have been proposed for implementing in-line SPE in flow analysis, generally involving the placement of a mini-column before, at or after the sample introduction port. Alternatively, by integrating SPE within the detection unit, moving beads, hyphenated techniques and/or commutation facilities can be exploited, as discussed at the end of this section.

MINI-COLUMN PRIOR TO SAMPLE INTRODUCTION INTO THE ANALYTICAL CHANNEL

In segmented flow analysis, SPE is generally used for selectivity enhancement and the spectrophotometric determination of sulphate in seawater [235] is a good example. The sample was passed through a cation-exchange resin mini-column where the potential interfering species (mainly Ca^{2+} ions) were retained. The modified sample zone eluting from the mini-column was directed to the main analytical channel (Fig. 8.15) where it was segmented and further processed. The flow system was rugged because the ion-exchange process did not require sample segmentation or retain the analyte. The column was relatively

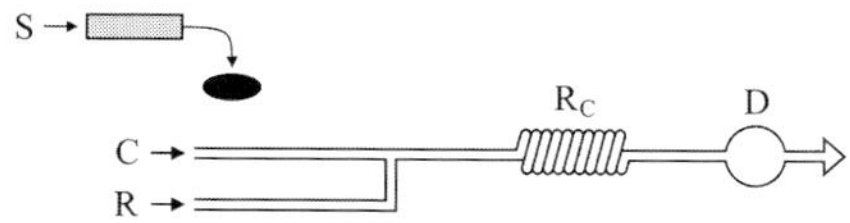

FIGURE 8.15 Didactic representation of a flow system with in-line SPE prior to sample introduction. S = sample; C = sample carrier stream; R = reagent; R_C = coiled reactor; D = detector; box = mini-column; black ellipse = depleted sample. Air stream and de-bubbler not shown, as the representation also holds for unsegmented flow analysis.

large, sample throughout was low and the concentration of potential interfering species was also low; hence, in-line elution of the retained cations to regenerate the column was not necessary.

In contrast, most applications involving SPE do require elution of the retained chemical species. To this end, the carrier/wash stream can also act as the eluent. Potential interfering species are eluted after the sample has passed through the mini-column, and the main stream is immediately segmented, thus avoiding any interaction between the modified sample and the separated potential interferents. Another strategy is to intercalate cups containing the eluent solution among the sample cups in the sampler tray. Removal of the retained chemical species is then accomplished after every sample assay (or after n samples have been run). Elution can also be accomplished off-line; from time to time, the column is removed from the manifold, the eluent is directed to flow through the mini-column, the retained species are eluted to waste, a wash solution is directed to flow through the mini-column for washing/conditioning purposes and the column is placed again in the flow manifold. This procedure is somewhat cumbersome, but the main advantage is that the eluted chemical species do not pass through the analytical path or the detector.

Although less frequently exploited, placement of a mini-column before the main analytical channel can also be used for in-line analyte concentration in segmented flow analysis. The sample flows through the mini-column, where the analyte is retained, and the carrier/wash stream also acts as the eluting/conditioning solution. Increasing the sample flow rate and/or the available time interval for analyte sorption improves the sensitivity.

Placement of the mini-column prior to the sample introduction port can also be useful in unsegmented flow analysis, as demonstrated in the flow injection spectrophotometric determination of molybdenum in plant digests [236]. This method relied on the influence of molybdate anions on the rate of iodide oxidation by hydrogen peroxide under acidic conditions, and some potential interfering cations needed to be separated. To this end, a cation-exchange resin mini-column was attached to the central sliding bar of an injector-commutator; hence, it could be displaced between two different streams. Initially, the sample solution was pumped

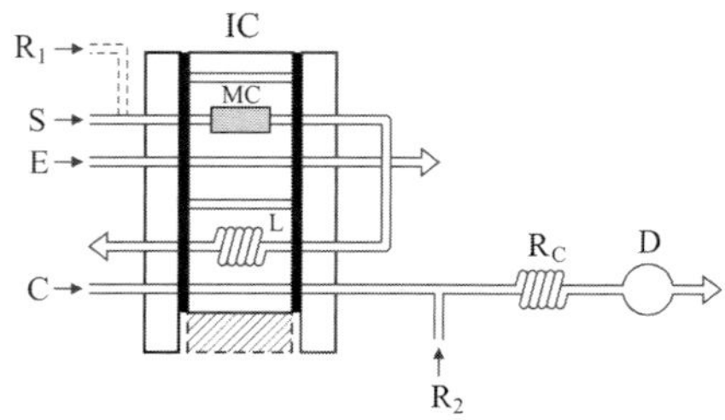

FIGURE 8.16 **Flow diagram of an unsegmented flow system with SPE prior to sample introduction.** S = sample; R_1 = optional make-up reagent; MC = mini-column; E = eluent; L = sampling loop; C = sample carrier stream; R_2 = reagent; R_C = coiled reactor; D = detector; IC = injector-commutator; shaded area = alternative IC position. For details, see text.

through this mini-column, where potential interfering species were retained, and the modified sample was directed to fill a sampling loop (Fig. 8.16). Switching the injector-commutator inserted the sample aliquot into the sample carrier stream and the resin mini-column into an eluent stream that directed the retained chemical species towards waste. With this innovation, the analyte did not participate in the ion-exchange process and selectivity was markedly improved.

An unsegmented conditioning stream (R_1, Fig. 8.16) can also be added by confluence before the mini-column, and this option is useful for procedures requiring an in-line sample conditioning step.

MINI-COLUMN AT THE SAMPLE INSERTION PORT

In unsegmented flow analysis, SPE can also be implemented by placing the mini-column at the injection port (Fig. 8.17), as demonstrated by

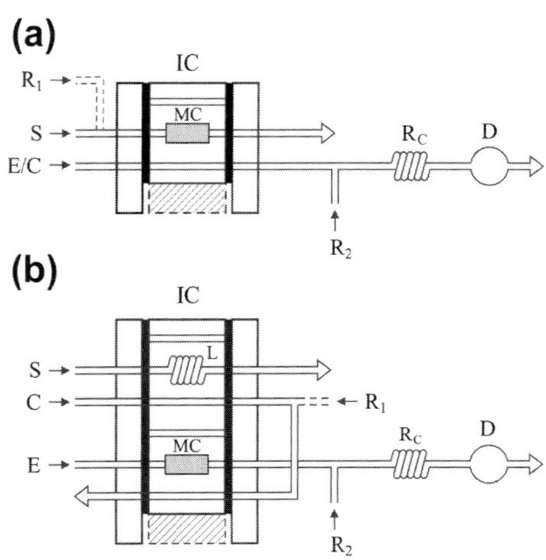

FIGURE 8.17 **Flow diagrams of unsegmented flow systems with SPE at the injection port involving (a) time-based or (b) loop-based sample introduction.** S = sample; R_1 = optional make-up reagent; MC = mini-column; L = sampling loop; E = eluent; C = sample carrier stream; R_2 = reagent; R_C = coiled reactor; D = detector; IC = injector-commutator; shaded area = alternative IC position; arrows = sites where pumping is applied. For details, see text.

pioneering work incorporating the use of ion-exchange resins in flow-injection analysis [65]. In order to improve sensitivity in the determination of ammonium in natural waters, the sample was pumped through a cation-exchange resin mini-column attached to the central sliding bar of an injector-commutator (Fig. 8.17a). The ammonium ions were retained and the depleted sample was directed to waste during the in-line concentration step. Thereafter, the injector-commutator was switched, thereby intercalating the mini-column into the sample carrier stream, which was an alkaline solution that also acted as the eluent. In this manifold geometry, the sample matrix does not pass through the detector and counter-current elution can be achieved. The washing/reconditioning step is directly accomplished as a consequence of the passage of the eluent/carrier/wash stream through the mini-column. The conditioning step can also be accomplished by the front portion of the next sample to be assayed. In this situation, some of the analyte in this sample portion is not retained and a larger sample volume is required.

As the sample volume through the mini-column is proportional to the sample flow rate and concentration time, the flow system should be computer-controlled. Better sensitivity is attained by increasing the sample flow rate and/or the concentration time, and the drop in sampling rate associated with longer concentration (or elution) times is minimised by the use of two sequentially operated mini-columns [237].

Analysis of complex samples (often coloured and/or turbid), *e.g.*, biological fluids, wines, soil extracts, juices, sediments and wastewaters, generally requires a clean-up or clarification procedure. An efficient approach for washing the mini-column after the concentration step is to take advantage of the manifold geometry in Fig. 8.17b which is analogous to that exploited to implement the zone sampling strategy (see 7.2.1). The spectrophotometric determination of zinc in plant digests [227] is a good example. The sample was inserted into the first carrier stream and the sample zone was directed towards a resin mini-column which was attached to another section of the central sliding bar of the injector-commutator. After the concentration step, the resin was washed (reconditioned) by the first sample carrier stream. Switching the commutator intercalated the mini-column into the eluent/second carrier stream and towards the detector. Manual operation of the flow system is possible using this strategy, which was also used for the in-line elimination of the spectral interference from iron in the determination of zinc by FAAS [238].

With this strategy, however, the efficiency of SPE is reduced as a consequence of broadening of the sample zone during emptying of the loop (see 6.2.2.3). The time interval needed for quantitative passage of the entire sample aliquot through the mini-column is thereby increased, thus impairing the sampling frequency and/or concentration efficiency, as shown in Fig. 8.18. As the regions below curves (a) and (b) are

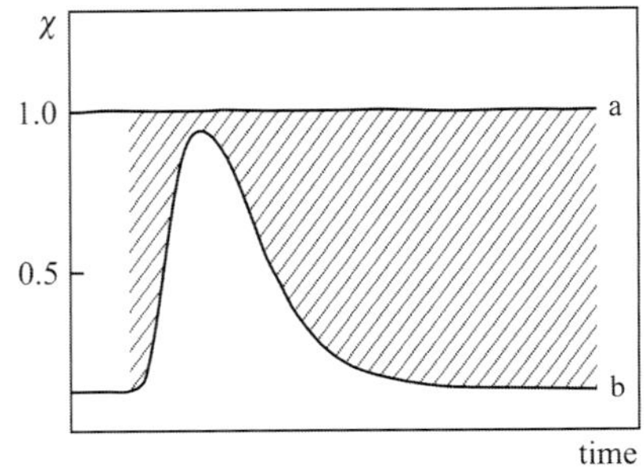

FIGURE 8.18 **Didactic comparison between time-based (upper) and loop-based (lower) mini-column loading.** The figure refers to the flow systems shown in Figs 8.17a, b. Shaded areas highlight the increased amount of sample available for SPE in flow system 8.17b relative to flow system 8.17a. The unshaded area on the left-hand side refers to the front portion of the sample that is not available for SPE but is used for reconditioning purposes. For details, see text.

proportional to the amounts of sample available for in-line SPE, the shaded area highlights the amount available for SPE in a system with time-based sample introduction and not for SPE in a system with loop-based sample introduction.

MINI-COLUMN AFTER THE SAMPLE INSERTION PORT

The mini-column can be placed between the injection port and the detection unit (Fig. 8.19), and the solutions involved (sample, wash, eluent and conditioning) are sequentially passed through the column. The flow system can be manual or computer-controlled. This manifold geometry is exemplified by the flow injection chemiluminometric determination of zinc and cadmium [207]. After sorption of the analytes as chloro-complexes, two different eluting solutions were injected, allowing sequential determinations; the resin was reconditioned by the water carrier stream. This geometry can also be implemented in sequential injection analysis and bead injection analysis, as demonstrated in the

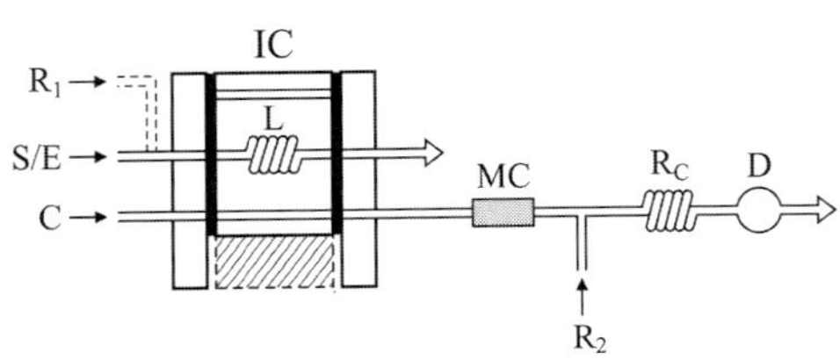

FIGURE 8.19 **Flow diagram of an unsegmented flow system with SPE after the injection port.** S/E = sample or eluent; R_1 = optional make-up reagent; L = sampling loop; C = sample carrier stream; MC = mini-column; R_2 = reagent; R_C = coiled reactor; D = detector; IC = injector-commutator; shaded area = alternative IC position; arrows = sites where pumping is applied. For details, see text.

determination of cadmium by ET-AAS and ICP-MS [234]. An advantage of having the mini-column in the analytical path is that the column behaves as a packed bed reactor (see 3.1.1.4), thus improving the mixing conditions. Although widely used, this manifold geometry is characterised by two intrinsic limitations, namely the passage of all the solutions involved through the detector and the drop in SPE efficiency associated with loop-based introduction of the solutions involved (see also Fig. 8.17). In some applications, it is necessary to check from time to time if elution is still quantitative. The eluent is inserted as a plug, so the available time for elution may not be enough for quantitative elution after long periods of system operation due to changes in particle packing density.

MINI-COLUMN AT THE DETECTOR

Some analytical procedures exploiting *e.g.*, chemiluminescence, fluorimetry and solid-phase spectrophotometry (see 4.1.1.4) require quantification of the product immediately after its formation. The analyte concentration/separation step should then be incorporated as close and possible to the detector, as exemplified by the speciation of iron in natural waters [239].

Another possibility is to take advantage of functionalised moving beads, which are inherent to the bead injection system (see 2.4.1). The analyte interacts with a specific reagent on the bead surface, thus forming a detectable species on the solid phase. Alternatively, an auxiliary reagent can be introduced into the system to perform this function [240]. By stopping the beads at the detector, the chemical species formed can be directly detected. The use of magnetic beads to achieve this has been comprehensively reviewed [201].

HYPHENATED TECHNIQUES

In view of the potential of the flow analyser and its compatibility with SPE, specific flow systems combining solid-phase extraction with flow analysis have been proposed.

In this context, the use of a flow system as a post-column detection system in liquid chromatography (Fig. 8.20) is attractive, and early applications focussed on the amperometric determination of amino acids [241] and the spectrophotometric determination of sugars [242]. The solution leaving the column is directed towards the sampling loop of a flow system; if necessary, a coupling element (open tank or pressure valve) with a small inner volume is placed between the column and the flow system for hydrodynamic pressure matching. The chromatogram is obtained with good resolution, because the flow system acts as a point sampler, inserting the eluate fractions at discrete time intervals, thus minimising mixing between the newly separated species. An injection frequency is chosen that allows a return to baseline between eluate

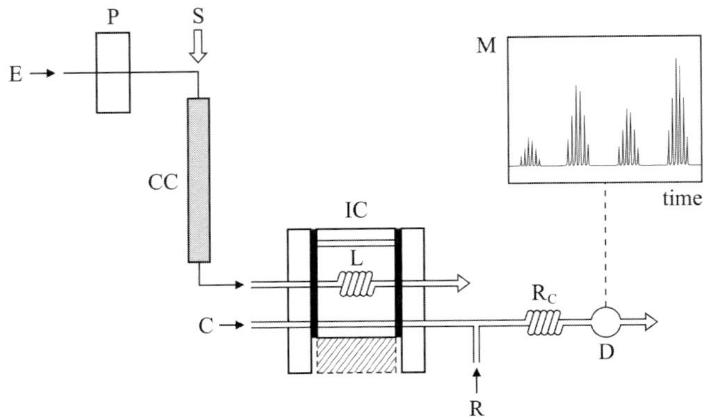

FIGURE 8.20 **A flow system for post-column monitoring in liquid chromatography.** S = sample; E = eluent; P = pump; CC = chromatographic column; L = sampling loop; C = sample carrier stream; R = reagent; R_C = coiled reactor; D = detector; M = monitored signal; IC = injector-commutator; shaded area = alternative IC position. For details, see text.

insertions to compensate for instrumental drift. The linear dynamic range can be expanded by implementing the strategies outlined in Chapter 7, and this is very important in relation to chromatography. The ease of simultaneous determinations in flow analysis opens up the possibility of monitoring several chemical species, thus expanding the range of chromatographic applications. This feature has not yet been fully exploited.

Alternatively, the flow system can be exploited for in-line sample conditioning prior to chromatographic separation, as demonstrated by the determination of benzoic and sorbic acids in food products [243]. The flow system comprised an electro-osmotic pump, five solenoid valves and a homemade SPE unit, combined with capillary zone electrophoresis. Tetrabutylammonium bromide was used as an ion pair reagent to improve analyte retention on a C_8-bonded silica sorbent.

OTHER MANIFOLD ARCHITECTURES

Other possibilities for implementing SPE without relying on packing the solid material inside a mini-column include:

- Membranes [244].
- Open tubular reactors with the solid phase immobilised on the inner walls of the tubing [245].
- Fluidised beads [246].
- Monolithic columns [211].

8.5.2.5. *Applications*

SPE is by far the most commonly used approach for in-line analyte separation/concentration in flow analysis. As already emphasised, it is more compatible with detection techniques unaffected by the Schlieren effect, such as atomic absorption/emission spectrometry, ICP-OES, ICP-MS, and related techniques. In spite of the relatively low number of reported *spectrophotometric* applications, the selected applications in Table 8.3 refer exclusively to flow systems with *spectrophotometric* detection.

8.5.3. Gas Diffusion

Gas diffusion (GD) involves the separation of gaseous species from the sample matrix, diffusion through a semi-permeable medium and collection into an acceptor solution for further analysis. In flow analysis, these steps are reproducibly accomplished but generally do not reach completion, and this expands the potential of gas diffusion. This approach has been exploited in both segmented and unsegmented flow analysis [247] but most applications relate to unsegmented flowing streams.

The analytes are generally volatile organic compounds (*e.g.*, phenols, ketones, alcohols andcarboxylic acids) or chemical species easily converted into simple gaseous species such as ammonia, sulphur dioxide or carbon dioxide [248]. The fluorimetric and spectrophotometric determinations of ethanol in fermentable medium to improve wine production [249] and urea in fresh waters [250] respectively are good examples. In the former application the analyte was directly separated from the sample matrix, whereas the latter application involved enzymatic analyte degradation yielding ammonia that was then separated from the sample. For metal determinations, the formation of volatile hydrides and/or metal vapours has been exploited, as demonstrated in the spectrophotometric flow injection determinations of mercury in river water [251] and arsenic in plant material [252]. In the former, Hg^{2+} ion was reduced to the volatile Hg^0 by tin(II) chloride under acidic conditions, separation was carried out in a GD chamber with a semi-permeable Teflon® membrane and the Hg^0 was collected in an acidic potassium permanganate stream that directed the analyte zone towards the spectrophotometer; the measurement basis was the transient colour reduction of the acceptor stream. In the latter situation, As^{3+} reacted with ammonium diethyl dithiophosphate, the complex formed was concentrated in a mini-column of C_{18}-bonded silica gel, the eluted complexes reacted with $NaBH_4$ and the generated arsine was collected in an alkaline iodine stream and determined by the arseno-molybdenum blue method.

Transport of the gaseous species from the donor to the acceptor stream depends on several parameters, *e.g.*, temperature, ionic strength, surface tension, contact time between the solutions and the semi-permeable medium, the characteristics of this medium (active surface, porosity, thickness), the partition coefficient between the fluid and the membrane,

TABLE 8.3 Selected Applications of Solid Phase Extraction

Analyte	Sample	Solid phase	Detection Technique	Concentration Range or Detection Limit	Remarks	Ref.
Al(III)	Drinking water	AmberliteXAD-4 resin modified by grafting salicylic acid	Fluorimetry	0.2 μg L^{-1}	Sequential injection system	[486]
Al(III), Fe(III)	Potable water	AmberliteXAD-4 resin modified by grafting salicylic acid	UV–Vis	4.9, 5.6 μg L^{-1}	Multi-syringe flow injection system; surfactant addition for improving sensitivity	[487]
Amiloride	Human urine	Bakerbond WCX (carboxy-group) ion-exchanger	Fluorimetry	0.05 μg mL^{-1}	Sequential injection system	[488]
Ammonium	Rain water	DowexAG1-X8 anion exchange resin	Turbidimetry	0.5–2.0 mg L^{-1}	Flow injection system with intermittent streams	[489]
Ascorbic acid, rutintrihydrate	Pharmaceutical formulations	Silica gel chemically modified with C_{18}	UV–Vis	1.0, 1.5 μg mL^{-1}	Simultaneous determinations in a sequential injection system; sequential elution with a single eluent	[490]
Benzo[a]pyrene	Drinking water	C_{18}	Fluorimetry	2.5 ng L^{-1}	Sequential injection system; variable angle fluorescence	[491]

Table 8.3 Selected Applications of Solid Phase Extraction—cont'd

Analyte	Sample	Solid phase	Detection Technique	Concentration Range or Detection Limit	Remarks	Ref.
Benzophenone-3, benzophenone-4	Urine from sunscreen users	C_{18}, diethylaminopropyl modified silica	UV–Vis (LC)	60, 30 ng mL^{-1}	Simultaneous determinations using two parallel cartridges in a sequential injection system hyphenated with liquid chromatography	[492]
β-Estradiol	Drinking, pond, well water	Molecularly imprinted polymer	Fluorimetry	1.12 μg L^{-1}	Flow injection system with intermittent carrier eluting streams; microwave-assisted extraction	[417]
Bismuth	Steels	Dowex 1-X8 anion exchange resin	UV–Vis	0.001 mg L^{-1} (injectate)	Flow injection system; analyte sorption as chlorocomplex	[493]
Boron	Steels	Sephadex G-25 gel	Fluorimetry	0.1 ng mL^{-1}	Flow injection system	[494]
Captopril	Tablets	Silica gel modified with quaternary ammonium	UV–Vis	2–50 μg mL^{-1}	Flow injection system; mini-column at the injection port	[495]
Cd(II), Cu(II), Pb(II)	River, well waters	Chemically modified Lewatit TP807'84 resin	UV–Vis	2.5, 2.0, 5.5 μg L^{-1}	Flow injection system; sequential elution, PLS multivariate calibration	[496]
Copper	Natural waters	Chelex-100	UV–Vis	0.5 μg L^{-1}	Bead injection system with a Jet Ring flow cell	[240]

(Continued)

Table 8.3 Selected Applications of Solid Phase Extraction—cont'd

Analyte	Sample	Solid phase	Detection Technique	Concentration Range or Detection Limit	Remarks	Ref.
Copper	Seawater digests	8-hydroxyquinoline	CL	1.0×10^{-7}–1×10^{-6} mol L^{-1}	Flow injection system; off-line UV-assisted sample digestion; SPE at the sampling loop	[138]
Cu(II), Hg(II)	River waters	Chelex-100	UV–Vis	0.63, 0.25 μg L^{-1}	Bead injection system; sequential determinations	[497]
Cr(VI), total Cr	Industrial hygiene samples	Strong anion exchange Dowex 1-X8 resin	UV–Vis	11 μg L^{-1} (both analytes)	Flow injection system; ultrasonic-assisted SPE	[498]
2,4-Dichlorophenol	River water	Molecularly imprinted polymer	CL	1.8×10^{-8} g mL^{-1}	Flow injection system	[499]
Dodecylamine (detergent)	Diesel fuels	200-mg solid-phase in 3-mL reservoir aminopropylisolute cartridges	CL	2.9 mg L^{-1}	Two mini-columns to improve sampling rate	[237]
Fe(III), total iron	River water, spiked seawater	Chelating disks (iminodiacetate groups)	UV–Vis	0.2 mg L^{-1} (both analytes)	Multi-pumping flow system; possibility to operate the system without SPE	[239]
Fluoride	Natural waters	Dowex 50 W-X8 cation-exchange resin	Fluorimetry	0.06 μg mL^{-1}	Flow injection system; SPE for interferent removal; resin mini-column prior to the sample injection port	[500]

Table 8.3 Selected Applications of Solid Phase Extraction—cont'd

Analyte	Sample	Solid phase	Detection Technique	Concentration Range or Detection Limit	Remarks	Ref.
Lead	Drinking water	AG1-X8 anion exchange resin	UV–Vis	Up to 100 μg L^{-1}	Sequential injection system; analyte sorption as chloro-complex; mini-column at the detection unit; catalytic procedure	[501]
Naptalam(pesticide)	River water	C_{18} modified silica	Fluorimetry	1.2×10^{-7} mol L^{-1}	Flow injection system; in-line analyte hydrolysis at 100°C	[502]
Nitrite	Tap, ground, harbour, aquarium waters	Octadecyl-bonded silica gel (C_{18}) disks	UV–Vis (diffuse reflectance; halogen-tungsten lamp)	0.1 ng mL^{-1}	Flow injection system; SPE at the detection unit; sandwich optrode	[503]
Nitrite	Tap, rain, aquarium waters	C_{18}-bonded silica end-capped material	UV–Vis (diffuse reflectance; green LED)	0.46 ng mL^{-1}	Flow injection system; SPE at the detection unit; sandwich optrode	[504]
Penicillins	Pharmaceuticals, milk	Methylene blue grafted polyurethane foam	UV–Vis	12, 15, 19 ng mL^{-1} for penicillin G, amoxicillin, ampicillin	Flow injection system with a mini-column at the injection port	[505]
Phenol	Waters, soil leachates	Amberlite XAD-4 absorbent (pH 2)	UV–Vis	0.2 ng mL^{-1}	Flow injection system; mini-column in the sampling loop; combination with LLE for improved sensitivity	[506]

(Continued)

Table 8.3 Selected Applications of Solid Phase Extraction—cont'd

Analyte	Sample	Solid phase	Detection Technique	Concentration Range or Detection Limit	Remarks	Ref.
Phenolic compounds	Waters, soils	Monolithic mini-column with polystyrene-divinylbenzene	UV–Vis	<1 ng mL^{-1}	Multi-syringe flow system coupled to HPLC; mini-column before the sampling loop	[507]
2-NP, 4-NP, 2,4-diNP	Mineral, tap, river, seawater	Anion exchange membrane	SPS	0.005, 0.003, 0.002 μmol L^{-1}	Sequential injection system; membrane as the optrode	[244]
Phenyl dibenzimidazoletetrasulfonate	Urine from sunscreen users	LiChrolut™ strong anion exchanger (trimethyl-aminopropylchoride)	Fluorimetry	10 ng mL^{-1}	Sequential injection system; standard addition method with a single solution	[508]
Phenylephrine hydrochloride	Pharmaceutical formulations	Dowex 50 W X8 ion-exchange resin	UV–Vis	5.8 mg L^{-1}	Multi-commuted flow injection system; sequential sample and eluent insertions	[509]
Orthophosphate	Seawater	Sep-Pak C_{18} cartridge	CL	0.002 μmol L^{-1}	Flow injection system	[510]
Ponceau 4R and its unsulfonated derivative (N2 N)	Food, cosmetic products	C_{18} silica gel	UV–Vis	0.052, 0.003 mg L^{-1}	Sequential determinations in a flow injection system; Ponceau 4R directly determined and N2 N determined after in-line concentration	[61]

Table 8.3 Selected Applications of Solid Phase Extraction—cont'd

Analyte	Sample	Solid phase	Detection Technique	Concentration Range or Detection Limit	Remarks	Ref.
Propoxur(insecticide)	Waters, fruits, vegetables	Octadecylsilica (DSC-18 cartridge)	CL	5 ng L^{-1}	Flow injection system with merging zones; GilsonAspecXLi device for implementing SPE	[511]
Salbutamol	Human urine, tablets	Silica gel modified with carboxylic acid	UV–Vis	0.1 μg mL^{-1}	Flow injection system; mini-column at the injection port	[512]
Salicylate	Urine, serum	Silica gel modified with quaternary ammonium	UV–Vis	0.05–20 μg mL–1	Flow injection system; mini-column at the injection port; injection of eluent	[513]
Selenium	Fresh, tap, seawater	C_{18} membrane disk	UV–Vis	1.7 μg L^{-1}	Multi-syringe flow system exploiting multi-commutation; acetonitrile as eluent	[514]
Soluble reactive phosphate	Sea water	Water™ hydrophilic-lipophilic balance (HLB) cartridge	UV–Vis	1.4 nmol L^{-1}	Sequential injection system; SPE for phosphomolybdenum blue enrichment	[515]
Sulfonated (WS), unsulfonated (SS) quinolineyellow	*Eau de Cologne*	C_{18} silica gel	UV–Vis	13.2 μg L^{-1}	Sequential determinations in a flow injection system; WS directly determined and SS determined after in-line concentration	[516]

(Continued)

Table 8.3 Selected Applications of Solid Phase Extraction—cont'd

Analyte	Sample	Solid phase	Detection Technique	Concentration Range or Detection Limit	Remarks	Ref.
Tetracycline	Fish samples	Molecularly imprinted polymer	CL	1×10^{-9} g mL^{-1}	Flow injection system; mini-column at the injection port	[517]
Thiabendazole	Natural waters	Nylon powder	Room-temperature phosphorescence	4.5 ng mL^{-1}	Flow injection system; solid phase at the detector (optode)	[518]
Thiamine	Pharmaceutical formulations	C_{18} – alkylated poly-(styrene/divinylbenzene)	Fluorimetry	31 ng mL^{-1}	Bead injection system	[519]
Trifluoperazine	Human urine	poly-(divinylbenzene)-co-N-vinylpyrrolidone	UV–Vis	18.2 ng mL^{-1}	Sequential injection system; extended optical path	[520]
Zinc	Biological materials	Polyurethane foam	UV–Vis	5–20 ng mL^{-1}	Flow injection system; mini-column at the injector port	[217]
Zinc, cadmium	Synthetic solutions	Amberlite IRA-400 anion exchange resin	CL	5, 3 ng mL^{-1}	Flow injection system; mini-column between injection port and detector; two eluting agents for sequential elution	[207]

UV–Vis = spectrophotometry; CL = chemiluminescence; SPS = solid phase spectrophotometry; LC = liquid chromatography; NP = nitrophenol.

diffusion coefficients and various experimental factors such as flow rates and composition of the donor and acceptor streams and the characteristics of the flow system. These parameters can be rigidly controlled during operation of the flow system, so the driving force for analyte transport is the difference in concentration of the diffusible gaseous analyte species (*i.e.* different gas pressures) in the donor and acceptor streams. Fast analyte conversion to simpler gaseous species (if required) and efficient interaction between the collected analyte and the acceptor stream are therefore important.

Semi-permeable membranes are generally used to select the chemical species to be diffused towards the acceptor stream [253]. Alternatively, other restrictive media can be used in a similar way to gas diffusion, *e.g.*, in pervaporation, membraneless gas diffusion and distillation, as discussed below.

8.5.3.1. *Perm-selective Membranes*

The main parameters influencing mass transference through membranes in flow analysis have been discussed in specialised texts that generally deal with both gas diffusion and dialysis [254–257].

CHARACTERISTICS

When using porous membranes, the pore size determines the efficiency of separation, and hence the selectivity, for both hydrophobic and hydrophilic membranes. Smaller pores favour the diffusion of small molecules and impair the passage of macromolecules, particulate and colloidal material [17]. Settling of these materials on the membrane surface can lead to "clogging" of the membrane which alters analyte transport efficiency and should be considered when designing the flow system, especially for the analysis of natural waters containing a high loading of suspended particulate matter or microorganisms.

Hydrophobic membranes, *e.g.*, PTFE, permit the efficient removal of volatile analytes from the sample matrix by diffusion though the micropores [257]. As these membranes have a high diffusion efficiency for many gaseous species, selectivity is usually low. For hydrophilic porous membranes, mass transference usually relies on dialysis, provided differences in donor and acceptor stream pressures are low [258]: the chemical species originally in the donor stream migrate through the solvent in the interstitial volume of the membrane. Ionic species are therefore efficiently separated from the macromolecules in the sample matrix. Increasing the difference in pressures of both streams favours the micro-filtration process; therefore, filtration and dialysis may occur simultaneously [259,260].

Nonporous membranes are homogeneous thin films that allow separation of chemical species based on different mass transport efficiencies, which is determined by the different solubilities of the chemical species and the different molecular diffusion (permeation) characteristics inside the membrane [248,257,261,262]. Silicone rubber is the most frequently used material. It is chemically inert and mechanically stable and has high permeation rates for a wide-range of chemical species, especially gases formed after sample acidification, *e.g.*, SO_2, CO_2 and NO_x.

Liquid membranes are characterised by enhanced selectivity because separation exploits specific carrier species that have a high affinity for certain compounds or compound classes [253].

MEMBRANE SHAPE

Most flow systems exploiting GD and related processes for in-line sample treatment make use of planar perm-selective membranes sandwiched between the donor and the acceptor streams (Fig. 8.21, upper) [17]. The membrane should be as thin as possible in order to permit fast analyte separation yet thick enough to be resistant to mechanical and/or pressure stresses. It is advisable to use a planar membrane with as small surface area but sufficient to cover the donor and acceptor channels. Planar membranes are fully compatible with both segmented and unsegmented flow systems. In the former, the gaseous analyte enters and leaves the membrane via the liquid phase (segments or thin film – see Fig. 5.2) adhering to the inner walls of the tubing whereas in the latter both liquid phases are intimately in contact with the planar membrane.

The use of a tubular membrane is very attractive due to its malleability and flexibility and the ease with which it can be directly incorporated in the flow manifold. The membrane is generally housed inside conventional tubing (made of plastic or metal), establishing a concentric geometry [263,264]. The donor and acceptor streams flow on either side of the tubular membrane (Fig. 8.21, lower). Analyte diffusion takes place in

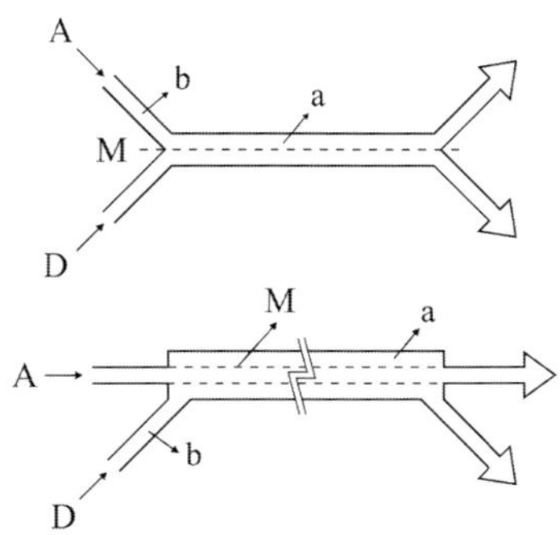

FIGURE 8.21 **Gas diffusion units with planar (upper) and tubular (lower) membranes.** D = donor stream flowing through the unit towards waste; A = acceptor stream flowing through the unit towards detection; M = membrane (traced lines); a = drilled channel (upper) or concentric tube (lower); b = manifold tube.

a continuous fashion along the entire analytical path. Moreover, the tubular membrane can be used as a sampling probe (see 8.1.2.2).

IMPLEMENTATION

The classical manifold architecture in Fig. 8.22, upper was exploited in the pioneering work incorporating GD in flow injection analysis for the spectrophotometric determination of total carbon dioxide in blood plasma [265]. Details of the separation unit are shown in Fig. 8.23, left. The donor stream with the sample zone was acidified and the released CO_2 diffused through the membrane towards the acceptor stream, which was an alkaline cresol-red indicator solution. Analyte collection resulted in a transient lowering of the pH of this stream and hence a transient modification to the monitored absorbance. The recorded peak height was proportional to the CO_2 content in the injectate.

To increase the available time interval for GD, thereby improving sensitivity, the donor stream can be stopped when the sample zone

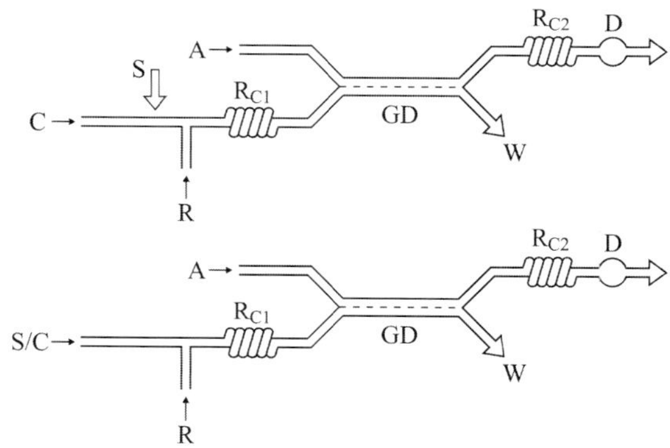

FIGURE 8.22 **Classical manifold architectures for implementing in-line gas diffusion involving intermittent (upper) or continuous (lower) sampling.** S = inserted sample; C = sample carrier stream; S/C = continuous inlet of sample or carrier stream; R = reagent; A = acceptor stream; R_{Ci} = coiled reactors; GD = gas diffusion unit; D = detector; W = waste.

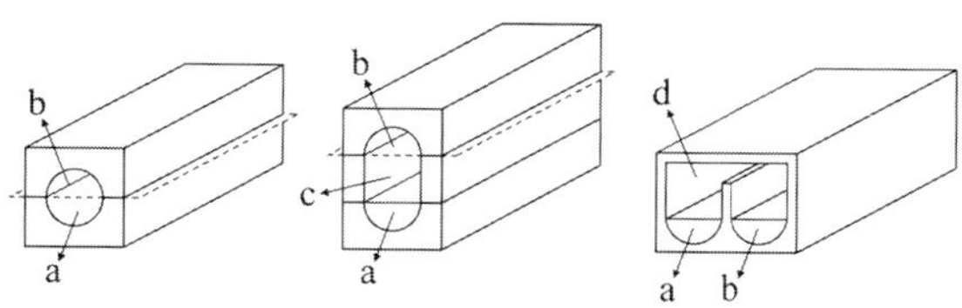

FIGURE 8.23 **Schematic representation of gas diffusion units involving the classical architecture (left), pervaporation (centre) and membraneless GD (right).** a = drilled channel for the donor stream; b = drilled channel for the acceptor stream; c = air-gap; d = air headspace; dashed lines = membrane. For details, see text.

reaches the GD unit. Alternatively, the sample can be inserted into the flow manifold using the approach of *infinite volume* [266] (see 2.1), as indicated in Fig. 8.22, lower. In both situations, stopping the acceptor stream also enhances GD efficiency.

Aspiration of a gaseous sample directly into the donor stream (Fig. 8.22, lower) is also feasible, an approach that is useful for gaseous samples (see 8.1.2).

8.5.3.2. Membraneless Diffusion

Gas diffusion can be accomplished using an air gap instead of a membrane between the donor and acceptor streams (Fig. 8.23, right). The feasibility of membraneless gas diffusion was demonstrated in the flow injection determination of ammonium in plant digests [267], and this innovation was originally referred to as *isothermal distillation*. The donor stream provided the alkaline conditions for analyte conversion to ammonia and flowed on an inclined silicone sheet inside the GD unit. Ammonia was released towards the headspace and collected by the acidic acceptor stream that flowed below another silicone sheet parallel to the first one. The air gap between the sheets was the barrier for gas diffusion and could be finely adjusted. The collected analyte was then directed towards the detector.

Another possibility is to exploit two parallel engraved channels separated from each other and sharing a common headspace and this innovation was originally proposed for the spectrophotometric determination of ethanol in alcoholic beverages [268]. This configuration of membraneless gas diffusion was also referred to as *thin layer distillation* and has been compared with the pervaporation process [269].

8.5.3.3. Pervaporation

Another possibility is to locate the semi-permeable membrane well above the donor stream (Fig. 8.23, centre), thus establishing an air gap between the sample and the membrane [270]. This approach has been referred to as *pervaporation*, and was defined as "a combination of continuous evaporation and gas diffusion through a gas permeable membrane" [271]. Both processes take place in a single step and in the same unit. The volatile chemical species present in the donor phase evaporates, crosses the air gap and the semi-permeable membrane and is collected in an acceptor stream.

The sample does not come into physical contact with the membrane, thus minimising potential matrix effects and increasing the range of possible applications. Complex matrices such as Kraft liquor [272], industrial effluents with high concentrations of surfactants, organic compounds, emulsions and/or suspended particulate matter [273] and fermentation media [274] can be analysed using this approach.

The donor and acceptor solutions are relatively far apart from each other. Therefore, it is possible to heat only the donor solution in order to speed up the evaporation process, and this aspect is useful for improving the analytical sensitivity.

Solid samples can also be analysed [275] by placing the sample at the bottom of the pervaporation unit and using the inlet stream as the digestion mixture.

8.5.3.4. Applications

The versatility of the flow analyser is of paramount importance for implementing gas diffusion (including pervaporation and membraneless gas diffusion) in flow analysis. High versatility is inherent to unsegmented flow analysis and therefore most of the applications listed in Table 8.4 refer to this type of flow analyser.

8.5.4. Dialysis

Dialysis involves the mass transference between two miscible liquid phases (the donor and acceptor solutions) separated by a liquid membrane through which some chemical species are likely to pass. Miscibility between the donor and acceptor solutions is inherent to dialysis and distinguishes it from *e.g.*, liquid–liquid extraction, osmosis and ultrafiltration [271]. These latter two membrane-based separation approaches tend to occur concomitantly with dialysis and involve the solvent rather than the solute crossing the membrane. In osmosis, the driving force towards separation is the concentration difference involved whereas in ultrafiltration, also called *reverse osmosis*, the driving force is an applied pressure that forces the solution across the membrane.

There are different types of dialysis, the most important of which are listed below [271].

- **Conventional or passive dialysis.** The driving force is the analyte concentration gradient across the membrane, which reflects the concentration difference between the donor and acceptor solutions. Ions and low-molecular-weight compounds are transferred, whereas dissolved and suspended material with high molecular mass is retained.
- **Donnan dialysis.** An ion-exchange membrane is used, allowing ions of a given charge (according to the membrane characteristics) to be transported as a result of an ionic strength gradient. Co-ions from the sample solution diffuse in the opposite direction in order to maintain electroneutrality. Donnan dialysis is generally faster than passive dialysis and efficiently promotes the transfer of charged compounds

TABLE 8.4 Selected Applications of Dialysis, Gas Diffusion, Pervaporation, Filtration, Distillation and Precipitation

Analyte	Sample	Separation/ Concentration Step	Detection Technique	Analytical Range or Detection Limit	Remarks	Ref.
Ammonia	Cigarettes	Pervaporation	UV–Vis	0.05 μg mL^{-1}	Flow injection system; ultrasound exploitation; iterative change of flow direction for improving extraction; Berthelot reaction	[415]
Ammonia	Surface, waste waters	Pervaporation	UV–Vis	0.03 mg L^{-1}	Flow injection system; enrichment cycle to attain wide-range spectrophotometry; heated donor solution; BTB acid–base indicator solution as the acceptor stream	[521]
Ammonium	Waters	GD	Fluorimetry	5.0 μg L^{-1}	Flow injection system; heating of the sample zone before the gas diffusion; Eosin-Bluish acid/base indicator solution as the acceptor stream	[522]
Ammonium	Waste waters, river waters	Membraneless GD	UV–Vis	2.20 mg L^{-1}	Multi-syringe flow system; BTB acid–base indicator solution as the acceptor stream; donor stream stopped during gas diffusion	[523]
Ammonium	Urban wastewaters, atmosphericaerosol extracts	GD	UV–Vis	0.5 mg L^{-1}	Sequential injection system; BTB acid–base indicator solution as the acceptor stream; flow reversals (donor stream) and flow stopping (acceptor stream) during gas diffusion.	[524]

Table 8.4 Selected Applications of Dialysis, Gas Diffusion, Pervaporation, Filtration, Distillation and Precipitation—cont'd

Analyte	Sample	Separation/ Concentration Step	Detection Technique	Analytical Range or Detection Limit	Remarks	Ref.
Ammonium	Industrial effluents	Pervaporation	UV–Vis	0.1 mg L^{-1}	Flow injection system; mixed cresol red/thymol blue acid–base indicator solution as the acceptor stream; results unaffected by surfactants, other organics and suspended solids	[273]
Ammonium, nitrate	Soil extracts	Distillation	UV–vis	0–15 mg L^{-1} N	Air-segmented flow system; 116 °C heating bath; indophenol blue reaction; Ti(III) addition for nitrate conversion to ammonium during distillation	[525]
Arsenic	Plant digests	GD	UV–Vis	5.00–50.0 μg L^{-1}	Flow injection system; for details, see text	[252]
B_{12} vitamin	Milk	Dialysis	UV–Vis	0.01 mg L^{-1}	Flow injection system; donor and acceptor streams recycled during dialysis; addition of trichloroaceticacid and sample centrifugation not required	[526]
Calcium carbonate	Calcium supplement tablets	Membraneless GD	UV–Vis	0.5 mg(solid sample)	Flow injection system; solid sample placed in a chamber into which HCl is added; released CO_2 collected after gas diffusion; cresol red acid–base indicator as the acceptor stream	[527]

(Continued)

Table 8.4 Selected Applications of Dialysis, Gas Diffusion, Pervaporation, Filtration, Distillation and Precipitation—cont'd

Analyte	Sample	Separation/ Concentration Step	Detection Technique	Analytical Range or Detection Limit	Remarks	Ref.
Carbamazepine-protein binding	Bovine serum albumin	Dialysis (for sampling)	Fluorimetry	6×10^{-6} mol L^{-1}	Flow injection system; *in vitro* study of the binding mechanism of the analyte with albumin; mini-column of PbO_2(s) for oxidising the analyte to a fluorescent product	[528]
Chloride	Industrial effluents	Electrodialysis	UV–Vis	200–3200 mg L^{-1}	Flow injection system; enhanced mass transference through passive neutral membranes	[288]
Chlorine	Waters	GD	CL	75 μg L^{-1} as Cl_2	Flow injection system; potassium hydroxide solution as the acceptor stream; reaction of hypochlorite with lophine generating the chemiluminescence emitting species	[529]
CO_2	Natural gaseous samples, headspace of milk containers	GD	UV–Vis	2.50×10^2 μL L^{-1}	Flow injection system; calibration relying on gaseous carbon dioxide or sodium carbonate standard solutions; mixed acid / base indicator solution as the acceptor stream	[530]
CO_2 (total)	Blood plasma	GD	UV–Vis	10–50 mmol L^{-1}	Flow injection system; original implementation of gas diffusion in flow injection analysis; buffered cresol-red indicator solution as the acceptor stream	[265]

Table 8.4 Selected Applications of Dialysis, Gas Diffusion, Pervaporation, Filtration, Distillation and Precipitation—cont'd

Analyte	Sample	Separation/ Concentration Step	Detection Technique	Analytical Range or Detection Limit	Remarks	Ref.
Cyanides	Tap, mineral waters	GD	UV–Vis	2.5 μg L^{-1}	Sequential injection system; analyte reaction with ninhydrin in carbonate medium to yielding the coloured product; standard addition method	[531]
Diacetyl	Beer	Pervaporation	UV–Vis	5.0 ng mL^{-1}	Flow injection system; pervaporation unit thermostated at 90°C; acceptor stream (alkaline α-naphthol + creatinesolution) halted during analyte collection	[532]
DIC	Seawaters	Dialysis	UV–Vis	1900–2300 μmol L^{-1}	Nitrogen-segmented flow system; semi-permeable silicone membrane; heated (35 °C) donor and acceptor streams	[533]
Ethanol	Spirits, wines	Membraneless GD	UV–Vis	0.27% v/v	Flow injection system; membraneless original proposal; acidic dichromate as the acceptor stream	[268]
Ethanol	Spirits, wines	GD	UV–Vis	0.5% v/v	Flow injection system; silicone tubular membrane; acidic dichromate as the acceptor stream	[264]
Ethanol	Gasohol	Membraneless GD	UV–Vis	0.9% (v/v)	Flow injection system; donor and acceptor streams stopped during pervaporation; gas segment inserted into the acceptor stream to improve system washing	[534]

(Continued)

Table 8.4 Selected Applications of Dialysis, Gas Diffusion, Pervaporation, Filtration, Distillation and Precipitation—cont'd

Analyte	Sample	Separation/ Concentration Step	Detection Technique	Analytical Range or Detection Limit	Remarks	Ref.
Ethanol	Fermentation medium for wine production	Pervaporation	Fluorimetry	0.0004% (v/v)	Flow injection system; ethanol monitoring; two-step enzymatic derivatisation involving alcohol oxidase and horseradish peroxidase; flow stopping during pervaporation.	[249]
Ethyl xanthate	Ore flotation process	Dialysis	UV–Vis	0.3 μmol L^{-1}	Flow injection system; sodium hydroxide solution as the acceptor stream; manual sample filtration not required even for sludge suspension samples	[535]
Fe(III)	Turbid pharmaceutical formulations	Dialysis	UV–Vis	45.0 mg L^{-1}	Sequential injection system; two subsystems each with a different holding coil, tiron as the colour-forming reagent; several flow reversals	[536]
Fluoride	Drainage water from water treatment plant	Distillation	UV–Vis	0.05–15 mg L^{-1}	Flow injection system; sulphuric and phosphoric acids for sample acidification; determination relies on the La(III)–fluoride–alizarin complex.	[311]
Formaldehyde	Indoor, ambient air	GD	UV–Vis	0.08 ppbv	Flow injection system; two parallel scrubbers for improving sampling rate; heated (50°C) main reactor	[537]

Table 8.4 Selected Applications of Dialysis, Gas Diffusion, Pervaporation, Filtration, Distillation and Precipitation—cont'd

Analyte	Sample	Separation/ Concentration Step	Detection Technique	Analytical Range or Detection Limit	Remarks	Ref.
Glucose, ethanol, ammonia, phosphate	Fermentation medium	Filtration	UV–Vis	5, 5, 50, 1 mg L^{-1} after 200-fold in-line sample dilution	Flow injection system; steriliseablecrossflowmicrofilter; process monitoring	[538]
Hg(II)	River water	GD	UV–Vis	4.0 μg L^{-1}	Flow injection system; for details, see text	[251]
Hydrogen peroxide	Lens care solutions	Dialysis	UV–Vis	Up to 342 mg L^{-1}	Sequential injection lab-on-valve system; low dialysis efficiency (1.25 %) to promote in-line dilution, thus sensitivity reduction; enzymatic assay	[60]
Iodide	Pharmaceutical formulations	GD	UV–Vis	1.0 mg I L^{-1}	Flow injection system; merging zones; exploitation of the iodine-starch reaction; possibility of using dialysis for in-line sample dilution	[539]
Iodide	Multi-vitamin tablets	Pervaporation	CL	0.5 mg L^{-1}	Flow injection system; prior SPE for ascorbic acid removal; analyte conversion to iodine; potassium iodide solution as the acceptor stream; luminol addition prior to detection	[540]
Malic acid, lactic acid	Wine	Dialysis	Fluorimetry, UV–Vis	0.015, 0.01 g L^{-1} (fluorimetry); 0.03, 0.05 g L^{-1} (UV–Vis)	Flow injection system; buffered β-NHD^{+} acceptor stream; dialysed divided into two portions each one directed towards a different enzymatic reactor; co-enzyme oxidation catalysed by malic or lactic dehydrogenase immobilised on controlled pore glass	[541]

(Continued)

Table 8.4 Selected Applications of Dialysis, Gas Diffusion, Pervaporation, Filtration, Distillation and Precipitation—cont'd

Analyte	Sample	Separation/ Concentration Step	Detection Technique	Analytical Range or Detection Limit	Remarks	Ref.
Nitrate, nitrite	Soil solutions	Filtration/dialysis	UV–Vis	0.02–0.60, 0.01–0.75 μg mL^{-1}	Unsegmented continuous flow system; filtration/dialysis probe for sampling; real-time monitoring of a leaching process; Griess reaction after optional nitrate conversion to nitrite	[542]
Nitrogen compounds degradable to ammonia	Synthetic solutions	Dialysis	Fluorimetry	1×10^{-6}–1.0×10^{-3} mol L^{-1}	Flow injection system; four parallel reactors for increasing sampling rate; enzymatic formation of ammonia; aqueous *o*-phthalaldehyde solution as the acceptor stream	[543]
Nitrous acid	Indoor environment	GD (for sampling)	CL	120 pptv (for 2-min sampling)	Flow injection system; *in situ* analysis involving a diffusion scrubber and peroxynitrite-induced luminol chemiluminescence	[18]
Peroxynitrite	Biological samples	Dialysis (for sampling)	CL	1×10^{-11} and 1×10^{-10} mol L^{-1} for systems without and with dialysis	Flow injection system; perm-selective cellulose acetate membrane sampler	[544]
pH, chloride, nickel	Electroplating baths	Dialysis (for chloride)	Potentiometry, UV–Vis	1–5, 0.10–1.00 mol L^{-1}, 0.10–1.60 mol L^{-1}	Sequential injection system; sample zone flows through a tubular pH electrode, a flow cell (for nickel determination as aquocomplex) and a dialysis unit; dialysed chloride flows through an ion-selective electrode	[545]

Table 8.4 Selected Applications of Dialysis, Gas Diffusion, Pervaporation, Filtration, Distillation and Precipitation—cont'd

Analyte	Sample	Separation/ Concentration Step	Detection Technique	Analytical Range or Detection Limit	Remarks	Ref.
Phosphorus (dissolved)	Waters, sediment extracts	Gel filtration	UV–Vis	1.1 (low range) or 12 (high range) $\mu g L^{-1}$	Flow injection system; speciation; Sephadex G-25 mini-column; UV-assisted in-line sample preparation; two dynamic concentration ranges	[546]
Reducing sugars	Wines	Dialysis	UV–Vis	2–140 $g\ L^{-1}$	Sequential injection system; dialysis to attain sample clean-up and high and variable dilution (wide-range spectrophotometry); sugar oxidation by cupric ions at 80 °C; formed cuprous ions react with neocuproine	[547]
Sulfide	Waste water	GD	UV–Vis	0.03 $mg\ L^{-1}$	Multi-syringe flow system; multi-commutation; acceptor stream stopped during gas diffusion	[548]
Sulfite	Wine	GD	UV–Vis	0.1 $mg\ L^{-1}$	Flow injection system; bromocresol green solution (pH 5.5) as the acceptor stream	[549]
Sulfite	Foods	Pervaporation	CL	0.2 $mg\ L^{-1}$	Flow injection system: rhodamine B solution (pH 2.5) as the acceptor stream	[550]
Sulphur dioxide (free and total)	Wines	Pervaporation	UV–Vis	1.2 $\mu g\ mL^{-1}$ (both analytes)	Flow injection system; *p*-rosaniline plus formaldehyde solution as the acceptor stream; sample alkalinisation prior to injection for total sulphur dioxide determination	[551]

(*Continued*)

Table 8.4 Selected Applications of Dialysis, Gas Diffusion, Pervaporation, Filtration, Distillation and Precipitation—cont'd

Analyte	Sample	Separation/ Concentration Step	Detection Technique	Analytical Range or Detection Limit	Remarks	Ref.
Sulphur dioxide (free and total)	Wines	GD	UV–Vis	2.0–40, 25–250 mg L^{-1}	Sequential injection system; *p*-rosaniline plus formaldehyde solution as the acceptor stream; sample alkalinisation prior to its insertion	[552]
Sulphur(IV)	Atmospheric water	Dialysis	UV–Vis	1.6×10^{-7} mol L^{-1}	Flow injection system; passive ion-exchange tubular membrane for reagent addition	[553]
Trimethylamine	Fish muscles (solid or extract)	Pervaporation	UV–Vis	0.08 mg g^{-1}, 1.6 mg L^{-1} (solid, extract)	Flow injection system; BTB solution (pH 6) as the	[554]
Urea	Freshwaters	GD	UV–Vis	2.40 μmol L^{-1}	Flow injection system; microwave-assisted N-urea reduction to N-ammonia prior to gas diffusion; BTB aqueous solution as the acceptor stream	[250]
Zinc	Synthetic solutions	Electrodialysis	UV–Vis	100 nmol L^{-1}	Flow injection system; electrodialysis for reagent addition via a cellulose membrane; LED detector	[289]
Zinc	Fertilisers	Dialysis	UV–Vis	4.75 mg L^{-1}	Sequential injection system; two holding coils related to the donor and acceptor channels; dialysis through a passive neutral membrane	[555]

GD = gas diffusion exploiting perm-selective membrane; UV–Vis = spectrophotometry; CL = chemiluminescence; BTB = bromothymol blue; DIC = dissolved inorganic carbon.

while retaining uncharged compounds. This type of dialysis has been reviewed elsewhere [276].

- **Electrodialysis.** The driving force is an external electric field that accelerates mass transference across the membrane. Charged compounds are transferred, whereas uncharged compounds are retained. As the dialysis efficiency depends on the external applied electric field, a potentiostat should be linked to both sides of the membrane. Details are given elsewhere [277,278].
- **Microdialysis.** This is similar to passive dialysis and is therefore used for separating ions and low-molecular-weight compounds from dissolved and suspended material with high molecular mass. The driving force is also a concentration gradient across the membrane, but the influence of the main physico-chemical aspects involved is different, as microfluidics is involved. A deeper discussion of this aspect is outside of the scope of this monograph and can be found elsewhere [259,260].

8.5.4.1. *Implementation*

Implementation of dialysis in flow analysis is analogous to gas diffusion using the classical manifold architecture (Fig. 8.22). As dialysis is generally a slow process, the mass-transfer equilibrium is frequently not reached. This feature can be exploited to adjust the degree of dialysis completion, thus matching the analytical signal with the dynamic concentration range of the detector, as demonstrated in the spectrophotometric flow injection determination of chloride in industrial effluents and plating bath solutions [151]. Samples with a very high chloride content (up to 60 g L^{-1}) were directly inserted into the flow system, which comprised two serial dialysis units. Precise results (r.s.d. = 0.8%) were obtained at a sampling rate of 100 h^{-1}.

To improve the dialysis efficiency, the donor and the acceptor streams can be stopped for a pre-set time interval and this strategy has often been exploited in unsegmented flow analysis [271]. In situations of relatively fast dialysis, the analytical sensitivity can be improved by stopping only the acceptor stream, thus continuously renewing the sample inside the separating unit and allowing the analyte to accumulate in the acceptor stream. To this end, large sample volumes or sample introduction by the infinite sample volume method (Fig. 8.22, lower) have been used.

8.5.4.2. *Applications*

Exploitation of dialysis in flow analysis dates back to the 1950's when the original air-segmented flow analyser was conceived (see 2.2). In 1963, a dialysis unit with a planar membrane was commercially available [279]. The first application of dialysis in unsegmented flow analysis dealt with the spectrophotometric determination of glucose in blood [280]. Regarding flow injection analysis, in-line dialysis was originally implemented for the spectrophotometric determination of inorganic phosphate and chloride in blood serum, as reported in the classical 10-article series on the proposal, concepts and applications of flow injection analysis [281] (see 2.3). The possibility of implementing dialysis in a sequential injection analyser was foreseen in 1993 [282].

Dialysis has been more frequently exploited in segmented flow analysis, as the related analyser is more widely used for clinical assays of complex and/or dirty matrices such as blood plasma and urine, which generally require dialysis. In addition, dialysis is generally a slow step requiring a longer mean sample residence time inside the separation unit, and this is better achieved in the segmented flow analyser. The vast majority of applications refer to passive dialysis (Table 8.4) and there are few examples of analytical applications exploiting Donnan dialysis or electrodialysis in unsegmented flow systems.

Donnan dialysis has been exploited in flow systems with flame atomic absorption spectrometry and ICP-OES, for which remarkable analytical figures of merit, including high enrichment factors, have been reported [283–285]. This type of dialysis is less compatible with flow-based *spectrophotometric* determinations, as the inherent concentration gradients can give rise to the occurrence of the Schlieren effect.

Electrodialysis is efficient for in-line sample clean-up prior to HPLC [286] or flow analysis, and the landmark work dealing with the determination of L-amino acids in a flow injection system with electrochemical detection [287] is a good example. Cation- and anion exchange membranes were used and separation relied on the large variation in the isoelectric point of the different amino acids. The buffered acceptor stream leaving the dialysis unit was directed towards the sampling loop of the flow-injection system, the inserted sampled aliquot then passed through two serial mini-columns containing catalase and L-amino acid oxidase and the hydrogen peroxide formed was monitored. Amino acid oxidation was nonspecific but good selectivity was achieved due to the use of in-line dialysis. With this innovation, arginine was determined in the presence of other amino acids.

This approach was further exploited in the spectrophotometric flow-injection determination of chloride in industrial effluents involving a passive neutral membrane [288]. The advantages inherent to passive dialysis were maintained and the mass transference was improved. Wide-range spectrophotometry was achieved by varying the applied voltage and/or the flow rates of donor and acceptor streams.

Electrodialysis is also useful for reagent addition without volumetric dilution, as demonstrated in the spectrophotometric determination of zinc involving 4-(2-pyridylazo) resorcinol [289]. The amount of added reagent was controlled by the applied voltage, and excellent mixing conditions were reported. A favourable aspect of this approach is that the gaseous species resulting from the electrolytic process, mainly hydrogen and oxygen, were generated outside of the analytical path.

8.5.5. Filtration

Filtration involves the removal of suspended particles and/or macromolecules from the sample by passing it through a selective material (porous material, membrane and knotted reactor) where they are retained. If only macromolecules are involved, the process is also called *gel filtration* or *ultrafiltration*, the latter term being more widely used for tangential flow filtration (see below). The material to be removed is either originally present in the assayed sample or formed during sample handling. This process is efficiently accomplished in flow analysis because the related sample manipulations (*e.g.*, addition of the precipitating agent, precipitate filtration and dissolution) are easily and reproducibly performed.

8.5.5.1. *Uses*

In-line filtration has been exploited in different ways as discussed below [290,291]:

PRECIPITATE REMOVAL FOR IMPROVING DETECTION

A coloured reagent stream interacts with the analyte, yielding a precipitate, and transient changes in the monitored signal reflect colour fading associated with a reduced reagent concentration (see 3.2.1). This transient modification, i.e. the recorded analytical signal, is monitored more easily if the suspended material is removed. This approach is suitable when the analyte is a major constituent of the sample, as originally demonstrated in the spectrophotometric determination of potassium in fertilisers in a segmented flow analyser [292]. The method involved analyte precipitation by sodium tetraphenylborate, in-line filtration of the slightly soluble potassium tetraphenylborate and spectrophotometric monitoring of the remaining reagent at 254 nm. Accumulation of particles on the walls of the main coiled reactor before the filtration unit was minimised by immersing it in an intermittently actuated ultrasonic bath.

In earlier applications using this approach, the filtration unit was periodically removed and cleaned. This limitation can be circumvented by placing the unit in the loop of an insertion valve [293], thus establishing a geometry analogous to that of multi-site detection (see 7.6.2.1). Switching this valve inserts the filtering unit into another stream that, in revere flow, solubilises the precipitate and directs it towards waste.

SAMPLE CLEAN-UP

In-line filtration can be accomplished before sample introduction into the main analytical channel, thus avoiding the introduction of undesirable particles into the flow manifold. As filtration occurs outside of the analytical path, it can be considered as complementary to the sampling step. This approach has been used mainly for *in situ* analysis of "dirty" samples, and the shipboard chemiluminometric determination of iron(II) in marine waters [13] is a good illustration (see 8.1.1 and Fig. 8.1). It is also useful for the analysis of leachates, partially dissolved alloys and partially digested samples (see Fig. 8.4). In this context, real-time monitoring of fulvic acid sorption onto modified vermiculites [294] is an interesting example. Fulvic acid was mixed with a vermiculite suspension inside a large (100 mL) beaker and the mixture was vigorously stired. A peristaltic pump continuously aspirated a small fraction of this mixture and directed it through the tangential filtration unit and then back to the main vessel. In this closed configuration, the process suspension was the donor stream and the acceptor stream was directly aspirated towards the holding coil and processed as in an ordinary sequential injection system. Fulvic acid was spectrophotometrically determined at 254 nm. With this innovation, important real-time information on the kinetics of the sorbent–sorbate interaction and the chemical equilibria involved was efficiently gathered.

ANALYTE SEPARATION/DILUTION

The sample aliquot is introduced into the main analytical channel and the established sample zone undergoes tangential flow filtration (see below). The unfiltered portion is directed towards waste while the filtrate is directed towards the detector. The manifold architecture is analogous to that for in-line dialysis, as both processes tends to occur simultaneously; consequently, the approach is also useful for in-line dilution.

ANALYTE SEPARATION/CONCENTRATION

The analyte is accumulated on the filter as a solid compound which is further dissolved and directed towards the detector. Matrix interferences are circumvented and sensitivity is improved, as demonstrated in the determination of copper in silicate rocks by flame atomic absorption spectrometry [295]. The analyte was precipitated by rubeanic acid, the precipitate was separated by continuous filtration and the depleted sample zone was directed towards waste. After precipitate collection, a potassium dichromate stream was allowed to pass through the filtering unit, thus dissolving the retained precipitate and transporting it towards the detector. The dynamic concentration range was expanded (0.3–200 $\mu g\,L^{-1}$ Cu) by passing different sample volumes through the filtration unit. Sensitivity associated with the highest volume (250 mL) improved 500-fold relative to the traditional operation of the spectrometer with direct sample aspiration into the nebuliser.

When the solubility product involved is not low enough for quantitative analyte precipitation, a high concentration of a chemical species with similar properties to the analyte is generally added, leading to analyte co-precipitation and quantitative analyte collection.

8.5.5.2. *Implementation*

The filtration unit is the key component for implementing in-line filtration in flow analysis. Different filtration strategies, each exploiting a different type of filtration unit, are discussed below.

CONVENTIONAL FILTRATION

The filter is a porous material [291] able to separate species by size exclusion. Large particles and macromolecules are retained when the sample passes through it. The main drawback is the possibility of permanent accumulation of the precipitate on the filtering element but this can be avoided by:

- Periodically removing the filter unit for washing. This procedure is generally adopted when in-line filtration is exploited for sampling.
- Passing a dissolution stream through the filtration unit after every analytical cycle. This procedure is normally associated with flow systems incorporating analyte separation/concentration by filtration.
- Placing the filtration unit in the loop of an insertion valve [293], as emphasised above.

An alternative approach exploits chemo-filtration and the spectrophotometric determination of gadolinium in biological materials [296] is a good example. The analyte was complexed by Arsenazo III at pH 2.5, and the complex was retained in a polyamide membrane; thereafter, the complex was removed by a buffered (pH 9) stream. With a 2.0 mL min^{-1} concentration flow rate and a 5 min sampling time, the detection limit was 15 μg L^{-1} Gd.

Another option is to place the filtration unit inside the flow cell, as demonstrated in the spectrophotometric flow injection determination of hydrogen peroxide [297]. The analyte interacted with titanium(IV) and 2-((5-bromopyridyl)azo)-5-(N-propyl-N-sulfopropylamino) phenol (PAPS) yielding a red-purple complex. After ion pairing with CTAB, the complex was adsorbed and concentrated on a very small area of the membrane filter positioned inside the flow cell. The analyte was quantified directly in the membrane by solid-phase spectrophotometry (see 4.1.1.4). Thereafter, ethanol was injected in order to solubilise the complex and transport it to waste.

TANGENTIAL FILTRATION

The filtering element, usually a membrane, is sandwiched between the donor and acceptor streams in a geometry analogous to the classical manifold architecture for gas diffusion (Fig. 8.22). An advantage inherent to this geometry is that there is no accumulation of insoluble materials on the filtration unit. Moreover, only a partial and reproducible amount of analyte passes through the membrane; consequently, the analytical sensitivity can be modified by altering the available time for filtration.

The spectrophotometric determination of Fe(II), Fe(III), nitrate and nitrite ions in natural and wastewaters in a sequential injection system [298] is an interesting example of this innovation. In-line tangential filtration was performed prior to the sample insertion port as part of the sampling step. Particulate material was not aspirated towards the holding coil and several samples could be analysed without the need for frequent cleaning of the filtration unit. Analytical figures of merit were similar to those for the corresponding flow-based analytical procedure with manual sample filtration prior to flow analysis.

THE FILTER-LESS APPROACH

In-line filtration without a filtering element is also feasible. To this end, a three-dimensional reactor [299], also called a knitted or knotted reactor (see 6.2.3.4), can be used, as emphasised in the landmark article reporting the flow injection determination of lead in blood and bovine liver by flame atomic absorption spectrometry [300]. The analyte was co-precipitated; the complex formed was retained on the inner walls of a knitted reactor and then released by isobutyl methyl ketone and transported to the detector. Interference from iron(III) at high concentrations was circumvented, sensitivity was markedly improved and precise results were obtained. This innovation was recently exploited to remove organic selenium and determine the speciation of inorganic selenium in a flow-injection system with atomic fluorescence spectrometric detection [301].

The availability of a large area for precipitate accumulation (a feature relevant to applications exploiting analyte co-precipitation), low back-pressure, the feasibility of using high flow rates, minimal contamination due to the use of chemically inert tubing, easy construction and long lifetime are the main advantages of this approach. This was emphasised in a review [291] that also discussed the mechanisms of this separation/concentration approach and presented relevant applications.

In-line filtration without using a filtering element can also be implemented by exploiting ultrasound irradiation, which is the most commonly used auxiliary energy for improving filtration through a semi-permeable medium. When ultrasonic standing waves and a laminar flowing stream are combined in a chamber, the *filtration chamber*, effective solid–liquid separation can be achieved [302], and this phenomenon was exploited in a multi-purpose flow system for the spectrophotometric analysis of waters, blood and other biological tissues [150]. To date, however, ultrasound filtration without a filtering element has not been further exploited in flow analysis.

8.5.5.3. *Applications*

The conventional geometry for in-line precipitate formation, separation from the sample matrix, accumulation on the filtering element, dissolution and transport to the detector has been more used with atomic absorption/emission spectrometric detection. This is probably because a concentrated solution is formed during the dissolution step which might lead to a pronounced Schlieren effect with *spectrophotometric* detection. Moreover, some particles could reach the detection unit without being fully dissolved and this would have a negative effect on *spectrophotometric* measurements. Regardless of the detection technique involved, tangential filtration has been preferentially exploited to complement the sampling step. The filter-less approach offers interesting possibilities but, to date, has rarely been used.

8.5.6. Other Processes

Flow systems exploiting other processes for in-line analyte separation/concentration have also been proposed. The characteristics, potential, limitations and analytical applications resulting from the implementation of these processes are briefly discussed below.

8.5.6.1. *Electrolytic Deposition*

In-line retention of electro-active chemical species, separation, accumulation at a specific manifold site and further release can be accomplished by taking advantage of electrolytic deposition, as originally demonstrated in the determination of lead in high-purity sodium chloride and sodium sulphate reagents by electrothermal atomic absorption spectrometry [303]. Selectivity was improved because the analyte was electro-deposited inside a 3 μL flow cell, thus separating it from the bulk matrix. By optimising the operation of a potentiostat linked to the flow cell, a deposition efficiency of 70% and analyte dissolution in 40 μLof eluent were reported. With a 3 min in-line concentration, a detection limit of 1.2 ng L^{-1} was achieved.

A good example of the combination of electrolytic deposition with *spectrophotometry* is the determination of cadmium in tap water and plant digests [304]. The sample zone was acidified and passed through an electrolytic cell comprised of two concentrically assembled Pt electrodes.

The potential interfering ions Cu^{2+} and Pb^{2+} were deposited on the cathode and anode, respectively, while Cd^{2+} was maintained in solution. The processed sample zone then passed through an AG1-X8 resin mini-column for in-line analyte concentration as $[CdCl_4]^{2-}$. The eluted analyte was spectrophotometrically determined using the malachite green-iodide method. With a 60 s electrolysis time and a 0.25 A applied current, Pb^{2+} and Cu^{2+}concentrations up to 50 and 250 mg L^{-1} respectively did not interfere and the detection limit for Cd was 0.23 μg L^{-1}. Precise results (r.s.d. = 3.85%) in agreement with those obtained by graphite furnace atomic absorption spectrometry were obtained at a sampling rate of 20 h^{-1}.

This approach was further modified in order to avoid the need for ion-exchange concentration after the electrolytic process [305]. Two electrolytic cells were placed in the main carrier stream, one for retaining the potential interferents and the other for in-line cadmium concentration. Cadmium was concentrated and eluted solely by modifying the applied potential in the second cell. This innovation was applied to the analysis of river waters using tetraphenylporphine tetrasulfonic acid as the colour-forming reagent and 2,2′-bipyridine as a catalyst. With a 29 min concentration time, precise results (r.s.d. = 2.1%) were obtained and the detection limit was 0.45 μg L^{-1}.

8.5.6.2. *Distillation*

Distillation is often used in industrial processes, *e.g.*, to remove volatile components for purification purposes, produce distilled water and separate substances through their different vapourisation temperatures. In analytical methods involving distillation, the volatile components (including the analyte) are generally separated from the heated sample through vapourisation, transported towards a cooled element and collected as a condensate for further analysis.

This process is not easily implemented in flow analysis for in-line analyte separation/concentration as the steps involved, especially solution heating, are relatively slow and difficult to achieve. As a consequence, most applications refer to segmented flow analysis, and eleven analytical applications exploiting in-line distillation had already been reported by 1973 [306]. The classical spectrophotometric determination of ammonium in plant digests [307] is a typical example. The air-segmented sample stream converged with an alkaline stream before reaching a steel coiled reactor immersed in a heated glycerine bath. An air stream blew the emergent heated aqueous/gaseous mixture into a series of splash heads where the liquid drops were collected, coalesced and were pumped towards waste. The steam (with ammonia) passing through the heads interacted with an acidic flowing stream where ammonia was absorbed as the ammonium ion. This stream was directed towards a condensing column and a phase-separation device from which the sample was continuously withdrawn for the spectrophotometric determination of ammonium.

For unsegmented flow analysis, the landmark article reporting the development of a continuous distillation unit for the flow injection amperometric determination of free and complexed cyanides in heavily contaminated wastewaters and effluents [308] should be highlighted. The sample stream merged with an alkaline stream before and reached the inlet of a distillation unit made from borosilicate glass, half-packed with glass helices and wrapped with a heating wire (Fig. 8.24). A nitrogen stream entering at the bottom of the distillation column carried the gaseous/liquid mixture with the released HCN through a condenser towards an alkaline absorbing stream near the top of the column. The gaseous/liquid mixture passed through a condenser, a gas–liquid separator and a cylindrical silver flow-through electrode, whereas the reflux from the condenser was cooled and pumped towards waste. Differentiation between total cyanide and strongly bound metal cyanide complexes was achieved by off-line UV-assisted decomposition of the complexes.

In-line distillation has rarely been used in flow analysis, but when it has been the basic design of the distillation unit has been maintained. Some advanced micro-distillation units have been proposed, mainly for analytical procedures with electrochemical detection [309–311]. A current tendency is to place several units in the manifold for successively performing different in-line analyte separation (and/or concentration) steps, *e.g.*, distillation, dialysis and filtration. It is also possible to perform a variant of in-line distillation under constant temperature. This approach has been called *isothermal distillation*, *membraneless gas diffusion* or *thin layer distillation*, and discussed in 8.5.3.2.

8.5.6.3. *Pyrolysis*

Pyrolysis was implemented in flow analysis in 1992 for the spectrophotometric determination of amino acids [312]. The sample was injected

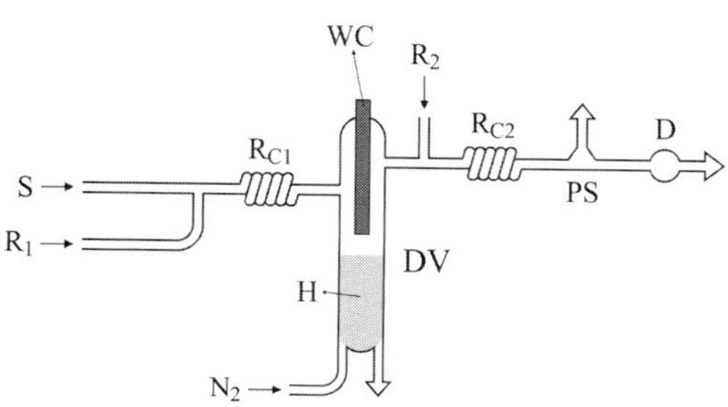

FIGURE 8.24 **Flow diagram of an unsegmented flow system with distillation.** S = sample; R_1 and R_2 = acidic and alkaline reagents; R_{C1} = coiled reactor; R_{C2} = coiled reactor for nitrogen absorption; DU = distillation unit; N_2 = nitrogen stream; H = heated glass helix; WC = water cooled condenser; PS = sample carrier stream; D = detector. For details, see text.

into a pyrolytic unit and the resulting degradation products were carried by an argon flow towards a heated PTFE capillary tube where these compounds underwent UV-assisted peroxydisulfate digestion. Thereafter, a colour-forming reagent stream was added to the handled sample and the specific amino acid was determined as nitrate. Proline, cysteine, alanine and tyrosine were discriminated from each other by modifying the pyrolysis temperature within the 220−350 °C range. In spite of the potential of this innovation, however, it has not been further exploited, probably because of the complexity of the flow manifold, including the pyrolyser, the programmable timer/heater and the phase separator.

Pyrolysis can also be used in flow-based determinations with electrothermal atomic absorption spectrometry, as demonstrated in the determination of nickel in environmental and biological reference materials using a sequential injection system with renewable beads [313]. After analyte sorption, the beads were directed towards the furnace of the spectrometer and stopped there; pyrolysis was accomplished as usual in order to release the analyte and destroy the beads. This innovation has often been exploited in the lab-on-valve system, but *spectrophotometric* applications have not been proposed to date.

8.5.7. Multiple Processes

A deeper examination of Sections 8.5.1−8.5.6 suggests that the analytical steps of analyte separation/concentration have particular nuances when carried out in a flow system. Their potential is often increased, especially when the temporal window inherent to the flow system is exploited and the geometry associated to the conditions for sample treatment is maintained.

For simultaneous determinations and/or speciation, different separation/concentration steps can be implemented in the same manifold, as in, *e.g.*, the determination of nitrogen, phosphorus and potassium in fertilisers [314]. The sample was inserted and passed successively through a dialysis unit and a gas diffusion unit to a flow cell for the spectrophotometric determination of phosphate. The dialysed potassium ions and the diffused gaseous ammonia were collected in specific streams and determined by flame photometry and potentiometry, respectively.

In some instances, the flow manifold is designed for sample preparation prior to, *e.g.*, HPLC or gas chromatography. Protocols for in-line sample filtration and dilution in sequential injection chromatography are presented elsewhere [315].

Some processes occur simultaneously, *e.g.*, gel filtration and dialysis; some processes are implemented in a similar manifold architecture, *e.g.*, supported membrane liquid extraction and solid-phase extraction; some processes involve the same material, *e.g.*, ion exchangers for SPE and for ultrafiltration. Consequently, it is possible to design multi-purpose units. In this context, a unit for implementing gas diffusion, dialysis, solid-phase extraction or precipitation/dissolution in continuous flow analysis was recently reported [316].

The discussion above demonstrates that in-line sample treatment in flow analysis presents endless possibilities.

8.6. MEASUREMENT STRATEGIES

The goal of the analytical steps highlighted in Sections 8.1–8.5 is to optimise the formation of a quantifiable chemical species, ideally reflecting the analyte concentration in the sample. As a rule, quantification is accomplished in real time during passage of the handled sample through the flow cell. The resulting analytical measurement relies on recording the transient modification in the monitored signal. The different measurement strategies for obtaining the final analytical result are presented below.

8.6.1. The Conventional Approach

The strategy relying on measuring the height, width, area or maximum value of the recorded peak resulting from the passage of the handled sample through the flow cell is by far the most commonly used, as discussed in Section 3.3. The analytical response curve is generally estimated *via* least-square regression analysis.

For multi-parametric determinations, successive measurements under different conditions should be made and the data treated using specific chemometric approaches. Flow systems provide an excellent tool for accomplishing this task (3.3.3).

8.6.2. Flow Titrations

Titration involves the reaction of the analyte (titrand) with another chemical species (titrant) and, in the classical version, the analytical result is obtained by considering the amount of titrant (mass or volume of a standardised solution) needed to attain the titration end point. This point ideally matches the equivalence point, i.e. the situation where the titrant and titrand equivalents are the same. This measurement strategy relies on several quantifications performed after successive titrant additions. As multiple additions are involved, titration is efficiently implemented in a flow system. Flow titration has been mostly exploited in unsegmented flow analysis, although recent applications have exploited mono-segmented flows.

8.6.2.1. *Initial Developments*

BATCHWISE TITRATIONS

Automation in the field of batchwise titrations was a reality in the early 1900's [317], and important improvements were made by the research group headed by Malmstadt (see *e.g.*, Refs. [318–320]). The setup involved placement of the sample solution in a titration vessel, addition of titrant by means of either an automatic burette or a syringe, and stirring of the reaction medium; concentration-orientated feedback mechanisms were generally exploited for assisting the operation of the titrant propeller. These automated batch titrators were characterised by a low sample throughput, a high consumption of sample and titrant, and a high implementation cost. These problems are circumvented in flow titrations.

FLOW TITRATIONS EXPLOITING MERGING STREAMS

The sample (titrand) and titrant streams merged with each other and the resulting stream passed through the detector. The potential of this innovation was demonstrated in the pioneering work of Blaedel and Laessig [321] who proposed flow titrations relying on the triangle programmed strategy.

The sample flow rate was maintained while the titrant flow rate was linearly increased to search for the end point. The titrant flow rate yielding a titrant amount chemically equivalent to the analyte was used to determine the analytical result. The potential of this innovation was demonstrated in a model system involving the titration of Fe(II) with Ce(IV) and electrochemical detection. In the 0.0007–0.06 mol L^{-1} Fe range, about 6 min were required per titration and highly precise results were obtained (r.s.d. 0.2–0.5%).

This innovation has been referred to a *triangle programmed titration*. However, a previous calibration step involving standard solutions was required to attain the analytical curve, a step that is not necessary in conventional titrations. Consequently, this innovation does not comply with the IUPAC definition of a titration, and can be regarded as a *pseudo-titration*. The flow analyser is simple, robust and performs titrations at a very high sampling rate.

FLOW TITRATIONS EXPLOITING CONCENTRATION GRADIENTS

After the advent of flow injection analysis, an alternative titration procedure based on gradient exploitation was proposed [322]. The sample was inserted into a titrant carrier stream, underwent high dispersion inside a mixing chamber placed in the analytical path and thereafter passed through the detector. The time interval elapsed between two measurement points with identical sample-to-titrant volumetric fractions was directly proportional to the logarithm of the analyte concentration [323]. This procedure is also considered a pseudo-titration because an analytical curve is involved.

FLOW TITRATIONS EXPLOITING MULTI-COMMUTATION

More recently, flow titrations have been implemented in multi-commuted flow systems. With these systems, the exact amounts of the solutions involved can be modified in real time according to a concentration-oriented feedback mechanism. The flow titration really mimics a *true titration* because an analytical curve is not needed. A flow system exploiting a binary search to define the end point of an acid–base titration [324] is a good example of this approach. Stream directing solenoid valves were used to modify the sample and titrant volumes after every measurement *via* a feedback mechanism. Samples with concentrations within a range of two orders of magnitude could be titrated without modifying the flow manifold.

Alternatively, the tandem streams inherent to multi-commuted flow systems can be exploited, and the titration of commercial hydrochloric acid and natural waters [325] is a good example. The sample and titrant streams converged towards a fast-switching three-way valve in order to establish a tandem stream. The frequency of the valve operation dictated the tritrand-to-titrant volumetric ratio of this stream, thus allowing the titration end point to be efficiently found.

8.6.2.2. Implementation

Several ingenious approaches for carrying out flow titrations have been proposed, some of which are summarised below.

BATCHWISE TITRATIONS

- Flow-batch system. A mixing/reaction chamber is the heart of the system (2.6) so titration is readily implemented [326].
- Micro-batch titration. Similar to flow-batch systems but the flow manifold does not include the sample carrier stream [327].
- Mono-segmented flow system. The sample mono-segment behaves as a titration vessel to which exact titrant amounts are added [328,329]. This innovation has also been implemented in the lab-on-valve system [330].
- Coulometric titration. The titrant is coulometrically generated inside a titration vessel and the amount is defined in terms of the electric current and elapsed time [103].

TITRATIONS EXPLOITING MERGING STREAMS

- Flow injection system with merging zones. The sample and titrant are simultaneously injected into convergent carrier streams and the titration is accomplished by adding different standard solutions via the titrant injection port [331] (see also 7.1.1.1).
- Flow ratiometry. Flow rates of the convergent sample and titrant streams are varied in real time according to a concentration-oriented feedback mechanism and the ratio of flow rates associated with the end point is used to estimate the analytical result. Flow rate variations are performed using, *e.g.*, stream directing valves or computer-controlled fluid delivery devices [332,333].

TITRATIONS EXPLOITING CONCENTRATION GRADIENTS

- Sample insertion into the titrant. Concentration gradients are established along the sample/titrant zone so that each fluid element corresponds to a different sample-to-titrant volumetric ratio [323].
- Sample insertion into a chemically inert carrier stream. A titrant stream is mixed with the sample zone by confluence; hence, a constant titrant concentration is used. The confluence configuration improves the sensitivity (3.2.2).
- Sequential injection system. The sample and titrant aliquots are sequentially aspirated towards the holding coil and, after flow reversal, the combined zone with its inherent analyte and titrant concentration gradients is directed towards the detector [334]. Exploitation of stream mono-segmentation [329] is valuable in this context.
- Zone sampling. This strategy (and related strategies) selects different titrant aliquots with known concentrations to be added to the sample zone [335] which expands the potential of flow titrations.

TITRATIONS EXPLOITING MULTI-COMMUTATION

- Management of the sample and titrant plugs. The number and length of the plugs of the solutions involved (carrier, wash, titrant and titrand) can be easily modified, allowing the development of true flow titrations [336,337]. These modifications can be performed in real time in accordance with a concentration-oriented feedback mechanism and this is the essence of expert titration flow systems.
- Tandem streams. The volumetric proportion between the different plugs of the tandem stream is taken into account to define the titration end point [325]; implementation of tandem streams is discussed in 3.1.1.2.

It should be emphasised that successive flow reversals are an important aspect for improving the different titration procedures exploiting gradient concentrations [338] and/or multi-commutation.

8.6.2.3. *Applications*

Recent applications have shown the potential of flow titration as a modern tool in analytical chemistry. As the required amount of titrand is associated with the analytical signal, important parameters, *e.g.*, oxidisability in wastewaters [339], bromine number in foodstuffs [340], bitterness of beers and similar [341], total acidity in wines and vinegars [342] and total alkalinity in natural waters [343], are efficiently determined. In addition, the total concentration of several analytes belonging to the same family, *e.g.*, amines [344], can be determined. The entire titration curve is generally available, allowing the determination of weak acids, complex stability constants and acid dissociation constants [345]. The determination of humidity by the Karl Fischer method [346] is another important application of flow titrations. For single analyte determinations, the analytical characteristics inherent to titrimetric procedures, such as enhanced accuracy and precision, should be emphasised.

8.6.3. The Standard Addition Method (SAM)

When the analytical procedure is susceptible to matrix effects, the standard addition method, SAM, can be applied to improve accuracy. SAM involves several additions of known amounts of the analyte (Δ_{Add} – Fig. 8.25) to the sample and measurement of the analytical signal (M – Fig. 8.25) after every addition. A specific "analytical curve" (M *vs* Δ_{Add}) is obtained for each assayed sample and the analytical result (C) is obtained by extrapolation.

When several analytes are involved, interferences due to both matrix and inter-analyte effects can occur. These interferences can be overcome by taking advantage of the generalised standard addition method, an

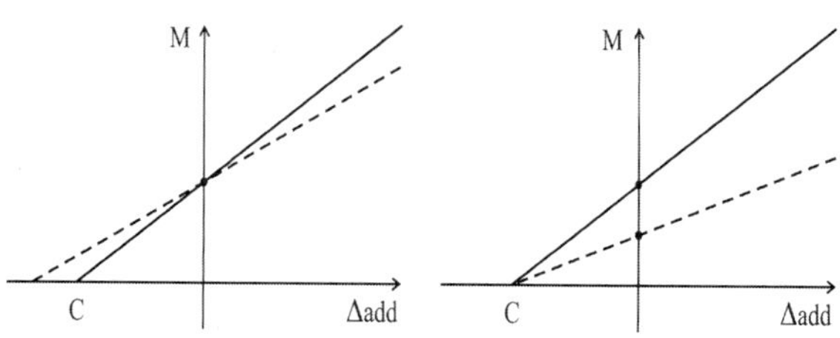

FIGURE 8.25 **Influence of the standard additions (Δ_{Add}) on the measurement (M) and the estimated analyte concentration (C).** Full and traced lines = curves affected and not affected, respectively, by a negative matrix effect. For details, see text.

expansion of SAM, which involves successive additions of known amounts of the analytes and multivariate calibration [7.1.2.7].

8.6.3.1. *General Approach*

Figure 8.25 shows hypothetical situations involving the analyses of two samples in the presence or absence of a pronounced negative matrix effect. In Fig. 8.25, left, the same initial measurements ($\Delta_{Add} = 0$) are obtained by both samples, but they become increasingly different from each other as successive standard additions are performed. By extrapolation, one can perceive that the analyte concentration of the sample associated with the pronounced matrix effect is relatively higher than the other sample. On the other hand, different measurements can be obtained for two samples with the same analyte concentration, depending on the matrix effect involved (Fig. 8.25, right).

In spite of the potential of SAM to minimise matrix effects, the analytical chemists generally avoid exploiting it in a batchwise manner, probably because its implementation is time-consuming, cumbersome and susceptible to errors due to the excessive number of sample manipulations involved. These drawbacks are minimised by implementing SAM in flow analysis, as originally demonstrated in the analysis of alloys by ICP-OES involving the generalised standard addition method [347] or by FAAS involving the injection of different standard solutions into the flowing sample [348]. Matrix effects due to spectral interferences (ICP-OES) or variations in nebulisation efficiency (FAAS) were efficiently circumvented.

8.6.3.2. *Implementation*

Implementations of SAM and titrations in flow analysis are analogous to each other, as both strategies involve successive additions to the sample. Consequently, mixing chambers [327,349], merging streams [350], concentration gradients [351] and/or multi-commutation [335] have been exploited.

Some strategies for implementing titrations, however, are less recommended for SAM, especially those requiring a high degree of sample dilution. In fact, the application of SAM presupposes that sensitivity is a limiting factor in the analysis and that sample dilution should therefore be avoided. Otherwise, the matrix effect could be circumvented simply by diluting the sample. In this context, the spectrophotometric flow injection determination of iron in milk samples [352] is a good example. A relatively large sample zone was pumped into a closed loop containing the flow cell; a relatively small volume of the standard solution was then inserted and successive measurements were obtained as a consequence of the repetitive passage of the overlapped sample/reagent zones through the flow cell. This innovation resembled the SAM batchwise procedure because a very small volume of standard solution was added to a relatively large volume of sample.

Another important aspect is that the sample matrix effect should ideally be maintained during the implementation of SAM. To this end, strategies involving a constant sample dilution during all of the measurement steps are preferred.

8.6.3.3. Applications

SAM has been more widely exploited with detection techniques involving nebulisation (*e.g.*, FAAS, ICP-OES and ICP-MS), as this step tends to be particularly affected by sample matrix effects, mainly salinity and viscosity. Relevant examples of SAM exploitation with *spectrophotometry* are given below.

- Nitrate determination in plant digests. With zone sampling one standard solution was sufficient to circumvent the influence of reducing agents in the extracts on the development of the colour-forming reaction [350].
- pH determination in blood cell suspensions. This pioneering SAM implementation relied on gradient exploitation. Asynchronous merging zones in a flow injection system were used in the single-line configuration [353].
- Analysis of synthetic solutions using a diode array spectrophotometer. This application of SAM involved gradient exploitation. The sample was kept undiluted ("infinite volume situation") and the standard zone was added by stream confluence; with two different standard zones, multi-parametric determinations involving the generalised standard addition method were achieved [354].
- Determination of orthophosphate in phosphatic rock digests and hydrochloric acid in the commercial reagent. The sample was injected into the flowing standard solution, and different fluid elements, each with a specific sample-to-reagent volumetric ratio, were considered for application of SAM [355].
- Spectrophotometric determination of the extractable content of chromium, lead, nickel and zinc in sewage sludges. A similar strategy to the above-mentioned [355] was used but different standard solution streams were involved. Matrix effects were successfully overcome [356].
- Spectrophotometric determination of iron in synthetic solutions. SAM was implemented in a sequential injection analyser with mono-segmentation. Concentration gradients were established at the interfaces between the different sequential zones [357].

These examples show that the flow systems used are rugged, yielding results unaffected by matrix effects, and the feasibility of using these systems to implement the generalised standard addition method has been demonstrated.

It is important to emphasise that the different manifold architectures originally designed for the application of SAM have also been used for spiking purposes to assess accuracy. This practice is however not recommended, as analyte addition for recovery experiments should always be done before the analytical procedure.

8.6.4. Time-dependent Aspects

Kinetic aspects are relevant to the different steps involved in the analytical process, which includes sampling, sample processing (chemical reactions, physical changes and separations), measurement, data processing and statistical evaluation [358]. The flow analyser is inherently associated with kinetic aspects (Chapter 1). Consequently, the time dependence of the measurement step in flow analysis is of the utmost relevance, especially for the exploitation of reaction rates.

8.6.4.1. *Methods Exploiting the Fixed-time Approach*

The analytical signal is evaluated immediately after sample handling is completed and re-evaluated after a pre-set fixed time interval. The difference in measurements reflects the production of the quantifiable chemical species during this interval or, in other words, the slope of the signal *vs* time curve, which ideally is proportional to the analyte concentration in the sample.

Exploitation of this approach is straightforward in flow analysis: the sample zone is handled, stopped inside the detector for a pre-set time interval and then directed towards waste. The initial and final measurements of the analytical signal are considered. The blank value is directly compensated by this approach and lag-phase effects become of minor concern, as demonstrated in the spectrophotometric determination of ethanol in whole blood [359]. The samples were injected without prior treatment other than dilution with a buffer solution and the stop period was 50–70 s. Sensitivity could be varied by selecting other stop periods.

8.6.4.2. *Differential Kinetic Methods*

Simultaneous determinations can be accomplished by exploiting the rates of reactions of different analytes with a common reagent, and this approach is efficiently carried out in flow analysis. To this end, different detection strategies associated with different manifold configurations have been proposed, *e.g.*, stream splitting/stream merging (7.3.2.1); stream splitting towards two different detectors [360]; sample incubation at the detector and monitoring of the analytical signal *vs* time curve

(3.3.3); double injection [361]; reactor replacement [362]; serial detection [363]; multi-site detection (7.6.2) [364]; sample stopping [365] and gradient exploitation (333) [366].

8.6.4.3. *Catalytic Methods*

Specific detection strategies and manifold configurations (8.3.3) are generally exploited for implementing analytical catalytic methods in the flow analyser. To this end, analytical signals related to different recorded peaks, to the signal increase during sample stopping periods [365] or to the measurement of multiple fluid elements along the dispersing zone [366] are used.

It is noteworthy to mention that, from a practical point of view, operation of the flow system for catalytic determinations is the same as for determinations involving the attainment of chemical equilibria. In fact, all of the temporal requirements inherent to catalytic analytical methods are efficiently fulfilled by the flow analyser.

8.7. EXPERT FLOW SYSTEMS

Flow analysers generally operate in a passive manner, *i.e.* the sample handling steps are defined prior to sample introduction into the analytical path. Precise timing is a cornerstone feature of flow-based analytical procedures that can be exploited to attain highly reproducible conditions for sample handling and, consequently, excellent analytical figures of merit. In ordinary flow analysers, internal timing, i.e. the inherent time window (see Chapter 1), is the key parameter for defining the sample handling conditions. In more advanced flow analysers, external timing associated with, *e.g.*, the time intervals for sample stopping or trapping, valve switching and the timing schedule associated with flow rate variations, can also be exploited to improve system versatility (see Chapter 7). In spite of the expanded potential and improved figures of merit of flow-based analytical procedures, these analysers are also operated in a passive manner, with the sample handling steps still defined prior to sample introduction into the system.

Alternatively, flow analysers can be operated in an active manner, and each sample is handled under specific conditions. To this end, real-time modifications of the sample handling conditions according to a concentration-oriented feedback mechanism are achieved. As a rule, a prior assay is carried out and the analytical result is used to feed the software with information for real-time decisions. The need for in-line modification of the degree of sample dilution, sample replacement, activation of pumping or stream directing devices, as well as further optional analysis, is then defined in real time.

This idea was embodied in the IUPAC definition of automatic systems: *"The use of combinations of mechanical and instrumental devices to replace, refine, extend, or supplement human effort and faculties in the performance of a given process, in which at least one major operation is controlled, without human intervention, by a feedback system"*[367]. The key point for an automatic system is therefore the incorporation of a feedback system, i.e. the *"combination of a sensing and a commanding device that can modify the performance of a given act"* [367]. In other words, the response of a sensing device is used for the real-time modification of system operation. These definitions also hold for flow analysis: automatic analysers are *expert flow systems* (Fig. 8.26), often referred to as *smart* or *intelligent* flow systems.

Expert flow systems have led to a paradigm shift in the practice of analytical chemistry by allowing response-oriented, real-time decisions based on preliminary results (prior assays). In addition to monitoring the sensor response, this innovation requires computer interfacing and the actuation of active devices (*e.g.*, valves and pumps) in the system manifold. The software in the expert system can initiate modifications during the analytical process, *e.g.*, by altering flow rates, sample volumes or added reagents. Another common strategy is stream redirection to submit the sample zone to different reaction conditions. Zone fluidics, i.e. the control and manipulation of flowing zones [101], is the underlying principle of these flow analysers and multi-commutation plays a decisive role in this context.

In view of the low costs of installation and operation, inherent flexibility, ease of operation and control, low consumption of reagents and short analysis times, as well as their compatibility with multi-commutation, most expert flow systems rely on unsegmented flow analysis. The

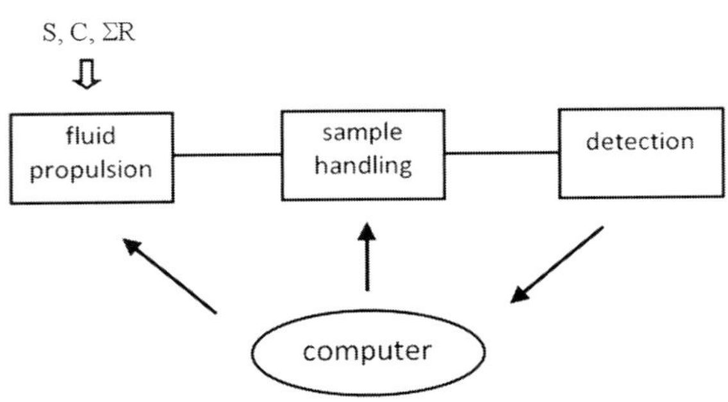

FIGURE 8.26 Block diagram of a typical expert flow system. Fluid propulsion, sample handling, detection = units of the flow analyser (Chapter 6); S, C, ΣR = sample, carrier, reagents solutions (inlet represented by the empty arrow); arrows = computer/unit interactions. For details, see text.

potential of these systems and examples of specific applications are discussed below.

8.7.1. Expansion of Analytical Laboratory Capacity

The result of a prior in-line assay is taken into account when deciding on the need for the determination of other parameters in the assayed sample. To this end, the result associated with the first analyte is compared with a pre-set threshold value specific for a given analytical problem and a binary answer (yes/no, present/absent, high/low) is obtained. Depending on the answer, other analytes may or may not be determined. A significant reduction in the total number of determinations is attained, thus expanding the analytical productivity of the laboratory.

This innovation generally involves modifications to the operation of the sampler and random access reagent selection, and can be implemented in both segmented and unsegmented flow analysers. For unsegmented flow analysis, the spectrophotometric determination of zinc and phosphate in soil extracts [368] is a good example. Zinc was determined only when phosphate was present at concentrations above a threshold level. The number of determinations required was reduced by 30%. Analogously, an expert flow system was proposed for the turbidimetric determination of chloride and sulphate in natural waters [369]. Both methods were implemented in the same manifold, and the need for sulphate determination was dependent on the chloride concentration determined.

In this context, expert flow systems are important tools for screening purposes, as they permit real-time sample classification [370]. It should be emphasised that ordinary flow analysers used with the objective of off-line classification of samples in different categories cannot be considered as expert flow systems.

Regarding industrial quality control, an expert system has been designed for monitoring a bio-chemical process, involving the determination of total acidity, reducing sugars, ethanol and pH during fermentation [371]. The system was able to decide when and which measurements should be carried out according to a predetermined set of rules relying on analytical results obtained in-line. The system was also able to take decisions concerning possible malfunctions.

8.7.2. Accuracy Assessment

Analytical procedures are sometimes susceptible to systematic errors that are difficult to detect. To a lesser extent, this also holds for flow analysis. A typical situation is the partial overlap of sample and reagent zones in sequential injection systems. The samples and reagents are sequentially inserted into the same analytical channel, and the analytical signal corresponds to the interfacial region with optimised sample and reagent volumetric fractions (5.3.1). This region is not necessarily associated with the best ruggedness, because interfering species may consume significant amounts of reagent, modifying the reagent concentration in that fluid element. The maximum response then corresponds to another fluid element with a higher reagent volumetric fraction but with a lower sample volumetric fraction. This was demonstrated in a model system for iron determination involving 1,10-phenanthroline as the colour-forming reagent [372]. A negative deviation of 19% was observed in the presence of copper, a potentially interfering chemical species (copper/iron ratio = 20% (m/m)). Inaccurate results can also be caused by differences in the rates of the chemical reactions involving the samples and the standard solutions (see Fig. 1.4) or by the decreased efficiency of a solid-phase reagent with continuous use. An expert multi-commuted flow system was effective for real-time characterisation and overcame these sources of inaccuracy for all assayed samples [372].

In addition, expert flow systems are able to assess the accuracy of analytical procedures. The high sampling rates, characteristic of flow analysis, can be exploited for the determination of a single analyte by two *quasi*-independent methods. Results are then checked by the control software and agreement between them is indicative of accuracy. If different results are obtained, in-line addition-recovery experiments can be performed in order to compensate for matrix effects, thus improving the accuracy. This approach was originally demonstrated in the flow injection determination of chloride in river waters with a highly variable sample matrix [373]. Spectrophotometric and turbidimetric methods were implemented in the same manifold, and the analyte concentration was determined after considering the results obtained by both methods. If agreement between results was attained, the mean value was considered as the final result. If not, the standard addition method was implemented in-line in order to compensate for matrix effects and the corrected values were compared with each other. If agreement was attained, the mean of these values was considered as the final result. If not, samples were re-analysed. Other applications involving sequential injection analysis with accuracy assessment, using two different detectors, include the spectrophotometric determinations of aluminium in plant digests [374], and captopril [375], diclofenac [376] and lansoprazole [377] in pharmaceutical formulations, the latter application being carried out in a single interface flow system.

8.7.3. System Optimisation

Optimisation is often a time-consuming step in the development and large-scale utilisation of an analytical procedure, as it requires evaluation of the effects of the main parameters involved, as well as the interactions between them, in order to achieve the best analytical response. The optimised conditions can however be altered due to uncontrolled experimental parameters, such as ageing of the pump tubing and changes in the ambient temperature; thus, re-optimisation is required. Although these deleterious effects can be reduced by acquiring several calibration graphs under different conditions, automation of this process can save time and yield more reliable analytical results.

Expert flow systems take advantage of iterative control devices which work in tandem and permit real-time alterations of the experimental parameters. They are designed to provide decisions for the real-time introduction of correction factors and, eventually, manifold self-optimisation. Enhanced system versatility and flexibility are also achieved.

For self-optimisation, a flow injection system able to implement a simplex algorithm in real time has been proposed [378]. The total flow rate and the volumes of sample and reagent were computer-controlled and permitted in-line application of the algorithm based on two experimental parameters (flow rate and sample-to-reagent volumetric ratio) and the detector response. The feasibility of this approach was demonstrated by the determination of chloride in natural waters, with wide variability in the analyte concentration. For multi-parametric analysis an ingenious approach relying on variations in flow rates was proposed for the sequential spectrophotometric determination of metal ions in synthetic solutions and natural waters [379]. The required modifications to ligand concentration and pH were achieved automatically in real time, allowing efficient system self-calibration and/or self-optimisation.

Due to its manifold geometry and system operation, the sequential injection analyser is more amenable to self-optimisation [380]. In this context, an expert sequential injection system with spectrophotometric detection and multivariate calibration was designed, validated and applied to the determinations of calcium and magnesium in natural waters [381]. The system was rugged and could efficiently define the parameters for system operation in real time. The information gathered was very consistent, demonstrating that sequential injection methodologies involving real-time optimisation could be developed for other analytes and/or other sample matrices.

8.7.4. Troubleshooting

Detecting and circumventing system failures caused by, *e.g.*, the ingress of undesirable air bubbles, that can give rise to spurious signals, system clogging and/or the poor performance of some manifold components (including the detection unit), is important, especially in the analysis of large sample batches or during process monitoring. To ensure reliable operation of the flow system, as well as providing system diagnosis, automated supervision is worthwhile [382], as emphasised by Bonastre *et al.* [383] who designed an expert system for monitoring the total acidity and peroxide value in olive oils. Feedback allowed the remote control of the system and in-line classification of the assayed samples. The analytical results were taken into account for estimating the overall quality of the oil and for identifying adulteration, age, storage conditions and the degree of oxidation.

Intelligent systems can also diagnose faults and find their causes. This innovation was demonstrated in the sequential injection analysis of calcium and magnesium in natural waters exploiting artificial neural networks [384]. As the diagnosis was carried out using information gathered in real time, the expert flow system was able to reconfigure itself to recognise analytical signals through the neural network and interpret the network output. Thus, faults were detected, the causes of failure were identified and samples were classified. Depending on the sample classification (low or high analyte concentration), real-time decisions were made to determine if it was necessary to re-analyse the sample under different experimental conditions.

Intelligent systems can also be designed for validation purposes, and the expert system proposed for the amperometric determination of penicillin in pharmaceutical formulations [385] is a good example. The input data required for in-line validation were obtained, allowing real-time evaluation of the calibration procedure, system drift and matrix effects.

8.7.5. Controlled Dilution

Sample dilution is often required to match the expected analyte concentration with the dynamic concentration range of the analytical procedure. Clinical analyses, where several parameters are routinely determined in large sample batches, is a good example of this requirement. Extensive dilution of biological fluids is common practice in view of the high analyte concentration and the need to minimise matrix effects. However, sample dilution is usually time-consuming, can degrade precision and lead to systematic errors.

The flow analyser is a good option for circumventing this drawback, as its operation presupposes precise control of the degree of sample dilution. On the other hand, difficulties can arise for sample batches with high variability in analyte concentration, i.e. when different samples need to be processed under different dispersion conditions.

Expert systems can be designed to achieve this goal by comparing the analytical response with previously defined upper and lower threshold values. Samples with analyte concentrations within these values can be directly quantified, whereas excessively concentrated or diluted samples should be handled under modified dispersion conditions. Feedback mechanisms can then be exploited to select the required degree of sample dilution in real time.

The concept of trial measurements [386] is important in this context. Different sample aliquots, each one corresponding to a different inserted volume, are successively analysed. The inserted sample volume is increased stepwise; thus, the analytical signal associated with each aliquot is continuously increased. If the analytical signal is lower than a threshold value, usually the quantification limit, the sample is run again but with a different dilution. The final analytical result is determined only when the analytical signal exceeds this value.

Matching the sample aliquot with the dynamic range of the detector also reduces the amount of sample reaching the detection unit, thus avoiding drawbacks such as clogging of the nebuliser of an atomic spectrometer when processing highly saline samples, impairing the liquid membrane of an ion-selective electrode, or undesirable precipitation effects. These beneficial aspects have been emphasised elsewhere [372,387,388].

Controlled variation of the degree of sample dilution is also a useful tool to widen the linear dynamic range, thus reducing the number of outlier samples, as demonstrated in the spectrophotometric determination of calcium in natural waters, plant materials, milk, fertilisers and calcareous rocks [389]. Zone splitting (see 7.4) and zone sampling (see 7.2) were involved, and the software was developed to control all of the sample processing steps and provide facilities to change the manifold configuration in response to the analytical signal. The required sample dilution was attained with up to three trials, allowing direct analysis of samples with a large variability in analyte concentrations. Precise results (r.s.d. = 0.83%) and a sampling rate of 60 h^{-1} were obtained. This innovation was recently applied to the flame atomic absorption spectrometric determination of potassium in soil extracts, parenteral solutions and wines [390]. Moreover, an expert sequential injection system with asynchronous merging zones was proposed for the spectrophotometric determination of glycerol in yeast cultivation media [57]. A water plug was placed between the sample and reagent aliquots, and its volume was selected in-line through a feedback mechanism. As each plug volume corresponded to a different analytical calibration graph, the dynamic range was expanded. After every tenth peak, a control standard was run and interrogation of the data allowed additional in-line decisions to be made, *e.g.*, the need for repeat of calibration or omission of cleaning protocols.

Other approaches that rely on real-time decisions to widen the dynamic range include:

- Automatic adjustment of the electronic gain of the detector, as demonstrated in the determination of mercury by cold-vapour atomic fluorescence [391].

- Controlled zone sampling prior to flame atomic absorption and flame emission spectrometry [387].
- Decision making based on the need for in-line analyte concentration in the determination and speciation of iron [392,393].
- Selection of the detection mode (transmittance or reflectance) and the need for in-line analyte concentration or dilution in the determination of sulphide in soil extracts with suspended particles [394].
- Use of sample/diluent convergent streams at variable flow rates prior to flame atomic absorption and flame emission spectrometry [395].

8.7.6. Matrix Matching

When the analytical signal is affected by the sample matrix, serious errors in the analytical results can arise due to improper calibration, especially for sample lots with highly variable matrix composition. The standard addition method is a powerful tool for compensating matrix effects, but its implementation in flow analysis usually involves a complex manifold and can impair the sampling rate (see 8.6.3). Matrix matching, i.e. matching the bulk compositions of the standard and sample solutions, is another way to achieve accurate results, but requires information about the sample matrix, which is not always available. To address this issue, expert systems have been exploited.

In a pioneering publication, a feedback mechanism was exploited to minimise matrix effects in the determination of metals in seawater by ICP-OES [396]. Direct analysis was not feasible, as variations in salinity affected the nebuliser efficiency and the intensity of the sodium spectral interference. For matrix matching, the sample salinity was estimated by a prior determination of sodium and the salinity of the standard solution used was selected by matching as closely as possible the salinity to that of the sample. This was done by modifying the t_{ZS} value in the sampling zone process (see 7.2.1), thus selecting a different amount of sodium chloride to be added to the working standard solutions.

Another possibility is to take advantage of the individual treatment of every assayed sample. For sample lots with high variability in, *e.g.*, pH or ionic strength, the required adjustments can be made in accordance with a prior assay performed in-line. This approach was proposed for the spectrophotometric determination of total nitrogen in Kjeldahl digests of plant leaves [397]. As the method was very susceptible to sample acidity, which varied from sample to sample, individual sample pH adjustment was needed. To this end, an alkaline solution was added to every sample and the amount added was determined according to a pH-orientated feedback mechanism. This innovation was further applied to the spectrophotometric determination of aluminium in plant digests [398] and iron in estuarine waters [399], characterised by pronounced variations in acidity and salinity, respectively. As the aluminium determination was very susceptible to the pH of the reaction medium, a feedback mechanism based on the colour change of an acid–base indicator was exploited for in-line adjustment of the sample acidity. The catalytic determination of iron was very susceptible to ionic strength, so a different amount of a sodium chloride solution was added to every assayed sample in order to give a constant salinity.

8.7.7. Flow Titrations

Titrimetric procedures have been carried out efficiently using expert flow systems because a different analytical signal can be obtained after every titrant addition. A feedback mechanism relying on these signals can then be exploited to modify the course of the titration in order to refine the end point search and improve the accuracy [400]. Fundamental aspects of the analytical potential, advantages/limitations and applications of the different types of flow titration (with or without expert flow systems) are presented in Section 8.6.2.

8.7.8. Localisation of the Sample Zone

Feedback mechanisms that are not typical expert systems, because all the assayed samples are handled in the same way, have also been used to improve the operation of specific flow systems. In general, the possibility of instant localisation of the sample zone expands the potential of these systems. A specific application is the determination of C_1–C_5 n–mercaptans in gasoline involving liquid–liquid extraction and zone sampling [199]. As emphasised in Section 7.2.2.4, a NaOH solution was inserted into the flowing sample and the analyte was extracted into it. The central portion of this alkaline aqueous zone was then re-sampled and inserted into another NaOH stream. For real-time definition of the t_{zs} value, the conductivity of the solution flowing inside the re-sampling loop was monitored and a sudden increase triggered the injector. With this innovation, Schlieren noise was minimised and the phase-separation step was simplified.

Instant localisation of the sample zone is also useful for improving analytical methods implemented in mono-segmented flow systems. These systems utilise sample insertion between two air bubbles to minimise sample dispersion, and thus carryover (see 5.5.1). Several sample zones can be handled simultaneously, which also improves the sampling rate. Precise and instant localisation of the flowing sample inside the analytical path can be achieved in the presence of the air bubbles, as optical switches can be used to sense the gas–liquid interfaces. Feedback mechanisms relying on changes in the logic status of the switches make selective reagent addition to only the sample zone feasible, thus minimising reagent consumption. This strategy was demonstrated by the determination of glucose, creatinine and urea in blood plasma and serum [401], involving the sequential addition of enzymatic and chromogenic reagents to the sample zone.

In the case of liquid–liquid extraction in mono-segmented flow systems, aliquots of organic solvents can be strategically added to the analytical path for single-phase [196] or dual-phase [195,402] separations.

The need for sample zone localisation in mono-segmented flow systems is also relevant for the analysis of gaseous samples. The sample replaces the air bubbles in the mono-segmented pattern, and either the volume contraction [403] or a change in a physico-chemical property of the aqueous sample zone due to the selective analyte sorption [404] can be used as the basis for measurement.

Segmented flow analysers using a bubble gating flow cell (Fig. 5.4) for monitoring the sample zone without the need for mechanical removal of the air phase [405] can also be considered as examples of flow systems exploiting a feedback mechanism that are not expert flow systems.

8.7.9. Trends

Expert flow systems play an important role in addressing the increasing demand for reliable analytical results, detectability, short analysis times and improved laboratory productivity. Accuracy assessment and sample conditioning for every assayed sample may become an essential requirement for accreditation of laboratories and/or for specific procedures. Regarding process monitoring, flow systems able to detect and circumvent failures during long and continuous working periods are also invaluable. Expert systems are therefore intrinsically more cost-effective and environmentally friendly.

The above applications show that feedback mechanisms to date have been devoted to specific applications. The next step is to incorporate real-time decision making in commercially available flow analysers but this will require software development that takes into account the needs and expertise of analytical chemists to deliver robust, effective and user-friendly tools.

8.8. FLOW METHODS AS TEACHING TOOLS

The relevance of flow analysis in the context of chemical education has three main strands: the strengthening of concepts, practical demonstrations and the practicality of flow chemography.

8.8.1. Strengthening Concepts

Manifold components such as reactors, chambers and connectors are generally made from transparent material (plastic and glass) and this imparts a unique characteristic to the flow analyser with *spectrophotometric* detection – the possibility of visually following the development of chemical reactions. Exploitation of this feature is an important pedagogical strategy for strengthening concepts, for both classroom demonstrations and large-scale analyses. The assimilation of concepts relies on the "*a single image is better than a thousand words*" premise. Moreover, the analytical peak-shaped signals are usually recorded in real time, which facilitates the assimilation and/or strengthening of concepts. Some illustrative examples are:

- Chemical kinetics. In a flow system designed in the confluence configuration, colour formation can be visually observed immediately after the confluence connector for reagent addition if the chemical reaction yielding the analytical signal is fast. Otherwise, a continuous increase in colour formation can be observed at the end of the main reactor. This is an important feature for distinguishing fast and slow chemical reactions.
- Beer's law. The intensity of the sample zone colour near the inlet of the detection unit increases with the analyte concentration in the injectate, and this proportionality can be visually observed. For concentrations that are too high, colour intensity appears less dependent on the analyte concentration, causing curvature of the analytical graph. These observations can be confirmed in real time by inspecting the recorded calibration peaks.
- Analytical repeatability. Visual inspection of the peaks recorded after successive replicate runs of the same sample emphasises the analytical repeatability. The skilled analytical chemist is then able to decide if the repeatability is acceptable, which is helpful for users less comfortable with mathematical calculations.
- Elution efficiency. In flow systems with SPE involving derivatisation immediately after elution, it is possible to visually evaluate the efficiency of the elution process. In the ideal situation, a very abrupt colour formation is observed immediately after the inlet of the make-up stream. When elution is a limiting factor for optimum system performance, a slow fading of the colour is observed.
- Mixing conditions. In a flow analyser comprising a mixing chamber *e.g.*, for analyte separation, a Schlieren pattern is observed inside the chamber inner volume under poor mixing conditions. This effect alters the recorded peak profile, which becomes noisy and less reproducible.

8.8.2. Practical Demonstrations

Pedagogical experiments can be efficiently implemented in flow systems to provide a powerful tool for assisting knowledge transfer in the field of chemical education. This may increase the interest of the students in practical demonstrations which, in traditional batchwise approaches, would be tedious and repetitive.

The flow system is characterised by its portability and low consumption of solutions, and these are additional positive factors in relation to its use in classroom experiments. A noteworthy feature is that the passage of the coloured sample zone through the detector can be displayed on a large screen, using *e.g.*, a slide projector [406] or a CCD digital camera [407]. The possibility of multiple repetitions of the experiments and improved student/teacher interactions during the demonstrations, even for larger classes, should be emphasised. Two strategies have been exploited in this context, the use of experiments to demonstrate the principles of flow analysis and use of the flow analyser to carry out pedagogical experiments.

In order to demonstrate the principles of flow analysis, classical spectrophotometric determinations, including a pseudo-flow titration, were carried out in the very first flow injection analyser [408]. More recently, the determination of hypochlorite in bleaching products using natural flower extracts [409] has also been exploited.

Regarding the use of the flow analyser to assist practical classes with specific examples in physical chemistry and/or analytical chemistry, reaction kinetics has been addressed in didactic experiments involving the Bertholet reaction with indophenol blue formation [410] and hexacyanoferrate(III) reduction by ascorbic acid [411]. Recently, video demonstrations have been successfully exploited aimed at the evaluation of stability constants and the classification of cations [412]. These experiments require combinations of a number of different solutions. The large number of flasks required for accomplishing these experiments in the traditional manner is not necessary, reagent consumption is low and the experiments can be indefinitely repeated.

8.8.3. Flow Chemography

The expression *flow chemography* was coined to "encompass all areas of flow chemistry in which readouts are graphically and mathematically represented" [413]. In this context, both flow analysis (flow injection analysis, chromatography, capillary electrophoresis, segmented flow analysis, flow titrations and flow optrodes) and flow synthesis (classical engineering, protein synthesis and protein sequencing) can be considered. It also applies to other analytical techniques yielding transient analytical signals, *e.g.*, atomic absorption/emission spectrometry. Concepts, such as smoothing, peak overlap, peak height or peak area, can also be effectively demonstrated.

In this context, the answer to the famous question: *Flow chemography – the future of chemical education?* [413] is YES!

References

[1] D.A. Skoog, F.J. Holler, T.A. Nieman, Principles of Instrumental Analysis, fifth ed., Harcourt Brace College Publishers, London, 1998.

[2] E.A.G. Zagatto, The renaissance of spectrophotometry, J. Flow Injection Anal. 19 (2002) 1.

[3] J. Thomsen, K.S. Johnson, R.L. Petty, Determination of reactive silicate in seawater by flow injection analysis, Anal. Chem. 55 (1983) 2378.

[4] P.J. Worsfold, J.R. Clinch, H. Casey, Spectrophotometric field monitor for water quality parameters. The determination of phosphate, Anal. Chim. Acta 197 (1987) 43.

[5] J.R. Clinch, P.J. Worsfold, H. Casey, An automated spectrophotometric field monitor for water quality parameters. Determination of nitrate, Anal. Chim. Acta 200 (1987) 523.

[6] J.R. Clinch, P.J. Worsfold, F.W. Sweeting, An automated spectrophotometric field monitor for water quality parameters. Determination of ammonia, Anal. Chim. Acta 214 (1988) 401.

[7] N.J. Blundell, A. Hopkins, P.J. Worsfold, H. Casey, A portable battery-powered flow injection monitor for the *in situ* analysis of nitrate in natural waters, J. Autom. Chem. 15 (1994) 159.

[8] N.J. Blundell, P.J. Worsfold, H. Casey, S. Smith, The design and performance of a portable, automated flow injection monitor for the *in-situ* analysis of nutrients in natural waters, Environ. Int. 21 (1995) 205.

[9] R.G.J. Bellerby, D.R. Turner, G.E. Millward, P.J. Worsfold, Shipboard flow injection determination of sea water pH with spectrophotometric detection, Anal. Chim. Acta 309 (1995) 259.

[10] L.K. Shpigun, I.Y. Kolotyrkna, Y.A. Zolotov, Experience with flow-injection analysis in marine chemical research, Anal. Chim. Acta 261 (1992) 307.

[11] K.N. Andrew, N.J. Blundell, D. Price, P.J. Worsfold, Flow injection techniques for water monitoring, Anal. Chem. 66 (1994) 916A.

[12] B.S. Gentle, P.S. Ellis, P.A. Fabera, M.R. Grace, I.D. McKelvie, A compact portable flow analysis system for the rapid determination of total phosphorus in estuarine and marine waters, Anal. Chim. Acta 674 (2010) 117.

[13] A.R. Bowie, E.P. Achterberg, P.N. Sedwick, S. Ussher, P.J. Worsfold, Real-time monitoring of picomolar concentrations of iron(II) in marine waters using automated flow injection-chemiluminescence instrumentation, Environ. Sci. Technol. 36 (2002) 4600.

[14] P.C.F.C. Gardolinski, A.R.J. David, P.J. Worsfold, Miniature flow injection analyser for laboratory, shipboard and *in situ* monitoring of nitrate in estuarine and coastal waters, Talanta 58 (2002) 1015.

[15] S. Gray, G. Hanrahan, I.D. McKelvie, A. Tappin, F. Tse, P. Worsfold, Flow analysis techniques for spatial and temporal measurement of nutrients in aquatic systems, Environ. Chem. 3 (2006) 3.

[16] C. Pasquini, M.V. Rebouças, Industrial and process analysis applications. Ch. 21, p.617, in: S.D. Kolev, I.D. McKelvie (Eds.), Advances in Flow Injection Analysis and Related Techniques, Wilson and Wilson's Comprehensive Analytical Chemistry, vol. 54, Elsevier, Amsterdam, 2008.

[17] M. Miro, W. Frenzel, Automated membrane-based sampling and sample preparation exploiting flow-injection analysis, Trends Anal. Chem. 23 (2004) 624.
[18] S.S. Park, J.H. Hong, J.H. Lee, Y.J. Kim, S.Y. Cho, S.J. Kim, Investigation of nitrous acid concentration in an indoor environment using an *in-situ* monitoring system, Atmos. Environ. 42 (2008) 6586.
[19] K. Toda, P.K. Dasgupta, Environmental applications: atmospheric trace gas analyses. Ch. 22, p. 640, in: S.D. Kolev, I.D. McKelvie (Eds.), Advances in Flow Injection Analysis and Related Techniques, Wilson and Wilson's Comprehensive Analytical Chemistry, vol. 54, Elsevier, Amsterdam, 2008.
[20] C. Pasquini, M.C.H. Silva, Determination of gaseous analytes in flow systems. A review, J. Braz. Chem. Soc. 10 (1999) 85.
[21] M.C.H. Silva, C. Pasquini, Bisegmented system for determination of low concentrations of gaseous constituents in gaseous samples, Anal. Chim. Acta 393 (1999) 121.
[22] S. Vicente, E.A.G. Zagatto, P.C.A.G. Pinto, M.L.M.E.S. Saraiva, J.L.F.C. Lima, E.P. Borges, Exploiting gas diffusion for non-invasive sampling in flow analysis: determination of ethanol in alcoholic beverages, An. Acad. Bras. Cienc. 78 (2006) 23.
[23] R.C. Chirnside, H.J. Cluley, P.M.C. Proffitt, The determination of traces of boron in nickel, Analyst 82 (1957) 18.
[24] S. Barabas, S.G. Lea, Application of anodic dissolution technique to automated analysis of metals. Determination of phosphorus in copper, Anal. Chem. 37 (1965) 1132.
[25] I.G. Souza, H. Bergamin-Filho, F.J. Krug, J.A. Nobrega, P.V. Oliveira, B.F. Reis, M.F. Gine, On-line electrolytic dissolution of alloys in flow-injection analysis. Part 3. Multi-elemental analysis of stainless steels by inductively coupled plasma atomic emission spectrometry, Anal. Chim. Acta 245 (1991) 211.
[26] C.A. Giacomozzi, R.R.U. Queiroz, I.G. Souza, J.A. Gomes-Neto, High current-density anodic electro-dissolution in flow-injection systems for the determination of aluminium, copper and zinc in non-ferroalloys by flame atomic absorption spectrometry, J. Autom. Method. Manag. Chem. 21 (1999) 17.
[27] C.A. Coutinho, E.C. Arruda, G.S.P. Etrusco, Análise rápida de alumínio solúvel no controle de fabricação de aço acalmado, Metalurgica ABM 39 (1983) 87.
[28] K. Ohls, K.H. Koch, Electrolytische Lösungshertellung zur Stahlanalyse mit der ICP-Spektrometrie, Fresenius Z. Anal. Chem. 326 (1987) 520.
[29] H. Bergamin-Filho, F.J. Krug, E.A.G. Zagatto, E.C. Arruda, C.A. Coutinho, On-line electrolytic dissolution of alloys in flow-injection analysis. Part 1. Principles and application in the determination of soluble aluminium in steels, Anal. Chim. Acta 190 (1986) 177.
[30] H. Bergamin-Filho, F.J. Krug, B.F. Reis, J.A. Nobrega, M. Mesquita, I.G. Souza, On-line electrolytic dissolution of alloys in flow injection analysis. Part 2. Spectrophotometric determination of molybdenum in steels, Anal. Chim. Acta 214 (1988) 397.
[31] J. Flock, K. Ohls, On-Line Stahlanalyse mit der ICP-Emissionsspektrometrie nach elektrolytischem Lösen, Fresenius J. Anal. Chem. 331 (1988) 408.
[32] A.P.G. Gervasio, G.C. Luca, A.A. Menegario, B.F. Reis, H. Bergamin-Filho, On-line electrolytic dissolution of alloys in flow injection analysis. Determination of iron, tungsten, molybdenum, vanadium and chromium in tool steels by inductively coupled plasma atomic emission spectrometry, Anal. Chim. Acta 405 (2000) 213.
[33] M. Aimoto, H. Kondo, A. Ono, Determination of silicon in high-silicon electrical steel by ICP-AES with on-line sample electrolytic dissolution, Anal. Sci. 23 (2007) 1367.
[34] A.P. Packer, A.P.G. Gervasio, C.E.S. Miranda, B.F. Reis, A.A. Menegario, M.F. Gine, On-line electrolytic dissolution for lead determination in high-purity copper by isotope dilution inductively coupled plasma mass spectrometry, Anal. Chim. Acta 485 (2003) 145.

[35] D.X. Yuan, X.R. Wang, P.Y. Yang, B.L. Huang, On-line electrolytic dissolution of solid metal samples and determination of copper in aluminium alloys by flame atomic absorption spectrometry, Anal. Chim. Acta 243 (1991) 65.
[36] J.B.B. Silva, M.B.O. Giacomelli, I.G. Souza, A.J. Curtius, Automated determination of tin and nickel in brass by on-line anodic electrodissolution and electrothermal atomic absorption spectrometry, Talanta 47 (1998) 1191.
[37] J.B.B. Silva, I.G. Souza, A.P.G. Gervasio, Anodic electrodissolution in flow injection system: a fast and efficient alternative for alloys dissolution, Quim. Nova 23 (2000) 244.
[38] W. Tiyapongpattana, P. Pongsakul, J. Shiowatana, D. Nacapricha, Sequential extraction of phosphorus in soil and sediment using a continuous-flow system, Talanta 62 (2004) 765.
[39] D.S. Ting, A. Appan, General characteristics and fractions of phosphorus in aquatic sediments of two tropical reservoirs, Water Sci. Technol. 34 (1996) 53.
[40] J. Buanuama, M. Miro, E.H. Hansen, J. Shiowatana, On-line dynamic fractionation and automatic determination of inorganic phosphorus in environmental solid substrates exploiting sequential injection microcolumn extraction and flow injection analysis, Anal. Chim. Acta 570 (2006) 224.
[41] M. Hulsman, M. Bos, W.E. van der Linden, Automated injection of slurry samples in flow-injection analysis, Anal. Chim. Acta 324 (1996) 13.
[42] C.E.C. Magalhaes, M.A.Z. Arruda, Slurry sampling: the technique employment in the direct sample analysis, Quim. Nova 21 (1998) 459.
[43] G.P. Meade, J.C.P. Chen, Cane Sugar Handbook: A Manual for Cane Sugar Manufacturers and their Chemists, tenth ed., John Wiley & Sons, New York, 1977.
[44] G.F. Pang, Y.Z. Cao, C.L. Fan, J.J. Zhang, X.M. Li, Z.Y. Li, G.Q. Jia, Liquid chromatography−fluorescence detection for simultaneous analysis of sulfonamide residues in honey, Anal. Bioanal. Chem. 376 (2003) 534.
[45] J.M.T. Carneiro, A.C.B. Dias, E.A.G. Zagatto, J.L.M. Santos, J.L.F.C. Lima, An improved sampling approach in multi-pumping flow systems applied to the spectrophotometric determination of glucose and fructose in syrups, Anal. Chim. Acta 531 (2005) 279.
[46] E.A.G. Zagatto, I.L. Mattos, A.O. Jacintho, Determination of sucrose in sugar-cane juice and molasses by flow-injection spectrophotometry, Anal. Chim. Acta 204 (1988) 259.
[47] C. Pasquini, W.A. Oliveira, Monosegmented system for continuous-flow analysis. Spectrophotometric determination of chromium(VI), ammonia, and phosphorus, Anal. Chem. 57 (1985) 2575.
[48] J. Ruzicka, E.H. Hansen, Recent developments in flow-injection analysis. Gradient techniques and hydrodynamic injection, Anal. Chim. Acta 145 (1983) 1.
[49] V.O. Brito, I.M. Raimundo-Junior, Simultaneous multiple injection in monosegmented flow analysis, Anal. Chim. Acta 371 (1988) 317.
[50] E.A.G. Zagatto, O. Bahia-Filho, M.F. Gine, H. Bergamin-Filho, A simple procedure for hydrodynamic injection in flow injection analysis applied to the atomic absorption spectrometry of chromium in steels, Anal. Chim. Acta 181 (1986) 265.
[51] S. Olsen, J. Ruzicka, E.H. Hansen, Gradient techniques in flow injection analysis. Stopped-flow measurements of the activity of lactate dehydrogenase with electronic dilution, Anal. Chim. Acta 136 (1982) 101.
[52] J. Ruzicka, Flow injection methods: A new tool for instrumental analysis, Philos. T. Roy. Soc. A 305 (1982) 645.
[53] B.F. Reis, A.O. Jacintho, J. Mortatti, F.J. Krug, E.A.G. Zagatto, H. Bergamin-Filho, L.C.R. Pessenda, Zone sampling processes in flow injection analysis, Anal. Chim. Acta 123 (1981) 221.

[54] G.D. Clark, J. Ruzicka, G.D. Christian, Split zone flow injection analysis: an approach to automated dilutions, Anal. Chem. 61 (1989) 1773.
[55] J. Mindegaard, Flow multi-injection analysis – a system for the analysis of highly concentrated samples without prior dilution, Anal. Chim. Acta 104 (1979) 185.
[56] J. Alonso, J. Bartroli, M. del Valle, M. Escalada, R. Barber, Sandwich techniques in flow injection analysis. Part 1. Continuous recalibration techniques for process control, Anal. Chim. Acta 199 (1987) 191.
[57] B. Horstkotte, E. Werner, A.K. Seresht, G. Cornelissen, O. Elsholz, V. Cerda, R. Luttmann, Sequential injection analyzer for glycerol monitoring in yeast cultivation medium, Talanta 71 (2007) 941.
[58] C.E. Lenehan, N.W. Barnett, S.W. Lewis, Sequential injection analysis, Analyst 127 (2002) 997.
[59] E.A.G. Zagatto, J.M.T. Carneiro, S. Vicente, P.R. Fortes, J.L.M. Santos, J.L.F.C. Lima, Mixing chambers in flow analysis: a review, J. Anal. Chem. (Russia) 64 (2009) 540.
[60] S.S.M.P. Vidigal, I.V. Toth, A.O.S.S. Rangel, Sequential injection lab-on-valve system for the on-line monitoring of hydrogen peroxide in lens care solutions, Microchem. J. 91 (2009) 197.
[61] L.F. Capitan-Vallvey, M.C. Valencia, E. Arana-Nicolas, Flow injection analysis with on-line solid phase extraction for spectrophotometric determination of Ponceau 4R and its subsidiary unsulfonated dye in sweets and cosmetic products, Mikrochim. Acta 138 (2002) 69.
[62] W.R. Melchert, F.R.P. Rocha, A green analytical procedure for flow-injection determination of nitrate in natural waters, Talanta 65 (2005) 461.
[63] R. Kuroda, T. Mochizuki, Continuous spectrophotometric determination of copper, nickel and zinc in copper-base alloys by flow-injection analysis, Talanta 28 (1981) 389.
[64] D.D. Cunningham, Fluidics and sample handling in clinical chemical analysis, Anal. Chim. Acta 429 (2001) 1.
[65] H. Bergamin-Filho, B.F. Reis, A.O. Jacintho, E.A.G. Zagatto, Ion exchange in flow injection analysis. Determination of ammonium ions at the μg/l level in natural waters with pulsed Nessler reagent, Anal. Chim. Acta 117 (1980) 81.
[66] A.P.G. Gervasio, P.R. Fortes, S.R.P. Meneses, C.E.S. Miranda, E.A.G. Zagatto, An improved flow-injection system for spectrophotometric determination of molybdenum and tungsten in tool steels, Talanta 69 (2006) 927.
[67] F.J. Krug, E.A.G. Zagatto, B.F. Reis, O. Bahia-Filho, A.O. Jacintho, S.S. Jørgensen, Turbidimetric determination of sulphate in plant digests and natural waters by flow injection analysis with alternating streams, Anal. Chim. Acta 145 (1983) 179.
[68] E.A.G. Zagatto, B.F. Reis, M. Martinelli, F.J. Krug, H. Bergamin-Filho, M.F. Gine, Confluent streams in flow injection analysis, Anal. Chim. Acta 198 (1987) 153.
[69] A.M. Pimenta, M.C.B.S.M. Montenegro, A.N. Araujo, J.M. Calatayud, Application of sequential injection analysis to pharmaceutical analysis, J. Pharm. Biomed. Anal. 40 (2006) 16.
[70] W.D. Basson, P.P. Pille, L. du Preez, Automated *in situ* preparation of Azomethine-H and the subsequent determination of boron in aqueous solution, Analyst 99 (1974) 168.
[71] W.D. Basson, R.G. Boehmer, D.A. Stanton, An automated procedure for the determination of boron in plant tissue, Analyst 94 (1969) 1135.
[72] R.A. Edwards, Automatic determination of boron (0.10–10.0 mg l^{-1}) in raw and waste waters, Analyst 105 (1980) 139.
[73] J.F. van Staden, T.A. van der Merwe, Automated *in situ* preparation of Azomethine-H and the subsequent determination of boron in fertilizer process and water effluent streams with sequential injection analysis, Analyst 125 (2000) 2094.

[74] R.C. Schothorst, J.M. Reijn, H. Poppe, G. den Boef, The application of strongly reducing agents in flow injection analysis. Part 1. Cr(II) and V(II), Anal. Chim. Acta 145 (1983) 197.
[75] R.C. Schothorst, G. den Boef, The application of strongly reducing agents in flow injection analysis. Part 2. Cr(II), Anal. Chim. Acta 153 (1983) 133.
[76] R.C. Schothorst, J.J.F. van Veen, G. den Boef, The application of strongly reducing agents in flow injection analysis. Part 3. V(II), Anal. Chim. Acta 161 (1984) 27.
[77] R.C. Schothorst, M. van Son, G. den Boef, The application of strongly reducing agents in flow injection analysis. Part 4. U(III), Anal. Chim. Acta 162 (1984) 1.
[78] R.C. Schothorst, G. den Boef, The application of strongly reducing agents in flow injection analysis. Part 5. Chromium(II) and vanadium(II) in acidic medium, Anal. Chim. Acta 175 (1985) 305.
[79] W.Th. Kok, D.T. Thuy, T.V. Nghi, G. den Boef, The application of strongly reducing agents in flow injection analysis. Part 6. Molybdenum(III), Anal. Chim. Acta 200 (1987) 533.
[80] R.C. Schothorst, G. den Boef, The application of strongly oxidizing agents in flow injection analysis. Part 1. Ag(II), Anal. Chim. Acta 169 (1985) 99.
[81] R.C. Schothorst, O.O. Schmitz, G. den Boef, The application of strongly oxidizing agents in flow injection analysis. Part 2. Mn(III), Anal. Chim. Acta 179 (1986) 299.
[82] R.C. Schothorst, G. den Boef, The application of strongly oxidizing agents in flow injection analysis. Part 3. Co(III), Anal. Chim. Acta 181 (1986) 235.
[83] W.E. van der Linden, G. den Boef, W. Ozinga, The application of strongly oxidizing agents in flow-injection analysis. Part 4. Manganese(VI) and copper(III), Anal. Chim. Acta 211 (1988) 99.
[84] G. den Boef, Unstable reagents in flow analysis, Anal. Chim. Acta 216 (1989) 289.
[85] G. den Boef, R.C. Schothorst, Unstable reagents in quantitative analysis? Why not apply FIA!, Anal. Chim. Acta 180 (1986) 1.
[86] X.W. Zheng, Z.J. Zhang, B.X. Li, Flow injection chemiluminescence determination of captopril with *in situ* electrogenerated Mn^{3+} as the oxidant, Electroanal. 13 (2001) 1046.
[87] B.X. Li, Z.J. Zhang, M.L. Wu, Flow-injection chemiluminescence determination of captopril using on-line electrogenerated silver(II) as the oxidant, Microchem. J. 70 (2001) 85.
[88] C.X. Zhang, J.C. Huang, Z.J. Zhang, M. Aizawa, Flow injection chemiluminescence determination of catecholamines with electrogenerated hypochlorite, Anal. Chim. Acta 374 (1998) 105.
[89] B.X. Li, Z.J. Zhang, W. Liu, Flow-injection chemiluminescence determination of chlortetracycline using on-line electrogenerated $[Cu(HIO_6)_2]^{5-}$ as the oxidant, Talanta 55 (2001) 1097.
[90] B.X. Li, Z.J. Zhang, M.L. Wu, Flow injection chemiluminescence determination of gentamycin using on-line electrogenerated Co(III) as the oxidant, Anal. Lett. 33 (2000) 1577.
[91] B.X. Li, Z.J. Zhang, X.W. Zheng, C.L. Xu, Flow injection chemiluminescence determination of isoniazid using on-line electrogenerated manganese(III) as oxidant, Microchem. J. 63 (1999) 374.
[92] X.W. Zheng, Z.J. Zhang, Flow-injection chemiluminescence determination of isoniazid using on-line electrogenerated BrO^- as oxidant, Analyst 124 (1999) 763.
[93] X.W. Zheng, Z.J. Zhang, Flow injection chemiluminescence determination of quinine using Mn^{3+} as the oxidant, Anal. Sci. 16 (2000) 1345.
[94] B.X. Li, Z.J. Zhang, M.L. Wu, Flow-injection chemiluminescence determination of quinine using on-line electrogenerated cobalt(III) as oxidant, Talanta 51 (2000) 515.

[95] X.W. Zheng, Y. Mei, Z.J. Zhang, Flow-injection chemiluminescence determination of tetracyclines with *in situ* electrogenerated bromine as the oxidant, Anal. Chim. Acta 440 (2001) 143.
[96] B.X. Li, Z.J. Zhang, M.L. Wu, Flow-injection chemiluminescence determination of sulfite using on-line electrogenerated silver(II) as the oxidant, Anal. Chim. Acta 432 (2001) 311.
[97] X.W. Zheng, Z.J. Zhang, Flow-injection chemiluminescence detecting sulfite with *in situ* electrogenerated Mn^{3+} as the oxidant, Sens. Actuators. B. Chem. 84 (2002) 142.
[98] F.C. Camargo, E.A.G. Zagatto, C.C. Oliveira, A flow-injection system for spectrophotometric catalytic determination of copper in plant digests and natural waters, Anal. Sci. 14 (1998) 565.
[99] A. Mellado-Romero, C. Gomez-Benito, J. Martinez-Calatayud, Continuous-flow spectrophotometric determination of sulfadiazine by diazotisation with *in situ* preparation of nitrite, Anal. Chim. Acta 308 (1995) 451.
[100] W.T. Suarez, H.J. Vieira, O. Fatibello-Filho, Generation and destruction of unstable reagent in flow injection system: determination of acetylcysteine in pharmaceutical formulations using bromine as reagent, J. Pharm. Biomed. Anal. 37 (2005) 771.
[101] G.D. Marshall, D. Wolcott, D. Olson, Zone fluidics in flow analysis: potentialities and applications, Anal. Chim. Acta 499 (2003) 29.
[102] S.C.B. Oliveira, E.C.S. Coelho, T.M.G. Selva, F.P. Santos, M.C.U. Araujo, F.C. Abreu, V.B. Nascimento, A coulometric flow cell for in-line generation of reagent, titrant or standard solutions, Microchem. J. 82 (2006) 220.
[103] R.H. Taylor, J. Ruzicka, G.D. Christian, Flow-injection coulometric titrations, Talanta 39 (1992) 285.
[104] A.D. Dakashev, V.T. Dimitrova, Pulse coulometric titration in continuous flow, Analyst 119 (1994) 1835.
[105] C. Pasquini, E.V. Aquino, M.V. Rebouças, F.B. Gonzaga, Robust flow-batch coulometric/biamperometric titration system: Determination of bromine index and bromine number of petrochemicals, Anal. Chim. Acta 600 (2007) 84.
[106] M.D. Luque de Castro, M.T. Tena, Solid interfaces as analytical problem solvers in flow injection analysis, Talanta 40 (1993) 21.
[107] J.M. Calatayud, Flow Injection Analysis of Pharmaceuticals: Automation in the Laboratory, Taylor & Francis, London, 1996, p. 263.
[108] H.A. Bruno, F.J. Andrade, P.C. Luna, M.B. Tudino, Kinetic control of reagent dissolution for the flow injection determination of iron at trace levels, Analyst 127 (2002) 990.
[109] M.F. Gine, H. Bergamin-Filho, E.A.G. Zagatto, B.F. Reis, Simultaneous determination of nitrate and nitrite by flow injection analysis, Anal. Chim. Acta 114 (1980) 191.
[110] M. Noroozifar, M. Khorasani-Motlagh, The application of the solid-phase Jones reagent as a reductant in the speciation flow injection analysis of Fe(III) and Fe(II) in real samples, Chem. Anal., (Warsaw) 49 (2004) 929.
[111] J.H. Lowry, K.H. Mancy, A rapid automated system for the analysis of dissolved total organic nitrogen in aqueous solutions, Water Res. 12 (1978) 471.
[112] J.F. van Staden, L.G. Kluever, Determination of manganese in natural water and effluent streams using a solid-phase lead(IV) dioxide reactor in a flow-injection system, Anal. Chim. Acta 350 (1997) 15.
[113] B.-X. Li, Z.-J. Zhang, W. Liu, Flow-injection chemiluminescence determination of manganese(II) in natural waters with solid bismuthate as an oxidant, Anal. Sci. 17 (2001) 1347.
[114] M. Kamaya, K. Nagashima, E. Ishii, Determination of sulphate and phosphate by flow-injection analysis using a barium chloranilate packed column, Fresenius' J. Anal. Chem. 347 (1993) 409.

[115] J.M.T. Carneiro, R.P. Sartini, E.A.G. Zagatto, Spectrophotometric determination of total nitrogen in plant materials using a flow-injection system with an AgCl(s) reactor, Anal. Chim. Acta 416 (2000) 185.
[116] F.J. Andrade, M.B. Tudino, O.E. Troccoli, Flow dissolution of 1,5-diphenylcarbazide for the determination of chromium(VI), Analyst 121 (1996) 613.
[117] S.R. Varma, J.M. Calatayud, H.A. Mottola, Continuous-flow determination of reserpine by oxidation with periodate ion and catalysis by manganese(II) in solution or by an MnO_2(s) reactor, Anal. Chim. Acta 233 (1990) 235.
[118] K.S. Seo, K.H. Choo, H.N. Chang, J.K. Park, A flow injection analysis system with encapsulated high-density *Saccharomyces cerevisiae* cells for rapid determination of biochemical oxygen demand, Appl. Microbiol. Biotechnol. 83 (2009) 217.
[119] P.J. Worsfold, J. Farrelly, M.S. Matharu, A comparison of spectrophotometric and chemiluminescence methods for the determination of blood glucose by flow injection analysis, Anal. Chim. Acta 164 (1985) 103.
[120] S.M.B. Brienza, R.P. Sartini, J.A. Gomes-Neto, E.A.G. Zagatto, Crystal seeding in flow-injection turbidimetry: determination of total sulfur in plants, Anal. Chim. Acta 308 (1995) 269.
[121] M.S. Nezio, M.E. Palomeque, B.S.F. Band, Automated flow-injection method for cadmium determination with preconcentration and reagent preparation on-line, Quim. Nova 28 (2005) 145.
[122] A. Ferrari, Nitrogen determination by a continuous digestion and analysis system, Ann. N. Y. Acad. Sci. 87 (1960) 792.
[123] T. Korenaga, H.Y. Ikatsu, The determination of chemical oxygen demand in waste-waters with dichromate by flow injection analysis, Anal. Chim. Acta 141 (1982) 301.
[124] Y. Hirai, N. Yoza, S. Ohashi, Flow injection analysis of inorganic polyphosphates, Anal. Chim. Acta 115 (1980) 269.
[125] H.M. Kingston, L.B. Jassie, Introduction to Microwave Sample Preparation: Theory and Practice, American Chemical Society, Washington, 1988.
[126] M. Burguera, J.L. Burguera, Microwave-assisted sample decomposition in flow analysis, Anal. Chim. Acta 366 (1998) 63.
[127] F.E. Smith, E.A. Arsenault, Microwave-assisted sample preparation in analytical chemistry, Talanta 43 (1996) 1207.
[128] A.F. Oliveira, O. Fatibello-Filho, J.A. Nobrega, Focused-microwave-assisted reaction in flow injection spectrophotometry: a new liquid-vapour separation chamber for determination of reducing sugar in wine, Talanta 55 (2001) 677.
[129] V. Cerda, J.M. Estela, On-line sample pretreatment: dissolution and digestion. Ch. 6, p. 129, in: S.D. Kolev, I.D. McKelvie (Eds.), Advances in Flow Injection Analysis and Related Techniques, Wilson and Wilson's Comprehensive Analytical Chemistry, vol. 54, Elsevier, Amsterdam, 2008.
[130] S.C. Stephen, D. Littlejohn, J.M. Ottaway, Evaluation of a slurry technique for the determination of lead in spinach by electrothermal atomic absorption spectrometry, Analyst 110 (1985) 1147.
[131] M.R. Cave, O. Butler, J.M. Cook, M.S. Cresser, L.M. Garden, D.L. Miles, Environmental analysis, J. Anal. At. Spectrom. 15 (2000) 181.
[132] J.L. Burguera, M. Burguera, Flow injection-electrothermal atomic absorption spectrometry configurations: recent developments and trends, Spectrochim. Acta B 56 (2001) 1801.
[133] M. Burguera, J.L. Burguera, O.M. Alarcon, Flow injection and microwave-oven sample decomposition for determination of copper, zinc and iron in whole blood by atomic absorption spectrometry, Anal. Chim. Acta 179 (1986) 351.
[134] M.L. Balconi, M. Borgarello, R. Ferraroli, F. Realini, Chemical oxygen demand determination in well and river waters by flow-injection analysis using a microwave oven during the oxidation step, Anal. Chim. Acta 261 (1992) 295.

[135] C.C. Oliveira, E.A.G. Zagatto, A.N. Araujo, J.L.F.C. Lima, Sample preparation in sequential injection analysis. Spectrophotometric determination of total phosphorus in food samples, Anal. Chim. Acta 371 (1998) 57.

[136] J. Golimowski, K. Golimowska, UV-photooxidation as pretreatment step in inorganic analysis of environmental samples, Anal. Chim. Acta 325 (1996) 111.

[137] A. Cavicchioli, I.G.R. Gutz, Use of UV radiation for sample pretreatment in inorganic analysis, Quim. Nova 26 (2003) 913.

[138] E.P. Achterberg, C.B. Braungardt, R.C. Sandford, P.J. Worsfold, UV digestion of seawater samples prior to the determination of copper using flow injection with chemiluminescence detection, Anal. Chim. Acta 440 (2001) 27.

[139] A. Cavicchioli, I.G.R. Gutz, In-line TiO_2-assisted photodigestion of organic matter in aqueous solution for voltammetric flow analysis of heavy metals in water samples, Anal. Chim. Acta 445 (2001) 127.

[140] S. Hinkamp, G. Schwedt, Determination of total phosphorus in waters with amperometric detection by coupling of flow-injection analysis with continuous microwave oven digestion, Anal. Chim. Acta 236 (1990) 345.

[141] M.T. Oms, A. Cerda, V. Cerda, Sequential injection system for on-line analysis of total nitrogen with UV-mineralization, Talanta 59 (2003) 319.

[142] I.D. McKelvie, B.T. Hart, T.J. Cardwell, R.W. Cattrall, Spectrophotometric determination of dissolved organic phosphorus in natural waters using in-line photo-oxidation and flow injection, Analyst 114 (1989) 1459.

[143] F. Priego-Capote, M.D. Luque de Castro, Analytical uses of ultrasound. I. Sample preparation, Trends Anal. Chem. 23 (2004) 9.

[144] P. Linares, F. Lazaro, M.D. Luque de Castro, M. Valcarcel, Effects of ultrasonic irradiation in flow-injection systems, Anal. Chim. Acta 200 (1987) 51.

[145] F. Priego-Capote, M.D. Luque de Castro, Ultrasound-assisted digestion: A useful alternative in sample preparation, J. Biochem. Biophys. Methods 70 (2007) 299.

[146] A. Caballo-Lopez, M.D. Luque de Castro, Continuous ultrasound-assisted leaching of phenoxyacid herbicides in soil and sediment with in-situ sample treatment, Chromatographia 58 (2003) 257.

[147] G.M. Greenway, T. Leelasattarathkul, S. Liawruangrath, R.A. Wheatley, N. Youngvises, Ultrasound-enhanced flow injection chemiluminescence for determination of hydrogen peroxide, Analyst 131 (2006) 501.

[148] A. Correia, M. Galesio, H. Santos, R. Rial-Otero, C. Lodeiro, A. Oehmen, A.C.L. Conceicao, J.L. Capelo, Can sample treatments based on advanced oxidation processes assisted by high-intensity focused ultrasound be used for toxic arsenic determination in human urine by flow-injection hydride-generation atomic absorption spectrometry? Talanta 72 (2007) 968.

[149] F. Priego-Capote, M.D. Luque de Castro, Ultrasound-assisted continuous liquid–liquid extraction without phase separation and hydrolysis of paracetamol in suppositories, Anal. Chim. Acta 489 (2003) 223.

[150] R.Y. Wang, J.A. Jarratt, P.J. Keay, J.J. Hawkes, W.T. Coakley, Development of an automated on-line analysis system using flow injection, ultrasound filtration and CCD detection, Talanta 52 (2000) 129.

[151] J.F. van Staden, Automated dilution in flow injection analysis with double on-line dialysis. A system for the determination of chloride in industrial effluents and plating bath solutions, Fresenius' J. Anal. Chem. 340 (1991) 415.

[152] M. Miro, E.H. Hansen, On-line processing methods in flow analysis, in: M. Trojanowicz (Ed.), Advances in Flow Methods of Analysis, Wiley-VCH, Weinhem, 2008, pp. 291–320.

[153] C.I.C. Silvestre, J.L.M. Santos, J.L.F.C. Lima, E.A.G. Zagatto, Liquid-liquid extraction in flow analysis: a critical review, Anal. Chim. Acta 652 (2009) 54.

[154] T.I. Zvarova, V.M. Shkinev, G.A. Vorob'eva, B.Ya. Spivakov, Yu.A. Zolotov, Liquid-liquid extraction in the absence of usual organic solvents. Application of two-phase aqueous systems based on a water-soluble polymer, Mikrochim. Acta 3 (1984) 449.
[155] L. Bulgariu, D. Bulgariu, Extraction of metal ions in aqueous polyethylene glycol–inorganic salt two-phase systems in the presence of inorganic extractants: correlation between extraction behaviour and stability constants of extracted species, J. Chromatogr. A 1196–1197 (2008) 117.
[156] R.E. Stanton, A.J. McDonald, Application of the Auto Analyzer to the determination of zinc in soils and sediments, Analyst 88 (1963) 608.
[157] E.I. Pentz, Liquid–liquid extraction in the autoanalyzer. Determination of taurine, Anal. Biochem. 8 (1964) 328.
[158] B. Karlberg, S. Thelander, Extraction based on flow-injection principle. Part 1. Description of extraction system 98 (1978) 1.
[159] H. Bergamin-Filho, J.X. Medeiros, B.F. Reis, E.A.G. Zagatto, Solvent extraction in continuous flow injection analysis. Determination of molybdenum in plant material, Anal. Chim. Acta 101 (1978) 9.
[160] K. Kina, K. Shiraishi, N. Ishibashi, Ultramicro solvent extraction and fluorimetry based on the flow-injection method, Talanta 25 (1978) 295.
[161] S. Motomizu, T. Sakai, On-line sample pretreatment: extraction and preconcentration, in: S.D. Kolev, I.D. McKelvie (Eds.), Advances in Flow Injection Analysis and Related Techniques, Wilson and Wilson's Comprehensive Analytical Chemistry, vol. 54, Elsevier, Amsterdam, 2008, pp. 159–201. Chapter 7.
[162] G. Audunsson, Aqueous/aqueous extraction by means of a liquid membrane for sample cleanup and preconcentration of amines in a flow system, Anal. Chem. 58 (1986) 2714.
[163] L. Nord, B. Karlberg, Extraction based on the flow-injection principle. Part 6. Film formation and dispersion in liquid liquid segmented flow extraction systems, Anal. Chim. Acta 164 (1984) 233.
[164] V. Kuban, Liquid–liquid extraction flow injection analysis, Crit. Rev. Anal. Chem. 22 (1991) 477.
[165] Z.-L. Fang, Flow Injection Separation and Preconcentration, VCH Pub., Weinheim, 1993, pp. 85–128.
[166] I. Facchin, C. Pasquini, Liquid–liquid extraction in flow systems, Quim. Nova 21 (1998) 60.
[167] J. Kawase, A. Nakae, M. Yamanaka, Determination of anionic surfactants by flow injection analysis based on ion pair extraction, Anal. Chem. 51 (1979) 1640.
[168] T. Sakai, Y.S. Chung, N. Ohno, S. Motomizu, Double-membrane phase separator for liquid–liquid extraction in flow-injection analysis, Anal. Chim. Acta 276 (1993) 127.
[169] L.N. Moskvin, J. Simon, P. Loffler, N.V. Michailova, D.N. Nicolaevna, Photometric determination of anionic surfactants with a flow-injection analyzer that includes a chromatomembrane cell for sample preconcentration by liquid-liquid solvent extraction, Talanta 43 (1996) 819.
[170] K. Bäckström, L.G. Danielsson, F. Ingman, H.Z. Zhao, Studies of two-phase equilibria by liquid-liquid segmented flow extraction of dithiocarbamic acids into various solvents, Talanta 34 (1987) 783.
[171] K.E. Rasmussen, S. Pedersen-Bjergaard, Developments in hollow fibre-based, liquid-phase microextraction, Trends Anal. Chem. 23 (2004) 1.
[172] J.F. Liu, J.B. Chao, G.B. Jiang, Continuous flow liquid membrane extraction: a novel automatic trace-enrichment technique based on continuous flow liquid–liquid extraction combined with supported liquid membrane, Anal. Chim. Acta 455 (2002) 93.

[173] C.A. Lucy, F.F. Cantwell, Mechanism of extraction and band broadening in solvent extraction-flow injection analysis, Anal. Chem. 61 (1989) 107.
[174] C. Thommen, A. Fromageat, P. Obergfell, H.M. Widmer, Application of a capillary flow cell to sophisticated flow-injection systems, Anal. Chim. Acta 234 (1990) 141.
[175] H.H. Liu, P.K. Dasgupta, Flow-injection extraction without phase separation based on dual-wavelength spectrophotometry, Anal. Chim. Acta 288 (1994) 237.
[176] A.N. Anthemidis, M. Miro, Recent developments in flow injection/sequential injection liquid-liquid extraction for atomic spectrometric determination of metals and metalloids, Appl. Spectrosc. Rev. 44 (2009) 140.
[177] J.F. Silva, W. Martins, Extraction of Fe(III), Cu(II), Co(II), Ni(II) and Pb(II) with thenoyltrifluoroacetone using the ternary solvent system water/ethanol/methylisobutylketone, Talanta 39 (1992) 1307.
[178] C.A. Lucy, K.C. Yeung, Solvent extraction-flow injection without phase separation through the use of the differential flow velocities within the segmented flow, Anal. Chem. 66 (1994) 2220.
[179] F. Canete, A. Rios, M.D. Luque de Castro, M. Valcarcel, Liquid-liquid extraction in continuous flow systems without phase separation, Anal. Chem. 60 (1988) 2354.
[180] F. Canete, A. Rios, M.D. Luque de Castro, M. Valcarcel, Kinetics of ion-pair extraction in continuous flow systems, Anal. Chim. Acta 224 (1989) 169.
[181] J.A.G. Mesa, P. Linares, M.D. Luque de Castro, M. Valcarcel, Direct automatic determination of polyphenols in olive oils in the aqueous phase of a flow-injection liquid-liquid extraction system without phase separation, Anal. Chim. Acta 235 (1990) 441.
[182] Y.Y. Luo, R. Al-Othman, J. Ruzicka, G.D. Christian, Solvent extraction-sequential injection without segmentation and phase separation based on the wetting film formed on a teflon tube wall, Analyst 121 (1996) 601.
[183] M. Miro, A. Cladera, J.M. Estela, V. Cerda, Dual wetting-film multi-syringe flow injection analysis extraction. Application to the simultaneous determination of nitrophenols, Anal. Chim. Acta 438 (2001) 103.
[184] C.D. Stalikas, Micelle-mediated extraction as a tool for separation and preconcentration in metal analysis, Trends Anal. Chem. 21 (2002) 343.
[185] Q. Fang, M. Du, C.W. Huie, On-line incorporation of cloud point extraction to flow injection analysis, Anal. Chem. 73 (2001) 3502.
[186] C. Ortega, M.R. Gomez, R.A. Olsina, M.F. Silva, L.D. Martinez, On-line cloud point preconcentration and determination of gadolinium in urine using flow injection inductively coupled plasma optical emission spectrometry, J. Anal. At. Spectrom. 17 (2002) 530.
[187] J.L. Burguera, M. Burguera, Analytical applications of organized assemblies for online spectrometric determinations: present and future, Talanta 64 (2004) 1099.
[188] X.S. Zhu, B. Hu, Z.C. Jiang, M.F. Li, Cloud point extraction for speciation of chromium in water samples by electrothermal atomic absorption spectrometry, Water Res. 39 (2005) 589.
[189] V. Kuban, F. Ingman, Design of a multichannel dropping segmenter for liquid—liquid extraction continuous flow-injection analysis, Anal. Chim. Acta 245 (1991) 251.
[190] R. Burakham, J. Jakmunee, K. Grudpan, Development of sequential injection-lab-at-valve (SI-LAV) micro-extraction instrumentation for the spectrophotometric determination of an anionic surfactant, Anal. Sci. 22 (2006) 137.
[191] M.J. Silva, E.V. Anjos, R.S. Honorato, M.F. Pimentel, A.P.S. Paim, Spectrophotometric cocaine determination in a biphasic medium employing flow-batch sequential injection analysis, Anal. Chim. Acta 629 (2008) 98.
[192] K. Carlsson, B. Karlberg, Determination of octanol-water partition coefficients using a micro-volume liquid-liquid flow extraction system, Anal. Chim. Acta 423 (2000) 137.

[193] E. Rodenas-Torralba, B.F. Reis, A. Morales-Rubio, M. de la Guardia, An environmentally friendly multicommutated alternative to the reference method for anionic surfactant determination in water, Talanta 66 (2005) 591.
[194] M. Agudo, A. Rios, M. Valcarcel, Continuous liquid–liquid extraction with on-line monitoring for the determination of anionic surfactants in waters, Analyst 119 (1994) 2097.
[195] I. Facchin, C. Pasquini, Two-phase liquid-liquid extraction in monosegmented flow analysis. Determination of cadmium with 1-(2′-pyridylazo) naphthol, Anal. Chim. Acta 308 (1995) 231.
[196] I. Facchin, J.W. Martins, P.G.P. Zamora, C. Pasquini, Single-phase liquid–liquid extraction in monosegmented continuous-flow systems, Anal. Chim. Acta 285 (1994) 287.
[197] R.D. Rogers, A.H. Bond, C.B. Bauer, J.-H. Zhang, S.T. Griffin, Metal ion separations in polyethylene glycol-based aqueous biphasic systems: correlation of partitioning behavior with available thermodynamic hydration data, J. Chromatogr. B 680 (1996) 221.
[198] C.I.C. Silvestre, S. Rodrigues, J.L.M. Santos, J.L.F.C. Lima, E.A.G. Zagatto, Exploitation of a single interface flow system for on-line aqueous biphasic extraction, Talanta 81 (2010) 1847.
[199] P.K. Dasgupta, W. Lei, Solvent extraction in continuous flow systems with intelligent zone sampling, Anal. Chim. Acta 226 (1989) 255.
[200] C.C. Lindgren, P.K. Dasgupta, Flow injection and solvent extraction with intelligent segment separation. Determination of quaternary ammonium ions by ion-pairing, Talanta 39 (1992) 101.
[201] M. Miro, S.K. Hartwell, J. Jakmunee, K. Grudpan, E.H. Hansen, Recent developments in automatic solid-phase extraction with renewable surfaces exploiting flow-based approaches, Trends Anal. Chem. 27 (2008) 749.
[202] V. Cerda, J.M. Estela, Automatic preconcentration and treatment for the analysis of environmental samples using non-chromatographic flow techniques, Int. J. Environ. Anal. Chem. 85 (2005) 231.
[203] J. Ruzicka, E.H. Hansen, Flow Injection Analysis, second ed., Wiley-Interscience, New York, 1988.
[204] S. Olsen, L.C.R. Pessenda, J. Ruzicka, E.H. Hansen, Combination of flow-injection analysis with flame atomic absorption spectrophotometry. Determination of trace amounts of heavy metals in polluted seawater, Analyst 108 (1983) 905.
[205] M. Valcarcel, M.D. Luque de Castro, Non-chromatographic Continuous Separation Techniques, Royal Society of Chemistry, Cambridge, 1991.
[206] A.C.B. Dias, E.P. Borges, E.A.G. Zagatto, P.J. Worsfold, A critical examination of the components of the Schlieren effect in flow analysis, Talanta 68 (2006) 1076.
[207] J.L. Burguera, M. Burguera, A. Townshend, Determination of zinc and cadmium by flow injection analysis and chemiluminescence, Anal. Chim. Acta 127 (1981) 199.
[208] Y. Narusawa, T. Hashimoto, Simultaneous determination of phosphate, silicate, and arsenateby on-line column flow injection analysis, Chem. Lett. 16 (1987) 1367.
[209] M. Sperling, S.K. Xu, B. Welz, Determination of chromium(III) and chromium(VI) in water using flow injection on-line preconcentration with selective adsorption on activated alumina and flame atomic absorption spectrometric detection, Anal. Chem. 64 (1992) 3101.
[210] O.B. Egorov, M.J. O'Hara, O.T. Farmer III, J.W. Grate, Extraction chromatographic separations and analysis of actinides using sequential injection techniques with on-line inductively coupled plasma mass spectrometry (ICP MS) detection, Analyst 126 (2001) 1594.
[211] D. Satinsky, P. Solich, P. Chocholous, R. Karlicek, Monolithic columns. A new concept of separation in the sequential injection technique, Anal. Chim. Acta 499 (2003) 205.

[212] P. Chocholous, P. Solich, D. Satinsky, An overview of sequential injection chromatography, Anal. Chim. Acta 600 (2007) 129.
[213] V. Camel, Solid phase extraction of trace elements, Spectrochim. Acta B 58 (2003) 1177.
[214] Q.-S. Pu, P. Liu, Q.Y. Sun, Z.X. Su, Flame atomic absorption spectrometric determination of gold and palladium using microcolumn on-line preconcentration and separation, Microchim. Acta 143 (2003) 45.
[215] J. Porath, P. Flodin, Gel filtration: a method for desalting and group separation, Nature 183 (1959) 1657.
[216] M. Gallego, Y.P. Pena, M. Valcarcel, Fullerenes as sorbent materials for metal preconcentration, Anal. Chem. 66 (1994) 4074.
[217] D.S. Jesus, R.J. Cassella, S.L.C. Ferreira, A.C.S. Costa, M.S. Carvalho, R.E. Santelli, Polyurethane foam as a sorbent for continuous flow analysis: preconcentration and spectrophotometric determination of zinc in biological materials, Anal. Chim. Acta 366 (1998) 263.
[218] A.N. Anthemidis, G.A. Zachariadis, J.A. Stratis, On-line solid phase extraction system using PTFE packed column for the flame atomic absorption spectrometric determination of copper in water samples, Talanta 54 (2001) 935.
[219] Z.H. Wang, Z.P. Zhang, Z.P. Wang, L.W. Liu, X.P. Yan, Acrylic acid grafted polytetrafluoroethylene fiber as new packing for flow injection on-line microcolumn preconcentration coupled with flame atomic absorption spectrometry for determination of lead and cadmium in environmental and biological samples, Anal. Chim. Acta 514 (2004) 151.
[220] X.P. Yan, Y. Li, Y. Jiang, Selective measurement of ultratrace methylmercury in fish by flow injection on-line microcolumn displacement sorption preconcentration and separation coupled with electrothermal atomic absorption spectrometry, Anal. Chem. 75 (2003) 2251.
[221] K. Pyrzynska, M. Trojanowicz, Functionalized cellulose sorbents for preconcentration of trace metals in environmental analysis, Crit. Rev. Anal. Chem. 29 (1999) 313.
[222] S.L. Lin, C.S. Zheng, G.Z. Yun, Determination of palladium by flame atomic absorption spectrometry combined on-line with flow injection preconcentration using a micro-column packed with activated carbon fibre, Talanta 42 (1995) 921.
[223] G.D. Matos, M.A.Z. Arruda, Online preconcentration/determination of cadmium using grape bagasse in a flow system coupled to thermospray flame furnace atomic absorption spectrometry, Spectrosc. Lett. 39 (2006) 755.
[224] C.R.T. Tarley, S.L.C. Ferreira, M.A.Z. Arruda, Use of modified rice husks as a natural solid adsorbent of trace metals: characterisation and development of an on-line preconcentration system for cadmium and lead determination by FAAS, Microchem. J. 77 (2004) 163.
[225] B. Godlewska-Zylkiewicz, J. Malejko, B. Lesniewska, A. Kojlo, Assessment of immobilized yeast for the separation and determination of platinum in environmental samples by flow-injection chemiluminescence and electrothermal atomic absorption spectrometry, Microchim. Acta 163 (2008) 327.
[226] V.A. Lemos, A.A. Jesus, E.M. Gama, G.T. David, R.T. Yamaki, On-line solid phase extraction and determination of copper in food samples using polyurethane foam loaded with Me-BTANC, Anal. Lett. 38 (2005) 683.
[227] J.R. Ferreira, E.A.G. Zagatto, M.A.Z. Arruda, S.M.B. Brienza, Determination of zinc in plants by flow injection spectrophotometry with ion-exchange separation, Analyst 115 (1990) 779.
[228] G. Theodoridis, C.K. Zacharis, P.D. Tzanavaras, D.G. Themelis, A. Economou, Automated sample preparation based on the sequential injection principle. Solid-phase extraction on a molecularly imprinted polymer coupled on-line to high-performance liquid chromatography, J. Chromatogr. A 1030 (2004) 69.

[229] A.C.B. Dias, E.C. Figueiredo, V. Grassi, E.A.G. Zagatto, M.A.Z. Arruda, Molecularly imprinted polymer as a solid phase extractor in flow analysis, Talanta 76 (2008) 988.
[230] C.F. Poole, New trends in solid-phase extraction, Trends Anal. Chem. 22 (2003) 362.
[231] C. Pons, R. Forteza, V. Cerda, The use of anion-exchange disks in an optrode coupled to a multi-syringe flow-injection system for the determination and speciation analysis of iron in natural water samples, Talanta 66 (2005) 210.
[232] S. Yalcin, X.C. Le, Speciation of arsenic using solid phase extraction cartridges, J. Environ. Monit. 3 (2001) 81.
[233] K.T. Lau, S.A.L. Fortescu, L.J. Murphy, J.M. Slater, Disposable glucose sensors for flow injection analysis using substituted 1,4-benzoquinone mediators, Electroanal. 15 (2003) 975.
[234] J.H. Wang, E.H. Hansen, M. Miro, Sequential injection–bead injection–lab-on-valve schemes for on-line solid phase extraction and preconcentration of ultra-trace levels of heavy metals with determination by electrothermal atomic absorption spectrometry and inductively coupled plasma mass spectrometry, Anal. Chim. Acta 499 (2003) 139.
[235] M.R. McSwain, R.J. Watrous, J.E. Douglass, Improved methylthymol blue procedure for automated sulfate determination, Anal. Chem. 46 (1974) 1329.
[236] L.C.R. Pessenda, A.O. Jacintho, E.A.G. Zagatto, Catalytic determination of molybdenum in plants by flow-injection spectrophotometry with ion-exchange separation, Anal. Chim. Acta 214 (1988) 239.
[237] P.J. Fletcher, K.N. Andrew, S. Forbes, P.J. Worsfold, Automated flow injection analyzer with on-line solid-phase extraction and chemiluminescence detection for the determination of dodecylamine in diesel fuels, Anal. Chem. 75 (2003) 2618.
[238] J.A. Sweileh, E.M. El-Nemma, On-line elimination of spectral interference of iron matrix in the flame atomic absorption determination of zinc by anion-exchange separation, Anal. Chim. Acta 523 (2004) 287.
[239] C. Pons, R. Forteza, V. Cerda, Multi-pumping flow system for the determination, solid-phase extraction and speciation analysis of iron, Anal. Chim. Acta 550 (2005) 33.
[240] C.C. Oliveira, E.A.G. Zagatto, J. Ruzicka, G.D. Christian, A bead injection system for copper determination, Anal. Lett. 33 (2000) 929.
[241] J.B. Kafil, C.O. Huber, Flow injection sample processing with nickel oxide electrode amperometric detection of amino acids separated by ion-exchange chromatography, Anal. Chim. Acta 139 (1982) 347.
[242] D. Betteridge, N.G. Courtney, T.J. Sly, D.G. Porter, Development of a flow-injection analyzer for the post-column detection of sugars separated by high-performance liquid chromatography, Analyst 109 (1984) 91.
[243] F. Han, Y.Z. He, L. Li, G.-N. Fu, H.-Y. Xie, W.E. Gan, Determination of benzoic acid and sorbic acid in food products using electrokinetic flow analysis-ion pair solid phase extraction-capillary zone electrophoresis, Anal. Chim. Acta 618 (2008) 79.
[244] M. Manera, M. Miro, M. Estela, V. Cerda, M.A. Segundo, J.L.F.C. Lima, Flow-through solid-phase reflectometric method for simultaneous multiresidue determination of nitrophenol derivatives, Anal. Chim. Acta 600 (2007) 155.
[245] I. Nukatsuka, S. Osanai, K. Ohzeki, Preparation of open tubular solid-phase extraction column with 5-amino-8-hydroxyquinoline-modified gold nanoparticle phase for the enrichment of heavy metal ions, Anal. Sci. 24 (2008) 267.
[246] M.F.T. Ribeiro, A.C.B. Dias, J.L.M. Santos, J.L.F.C. Lima, E.A.G. Zagatto, Fluidized beds in flow analysis: use with ion-exchange separation for spectrophotometric determination of zinc in plant digests, Anal. Bioanal. Chem. 384 (2006) 1019.
[247] W.E. van der Linden, Membrane separation in flow injection analysis. Gas diffusion, Anal. Chim. Acta 151 (1983) 359.

[248] L.N. Moskvin, T.G. Nikitina, Membrane methods of substance separation in analytical chemistry, J. Anal. Chem. 59 (2004) 2.
[249] F. Delgado-Reyes, I. Papaefstathiou, J.M. Fernandez-Romero, M.D. Luque de Castro, Monitoring ethanol production during wine fermentation processes by pervaporation-enzymic derivatisation approach, Analyst 123 (1998) 2367.
[250] A. Schmitt, L. Buttle, R. Uglow, K. Williams, S. Haswell, An integrated reduction method for the determination of urea as ammonia in fresh water samples, Anal. Chim. Acta 284 (1993) 249.
[251] N. Amini, T.J. Cardwell, R.W. Cattrall, R.J.S. Morrison, S.D. Kolev, On-line determination of mercury(II) by membrane separation flow injection analysis, Talanta 63 (2004) 1069.
[252] J.A. Gomes-Neto, R. Montes, A.A. Cardoso, Spectrophotometric detection of arsenic using flow-injection hydride generation following sorbent extraction preconcentration, Talanta 50 (1999) 959.
[253] L. Chimuka, E. Cukrowska, J.A. Jonsson, Why liquid membrane extraction is an attractive alternative in sample preparation, Pure Appl. Chem. 76 (2004) 707.
[254] S.D. Kolev, W.E. van der Linden, Mathematical-modeling of a flow-injection system with a membrane separation module, Anal. Chim. Acta 268 (1992) 7.
[255] S.D. Kolev, W.E. van der Linden, Influence of the main parameters on a parallel plate dialyzer under laminar flow conditions, Anal. Chim. Acta 257 (1992) 317.
[256] A.B. Shelekhin, A.G. Dixon, Y.H. Ma, Theory of gas-diffusion and permeation in inorganic molecular-sieve membranes, AICHE J. 41 (1995) 58.
[257] V. Kuban, Gas diffusion/permeation flow injection analysis. Part 1. Principles and instrumentation, Crit. Rev. Anal. Chem. 23 (1992) 323.
[258] Z. Fang, Flow Injection Separation and Preconcentration, VCH Pub., Weinheim, 1993, p. 251.
[259] T. Buttler, C. Nilsson, L. Gorton, G. Marko-Varga, T. Laurell, Membrane characterisation and performance of microdialysis probes intended for use as bioprocess sampling units, J. Chromatogr. A 725 (1996) 41.
[260] S. Kjellstrom, S. Lindberg, T. Laurell, G. Marko-Varga, Development of a push-pull microdialysis sampling technique for the quantitative determination of proteins, Chromatographia 52 (2000) 334.
[261] W. Frenzel, Permeation denuder for sampling and continuous analysis of gases. Part 1. System configuration, basic studies and application to atmospheric ammonia and sulphur dioxide, Anal. Chim. Acta 291 (1994) 305.
[262] J.F. van Staden, Membrane separation in flow injection systems. Part 1. Dialysis, Fresenius J. Anal. Chem. 52 (1995) 271.
[263] S. Motomizu, K. Toei, T. Kuwaki, M. Oshima, Gas-diffusion unit with tubular microporous poly (tetrafluoroethy1ene) membrane for flow-injection determination of carbon dioxide, Anal. Chem. 59 (1987) 2930.
[264] I.L. Mattos, R.P. Sartini, E.A.G. Zagatto, B.F. Reis, M.F. Gine, Spectrophotometric flow injection determination of ethanol in distilled spirits and wines involving permeation through a silicon tubular membrane, Anal. Sci. 14 (1998) 1005.
[265] H. Baadenhuijsen, H.E. Seuren-Jacobs, Determination of total CO_2 in plasma by automated flow-injection analysis, Clin. Chem. 25 (1979) 443.
[266] F.J. Krug, J. Mortatti, L.C.R. Pessenda, E.A.G. Zagatto, H. Bergamin-Filho, Flow injection spectrophotometric determination of boron in plant material with Azomethine-H, Anal. Chim. Acta 125 (1981) 29.
[267] E.A.G. Zagatto, B.F. Reis, H. Bergamin-Filho, F.J. Krug, Isothermal distillation in flow injection analysis. Determination of total nitrogen in plant material, Anal. Chim. Acta 109 (1989) 45.

[268] N. Choengchan, T. Mantim, P. Wilairat, P.K. Dasgupta, S. Motomizu, D. Nacapricha, A membraneless gas diffusion unit: design and its application to determination of ethanol in liquors by spectrophotometric flow injection, Anal. Chim. Acta 579 (2006) 33.

[269] P. Mornane, J. van den Haak, T.J. Cardwell, R.W. Cattrall, P.K. Dasgupta, S.D. Kolev, Thin layer distillation for matrix isolation in flow analysis, Talanta 72 (2007) 741.

[270] M.D. Luque de Castro, I. Papaefstathiou, Analytical pervaporation: a new separation technique, Trends Anal. Chem. 17 (1998) 41.

[271] M.D. Luque de Castro, Membrane-based separation techniques: dialysis, gas diffusion and pervaporation. Ch. 8, p. 204, in: S.D. Kolev, I.D. McKelvie (Eds.), Advances in Flow Injection Analysis and Related Techniques, Wilson and Wilson's Comprehensive Analytical Chemistry, vol. 54, Elsevier, Amsterdam, 2008.

[272] I. Papaefstathiou, M.D. Luque de Castro, M. Valcarcel, Flow injection/pervaporation coupling for the determination of sulphite in Kraft liquours, Fresenius J. Anal. Chem. 354 (1996) 442.

[273] L.-J. Wang, T.J. Cardwell, R.W. Cattrall, M.D. Luque de Castro, S.D. Kolev, Pervaporation-flow injection determination of ammonia in the presence of surfactants, Anal. Chim. Acta 416 (2000) 177.

[274] I. Papaefstathiou, U. Bilitewiski, M.D. Luque de Castro, Pervaporation: an interface between fermenters and monitoring, Anal. Chim. Acta 330 (1996) 265.

[275] B. Vallejo, P. Richter, I. Toral, C. Tapia, M.D. Luque de Castro, Determination of sulphide in liquid and solid samples by integrated pervaporation-potentiometric detection, Anal. Chim. Acta 436 (2001) 301.

[276] K. Pyrzynska, Preconcentration and recovery of metal ions by donnan dialysis, Microchim. Acta 153 (2006) 117.

[277] P. Kuban, A. Slampova, P. Bocek, Electric field-enhanced transport across phase boundaries and membranes and its potential use in sample pretreatment for bioanalysis, Electrophoresis 31 (2010) 768.

[278] G. Morales-Cid, S. Cardenas, B.M. Simonet, M. Valcarcel, Sample treatments improved by electric fields, Trends Anal. Chem. 29 (2010) 158.

[279] N. Gochman, Adaptation of the autoanalyzer to routine clinical microdeterminations, Microchem. J. 11 (1966) 358.

[280] A.H. Kadish, D.A. Hall, A new method for continuous monitoring of blood glucose by measurement of dissolved oxygen, Clin. Chem. 11 (1965) 869.

[281] J. Ruzicka, E.H. Hansen, Flow injection analysis. Part VI. The determination of phosphate and chloride in blood serum by dialysis and sample dilution, Anal. Chim. Acta 87 (1976) 353.

[282] A. Ivaska, J. Ruzicka, From flow injection to sequential injection: comparison of methodologies and selection of liquid drives, Analyst 118 (1993) 885.

[283] A. Koropchak, L. Allen, Flow injection donnan dialysis preconcentration of cations for flame atomic absorption spectrophotometry, Anal. Chem. 61 (1989) 1410.

[284] N. Kasthurikrishnan, J.A. Koropchak, Flow injection Donnan dialysis preconcentration of trace metal cations for inductively coupled plasma atomic emission spectrometry, Anal. Chem. 65 (1993) 857.

[285] A. Antonia, L.B. Allen, Extraction and analysis of lead in sweeteners by flow-injection Donnan dialysis with flame atomic absorption spectroscopy, J. Agric. Food Chem. 49 (2001) 4615.

[286] A.J.J. Debets, W.T. Kok, K.P. Hupe, U.A.T. Brinkman, Electrodialytic sample treatment coupled online with high-performance liquid chromatography, Chromatographia 30 (1990) 361.

[287] J.C. Cooper, J. Danzer, H.-L. Schmidt, Enhanced selectivity in flow-injection analysis for L-aminoacids using electrodialysis with amino acid oxidation, Anal. Chim. Acta 282 (1993) 369.

[288] J.F. van Staden, C.J. Hattingh, Incorporation of electrodialysers into the conduits of FIA systems: enhancement of the mass transfer of chloride anions through passive neutral membranes, Talanta 45 (1998) 485.
[289] S.K. Mishra, P.K. Dasgupta, Electrodialytic reagent introduction in flow systems, Anal. Chem. 82 (2010) 3981.
[290] M. Valcarcel, M. Gallego, Automatic precipitation-dissolution in continuous flow systems, Trends Anal. Chem. 8 (1989) 34.
[291] M.D. Luque de Castro, F. Priego-Capote, Continuous filtration as a separation technique, Trends Anal. Chem. 27 (2008) 101.
[292] J.M. Skinner, A.C. Docherty, Determination of potassium by an automatic ultraviolet absorptiometric method, Talanta 14 (1967) 1393.
[293] O. Fatibello-Filho, J.A. Nobrega, A.J.M. Guarita-Santos, Flow injection potentiometric determination of saccharin in dietary products with relocation of filtration unit, Talanta 41 (1994) 731.
[294] G. Abate, L.B.O. Santos, S.M. Colombo, J.C. Masini, Sequential injection analysis as a tool for on-line monitoring the sorption of fulvic acid onto modified vermiculite, J. Braz. Chem. Soc. 17 (2006) 491.
[295] R.E. Santelli, M. Gallego, M. Valcarcel, Atomic absorption determination of copper in silicate rocks by continuous precipitation preconcentration, Anal. Chem. 61 (1989) 1427.
[296] I.E. Vito, R.A. Olsina, J. Raba, A.N. Masi, Flow injection preconcentration by chemofiltration and spectrophotometric determination of Gd, Anal. Chim. Acta 501 (2004) 11.
[297] C. Matsubara, K. Sakai, K. Takamura, Collection by membrane filtration/spectrophotometric determination of trace hydrogen peroxide, Anal. Sci. 7 (1991) 311.
[298] C.X. Galhardo, J.C. Masini, Sequential injection analysis as a tool for *in situ* monitoring of Fe(II), Fe(III), NO_3^- and NO_2^- in natural and waste waters, Anal. Chim. Acta 438 (2001) 39.
[299] U.D. Neue, H. Engelhardt, Reaction detector with three-dimensional coiled open tubes in HPLC, Chromatographia 15 (1982) 403.
[300] Z.-L. Fang, M. Sperling, B. Welz, Flame atomic-absorption spectrometric determination of lead in biological samples using a flow-injection system with online preconcentration by coprecipitation without filtration, J. Anal. At. Spectrom. 6 (1991) 301.
[301] H. Wu, Y. Jin, Y.-Q. Shi, S.-P. Bi, On-line organoselenium interference removal for inorganic selenium species by flow injection coprecipitation preconcentration coupled with hydride generation atomic fluorescence spectrometry, Talanta 71 (2007) 1762.
[302] J.J. Hawkes, W.T. Coakley, Force field particle filter, combining ultrasound standing waves and laminar flow, Sens. Actuators B 75 (2001) 213.
[303] E. Bulska, W. Jedral, ETAS determination of lead with on-line preconcentration using a flow-through electrochemical microcell, At. Spectros. 18 (1997) 202.
[304] J.A. Gomes-Neto, A.P. Oliveira, G.P.G. Freshi, C.S. Dakuzaku, M. Moraes, Minimization of lead and copper interferences on spectrophotometric determination of cadmium using electrolytic deposition and ion-exchange in multi-commutation flow system, Talanta 53 (2000) 497.
[305] M. Ito, M. Itagaki, K. Watanabe, Flow-injection spectrophotometry of cadmium in river water with tetraphenylporphinetetrasulfonic acid using on-line electrolytic preconcentration, Bunzeki Kagaku 52 (2003) 873.
[306] J.T. van Gemert, Automated wet chemical analysers and their applications, Talanta 20 (1973) 1045.
[307] J. Keay, P.M.A. Menage, Automated distillation procedure for determination of nitrogen, Analyst 94 (1969) 895.
[308] B. Pihlar, L. Kosta, Determination of cyanides by continuous distillation and flow analysis with cylindrical amperometric electrodes, Anal. Chim. Acta 114 (1980) 275.

[309] C.W.K. Chow, R. Lane, T.C.W. Yeow, D.E. Davey, D.E. Mulcahy, Development of a fully integrated microdistillation flow injection system for the determination of trace level ammonia, Lab. Autom. Info. Manag. 33 (1998) 199.

[310] S. McLeod, Micro-distillation unit for use in continuous flow analyzers. Its construction and use in determination of ammonia and nitrate in soils, Anal. Chim. Acta 256 (1992) 107.

[311] K. Shimada, T. Shimoda, H. Kokusen, S. Nakano, Automatic microdistillation flow-injection system for the spectrophotometric determination of fluoride, Talanta 66 (2005) 80.

[312] J.L. Burguera, M. Burguera, M. de la Guardia, A. Salvador, Pyrolysis-flow-injection analysis-spectrophotometric determination of amino acids in aqueous solutions, Anal. Chim. Acta 261 (1992) 23.

[313] J.-H. Wang, E.H. Hansen, Coupling on-line preconcentration by ion-exchange with ETAAS. A novel flow injection approach based on the use of a renewable micro-column as demonstrated for the determination of nickel in environmental and biological samples, Anal. Chim. Acta 424 (2000) 223.

[314] P.B. Martelli, J.A. Gomes-Neto, E.A.G. Zagatto, S.M.B. Brienza, M.C.B.S.M. Montenegro, J.L.F.C. Lima, Sequential analyte removal in flow analysis: determination of nitrogen, phosphorus and potassium in fertilizers, Anal. Chim. Acta 317 (1995) 239.

[315] C.K. Zacharis, A. Verdoukas, P.D. Tzanavaras, D.G. Themelis, Automated sample preparation coupled to sequential injection chromatography: on-line filtration and dilution protocols prior to separation, J. Pharm. Biomed. Anal. 490 (2009) 726.

[316] R. Lucena, S. Cardenas, M. Gallego, M. Valcarcel, Multipurpose chamber for the implementation of gas diffusion, dialysis, solid-phase extraction and precipitation/dissolution in continuous flow analyzers, Anal. Chim. Acta 509 (2004) 47.

[317] H. Ziegel, Panel for electro-analysis, J. Am. Chem. Soc. 36 (1914) 1450.

[318] H.V. Malmstadt, E.C. Gohrbandt, Automatic spectrophotometric titrations. Determination of milligram quantities of thorium, Anal. Chem. 26 (1954) 442.

[319] H.V. Malmstadt, C.B. Roberts, Automatic derivative spectrophotometric titrations, Anal. Chem. 28 (1956) 1408.

[320] H.V. Malmstadt, Rapid and accurate automatic titration of calcium and magnesium in dolomites and limestones. Use of EDTA titrant and automatic derivative spectrophotometric end-point termination, Anal. Chim. Acta 19 (1958) 563.

[321] W.J. Blaedel, R.H. Laessig, Continuous automated buretless titrator with direct readout, Anal. Chem. 36 (1964) 1617.

[322] J. Ruzicka, E.H. Hansen, H. Mosbaek, Flow Injection Analysis. Part IX. A new approach to continuous flow titrations, Anal. Chim. Acta 92 (1977) 235.

[323] A.U. Ramsing, J. Ruzicka, E.H. Hansen, The principles and theory of high-speed titrations by flow injection analysis, Anal. Chim. Acta 129 (1981) 1.

[324] M. Korn, L.F.B.P. Gouveia, E. Oliveira, B.F. Reis, Binary search in flow titration employing photometric end-point detection, Anal. Chim. Acta 313 (1995) 177.

[325] J.-Y. Zhu, L.-Y. Shi, C. Dallner, H. Niebauer, C. Muenzel, H. Bauer, Study and application of a new on-line titration technique, (English edn) J. Shanghai Univ. 6 (2002) 181.

[326] E.V. Aquino, J.J.R. Rohwedder, C. Pasquini, A new approach to flow-batch titration. A monosegmented flow titrator with coulometric reagent generation and potentiometric or biamperometric detection, Anal. Bioanal. Chem. 386 (2006) 1921.

[327] J.A. Sweileh, P.K. Dasgupta, Applications of in situ detection with an automated micro batch analyzer, Anal. Chim. Acta 214 (1988) 107.

[328] R.S. Honorato, M.C.U. Araujo, G. Veras, E.A.G. Zagatto, R.A.S. Lapa, J.L.F.C. Lima, Monosegmented flow titration for the spectrophotometric determination of total acidity in vinegar, Anal. Sci. 15 (1999) 665.

[329] J. Kozak, M. Wojtowicz, N. Gawenda, P. Koscielniak, An automatic system for acidity determination based on sequential injection titration and the monosegmented flow approach, Talanta 84 (2011) 1379.
[330] J. Jakmunee, L. Pathimapornlert, S.K. Hartwell, K. Grudpan, Novel approach for mono-segmented flow micro-titration with sequential injection using a lab-on-valve system: a model study for the assay of acidity in fruit juices, Analyst 130 (2005) 299.
[331] M. Wojtowicz, J. Kozak, K. Danielewska, P. Koscielniak, Flow injection titration basing on the merging-zone technique, Talanta 79 (2009) 1006.
[332] H. Tanaka, P.K. Dasgupta, J.-M. Huang, Continuous on line true titrations by feed-back-based flow ratiometry. The principle of compensating errors, Anal. Chem. 72 (2000) 4713.
[333] J. Marcos, A. Rios, M. Valcarcel, Automatic titrations in unsegmented flow systems based on variable flow-rate patterns. Part 1. Principles and applications to acid-base titrations, Anal. Chim. Acta 261 (1992) 489.
[334] H. du Plessis, J.F. van Staden, Determination of sulphuric acid in process effluent streams using sequential injection titration, Talanta 52 (2000) 83.
[335] F.R.P. Rocha, C.M.C. Infante, W.R. Melchert, A multi-purpose flow system based on multi-commutation, Spectrosc. Lett. 39 (2006) 651.
[336] M.A. Feres, P.R. Fortes, E.A.G. Zagatto, J.L.M. Santos, J.L.F.C. Lima, Multi-commutation in flow analysis: recent developments and applications, Anal. Chim. Acta 618 (2008) 1.
[337] A.P.S. Paim, B.F. Reis, An automatic spectrophotometric titration procedure for ascorbic acid determination in fruit juices and soft drinks based on volumetric fraction variation, Anal. Sci. 16 (2000) 487.
[338] G.D. Clark, J. Zable, J. Ruzicka, G.D. Christian, Flow-reversal flow-injection analysis. Enhancement of flow-injection titrations, Talanta 38 (1991) 119.
[339] C.M.N.V. Almeida, R.A.S. Lapa, J.L.F.C. Lima, Oxidability determination in waste waters using an automatic titrator based on a multicommutated unsegmented flow system, Int. J. Environ. Anal. Chem. 78 (2000) 315.
[340] R.H. Taylor, C. Winbo, G.D. Christian, J. Ruzicka, Bromine number determination by coulometric flow-injection titration, Talanta 39 (1992) 789.
[341] Y. Sahlestroem, S. Twengstroem, B. Karlberg, An automated flow-injection extraction method for determination of bittering compounds in beer, Anal. Chim. Acta 187 (1986) 339.
[342] T.J. Cardwell, R.W. Cattral, G.J. Cross, G.R. O'Connell, J.D. Petty, G.R. Scollary, Determination of titratable acidity in wines and total acidity of vinegars by discontinuous flow analysis using photometric end-point, Analyst 116 (1991) 1043.
[343] D.R. Turner, S. Knox, M. Whitfield, M.M.C. Santos, Flow injection titration of alkalinity in natural waters, Anal. Proc. 24 (1987) 360.
[344] C.A. Georgiou, M.A. Koupparis, Automated flow injection spectrophotometric non-aqueous pseudotitrations of amines and their hydrohalide salts, Analyst 113 (1988) 755.
[345] D.R. Turner, M.M.C. Santos, P. Coutinho, M.L. Goncalves, S. Knox, Rapid pK measurements for multibasic weak acids by gradient flow injection titration, Anal. Chim. Acta 258 (1992) 259.
[346] Y.Y. Liang, Automation of Karl Fischer water titration by flow injection sampling, Anal. Chem. 62 (1990) 2504.
[347] E.A.G. Zagatto, A.O. Jacintho, F.J. Krug, B.F. Reis, R.E. Bruns, M.C.U. Araujo, Flow-injection systems with inductively-coupled argon plasma atomic emission spectrometry. Part 2. The generalized standard addition method, Anal. Chim. Acta 145 (1983) 169.
[348] J.F. Tyson, J.M.H. Appleton, A.B. Idris, Flow injection calibration methods for atomic absorption spectrometry, Anal. Chim. Acta 145 (1983) 159.

[349] I.P.A. Moraes, M.R.S. Souto, A.O.S.S. Rangel, Sequential injection standard addition system with a mixing chamber: determination of orthophosphate in waters, J. Flow Injection Anal. 20 (2003) 187.
[350] M.F. Gine, B.F. Reis, E.A.G. Zagatto, F.J. Krug, A.O. Jacintho, A simple procedure for standard additions in flow injection analysis. Spectrophotometric determination of nitrate in plant extracts, Anal. Chim. Acta 155 (1983) 131.
[351] E.C. Silva, M.C.U. Araujo, R.S. Honorato, J.L.F.C. Lima, E.A.G. Zagatto, S.M.B. Brienza, Standard additions in flow injection analysis based on merging zones and gradient exploitation. Application to copper determination in spirits, Anal. Chim. Acta 319 (1996) 153.
[352] I. Lugo, H. Carrero, L.E. Leon, The generalized standard addition method in a closed-loop flow system, Anal. Lett. 34 (2001) 2735.
[353] J.J. Harrow, J. Janata, Heterogeneous samples in flow-injection systems. Part 2. Standard addition, Anal. Chim. Acta 174 (1985) 123.
[354] I.E. Bechmann, L. Nørgaard, C. Ridder, Generalized standard addition in flow-injection analysis with UV-visible photodiode array detection, Anal. Chim. Acta 304 (1995) 229.
[355] Y. Israel, R.M. Barnes, Flow injection sample-to-standard additions method. Spectrophotometric determination of hydrochloric acid and orthophosphate, Analyst 114 (1989) 843.
[356] I. Lavilla, B. Perez-Cid, C. Bendicho, Use of flow-injection sample-to-standard addition methods for quantification of metals leached by selective chemical extraction from sewage sludge, Anal. Chim. Acta 381 (1999) 297.
[357] M.S.P. Silva, J.C. Masini, Exploiting monosegmented flow analysis to perform in-line standard additions using a single stock standard solution in spectrophotometric sequential injection procedures, Anal. Chim. Acta 466 (2002) 345.
[358] H.L. Pardue, Kinetic aspects of analytical chemistry, Anal. Chim. Acta 216 (1989) 69.
[359] P.J. Worsfold, J. Ruzicka, E.H. Hansen, Rapid automated enzymatic method for the determination of alcohol in blood and beverages using flow injection analysis, Analyst 106 (1981) 1309.
[360] A. Fernandez, M.D. Luque de Castro, M. Valcarcel, Comparison of flow injection analysis configurations for differential kinetic determination of cobalt and nickel, Anal. Chem. 56 (1984) 1146.
[361] D.A. Whitman, M.B. Seasholtz, G.D. Christian, J. Ruzicka, B.R. Kowalski, Double-injection flow-injection analysis using multivariate calibration for multicomponent analysis, Anal. Chem. 63 (1991) 775.
[362] M.A.Z. Arruda, E.A.G. Zagatto, N. Maniasso, Kinetic determination of cobalt and nickel by flow-injection spectrophotometry, Anal. Chim. Acta 283 (1993) 476.
[363] H. Kagenow, A. Jensen, Differential kinetic analysis and flow injection analysis, Part 3. The (2.2.2) cryptates of magnesium, calcium and strontium, Anal. Chim. Acta 114 (1980) 227.
[364] D. Vendramini, V. Grassi, E.A.G. Zagatto, Spectrophotometric flow injection determination of copper and nickel in plant digests exploiting differential kinetic analysis and multi site detection, Anal. Chim. Acta 570 (2006) 124.
[365] J.-H. Wang, R.-H. He, Stopped-flow kinetic spectrophotometric method for the simultaneous determination of iron, titaniumand vanadium, Anal. Chim. Acta 276 (1993) 419.
[366] P.R. Fortes, S.R.P. Meneses, E.A.G. Zagatto, A novel flow-based strategy for implementing differential kinetic analysis, Anal. Chim. Acta 572 (2006) 316.
[367] H.M. Kingston, M.L. Kingston, Nomenclature in laboratory robotics and automation (IUPAC recommendations 1994), J. Autom. Chem. 16 (1994) 43.

[368] V. Grassi, A.C.B. Dias, E.A.G. Zagatto, Flow systems exploiting in-line prior assays, Talanta 64 (2004) 1114.
[369] P.R. Fortes, M.A. Feres, E.A.G. Zagatto, An expert flow system involving in-line prior assay for turbidimetric determination of chloride and sulphate in natural waters, Talanta 77 (2008) 571.
[370] M. Valcarcel, S. Cardenas, M. Gallego, Sample screening systems in analytical chemistry, Trends Anal. Chem. 18 (1999) 685.
[371] M. Peris, A. Maquieira, R. Puchades, V. Chirivella, R. Ors, J. Serrano, An expert system as a tool for the specification and intelligent control of a flow injection analysis system, Chemometr. Intell. Lab. Syst. 234 (1990) 207.
[372] E.A.G. Zagatto, F.R.P. Rocha, P.B. Martelli, B.F. Reis, Detecting and circumventing sources of inaccuracy in flow analysis, Pure Appl. Chem. 73 (2001) 45.
[373] C.C. Oliveira, R.P. Sartini, E.A.G. Zagatto, J.L.F.C. Lima, Flow analysis with accuracy assessment, Anal. Chim. Acta 350 (1997) 31.
[374] E.A.G. Zagatto, S. Vicente, C.C. Oliveira, R.P. Sartini, J.L.F.C. Lima, Sequential injection analysis with accuracy assessment, J. Flow Injection Anal. 15 (1998) 226.
[375] A.M. Pimenta, A.N. Araujo, M.C.B.S.M. Montenegro, Sequential injection analysis of captopril based on colorimetric and potentiometric detection, Anal. Chim. Acta 438 (2001) 31.
[376] A.M. Pimenta, A.N. Araujo, M.C.B.S.M. Montenegro, Simultaneous potentiometric and fluorimetric determination of diclofenac in a sequential injection analysis system, Anal. Chim. Acta 470 (2002) 185.
[377] C.I.C. Silvestre, J.L.M. Santos, J.L.F.C. Lima, M.A. Feres, E.A.G. Zagatto, Single interface flow analysis with accuracy assessment, Microchem. J. 94 (2010) 60.
[378] M.F. Gine, R.L. Tuon, A.A. Cesta, A.P. Packer, B.F. Reis, Real-time simplex optimization of flow-injection systems for chemical analysis, Anal. Chim. Acta 366 (1998) 313.
[379] J. Marcos, A. Rios, M. Valcarcel, Automated simultaneous determination of metal ions by use of variable flow rates in unsegmented systems, Analyst 117 (1992) 1629.
[380] M. Valcarcel, M.D. Luque de Castro, Automatic methods of analysis, in: Techniques and Instrumentation in Analytical Chemistry vol. 9, Elsevier Sci. Pub., Amsterdam, 1988.
[381] A. Rius, M.P. Callao, F.X. Rius, Self-configuration of sequential injection analytical systems, Anal. Chim. Acta 316 (1995) 27.
[382] J. Brandt, B. Hitzmann, Knowledge-based fault detection and diagnosis in flow-injection analysis, Anal. Chim. Acta 291 (1994) 29.
[383] A. Bonastre, R. Ors, M. Peris, Advanced automation of a flow injection analysis system for quality control of olive oil through the use of a distributed expert system, Anal. Chim. Acta 506 (2004) 189.
[384] I. Ruisanchez, J. Lozano, M.S. Larrechi, F.X. Rius, J. Zupan, On-line automated analytical signal diagnosis in sequential injection analysis using artificial neural networks, Anal. Chim. Acta 348 (1997) 113.
[385] R. Wolters, M.A.J. van Opstal, G. Kateman, Quantitative validation of a flow-injection determination of penicillin in pharmaceutical formulations by means of a validation program based on an expert system, Anal. Chim. Acta 233 (1990) 65.
[386] B.F. Reis, E.A.G. Zagatto, P.B. Martelli, S.M.B. Brienza, Trial measurements in flow analysis, Analyst 118 (1993) 719.
[387] J.M.P.J. Garrido, R.A.S. Lapa, J.L.F.C. Lima, C. Delerue-Matos, J.L.M. Santos, FIA automatic dilution system for the determination of metallic cations in waters by atomic absorption and flame emission spectrometry, J. Autom. Chem. 18 (1996) 17.
[388] M.F. Gine, A.P. Packer, T. Blanco, B.F. Reis, Flow system based on a binary sampling process for automatic dilutions prior to flame atomic spectrometry, Anal. Chim. Acta 323 (1996) 47.

[389] F.R.P. Rocha, P.B. Martelli, R.M. Frizzarin, B.F. Reis, Automatic multicommutation flow system for wide range spectrophotometric calcium determination, Anal. Chim. Acta 366 (1998) 45.
[390] M.I.G.S. Almeida, M.A. Segundo, J.L.F.C. Lima, A.O.S.S. Rangel, Interfacing multi-syringe flow injection analysis to flame atomic emission spectrometry: an intelligent system for automatic sample dilution and determination of potassium, J. Anal. At. Spectrom. 24 (2009) 340.
[391] A.M. Serra, J.M. Estela, V. Cerda, MSFIA system for mercury determination by cold vapour technique with atomic fluorescence detection, Talanta 77 (2008) 556.
[392] C. Pons, M. Miro, E. Becerra, J.M. Estela, V. Cerda, An intelligent flow analyser for the in-line concentration, speciation and monitoring of metals at trace levels, Talanta 62 (2004) 887.
[393] C. Pons, R. Forteza, V. Cerda, Expert multi-syringe flow-injection system for the determination and speciation analysis of iron using chelating disks in water samples, Anal. Chim. Acta 524 (2004) 79.
[394] L. Ferrer, J.M. Estela, V. Cerda, A smart multisyringe flow injection system for analysis of sample batches with high variability in sulfide concentration, Anal. Chim. Acta 573 (2006) 391.
[395] I. Lopez-Garcia, J. Arroyo, P. Vinas, M. Hernandez-Cordoba, A manifold for the automatic dilution of concentrated solutions in flame atomic absorption spectrometry, Anal. Bioanal. Chem. 372 (2002) 587.
[396] M.F. Gine, H. Bergamin-Filho, B.F. Reis, R.L. Tuon, Intelligent flow-injection-inductively coupled plasma system for matrix matching, Anal. Chim. Acta 234 (1990) 207.
[397] J.M.T. Carneiro, R.S. Honorato, E.A.G. Zagatto, Individual sample conditioning in flow analysis. Determination of N-total in plant materials, Fresenius J. Anal. Chem. 368 (2000) 496.
[398] R.S. Honorato, J.M.T. Carneiro, E.A.G. Zagatto, Spectrophotometric flow-batch determination of aluminum in plant tissues exploiting a feedback mechanism, Anal. Chim. Acta 441 (2001) 309.
[399] J.M.T. Carneiro, A.C.B. Dias, E.A.G. Zagatto, R.S. Honorato, Spectrophotometric catalytic determination of Fe(III) in estuarine waters using a flow-batch system, Anal. Chim. Acta 455 (2002) 327.
[400] R.S. Honorato, M.C.U. Araujo, R.A.C. Lima, E.A.G. Zagatto, R.A.S. Lapa, J.L.F.C. Lima, A flow-batch titrator exploiting a one-dimensional optimisation algorithm for end point search, Anal. Chim. Acta 396 (1999) 91.
[401] I.M. Raimundo-Junior, C. Pasquini, Automated monosegmented flow analyser. Determination of glucose, creatinine and urea, Analyst 122 (1997) 1039.
[402] I. Facchin, J.J.R. Rohwedder, C. Pasquini, An automated system for liquid-liquid extraction in monosegmented flow analysis, J. Autom. Chem. 19 (1997) 33.
[403] M.C.H. Silva, C. Pasquini, Determination of gaseous species by monosegmented flow systems. Volumetric determination of oxygen and carbon dioxide, Anal. Chim. Acta 349 (1997) 377.
[404] M.C.H. Silva, J.J.R. Rohwedder, C. Pasquini, Determination of CO_2 in gaseous samples using a monosegmented flow system and conductimetric detection, Anal. Chim. Acta 366 (1998) 223.
[405] R.L. Habig, B.W. Schlein, L. Walters, R.E. Thiers, A bubble-gating flow cell for continuous-flow analysis, Clin. Chem. 15 (1969) 1045.
[406] I.D. McKelvie, T.J. Cardwell, R.W. Cattrall, A microconduit flow injection analysis demonstration using a 35-mm slide projector, J. Chem. Educ. 67 (1990) 262.
[407] N. Maleki, A. Safavi, F. Sedaghatpour, Single-step calibration, prediction and real samples data acquisition for artificial neural network using a CCD camera, Talanta 64 (2004) 830.

[408] E.H. Hansen, J. Ruzicka, Principles of flow injection analysis as demonstrated by three lab exercises, J. Chem. Educ. 56 (1979) 677.
[409] L.A. Ramos, K.R. Prieto, E.T.G. Cavalheiro, C.C.S. Cavalheiro, Determination of hypochlorite in bleaching products with flower extracts to demonstrate the principles of flow injection analysis, J. Chem. Educ. 82 (2005) 1815.
[410] J.A. Nobrega, A.A. Mozeto, R.M. Alberici, Reaction kinetics in a flow-injection spectrophotometric experiment. A laboratory exercise, J. Chem. Educ. 68 (1991) 966.
[411] J.A. Nobrega, F.R.P. Rocha, Ionic strength effect on the rate of reduction of hexacyanoferrate(III) by ascorbic acid. A flow injection kinetic experiment, J. Chem. Educ. 74 (1997) 560.
[412] M.K. Sasaki, C.R. Silva, P.R. Fortes, M.A. Feres, E.A.G. Zagatto, A flow system for video demonstrations: comparative evaluation of the stability of some cuprate(II) complexes, J. Flow Injection Anal. 28 (2011) 3.
[413] B.R. Kowalski, J. Ruzicka, G.D. Christian, Flow chemography – the future of chemical education? Trends Anal. Chem. 9 (1990) 8.
[414] A.V. Pereira, C. Aniceto, O. Fatibello-Filho, Flow injection spectrophotometric determination of acetylsalicylic acid in tablets after on-line microwave-assisted alkaline hydrolysis, Analyst 123 (1998) 1011.
[415] A. Caballo-Lopez, M.D. Luque de Casto, Continuous ultrasound-assisted extraction coupled to flow injection-pervaporation, derivatization, and spectrophotometric detection for the determination of ammonia in cigarettes, Anal. Chem. 78 (2006) 2297.
[416] A. Chivulescu, M. Catala-Icardo, J.V.G. Mateo, J.M. Calatayud, New flow-multicommutation method for the photo-chemiluminometric determination of the carbamate pesticide asulam, Anal. Chim. Acta 519 (2004) 113.
[417] J.C. Bravo, P. Fernandez, J.S. Durand, Flow injection fluorimetric determination of β-estradiol using a molecularly imprinted polymer, Analyst 130 (2005) 1404.
[418] Z. Pawlicova, J.R. Albert-Garcia, I. Sahuquillo, J.V.G. Mateo, M.C. Icardo, J.M. Calatayud, Chemiluminescent determination of the pesticide bromoxynil by on-line photodegradation in a flow-injection system, Anal. Sci. 22 (2006) 29.
[419] D.-H. Chen, F. Lazaro, M.D. Luque de Castro, M. Valcarcel, Direct spectrophotometric determination of total boron in soils with ultrasonic leaching in automatic flow systems, Anal. Chim. Acta 226 (1989) 221.
[420] D.-Z. Dan, R.C. Sandford, P.J. Worsfold, Determination of chemical oxygen demand in fresh waters using flow injection with on-line UV-photocatalytic oxidation and spectrophotometric detection, Analyst 130 (2006) 227.
[421] M. Catala-Icardo, M. Misiewicz, A. Ciucu, J.V. Garcia-Mateo, J.M. Calatayud, FI-on line photochemical reaction for direct chemiluminescence determination of photodegradated chloramphenicol, Talanta 60 (2003) 405.
[422] S. Morales-Munoz, J.L. Luque-Garcia, M.D. Luque de Castro, A continuous approach for the determination of Cr(VI) in sediment and soil based on the coupling of microwave-assisted water extraction, preconcentration, derivatization and photometric detection, Anal. Chim. Acta 515 (2004) 343.
[423] J.L. Luque-Garcia, M.D. Luque de Castro, Continuous ultrasound-assisted extraction of hexavalent chromium from soil with or without on-line preconcentration prior to photometric monitoring, Analyst 127 (2002) 1115.
[424] O. Tue-Ngeun, R.C. Sandford, J. Jakmunee, K. Grudpan, I.D. McKelvie, P.J. Worsfold, Determination of dissolved inorganic carbon (DIC) and dissolved organic carbon (DOC) in freshwaters by sequential injection spectrophotometry with on-line UV photo-oxidation, Anal. Chim. Acta 554 (2005) 17.
[425] C.B. Lopes, S. Abreu, M. Valega, R.M.B.O. Duarte, M.E. Pereira, A.C. Duarte, The assembling and application of an automated segmented flow analyzer for the

determination of dissolved organic carbon based on UV-persulphate oxidation, Anal. Lett. 39 (2006) 1979.
[426] J.A. Sweileh, On-line flow injection solid sample introduction digestion and analysis: spectrophotometric and atomic absorption determination of iron, copper and zinc in multi-vitamin tablets, Microchem. J. 65 (2000) 87.
[427] X.-L. Hu, J.-G. You, C.-G. Bao, H.-R. Zhang, X.-Z. Meng, T.-T. Xiao, K. Zhang, Y.-T. Wang, H.-M. Wang, H.-Q. Zhang, A.-M. Yu, Determination of total flavonoids in *Scutellaria barbata* D. Don by dynamic ultrasonic extraction coupled with on-line spectrophotometry, Anal. Chim. Acta 610 (2008) 217.
[428] S. Morales-Munoz, J.L. Luque-Garcıa, M.D. Luque de Castro, Continuous microwave-assisted extraction coupled with derivatization and fluorimetric monitoring for the determination of fluoroquinolone antibacterial agents from soil samples, J. Chromatogr. A 1059 (2004) 25.
[429] X.-Q. Zhao, Z.-Q. Zhang, Microwave-assisted on-line derivatization for sensitive flow injection fluorometric determination of formaldehyde in some foods, Talanta 80 (2009) 242.
[430] M.J. Escuriola, A. Morales-Rubio, M. de La Guardia, A clean analytical method for the spectrophotometric determination of formetanate incorporating an on-line microwave assisted hydrolysis step, Anal. Chim. Acta 390 (1999) 147.
[431] S. Pinzi, F. Priego-Capote, J. Ruiz-Jimenez, M.P. Dorado, M.D. Luque de Castro, Flow injection analysis-based methodology for automatic on-line monitoring and quality control for biodiesel production, Bioresource Technol. 100 (2009) 421.
[432] C.C. Oliveira, R.P. Sartini, E.A.G. Zagatto, Microwave-assisted sample preparation in sequential injection: spectrophotometric determination of magnesium, calcium and iron in food, Anal. Chim. Acta 413 (2000) 41.
[433] A. Caballo-Lopez, M.D. Luque de Castro, Continuous ultrasound-assisted extraction coupled to on line filtration—solid-phase extraction—column liquid chromatography—post column derivatisation—fluorescence detection for the determination of N-methylcarbamates in soil and food, J. Chromatogr. A 998 (2003) 51.
[434] A. Cerda, M.T. Oms, R. Forteza, V. Cerda, Total nitrogen determination by flow injection using on-line microwave-assisted digestion, Anal. Chim. Acta 351 (1997) 273.
[435] I.D. McKelvie, M. Mitri, B.T. Hart, L.C. Hamilton, A.D. Stuart, Analysis of total dissolved nitrogen in natural waters by on-line photooxidation and flow injection, Anal. Chim. Acta 293 (1994) 155.
[436] Z. Bouhsain, S. Garrigues, A. Morales-Rubio, M. de la Guardia, Flow injection spectrophotometric determination of paracetamol in pharmaceuticals by means of on-line microwave-assisted hydrolysis and reaction with 8-hydroxyquinoline (8-quinolinol), Anal. Chim. Acta 330 (1996) 59.
[437] A. Criado, S. Cardenas, M. Gallego, M. Valcarcel, Continuous flow spectrophotometric determination of paracetamol in pharmaceuticals following continuous microwave assisted alkaline hydrolysis, Talanta 53 (2000) 417.
[438] J. Ruiz-Jimenez, M.D. Luque de Castro, Flow injection manifolds for liquid-liquid extraction without phase separation assisted by ultrasound, Anal. Chim. Acta 489 (2003) 1.
[439] R.L. Benson, I.D. McKelvie, B.T. Hart, I.C. Hamilton, Determination of total phosphorus in waters and wastewaters by on-line microwave-induced digestion and flow-injection analysis, Anal. Chim. Acta 291 (1994) 233.
[440] K.E. Williams, S.J. Haswell, D.A. Barclay, G. Preston, Determination of total phosphate in waste waters by on-line microwave digestion incorporating colorimetric detection, Analyst 118 (1993) 245.
[441] M.I.G.S. Almeida, M.A. Segundo, J.L.F.C. Lima, A.O.S.S. Rangel, Multi-syringe flow injection system with in-line microwave digestion for the determination of phosphorus, Talanta 64 (2004) 1283.

[442] C. Pons, I.V. Toth, A.O.S.S. Rangel, R. Forteza, V. Cerda, Multi-pumping flow system for the determination of dissolved orthophosphate and dissolved organic phosphorus in wastewater samples, Anal. Chim. Acta 572 (2006) 148.
[443] A.F. Oliveira, O. Fatibello-Filho, Flow injection spectrophotometric determination of reducing sugars using a focalized coiled reactor in a domestic microwave oven, Talanta 50 (1999) 899.
[444] L.-G. Chen, L. Ding, H.-R. Zhang, J. Li, Y.-T. Wang, X.-P. Wang, C.-L. Qu, H.-Q. Zhang, Dynamic microwave-assisted extraction coupled with on-line spectrophotometric determination of safflower yellow in *Flos Carthami*, Anal. Chim. Acta 580 (2006) 75.
[445] A. Caballo-Lopez, M.D. Luque de Castro, Fast microwave-assisted free sugars washing and hydrolysis pre-treatment for the flow injection determination of starch in food, Talanta 59 (2003) 837.
[446] A. Cladera, M. Miro, J.M. Estela, V. Cerda, Multicomponent sequential injection analysis determination of nitro-phenols in waters by on-line liquid-liquid extraction and preconcentration, Anal. Chim. Acta 421 (2000) 155.
[447] L. Fossey, F.F. Cantwell, Simultaneous monitoring of both phases in the solven textraction/flow injection analysis of dramamine tablets, Anal. Chem. 55 (1983) 1882.
[448] M. Miro, E. Gomez, J.M. Estela, M. Casas, V. Cerda, Sequential injection ^{90}Sr determination in environmental samples using a wetting-film extraction method, Anal. Chem. 74 (2002) 826.
[449] A. Alonso, M.J. Almendral, M.J. Porras, Y. Curto, C.G. Maria, Flow-injection solvent extraction with and without phase separation. Fluorimetric determination of aluminium in water, Anal. Chim. Acta 447 (2001) 211.
[450] T. Sakai, H. Harada, X.-Q. Liu, N. Ura, K. Takeyoshi, K. Sugimoto, New phase separator for extraction-spectrophotometric determination of anionic surfactants with Malachite Green by flow injection analysis, Talanta 45 (1998) 543.
[451] M.I. Acedo-Valenzuela, T. Galeano-Diaz, N. Mora-Diez, A. Silva-Rodriguez, Response surface methodology for the optimisation of flow-injection analysis with in situ solvent extraction and fluorimetric assay of tricyclic antidepressants, Talanta 66 (2005) 952.
[452] K.L. Peterson, B.K. Logan, G.D. Christian, J. Ruzicka, Sequential-injection extraction for sample preparation, Anal. Chim. Acta 337 (1997) 99.
[453] T. Sakai, Spectrophotometric determination of quaternary ammonium salts by flow injection method coupled with termochromism of ion associates, Analyst 117 (1992) 211.
[454] T. Sakai, N. Ohno, Y.S. Chug, H. Nishikawa, Spectrofluorimetric determination of berberine in oriental pharmaceutical preparations by flow-injection analysis coupled with liquid-liquid extraction, Anal. Chim. Acta 308 (1995) 329.
[455] C. Lu, G.-Q. Song, J.-M. Lin, C.-W. Huie, Enhancement in sample preconcentration by the on-line incorporation of cloud point extraction to flow injection analysis inside the chemiluminescence cell and the determination of total serum bilirubin, Anal. Chim. Acta 590 (2007) 159.
[456] O. Klinghoffer, J. Ruzicka, E.H. Hansen, Flow-injection analysis of traces of lead and cadmium by solvent extraction with dithizone, Talanta 27 (1980) 169.
[457] S. Motomizu, M. Oshima, N. Yoneda, T. Iwachido, Spectrophotometric determination of calcium with dicyclohexano-24-crown-8 and propyl orange by solvent-extraction flow-injection method, Anal. Sci. 6 (1990) 215.
[458] Y. He, H.-K. Lee, Liquid-phase microextraction in a single drop of organic solvent by using a conventional microsyringe, Anal. Chem. 69 (1997) 4634.
[459] T. Blanco, N. Maniasso, M.F. Gine, A.O. Jacintho, Liquid-liquid extraction in flow injection analysis using an open-phase separator for the spectrophotometric determination of copper in plant digests, Analyst 123 (1998) 191.
[460] M.C.T. Diniz, O. Fatibello-Filho, J.J.R. Rohwedder, An automated system for liquid–liquid extraction based on a new micro-batch extraction chamber with

on-line detection. Preconcentration and determination of copper(II), Anal. Chim. Acta 525 (2004) 281.

[461] Y.-Y. Luo, S. Nakano, D.A. Holman, J. Ruzicka, G.D. Christian, Sequential injection wetting film extraction applied to the spectrophotometric determination of chromium(VI) and chromium(III) in water, Talanta 44 (1997) 1563.

[462] T. Kato, Flow-injection spectrophotometric determination of enalapril in pharmaceuticals with bromothymol blue, Anal. Chim. Acta 175 (1985) 339.

[463] A. Sanz, V. Tomas, C. Martinez-Lozano, T. Perez-Ruiz, Extraction-spectrofluorimetric method for the determination of erythromycin and its esters in pharmaceutical formulations using manual and flow-injection procedures, Analyst 118 (1993) 567.

[464] N. Clarke, L.-G. Danielsson, The simultaneous speciation of aluminum and iron in a flow-injection system, Anal. Chim. Acta 306 (1995) 5.

[465] M.J. Almendral, A. Alonso, M.J. Porras, M.A. Garcia, Y. Curto, Determination of iron in tap and waste water using liquid-liquid extraction in a flow-injection system, Microchim. Acta 147 (2004) 117.

[466] D.G. Themelis, P.D. Tzanavaras, Solvent extraction flow-injection manifold for the simultaneous spectrophotometric determination of free cyanide and thiocyanate ions based upon on-line masking of cyanides by formaldehyde, Anal. Chim. Acta 452 (2002) 295.

[467] H. Koshima, H. Onishi, Flow-injection method for the separation of thallium and gold on a Sephadex column and their extraction-photometric determination with Rhodamine-B, Anal. Sci. 6 (1990) 421.

[468] R.P. Siraganian, An automated continuous-flow system for the extraction and fluorometric analysis of histamine, Anal. Biochem. 57 (1974) 383.

[469] A.L.D. Comitre, B.F. Reis, Automatic flow procedure based on multicommutation exploiting liquid−liquid extraction for spectrophotometric lead determination in plant material, Talanta 65 (2005) 846.

[470] I. Nemcova, P. Rychlovsky, V. Tomankova, L. Zivanovic, Extraction spectrophotometric determination of lidocaine using flow injection analysis, Anal. Lett. 34 (2001) 2457.

[471] S. Nakano, Y.-Y. Luo, D. Holman, J. Ruzicka, G.D. Christian, Sequential injection photometric determination of molybdenum using organic wetting film extraction, Microchem. J. 55 (1997) 392.

[472] T. Sakai, X.-Q. Liu, Y. Maeda, On-line extraction-spectrophotometric determination of neostigmine in pharmaceuticals using double membrane phase separator and monovalent dyestuff, Talanta 49 (1999) 997.

[473] M.J. Whitaker, Spectrophotometric determination of nonionic surfactants by flow injection analysis utilizing ion-pair extraction and an improved phase separator, Anal. Chim. Acta 179 (1986) 459.

[474] A.L. Moskvin, L.N. Moskvin, A.V. Moszhuchin, V.V. Fomin, Extraction-chromatographic preconcentration with chromatomembrane separation of extract from aqueous phase for luminescence determination of oil products and phenols in natural water by flow injection analysis, Talanta 50 (1999) 113.

[475] A.N. Anthemidis, D.G. Themelis, J.A. Stratis, Stopped-flow injection liquid-liquid extraction spectrophotometric determination of palladium in airborne particulate matter and automobile catalysts, Talanta 54 (2001) 37.

[476] A.A. Ensafi, B. Rezaei, Automatic liquid-liquid extraction flow injection analysis determination of trace amounts of perchlorate with spectrophotometric detection, Anal. Lett. 31 (1998) 167.

[477] S. Motomizu, M. Oshima, Spectrophotometric determination of phosphorus as orthophosphate based on solvent extraction of the ion associate of molybdophosphate with Malachite Green using flow-injection, Analyst 112 (1987) 295.

[478] J. Skrlikova, V. Andruch, H. Sklenarova, P. Chocholous, P. Solich, I.S. Balogh, A novel dual-valve sequential injection manifold (DV-SIA) for automated liquid−liquid extraction. Application for the determination of picric acid, Anal. Chim. Acta 666 (2010) 55.
[479] S. Motomizu, M. Onoda, M. Oshima, T. Iwachido, Spectrophotometric determination of potassium in river water based on solvent extraction of the complex formed with a crown ether and an anionic azo dye using flow injection, Analyst 113 (1988) 743.
[480] H. Koshima, H. Onishi, Separation of rhenium by extraction with crown ethers and flow-injection extraction spectrophotometric determination with Brilliant Green, Anal. Chim. Acta 232 (1990) 287.
[481] L.A. Trivelin, J.J.R. Rohwedder, S. Rath, Determination of pentavalent antimony in antileishmaniotic drugs using an automated system for liquid-liquid extraction with on-line detection, Talanta 68 (2006) 1536.
[482] W. Thongchai, B. Liawruangrath, S. Liawruangrath, Sequential injection analysis with lab-at-valve (SI-LAV) for the determination of solasodine in Solanum species, Talanta 81 (2010) 565.
[483] M. Maeda, A. Tsuji, Fluorescence and chemiluminescence determination of steroid and bile acid sulphates with lucigenin by flow injection analysis based on ion-pair solvent extraction, Analyst 110 (1985) 665.
[484] A. Alonso, M.J. Almendral, J.J. Porras, Y. Curto, Flow-injection solvent extraction without phase separation. Fluorimetric determination of thiamine by the thiochrome method, J. Pharm. Biomed. Anal. 42 (2006) 171.
[485] T.P. Lynch, A.F. Taylor, J.N. Wilson, Fully automatic flow injection system for the determination of uranium at trace levels in ore leachates, Analyst 108 (1983) 470.
[486] C. Brach-Papa, B. Coulomb, C. Branger, A. Margaillan, F. Theraulaz, P. van Loot, J.L. Boudenne, Fluorimetric determination of aluminium in water by sequential injection through column extraction, Anal. Bioanal. Chem. 378 (2004) 1652.
[487] P. van Loot, C. Branger, A. Margaillan, C. Brach-Papa, J.L. Boudenne, B. Coulomb, On-line solid-phase extraction and multisyringe flow injection analysis of Al(III) and Fe(III) in drinking water, Anal. Bioanal. Chem. 389 (2007) 1595.
[488] J. Huclova, D. Satínsky, O. Pavlicek, L. Vedralova, R. Karlicek, Using on-line solid phase extraction for determination of amiloride in human urine by sequential injection technique, Anal. Chim. Acta 573-574 (2006) 376.
[489] M.M. Santos-Filha, B.F. Reis, F.J. Krug, C.H. Collins, N. Baccan, Sulphate preconcentration by anion exchange resin in flow injection and its turbidimetric determination in water, Talanta 40 (1993) 1529.
[490] Z. Legnerova, D. Satínsky, P. Solich, Using on-line solid phase extraction for simultaneous determination of ascorbic acid and rutintrihydrate by sequential injection analysis, Anal. Chim. Acta 497 (2003) 165.
[491] J.A. Erustes, A. Andrade-Eiroa, A. Cladera, R. Forteza, V. Cerda, Fast sequential injection determination of benzo[A]pyrene using variable angle fluorescence with on-line solid-phase extraction, Analyst 126 (2001) 451.
[492] Z. Leon, A. Chisvert, T. Balaguer, A. Salvador, Development of a fully automated sequential injection solid-phase extraction procedure coupled to liquid chromatography to determine free 2-hydroxy-4-methoxybenzophenone and 2-hydroxy-4-methoxybenzophenone-5-sulphonic acid in human urine, Anal. Chim. Acta 664 (2010) 178.
[493] T. Yamane, H. Ishikawa, T. Tanaka, Simple, rapid and sensitive determination of bismuth in iron and steel based on in-line preconcentration/separation directly coupled with spectrophotometric detection in a continuous flow system, ISIJ Int. 44 (2004) 698.

[494] T. Yamane, Y. Kouzaka, M. Hirakawa, Continuous flow system for the determination of trace boron in iron and steel utilizing in-line preconcentration/separation by Sephadex column coupled with fluorimetric detection, Talanta 55 (2001) 387.

[495] R. Karlicek, P. Solich, Flow-injection spectrophotometric determination of captopril using on-line solid phase extraction, Pharmazie 53 (1998) 549.

[496] E. Castillo, J.L. Cortina, J.L. Beltran, M.D. Prat, M. Granados, Simultaneous determination of Cd(II), Cu(II) and Pb(II) in surface waters by solid phase extraction and flow injection analysis with spectrophotometric detection, Analyst 126 (2001) 1149.

[497] E.C. Vidotti, V.C. Almeida, C.C. Oliveira, Exploiting the bead injection concept for sequential determination of copper and mercury ions in river-water samples, Talanta 64 (2004) 993.

[498] J. Wang, K. Ashley, E.R. Kennedy, C. Neumeister, Determination of hexavalent chromium in industrial hygiene samples using ultrasonic extraction and flow injection analysis, Analyst 122 (1997) 1307.

[499] Q.Z. Feng, L.X. Zhao, W. Yan, F. Ji, Y.L. Wei, J.M. Lin, Molecularly imprinted solid-phase extraction and flow-injection chemiluminescence for trace analysis of 2,4-dichlorophenol in water samples, Anal. Bioanal. Chem. 391 (2008) 1073.

[500] M. Garrido, A.G. Lista, M. Palomeque, B.S. Fernandez-Band, Fluorimetric determination of fluoride in a flow assembly integrated on-line to an open/closed FIA system to remove interference by solid phase extraction, Talanta 58 (2002) 849.

[501] N.Z. Aracama, A.N. Araujo, R. Perez-Olmos, Sequential injection analysis of lead using time-based colorimetric detection and preconcentration on an anionic-exchange resin, Anal. Sci. 20 (2004) 679.

[502] T. Galeano-Díaz, M.I. Acedo-Valenzuela, F. Salinas, Determination of the pesticide Naptalam, at the ppb level, by FIA with fluorimetric detection and on-line preconcentration by solid-phase extraction on C18 modified silica, Anal. Chim. Acta 384 (1999) 185.

[503] M. Miro, W. Frenzel, J.M. Estela, V. Cerda, A novel flow-through disk-based solid-phase extraction diffuse reflectance optrode. Application to preconcentration and determination of trace levels of nitrite, Analyst 126 (2001) 1740.

[504] M. Miro, W. Frenzel, V. Cerda, J.M. Estela, Determination of ultratraces of nitrite by solid-phase preconcentration using a novel flow-through spectrophotometric optrode, Anal. Chim. Acta 437 (2001) 55.

[505] M.F. El-Shahat, N. Burham, S.M.A. Azeem, Flow injection analysis-solid phase extraction (FIA-SPE) method for preconcentration and determination of trace amounts of penicillins using methylene blue grafted polyurethane foam, J. Hazard. Mater. 177 (2010) 1054.

[506] Z.-L. Zhi, A. Rios, M. Valcarcel, Continuous-flow method for the determination of phenols at low levels in water and soil leachates using solid-phase extraction for simultaneous preconcentration and separation, Analyst 121 (1996) 1.

[507] H.M. Oliveira, M.A. Segundo, J.L.F.C. Lima, V. Cerda, Multisyringe flow injection system for solid-phase extraction coupled to liquid chromatography using monolithic column for screening of phenolic pollutants, Talanta 77 (2009) 1466.

[508] A. Balaguer, A. Chisvert, A. Salvador, Sequential-injection determination of traces of disodium phenyl dibenzimidazoletetrasulphonate in urine from users of sunscreens by on-line solid-phase extraction coupled with a fluorimetric detector, J. Pharm. Biomed. Anal. 40 (2006) 922.

[509] M. Knochen, J. Giglio, Flow-injection determination of phenylephrine hydrochloride in pharmaceutical dosage forms with on-line solid-phase extraction and spectrophotometric detection, Talanta 64 (2004) 1226.

[510] Y. Liang, D.X. Yuan, Q.L. Li, Q.M. Lin, Flow injection analysis of ultratrace orthophosphate in seawater with solid-phase enrichment and luminol chemiluminescence detection, Anal. Chim. Acta 571 (2006) 184.
[511] T. Perez-Ruiz, C. Martinez-Lozano, M.D. Garcia, Determination of propoxur in environmental samples by automated solid-phase extraction followed by flow-injection analysis with tris(2,2'-bipyridyl)ruthenium(II) chemiluminescence detection, Anal. Chim. Acta 584 (2007) 275.
[512] D. Satinsky, R. Karlıcek, A. Svoboda, Using on-line solid phase extraction for flow-injection spectrophotometric determination of salbutamol, Anal. Chim. Acta 455 (2002) 103.
[513] R. Karlicek, M. Gargos, P. Solich, Flow-injection spectrophotornetric determination of salicylate using on-line solid phase extraction, J. Flow Injection Anal. 13 (1996) 45.
[514] A.M. Serra, J.M. Estela, B. Coulomb, J.L. Boudenne, V. Cerda, Solid phase extraction – multisyringe flow injection system for the spectrophotometric determination of selenium with 2,3-diaminonaphthalene, Talanta 81 (2010) 572.
[515] J. Ma, D.X. Yuan, Y. Liang, Sequential injection analysis of nanomolar soluble reactive phosphorus in seawater with HLB solid phase extraction, Mar. Chem. 111 (2008) 151.
[516] L.F. Capitan-Valvey, M.C. Valencia, E. Arana-Nicolas, Flow injection analysis with in-line solid phase extraction for the spectrophotometric determination of sulfonated and unsulfonated Quinoline Yellow in Cologne, Fresenius' J. Anal. Chem. 367 (2000) 672.
[517] Y. Xiong, H.J. Zhou, Z.J. Zhang, D.Y. He, C. He, Molecularly imprinted on-line solid-phase extraction combined with flow-injection chemiluminescence for the determination of tetracycline, Analyst 131 (2006) 829.
[518] G.N. Piccirilli, G.M. Escandar, Flow injection analysis with on-line nylon powder extraction for room-temperature phosphorescence determination of thiabendazole, Anal. Chim. Acta 646 (2009) 90.
[519] H.L. Zhu, H.W. Chen, Y.L. Zhou, Determination of thiamine in pharmaceutical preparations by sequential injection renewable surface solid-phase spectrofluorimetry, Anal. Sci. 19 (2003) 289.
[520] A.M. Idris, On-line coupling of solid-phase extraction, derivatization reaction and spectrophotometry by sequential injection analysis: application to trifluoperazine assay in human urine, J. Pharmacol. Toxicol. Method 56 (2007) 330.
[521] L.C. Hong, X.Y. Sun, L.Q. Wang, Determination of ammonia in water using flow injection analysis with automatic pervaporation enrichment, Anal. Lett. 42 (2009) 2364.
[522] M.J. Almendral-Parra, A. Alonso-Mateos, M.S. Fuentes-Prieto, A gas diffusion technique coupled with flow injection systems. Optimization of the process in its application to the fluorimetric determination of ammonium in water samples, J. Fluoresc. 20 (2010) 55.
[523] M.I.G.S. Almeida, J.M. Estela, M.A. Segundo, V. Cerda, A membraneless gas-diffusion unit – multisyringe flow injection spectrophotometric method for ammonium determination in untreated environmental samples, Talanta 84 (2011) 1244.
[524] M.T. Oms, A. Cerda, A. Cladera, V. Cerda, R. Forteza, Gas diffusion techniques coupled sequential injection analysis for selective determination of ammonium, Anal. Chim. Acta 318 (1996) 251.
[525] J. Keay, P.M.A. Menage, Automated determination of ammonium and nitrate in soil extracts by distillation, Analyst 95 (1970) 379.
[526] G. Medina-Alonso, M. Carrasco-Fuentes, M.P. Canizares-Macias, Coupling on-line of a dialyser with a flow-continuous system to separate vitamin B_{12} from milk, Talanta 68 (2005) 292.
[527] K. Sereenonchai, P. Saetear, N. Amornthammarong, K. Uraisin, P. Wilairat, S. Motomizu, D. Nacapricha, Membraneless vaporization unit for direct analysis of solid sample, Anal. Chim. Acta 597 (2007) 157.

[528] Z.Q. Zhang, G.-X. Liang, Flow injection on-line oxidizing fluorometry coupled to dialysis sampling for the study of carbamazepine-protein binding, Anal. Chim. Acta 536 (2005) 145.
[529] J.R. Gord, G. Gordon, G.E. Pacey, Selective chlorine determination by gas-diffusion flow injection analysis with chemiluminescent detection, Anal. Chem. 60 (1988) 2.
[530] S. Satienperakul, T.J. Cardwell, R.W. Cattrall, I.D. McKelvie, D.M. Taylor, S.D. Kolev, Determination of carbon dioxide in gaseous samples by gas diffusion-flow injection, Talanta 62 (2004) 631.
[531] D.G. Themelis, S.C. Karastogianni, P.D. Tzanavaras, Selective determination of cyanides by gas diffusion-stopped flow-sequential injection analysis and an on-line standard addition approach, Anal. Chim. Acta 632 (2009) 93.
[532] J.M. Izquierdo-Ferrero, J.M. Fernandez-Romero, M.D. Luque de Castro, On-line flow injection-pervaporation of beer samples for the determination of diacetyl, Analyst 122 (1997) 119.
[533] M.H.C. Stoll, K. Bakker, G.H. Nobbe, R.R. Haese, Continuous-flow analysis of dissolved inorganic carbon content in seawater, Anal. Chem. 73 (2001) 4111.
[534] S. Munchareoen, J. Sitanurak, W. Tiyapongpattana, N. Choengchan, N. Ratanawimarmwong, S. Motomizu, P. Wilairat, D. Nacapricha, Quality control of gasohol using a micro-unit for membraneless gas diffusion, Microchim. Acta 164 (2009) 203.
[535] R.S. Fontenele, P. Hidalgo, I.G.R. Gutz, J.J. Pedrotti, Flow injection analysis of ethyl xanthate by in-line dialysis and UV spectrophotometric detection, Talanta 72 (2007) 1017.
[536] J.F. van Staden, H. du Plessis, R.E. Taljaard, Determination of iron(III) in pharmaceutical samples using dialysis in a sequential injection analysis system, Anal. Chim. Acta 357 (1997) 141.
[537] K. Toda, K.-I. Yoshioka, K. Mori, S. Hirata, Portable system for near-real time measurement of gaseous formaldehyde by means of parallel scrubber stopped-flow absorptiometry, Anal. Chim. Acta 531 (2005) 41.
[538] M. Garn, M. Gisin, C. Thommen, P. Cevey, A flow injection analysis system for fermentation monitoring and control, Biotechnol. Bioeng. 34 (1989) 423.
[539] D. Nacapricha, K. Uraisin, N. Ratanawimarnwong, K. Grudpan, Simple and selective method for determination of iodide in pharmaceutical products by flow injection analysis using the iodine-starch reaction, Anal. Bioanal. Chem. 378 (2004) 816.
[540] D. Nacapricha, P. Sangkarn, C. Karuwan, T. Mantim, W. Waiyawat, P. Wilairat, T. Cardwell, I.D. McKelvie, N. Ratanawimarnwong, Pervaporation-flow injection with chemiluminescence detection for determination of iodide in multivitamin tablets, Talanta 72 (2007) 626.
[541] E. Mataix, M.D. Luque de Castro, Determination of L-(-)-malic and L-(+)-lactic acid in wine by a flow injection-dialysis-enzymic derivatisation approach, Anal. Chim. Acta 428 (2001) 7.
[542] Z. Zhi, A. Rios, M. Valcarcel, Direct determination of nitrate and nitrite in soils by use of a hydrodynamic injection probe based on filtration-dialysis processes, Int. J. Environ. Anal. Chem. 57 (1994) 279.
[543] H. Mana, U. Spohn, Selective flow injection procedures for the determination of nitrogen containing analytes by gas dialytic-fluorimetric detection of enzymatically generated ammonia, Anal. Chim. Acta 325 (1996) 93.
[544] K. Dai, A.G. Vlessidis, N.P. Evmiridis, Dialysis membrane sampler for on-line flow injection analysis/chemiluminescence detection of peroxynitrite in biological samples, Talanta 59 (2003) 55.
[545] J.E. Silva, M.F. Pimentel, V.L. Silva, M.C.B.S.M. Montenegro, A.N. Araujo, Simultaneous determination of pH, chloride and nickel in electroplating baths using sequential injection analysis, Anal. Chim. Acta 506 (2004) 197.

[546] I.D. McKelvie, B.T. Hart, T.J. Cardwell, R.W. Cattrall, Speciation of dissolved phosphorus in environmental samples by gel filtration and flow-injection analysis, Talanta 40 (1993) 1981.

[547] A.N. Araujo, J.L.F.C. Lima, A.O.S.S. Rangel, M.A. Segundo, Sequential injection system for the spectrophotometric determination of reducing sugars in wines, Talanta 52 (2000) 59.

[548] G. Armas, L. Ferrer, M. Miro, J.M. Estela, V. Cerda, In-line membrane separation method for sulfide monitoring in wastewaters exploiting multisyringe flow injection analysis, Anal. Chim. Acta 524 (2004) 89.

[549] L.G. Decnop-Weever, J.C. Kraak, Determination of sulphite in wines by gas-diffusion flow injection analysis utilizing spectrophotometric pH-detection, Anal. Chim. Acta 337 (1997) 125.

[550] S. Satienperakul, P. Phongdong, S. Liawruangrath, Pervaporation flow injection analysis for the determination of sulphite in food samples utilising potassium permanganate–rhodamine B chemiluminescence detection, Food Chem. 121 (2010) 893.

[551] E. Mataix, M.D. Luque de Castro, Determination of total and free sulfur dioxide in wine by pervaporation-flow injection, Analyst 123 (1998) 1547.

[552] M.A. Segundo, A.O.S.S. Rangel, A gas diffusion sequential injection system for the determination of sulphur dioxide in wines, Anal. Chim. Acta 427 (2001) 279.

[553] P.K. Dasgupta, V.K. Gupta, Membrane-based flow injection system for determination of sulfur(IV) in atmospheric water, Environ. Sci. Technol. 20 (1986) 524.

[554] J.A. Garcia-Garrido, M.D. Luque de Castro, Determination of trimethylamine in fish by pervaporation and photometric detection, Analyst 122 (1997) 663.

[555] J.F. van Staden, S.I. Tlowana, On-line separation, simultaneous dilution and spectrophotometric determination of zinc in fertilizers with a sequential injection system and xylenol orange as complexing agent, Talanta 58 (2002) 1115.

CHAPTER

9

Future Trends

We dance around a circle and suppose,
while the answer sits in the middle and knows. **Robert Frost**

Major trends in the future growth of flow analysis are likely to include the evolution of instrument design (including miniaturisation and expert flow systems), the recognition of more flow-based standard methods, hyphenation with other detection systems and an impact on chemical measurements in new and emerging areas of science.

One of the inherent strengths of flow analysers is the range of temporal windows that can be exploited within the elution profile of the sample zone (see Chapter 1). The strategies that can be used to maximise the information from these windows are described in detail in Chapter 7 and include kinetic and catalytic procedures. However, the wider availability of reliable and fully automated commercial instrumentation will facilitate the implementation of these strategies and hence maximise the potential for information retrieval.

Fully automated instrumentation will also allow a more systematic application of expert flow systems as discussed in Chapter 8.

Another trend in analytical instrumentation that will continue to impact on the design and operation of flow analysers is miniaturisation. This will not only reduce reagent consumption/waste minimisation and open up new areas for deployment in remote and/or hazardous locations, *e.g.*, in situ and in vivo assays, but also lead to new chemistries and chemical processes that take advantage of the microenvironment of miniaturised "lab-on-a-chip" type devices. A note of caution, however, is that whilst flow analysis will generally benefit from miniaturisation, there will be a point for most applications at which the benefits of scaling down will be outweighed by the practical challenges, particularly relating to sample introduction and conditioning.

Flow analysers are also able to deal effectively with hazardous and/or toxic reagents and samples due to their enclosed nature, low reagent volumes and in-line waste storage. These features will become increasingly attractive to both analytical chemists and legislators in the future.

The capacity of flow analysers to be used as a teaching aid to visually demonstrate analytical chemistry applications and general chemical principles (Chapter 8) will expand significantly.

A key feature of quality assurance in analytical chemistry is the adoption and wider implementation of standard methods. Flow analysis has a major role to play in this area due to the inherent attractions of flow-based instrumentation, not least the reproducibility of sample transport and manipulation. This in turn will generate a greater awareness of, and interest in, flow analysis. The challenge is to get such methods adopted by the relevant regulatory agencies.

Hyphenation of flow analysers with other detection systems will become increasingly common, *e.g.*, using flow manifolds for sample conditioning at the front end of other detection systems.

Flow-based analytical procedures are already used in many fields, *e.g.*, environment, food and health. In tandem with the developments described above, flow analysers will also be used in new and emerging fields of application such as the "omics" (metallomics, genomics and proteomics), biotechnology (enzymatic assays, platforms for bio-sensors and flow immunoassays) and quality-control applications in general. Economic resources will also be a driver for more multi-parametric flow-based methods, particularly with spectrophotometric detection.

Human resources will play a fundamental role in making the trends highlighted above come true. Effort will also be required to transmit the know-how related to flow analysis to the wider scientific community. Enhanced knowledge has to be created, expanded, transmitted and learnt. In this context,

It is a happy day when we can learn something. **H. Bergamin-Filho**

Index

Note: Page numbers followed by *f* indicate figures and *t* indicate tables.

B

C

E

G

Q

R

S

T